Biodiversity Conservation and Livelihood Management

THE EDITORS

Dr. Hrudayanath Thatoi is a Professor, Department of Biotechnology, North Orissa University, Odisha, India. He is also presently working as Director of Centre for Similipal Studies and Director, HRDC, North Orissa University. He has also served the University as Reader and Head, Department of Biotechnology, Biju Patnaik University of Technology, Odishar and Sr. Scientist, M.S. Swaminathan Research Foundation, Chennai. Prof. Thatoi obtained his M. Sc, M. Phil. and Ph. D. from Utkal University, Odisha, and has more than 15 years of teaching and 25 years of research experience. He has contributed immensely to the field of microbiology and biotechnology throughout his research and teaching career. His research activities are mainly based on exploration microbial diversity, bioremediation, biofuel production, mangrove biology, bioactivity studies in medicinal plants and mushrooms, ethnopharmacology and *in silico* drug designing and modeling *etc.* Prof. Thatoi has handled many research projects funded by DST, Govt. of Odisha, UGC-DAE, Govt. of India, Department of Forests, Govt. of Odisha, *etc.* So far, 16 students have obtained Ph.D.s under his guidance and supervision and some 7 students are still continuing. Besides, several M.Sc., M.Tech and B.Tech students have received his guidance for their dissertation works. Prof. Thatoi has published more than 200 research papers in various nationally and internationally reputed journals/proceedings and more than 30 book chapters in National and International publications. Prof. Thatoi has also authored 18 reference books and Seminar Proceedings published by various notable publishers like, Elsevier, Springer, Taylor and Francis (CRC and Apple Press), Studium Press LLC USA, IK International, Narosa, Biotech Books, APH Publication, *etc.* He has also authored two text books on Microbiology and Immunology, published by India Tech Publication, New Delhi, and Practical Biotechnology: Principles and Protocols by IK International Publication, New Delhi for Graduation and Post-Graduation students in the field of Life Sciences and Biotechnology.

Dr. Priya Ranjan Debata presently working as an Assistant Professor in the Department of Zoology, North Orissa University, Odisha. He passed M.Sc. in Zoology from Berhampur University in 1998. He qualified CSIR-JRF-NET from the Council of Scientific and Industrial Research, New Delhi and joined the Ph.D. programme in Life Sciences at the Institute of Life Sciences, Bhubaneswar. He was awarded Ph.D. degree from Utkal University, Bhubaneswar in 2008. He has more than five years of Postdoctoral Research experience at City University of New York, USA. Dr. Debata joined at Utkal University, Bhubaneswar as Ramalingaswami Fellow awarded by Department of Biotechnology, Govt. of India during 2013-14. He joined the North Orissa University as faculty in 2014. He has published 21 research papers in international journals of repute and four book chapters. He has guided seven M.Phil. students and four research scholars are continuing for their Ph.D. under his guidance. He has implemented research projects from DBT Govt. of India, DST Govt. of Odisha, and DST Govt. of India.

Biodiversity Conservation and Livelihood Management

— *Editors* —

Prof. Hrudayanath Thatoi

Professor,
Department of Biotechnology, and
Director, Centre for Similipal Studies

Dr. Priya Ranjan Debata

Asst. Professor,
Department of Zoology

North Orissa University,
Baripada – 757 003, Odisha

2024

Daya Publishing House®
A Division of
Astral International Pvt. Ltd.
New Delhi – 110 002

Publisher's Note:

Every possible effort has been made to ensure that the information contained in this book is accurate at the time of going to press, and the publisher and author cannot accept responsibility for any errors or omissions, however caused. No responsibility for loss or damage occasioned to any person acting, or refraining from action, as a result of the material in this publication can be accepted by the editor, the publisher or the author. The Publisher is not associated with any product or vendor mentioned in the book. The contents of this work are intended to further general scientific research, understanding and discussion only. Readers should consult with a specialist where appropriate.

Every effort has been made to trace the owners of copyright material used in this book, if any. The author and the publisher will be grateful for any omission brought to their notice for acknowledgement in the future editions of the book.

Cataloging in Publication Data--DK

Courtesy: D.K. Agencies (P) Ltd. <docinfo@dkagencies.com>

National Seminar "Biodiversity Conservation and Livelihood Management in Similipal Biosphere Reserve" (2017 : Baripada, India)

Biodiversity conservation and livelihood management / editors, Prof. Hrudayanath Thatoi, Dr. Priya Ranjan Debata.

pages cm

"... a National Seminar "Biodiversity Conservation and Livelihood Management in Similipal Biosphere Reserve" was organized by the Centre for Similipal Studies, North Orissa University on 26-27 March 2017..."--Preface.

Includes bibliographical references and index.

ISBN 9789359190983

1. Biodiversity conservation--India--Similipal Biosphere Reserve--Congresses. 2. Tribes--India--Similipal Biosphere Reserve--Economic conditions--Congresses. 3. Community forestry--India--Similipal Biosphere Reserve--Congresses. 4. Forestry and community--India--Similipal Biosphere Reserve--Congresses. I. Thatoi, H. N., editor. II. Debata, Priya Ranjan, editor. III. North Orissa University. Centre for Similipal Studies, organizer. IV. Title.

LCC QH77.I4N38 2017 | DDC 333.95160954133 23

Published by : **Daya Publishing House®**
A Division of
Astral International Pvt. Ltd.
– ISO 9001:2015 Certified Company –
4736/23, Ansari Road, Darya Ganj
New Delhi-110 002
Ph. 011-43549197, 23278134
e-mail: info@astralint.com
Website: www.astralint.com

Foreword

India is one among, twelve countries that collectively account for 60–70 per cent of the world's biological diversity. India, with only 2.4 per cent of the world's landmass, supports an astounding 8.1 per cent of the world's biodiversity. In a country like India, millions of people rely on products from natural ecosystems to sustain their livelihood. For these reasons, rapid, often irreversible, loss of species and ecosystems is of more than just academic concern. Indeed, our understanding of biodiversity in natural ecosystems remains so woefully inadequate that we are unable to fully comprehend the consequences of its loss. With impending climate change and increasing spread of invasive species, the biodiversity crisis is likely to get worse, with far-reaching effects on human societies. In this context, the present book "Biodiversity Conservation and Livelihood Management", being published by North Orissa University will be a valuable source of information for all those who are associated in the conservation of biodiversity and management of livelihood of the local inhabitants. I thank the efforts made by Prof. H.N. Thatoi, Director, Centre for Similipal Studies and Dr. P.R. Debata, Asst. Professor, Department of Zoology, North Orissa University for editing and taking responsibility for the publication of the book. The book deals with 32 interesting articles on the extent of diversity among biological species including plants, animals and microbes and their conservation in addition to the livelihood of the native populations. I congratulate the authors and editors for this fruitful effort and strongly hope that the book would be an important source of information and referral document for conservation scientists, researchers and academicians working in the areas of biodiversity conservation and management as well as the scientific community and entire humanity.

Prof. Pradeep Kumar Chand

Vice-Chancellor

North Orissa University

Preface

The world population has been increasing at an alarming rate. There is a need to provide food and livelihood support to the growing human population. Forest has been serving livelihood to forest dependant people from ancient times. Unfortunately, we continued to exploit these resources without thinking about its sustainability. The forest has tremendous opportunities for alternate energy, ecotourism, natural resource management, and marketing-networking and alternate livelihood.

Currently, issues related to biodiversity is of prime focus as unprecedented degradation of habitat throughout the globe in parallel with human development. The rate at which the species are disappearing as a result of human activity is high. In this context, a national seminar "Biodiversity conservation and livelihood management in Similipal Biosphere Reserve" was organized by the Centre for Similipal Studies, North Orissa University on 26-27 March 2017 to discuss the present issue related to Biodiversity and livelihood management. The present book is an outcome of the national seminar and comprised of the articles contributed by the researchers, academicians, field level experts involved in biodiversity conservation and management. Apart from the participants, some articles were also contributed by some renowned experts in the relevant field.

The present book is intended to reach a broad audience with the latest thinking about maintaining biodiversity management and lively hood management. The focus of this book is on the relevance of biodiversity conservation and livelihood management. The chapters covered in this book deals with the biodiversity of plants, animals, and microbes from different angles like genetic, species, and ecosystems.

This book is divided into two sections; section 1 Biodiversity conservation and section 2 Livelihood management. The book is published with the support of UGC grant of North Orissa University and printed by Astral Publication, New Delhi. We thank our Hon'ble Vice-Chancellor, North Orissa University, for his wholehearted support and writing foreword for the book. We appreciate the efforts of all the contributors who have promptly agreed to contribute their valuable articles and reviewers for critically reviewing the chapters to enhance the quality of the publication. Our special thanks are due to the publisher of this book Astral International (P) Ltd., New Delhi for taking the responsibility to print the book well in time. We hope the book will be beneficial to all the scientists, research scholars, students, and policymakers who are directly or indirectly related to the subject.

Prof. Hrudayanath Thatoi

Dr. Priya Ranjan Debata

Contents

— *Section 1* —

BIODIVERSITY CONSERVATION

Chapter 1

Biodiversity Conservation, Environmental Challenges and Sustainable Livelihood

Sharat Kumar Palita* and Debabrata Panda

Department of Biodiversity and Conservation of Natural Resources, Central University of Odisha, Koraput – 764 021, Odisha
**e-mail: skpalita@gmail.com*

Biodiversity is the variety and variability of Life on Earth, which includes variety of genes, populations, species, communities, ecosystems, and ecological processes. From genes to micro-organisms to top predators and even whole ecosystems, humanity depend on biodiversity for everything from clean air and water to medicines and secure food supplies. In spite of the growing awareness of the importance of ecosystems and biodiversity to human welfare, loss of biodiversity and degradation of ecosystems still continue on a large scale. We are at risk of losing much of biodiversity and the benefits it provides to humanity.

There is increasing recognition that a deteriorating global environment poses not only challenges for survival of biodiversity, it also posses significant threats to future economic growth and development. Strategic approaches that integrate biodiversity conservation and livelihood goals are necessity of the time. Creating or highlighting explicit interdependencies between socioeconomic benefits and biodiversity conservation is expected to reduce threats emanating from unsustainable natural resource exploitation and contribute to human well-being by improving human quality of life in areas of conservation concern.

Keywords: *Biodiversity, Environmental challenges, Climate change, Sustainable livelihood.*

INTRODUCTION

The Convention of Biological Diversity (CBD) (1992, Rio de Janeiro, Brazil) defined "Biodiversity" as variety and variability of life forms on Earth including;

terrestrial, marine, and other aquatic ecosystems. Biodiversity includes variety of genes, populations, species, communities, ecosystems, and ecological processes. The Convention has three objectives, namely,conservation of biodiversity, sustainable use of the components of biodiversity, and fair and equitable sharing of benefits arising out of the use of genetic resources. Biodiversity have multiple values and is very important for the production of food and to preserve the ecological foundations needed to maintain people's livelihood. All species on our Earth are connected in the web of life and "rely on each other to survive and thrive". Biodiversity within an area is a unique community of organisms interacting with the air, water, rocks, and soil to form an ecosystem. Microbes in soils, phytoplankton in oceans and top predators in the wilderness all interact in their ecosystem to regulate the state of the planet (Abrahms *et al.*, 2017).

Life on Earth is believed to have originated around 3.5 billion years ago. Over time, it evolved into myriad fascinating forms through the process of speciation. Some of these life forms were, however, lost along the way through extinction. The current stock of species is a product of these two processes occurring simultaneously over a long period of time. To date, approximately 1.75 million species (Plants, Animals and Microorganisms) have been formally identified. While estimates of the total numbers vary widely, some report predicted that there may be as many as 8-9 million and possibly up to 100 million species living on Earth. However, the fossil record prompts us to think bigger as it indicates more than 50 billion species might have existed on Earth at varying points in history, and more than 98 per cent have disappeared. Extinction of species is a global issue and a natural phenomena in geographical history but main intervention speed up the extinction rate. Between 1600-1950,the rate of extinction was 1 species in every 10 year. But currently the rate of extinction was 1 species in every year. Over last 100 yrs, human-caused species extinctions have multiplied ~ 1,000 times. 12 per cent of birds, 23 per cent of mammals, and 32 per cent of amphibians are threatened with extinction. Current estimates of species loss range between 10-50 per cent over next 20-50 years.

The Millennium Ecosystem Assessment (MA 2005) recognized four categories of services rendered by biodiversity: supporting (*e.g.* nutrient cycling, soil formation and primary production); provisioning (*e.g.* food, fresh water, wood and fibber and fuel); regulating (*e.g.* climate regulation, flood and disease regulation and water purification); and cultural (aesthetic, spiritual, educational and recreational). The introduction of the concept of ecosystem services on the global agenda by the MA provides an important bridge between the imperatives of maintaining biodiversity and the challenges in meeting to meet the Millennium Development Goals.

In the present chapter, an attempt has been made to highlight importance of biodiversity conservation, environmental challenges confronting biodiversity and livelihoods of millions of people in the world dependent on biodiversity and threatened by environmental challenges including climate change.

Importance of Biodiversity

Provisioning of Food

World over, agro-ecosystems provide food for human consumption and, together with the associated ecosystems supporting marine and freshwater fisheries, underpin global food security. Presently, 35 per cent of the Earth's surface is used for growing crops or rearing livestock (MA 2005). Among these grazing land alone accounts for 26 per cent of the Earth's surface, and animal feed crops account for a third of all cultivated land (FAO 1999). It has been estimated that well over 6,000 species of plants are known to have been cultivated at some time or another, and many thousands that are grown locally are scarcely or only partly domesticated, whilst as many, if not more, are gathered from the wild (Heywood, 1999). Despite this, 95 per cent of humanity's food comes from only about 30 crop species (Williams and Haq, 2002) and it has been argued that the world is currently over-dependent on a few plant species.

The increasing dependence on fewer species for crops has lead to the loss of native genetic resources, and higher yielding modern varieties were displacing "landraces" uniquely adapted to local conditions (Harlan 1975). Failure to maintain sufficient genetic diversity in crops can incur high economic and social costs. The potato famine in Ireland in the nineteenth century is generally attributed to the low genetic diversity of the potatoes cultivated there, making the entire crop susceptible to potato blight fungus, a problem resolved by using resistant varieties from original gene pools in South America. Mixtures of varieties may successfully reduce disease incidence and increase yields as for example with the case of barley in Europe (Hajjar *et al.*, 2008). In the present juncture, biodiversity could be promoted by using intensive agriculture and devoting spare land to biodiversity or by extending „organic or integrated farming systems that promote biodiversity (Fischer *et al.*, 2008), but the outcomes of these two approaches would be very different.

The importance of biodiversity extends far beyond the species for which we see direct benefit to us as food, fuel, medicines, or other uses. World Wildlife Fund has produced reports showing that, globally, we are using 25 per cent more natural resources than the planet can sustain, which puts habitats and species survival under severe pressure. By not protecting biodiversity, we rob ourselves and other species of choices for the future. Ecosystems with high biodiversity are characterized by complex interactions between different species, which can help the ecosystem remain intact and healthy in the face of disturbance and environmental change. For this reason, looking at biodiversity is a good parameter for assessing the overall health of an ecosystem. Beside habitat loss and fragmentation, overexploitation, pollution, and the impact of invasive alien species, climate change is serious environmental challenge that could undermine the drive for sustainable development. A holistic approach to conservation of biodiversity and conservation of natural habitat are crucial for their sustainability.

At current levels of consumption, global food production will need to increase by 50 per cent within the next four decades to meet the demands of a growing human population (UN 2009) and as consumption levels and world food prices rise, pressure to maximize the area under production will grow. Given the rapidly growing demands on the planetary ecosystems (Rockström *et al.*, 2009), it is becoming critical to understand how a dramatic increase in agricultural production and shifting land use in combination with climate change will affect natural processes of the biosphere and levels of key regulating ecosystem services (*e.g.*, CO_2, nitrogen flow, freshwater consumption).

Provisioning of Water

Ecosystems play important roles, In the global hydrological cycle, contributing to water provision (quantity, defined as total water yield), regulation (timing, the seasonal distribution of flows) and purification (quality, including biological purity as well as sediment load) (Dudley and Stolton 2003; Bruijnzeel 2004; Brauman *et al.*, 2007). Global water use is dominated by agricultural withdrawals (70 per cent of all use and 85 per cent of consumptive use), including livestock production, followed by industrial and domestic applications. Vegetation, particularly forests, significantly influences the quantity of water circulating in a watershed. It is commonly assumed that forests generate rainfall and, in comparison with pasture and agriculture, promote higher rates of evapotranspiration and greater aerodynamic roughness, leading to increased atmospheric humidity and moisture convergence, and thus to higher probabilities of cloud formation and rainfall generation. Although evidence is increasing (Bruijnzeel, 2004) that large-scale land use conversions affect cloud formation and rainfall patterns, this effect is highly variable and specific. Ecosystems such as forest and wetlands with intact groundcover and root systems are considered very effective at regulating water flow and improving water quality.

Provisioning of Fuel and Fibre

Historically, the provision of fuels and fibres – such as timber, cotton, jute, sisal, sugars and oils - has been regarded as a highly important ecosystem service. Natural systems provide a great diversity of materials for construction and fuel, notably oils and wood that are derived directly from wild or cultivated plant species. Wood and non-wood production is the primary commercial function of 34 per cent of the world's forests, while more than half of all forests are used for such production in combination with other functions, however, only 3.8 per cent of global forest cover corresponds to forest plantations, indicating that a substantial fraction of natural forests is used for productive uses (FAO, 2006).

Currently there is an intense interest and strong policy direction to increase the proportion of energy derived from renewable sources, mostly from biological materials. This is being achieved partly by the cultivation of biomass crops and partly by diversion of materials otherwise usable as food for people or animals, including wheat and maize, to manufacture ethanol as a replacement for petrol and other oil-derived fuels. Further a big effort has been put into the cultivation of algae for bio fuels (Ross *et al.*, 2008; Adams *et al.*, 2009).

It is likely that a decline in the provision of wild timber, plant fibres and fuel wood will take place in proportion to the decline in the forested area. Fragmentation, however, may result in a much quicker decline in forest productivity than what would be expected given the total area of remaining forest (Laurance *et al.*, 2001).

Genetic Resource Provisioning

Genetic diversity of crops is responsible for increase in production and decrease in susceptibility to pests and climate variation (Ewel 1986; Altieri 1990; Zhu *et al.*, 2000). In low-input systems especially, locally adapted varieties often produce higher yields or are more resistant to pests than varieties bred for high performance under optimal conditions (Joshi *et al.*, 2001). In agriculture, the diversity of genetic resources comprises the traditional resources (wild types and the older domesticated landraces) together with modern cultivars. In the time to come, genetic resources will be increasingly important in support of improved breeding programs (*e.g.* for crop plants, farm animals, fisheries and aquaculture), with a wide range of objectives for increasing yield, resistance to disease, optimization of nutritional value, and adaptation to local environment and climate change. Advances in genomics research are opening up a new era in breeding, where the linkage of genes to traits (marker-assisted selection) provides a more efficient and predictable route than conventional breeding programs to improved strains (Elmqvist, 2010).

Environmental Challenges and Biodiversity

The environment has an intense effect on all the living organism of the world. Environmental challenges are universal and many of them are damaging for us and also for the future generations. Seven biggest threats to environment are climate change, species extinction and biodiversity loss, air and water pollution, depletion of natural resources, deforestation and soil degradation. Presently there is overwhelming pressure on global natural resources. The world population is rapidly rising and in the same time the world's forest resources depleting by a considerable amount. Over exploitation of selected species for food and medicine results in drastic decline in size of populations. In Primitive Agriculture age man used about 5000 species of edible plants. Now nearly 150 species of plants cultivated in large scale. Only 4 species such as rice, wheat, corn and potato provides 90 per cent of food supply of the world. Rapid continuous pollution of air, water and land threatens our health and also our future. Loss of biodiversity and degradation of ecosystems still continue on a large scale for that countless species of animals and plants have been wiped out from earth, many of them are valuable for human beings. Increase of greenhouse gases released around the globe by industrialization, urbanisation, automobiles and burning forests affect many billion dollar damage to our health and our climate.

Population Growth

It is projected that by the year 2050 the global population will be approximately between 7.9 to 12 billion. Now it is seen that faster growth is taking place in urban areas, with a rapid expansion in the number of mega-cities with population of 10 to 20 million or more (Flower, 2006). Over population pressures will tax more on the natural and institutional resources. Growing demands on resources, including water, electricity, oil, timber, hydrocarbons, and food will harm to people and the environment. High population growth and urbanization results in more and more energy demand, economic expansion and waste generation which will ultimately damage the environments. Providing adequate food, nutrition, environment and health security to the growing population is the greatest challenge of the new millennium.

Pollution

Air, water and land are polluted by various pollutants. Ninety-two per cent of the world population lives in areas with polluted air which causes 11.6 per cent of global recorded deaths, points out WHO. Air quality is particularly bad in cities, and this situation is going to get worse as more people move to cities. Chemical pollution from agriculture and industries is another problem where plants and animals are killed or affected by toxins. In addition nutrient pollution from fertilizers, households, and other sources end up in lakes, ponds, and the oceans to cause eutrophication. In the oceans nutrient pollution together with global warming has caused 500 dead zones where there are no oxygen reports.

Global Warming and Climate Change

With the recent expansion of human activities, a large amount of greenhouse gases such as carbon dioxide and methane gas is emitted into the air, enhancing greenhouse effects and increasing the possibility of global warming. The Intergovernmental Panel on Climate Change (IPCC) predicted that the temperature rises during this century of 1.4 to 5.8 degree Celsius. Constant changes in the environment significantly change the chemical composition of the atmosphere. By 2030, the carbon dioxide concentrations will have doubled as compared to pre-industrial levels (IPCC, 2018). The multiple components of climate change are anticipated to affect all the levels of biodiversity, from organism to biome levels. The effect of climate change might result to name a few such as flooding, cyclone, sea level rise, heat or cold waves, water shortage, fires, hail, windstorm, permafrost melting, *etc.* To stop the dangerous trends of climate change, IPCC recommends a reduction of green house gas emission by around 50-80 per cent by the middle of our century.

Climate change is having widespread impacts across multiple scales of biodiversity including genes, species, communities, and ecosystems (Parmesan 2006; Bellard *et al.*, 2012). Biological responses to climate change vary widely among species and populations; some responses are positive, leading to increased growth rates or range expansions, while others are negative, resulting in localized or widespread declines (Miller-Rushing *et al.*, 2010; Montoya and Raffaelli 2010; Dawson *et al.*, 2011; Geyer *et al.*, 2011). Many species have already shifted their geographical ranges, generally pole ward, towards higher elevations, or to deeper depths in marine environments (Nye *et al.*, 2009; Burrows *et al.*, 2011; Chen *et al.*, 2011; Doney *et al.*, 2012). Species have altered the temporal patterns of seasonal migrations and other life cycle events (phenology), showed changes in population demographics (Doak and Morris 2010; Miller-Rushing *et al.*, 2010; Dawson 2011), or in some cases are adapting in place to the new environmental conditions (Bellard *et al.*, 2012).

Climate change has taken its toll on biodiversity and as a development challenge it requires an integrated and strategic approach such as the integration of biodiversity based strategies in climate change mitigation. The United Nations Framework Convention on Climate Change (UNFCCC) has introduced targeted measures such as REDD (reducing emissions from deforestation and forest degradation), which now includes "conservation, sustainable management of forests and enhancement of forest carbon stocks" (REDD+). Thus, biodiversity paved the way to provide a mitigation solution for reducing carbon emissions from forestlands and invest in low-carbon paths to achieve sustainable development.

India is a large developing country with nearly 700 million rural population directly depending on climate-sensitive sectors (agriculture, forests and fisheries) and natural resources (such as water, biodiversity, mangroves, coastal zones, grasslands) for their subsistence and livelihoods. Further, the adaptive capacity of dry land farmers, forest dwellers, fisher folk, and nomadic shepherds is very low (Ravindranath and Sathaye, 2002). Climate change is likely to impact all the natural ecosystems as well as socio-economic systems as shown by the National Communications Report of India to the UNFCCC (MoEF 2004).

Agriculture is one of the largest contributors to India's GDP, approximately 20 per cent. It is the main source of livelihood for almost 60 per cent of the country's total population. The impacts of climate change on agriculture will therefore be severely felt in India. It has been projected that under the scenario of a 2.5 °C to 4.9 °C temperature rise in India, rice yields will drop by 32 per cent -40 per cent and wheat yields by 41 per cent -52 per cent. This would cause GDP to fall by 1.8 per cent -3.4 per cent (OECD, 2002; Guiteras, 2007; GoI 2011).

Despite the gloomy predictions about the negative impacts for India's agricultural sector, climate change is generally also expected to bring opportunities, *e.g.* production gains through the 'CO_2 fertilization effect' or the expansion of cultivated land to higher altitudes and northern latitudes. However, it must be noted that to date all climate change projections have been accompanied by uncertainty – not primarily concerning trends but extent (IFPRI 2009; UNFCCC, 2009).

India is a large developing country with diverse climatic zones. The livelihood of vast population depends on climate-sensitive economic sectors like agriculture, forestry and fisheries. The climate change vulnerability and impact studies in India (Shukla *et al.*, 2003) assume high degree of uncertainty in the assessment due to 'limited understanding of many critical processes in the climate system, existence of multiple climatic and non-climatic stresses, regional-scale variations and non-linearity'. The costs of not addressing climate change or to adapt to it are very uncertain, but their welfare consequences are enormous. Early actions on adaptation therefore are prudent and consistent from the viewpoint of 'precautionary principle'(Sathe *et al.*, 2006). The projected climate change under various scenarios is likely to have implications on food production, water supply, biodiversity and livelihoods. Thus, India has a significant stake in scientific advancement as well as an international understanding to promote mitigation and adaptation. This requires improved scientific understanding, capacity building, networking and broad consultation processes (Sathe *et al.*, 2006).

Deforestation

Despite global increases in protected area coverage and the amount of community-managed forest lands (Zimmerer *et al.*, 2004; Molnar *et al.*, 2007), natural tropical forests are still declining worldwide (Hansen *et al.*, 2013). Over the last few decades, forest landscapes have become increasingly fragmented (Koehler *et al.*, 2003). The resulting mosaic landscapes that have fragments of natural or semi-natural habitats have long been poorly valued in terms of biodiversity conservation (Hanski, 2005). Deforestation threatens animal and plant biodiversity, water and air quality, the livelihoods and food security of forest-dependent communities that are the poorest and most vulnerable (Kummer 1992; Kummer and Turner 1994; Saastamoinen 1996).

According to projections based on a global dynamic vegetation model, the impacts of climate change on forests in India (Gopalakrishnan *et al.*, 2011) are the following:

- ✰ About 45 per cent of the forested area is likely to undergo change by 2035. The changes comprise an expansion of tropical evergreen forests in the eastern India plateau and the Western Ghats, slight expansion of forests into the western part of central India, and almost no change in vegetation type in the north-east.
- ✰ Vulnerability assessments show that the vulnerable forest areas are spread across India, with a higher concentration in upper Himalayan stretches, parts of central India, the northern Western Ghats and Eastern Ghats. North-eastern forests, the southern Western Ghats and the forested regions of eastern India are estimated to be the least vulnerable.
- ✰ Temperate forests (over 50 per cent of India's forest area) are the most vulnerable. They are followed by tropical semi-evergreen forests and dry and moist deciduous forests.

Livelihood

Definition of 'Livelihood' suggested by Chambers and Conroy:

*A **livelihood** comprises the capabilities, assets (including both material and social resources) and activities required for a means of living. A livelihood is sustainable when it can cope with and recover from stress and shocks and maintain or enhance its capabilities and assets both now and in the future, while not undermining the natural resource base* (Chambers and Conway 1991)

In order to better understand how people develop and maintain livelihoods, the UK Department for International Development (DFID), building on the work of practitioners and academics, developed the Sustainable Livelihoods Framework (SLF).

The SLF views livelihoods as systems and provides a way to understand:

1. The assets people draw upon
2. The strategies they develop to make a living

3. The context within which a livelihood is developed
4. Those factors that make a livelihood more or less vulnerable to shocks and stresses

People living in poor countries are disproportionately vulnerable to the loss of biodiversity and reduced ecosystem services. While links between climate change, biodiversity and poverty are clear, and there is an obvious imperative to support projects and activities that meet the objectives of all three agreements, in practice these tend to be thin on the ground. And in some cases, activities intended to meet the goals of one agreement may negatively affect the goals of another. As such the integration of objectives can be improved through stronger linkages between Conventions, better research and inter-disciplinary thinking and a wider engagement between the public, governments, agencies and other stakeholders.

Geographic location is a key factor in the vulnerability of poor people and poor nations. Many of these countries lie in the regions most at risk from climate change (such as drought-prone sub-Saharan Africa); many of the poor may live in marginal areas such as floodplains or at the foot of unstable hillsides. Poor people also have the fewest choices available to them, and the lowest capacity, for example because of a lack of resources and mobility, to cope with climate change related shocks such as cyclones (Huq *et al.*, 2004).

Poverty is both complex and multidimensional (OECD-DAC 2001; World Bank 2000-1), but in general, it represents people's inability to achieve acceptable standards of well-being. Well-being includes economic capabilities (the ability to earn an income, access to land and resources, decent work), human capabilities (health, education, nutrition), political capabilities (empowerment, rights, voice), socio-cultural capabilities (status, dignity) and protective capabilities (to address security, risk and vulnerability) (OECD-DAC 2001).

Given the diverse linkages between livelihoods and natural resources, there are numerous ways of incorporating ecological and socioeconomic goals at the community level (Levang *et al.*, 2005). For example, if communities passively rely on forest resources such as clean water and soil qualities, then conservation in itself can be seen as beneficial with minimal attention given to livelihood development. But in communities that rely heavily on forest resources, such as logging or hunting, strict protection will increase poverty. In these settings, providing alternative livelihoods or pursuing other strategies of poverty reduction may also benefit conservation objectives (Adams *et al.*, 2004). In many instances of very remote and poor places, providing basic healthcare, education facilities, and access to markets can deter communities from extracting natural forest for income (Levang *et al.*, 2005). This approach also shows that well-being is not always measured by poverty rates. Rural households often lack earned income, but if they have land to farm, access to natural resources and basic amenities, their well-being would be defined by these factors rather than by income alone (Levang *et al.*, 2005). As such, within subsistence economies, one cannot infer the level of well-being within a household based on income.

Forest dwelling communities and indigenous tribes have for centuries depended on forest resources for their livelihoods. There has been increasing interest in the contribution that natural forests make to local employment and income generating activities (Arnold and Towns 1998; Chileshe 2005). The people inhabited in and around the protected areas (PAs) live in a subsistence economy with little or no access to market, education, health and other sanitation services and try to improve their living standards by extracting more forest resources (Godoy *et al.*, 1998; Fisher 2004) that may result in serious implications on conservation of biodiversity and natural habitats. Nevertheless, the PAs are subjected to pressure from these human extractions/dependencies in terms of grazing, cutting trees for firewood and timber, extraction of non-timber forest products (NTFPs), hunting, *etc.* (World Bank 1995; Amacher *et al.*, 1999; Heltberg *et al.*, 2000; Linde-Rahr 2003).

Forest-dependent communities in the tropics typically rank lower in socioeconomic status than agricultural and urban communities, and improving livelihood choices while protecting forest resources can be a difficult task. Conflicts can arise where biodiversity conservation objectives restrict resource access to forest communities (Chechinaa *et al.*, 2018).

Increasing natural forest protection and reforestation efforts are required to offset the effects of deforestation. However, most natural forests, even strictly protected, are inhabited by human communities (Nagendra *et al.*, 2009). In the absence of other options for employment and income, these communities are often the poorest and the most depend on natural resources for their livelihoods (Adams *et al.*, 2004; Levang *et al.*, 2005).

Biodiversity Conservation and Non-Timber Forest Products

Forest dwelling communities and indigenous tribes have for centuries depended on forest resources for their livelihoods. There has been increasing interest in the contribution that natural forests make to local employment and income generating activities (Arnold and Towns 1998; Chileshe 2005). The linkages between biodiversity conservation and indigenous communities world over and their livelihood dependent on forest resources was recognised by United Nations. The Agenda 21, Chapter 11 adopted during the United Nations Conference on Environment and Development (UNCED) in 1992 strives for combating deforestation, includes the promotion and development of NTFPs through value addition, domestic processing, and promotion of small-scale forest-based enterprises for rural income and employment. On December 17, 2007, the UN General Assembly adopted the Non-Legally Binding Instrument on All Types of Forests (2007). Among the four global forest objectives, the second objective speaks about - enhancing forest-based economic, social and environmental benefits, including by improving the livelihoods of forest dependent people.

NTFP are defined as "Non-timber forest products consist of goods of biological origin other than wood, derived from forests, other wooded land and trees outside forests" (FAO 1999). Non-Timber Forest Products (NTFPs) have been welfare, subsistence or livelihood commodity since long and are traditional source of food, fibre, medicine, *etc.* for the forest dwellers. The NTFPs are bioresources which include fruits, nuts, vegetables, medicinal plants, gum, resins, essences, bamboo, rattans and palms, fibres and flosses, grasses, leaves, seeds, honey and lac, *etc.* The forest dwellers mainly belong to tribal communities. There is strong evidence that forest products play a significant role in the livelihoods of the world's rural poor. Forest products are the main source of income for the forest dwelling population in many countries including India. It is estimated that of the 6.2 billion people on the planet, 25 per cent depend to varying degrees on the forest's resources for their livelihood and 350 million people living in or near dense forest depend highly on them for their subsistence or livelihood (Killman, 2003). Some 80 per cent of the people living in developing countries depend on non-timber forest products, such as fruits and herbs, for their primary health and nutritional needs (FAO 2008; www.agroforestry.net). The most used categories of NTFPs are medicine, edible, miscellaneous and fuel wood (Bouri and Mukharjee, 2013).

It is estimated that 275 million poor rural people in India - 27 per cent of the total population - depend on NTFPs for at least part of their subsistence and cash livelihoods (Malhotra and Bhattacharya 2010). This dependency is particularly intense for half of India's 89 million tribal people, the most disadvantaged section of society, who live in forest fringe areas. According to an estimate the NTFP sector alone is able to create about 10 million workdays annually in the country. Sizable populations of rural and tribal poor are involved in NTFP collection, processing and marketing (Yadav and Dugaya, 2013).

Non-timber forest products (NTFPs) are of socio-economic and cultural importance for the forest dwelling communities, particularly for the tropical countries like India (Yadav and Dugaya 2013). More recently, NTFPs have received international attention with a steady growth of green consumerism; awareness of biodiversity conservation and sustainable and protective use of forest resources; paradigm shift in forest management from 'sustained yield' to conservation focused; importance of joint forest management for sustainable management of forest resources; more open international markets; *etc.* (Yadav and Durgaya, 2013).

Biodiversity and Livelihood Opportunities: Exampls from Odisha

Odisha, an eastern state of India is the second largest state producing NTFPs and 37 per cent of its population depend on NTFPs for their day to day requirements (Kandari *et al.*, 2012). In Odisha, up to 40 per cent of the income of the rural people comes from the collection of forest products (Dash 2001). A variety of NTFPs from forests of Odisha are presented in Table 1.1. Many major and minor NTFPs are collected in Odisha from various parts of forest trees such as sal seed, kendu leaves, lac, broom grass, bahada, harida, amla, karanj, *etc.* The collected NTFPs were used in different purposes *i.e.* medicines, vegetables, essential oils, dye yielding, food items and other miscellaneous items by the tribal people (Panigrahi *et al.*, 2019). They are also important in generating income through harvesting, processing and trading of products such as gums, resins, oils, leaves, fruits, fodder, honey, mushrooms, tanning materials, bush meat, medicine (Anup, 2014). Fuel wood, fodder and tooth brush are collected and marketed throughout the year while sal leaf plates, lac, fruits and vegetables are sold for 9-10 months by the tribes. Oilseeds and ethno-medicine are vended for 5-6 months whereas mahua flower is traded only during the months of summer, spring and autumn its abundance is seen against to rainy season (Islam *et al.*, 2017).

Table 1.1. Common Non-Timber Forest Produce (NTFP) of Odisha

Sl. No.	*Scientific Name*	*Common Name/ Local Name*	*Uses of Plant parts*
1	*Albizia procera*	Silver bark Rain Tree/Siris	Whole plant is used as wood
2	*Andrographis paniculata*	Green Chiteta/Bhuineema	Whole plant is used in ayurverdic medicine
3	*Bamboo* sp.	Bamboo/Baunsha	Variety of uses. Used in construction, decorations, cooking, weapon
4	*Boswellia serrata*	Indian Francincense/Siali Gogul	Fruits are used as medicines
5	*Buchanania lanzan*	Chironji/Char	Seeds are eaten in raw or roasted. Often used in sweets
6	*Cinnamomum tamala*	Indian bay leaf/Tejpatta	Leaves are used as spices
7	*Diospyros melanoxylon*	Coromandel Ebony/Kendu	Leaves are used for making of beedi (mini cigar)
8	*Eulalia trispicata*	Sabai grass	Whole plant is used for rope makmimg. Also used in paper industry
9	*Madhuca longifolia*	Mahua/Mahulo	Flower is used to produce alcoholic drink
10	*Phylanthus embalica*	Indian Gooseberry/Amla	Fruits are used for medicinal purpose
11	*Pongamia pinnata.*	Indian Beech Tree/Karanj	Fruits used as herbal medicines and oil is used for a variety of purposes
12	*Rauvolfia serpentina*	The Indian Snakeroot/ Patalgaruda	It is a source of the phytochemical, reserpine, which has been used in the treatment of systolic hypertension
13	*Salvia hispanica*	Chia	Seeds are used as food
14	*Schleichera oleosa*	Gum Lac Tree/Kusum	Seeds are used as oil seed and medicines
15	*Shorea robusta*	Sal Tree/Sal	Laeves are used for making and resin is used as astringent in ayurvedic medicines
16	*Sterculia urens*	Indian Tragacanth/Genduli	Gums and resins are used for medicinal purpose
17	*Tamarindus indica*	Tamarind/Tentuli	Fruit contains a brown, edible pulp used in cuisines around the world.
18	*Terminalia bellirica*	Bastard Myrobaln/Bahada	Fruits are used as food and medicine for stomach problem
19	*Terminalia chebula*	Black Myrobalan/Harida	Used as food and medicine
20	*Thysanolaena maxima*	Tiger Grass/Broom grass	Used to feed livestock, soil conservation, staking in vegetable growing fields.
21	*Woodfordia frusticosa*	Dhataki	Flower is used for medicinal purpose

Source: Rout *et al.* (2010) and Panigrahi *et al.* (2019).

The Similipal Biosphere Reserve (SBR) in the Indian State of Odisha is the sixth largest biosphere reserve in the country. It is situated in the district of Mayurbhanj in which tribes occupy a bigchunk of the population constituting 52 per cent of it (Naik , 1998). The major tribes are Santal, Kolhs, Bathudi, Kharia, Mankidia,Gond, Ho, Bhuyan, Paudi and Bhuyan. Some ofthe tribes namely Kharia, Mankdias and Saharas are still in primitive state of living. They depends only on their surrounding plants for more of their requirements from food to medicines (Rout 2004; Saxena *et al.*, 1988). Kharias mostly collect non-timber forest products like honey, resin, lac and arrowroot *etc.* Mankidias collect mostly Siali fibre and use it for rope making. Some of the tribes like Gond and Bathudi spend their time in collecting Sabai grass from the forestland. They hardly come out of the forests to the plain area only to collect salt, potato, or onion in exchange of honey and resin collected by them (Pattnaik, 1997).

Bhitarkanika Wildlife Sanctuary located in the east coast of India in the state of Odisha harbours one of India's largest mangrove forest in terms of number of species and levels of variation in ecological conditions. The conservation values for Bhitarkanika are very high, since it the only site in mainland India that's has enough mangrove and associates to provide a secure habitat to many endangered species, especially reptiles. The protected area also provides an undisturbed mass-nesting beach for Olive Ridley Turtles. In the cyclone in 1999, the entire Paradeep area, south of this site was devastated, while at this site the damage was minimal in the peripheral areas. In fact the locals spoke of no damage at all in the core areas of the protected zones. The livelihood benefits of Bhitarkanika mangrove has been described under two categories- consumptive use value and productive use value (Kanvinde 1997). Consumptive use value are the produce from the forest that are used are: timber, fuel-wood, bark, grasses, vegetables, medicines, fish and crabs and prawn seeds and productive use values are the ecosystem produces fish, prawn,honey, fuel wood and timber. Kanvinde (2003) reported two NTFPs, honey and a grass used to make baskets.

Honey

The local name for honey is *'Mahu'*. There are a group of 35 adivasi households called 'Dalai' living in Dangmal village sell honey to the Forest Department and distribute the profit amongst themselves (Chadha and Kar, 1999). Good quality honey is collected mainly in the flowering season of *'Khorsi'* (*Aegiceras corniculatum*), and of *'Garani'* (*Ceriops decandra*).

Basket and Mat Making

Every household makes baskets, mats, hand fans, and various other items of the leaves and bark of mangrove trees. The bark fibres of 'Bahumurga' (*Flagellaria indica*), is used for rope making. The leaves of 'Hental' (*Phoenix paludosa*) are used for mat making. Harijan communities (Scheduled Caste) from the sanctuary area make mats and baskets out of 'Naalia grass'(*Myriostachya wightiana*).

Under-utilized Crops Resources of Koraput and Tribal Livelihood

Koraput district of Indian state of Odisha is the home for a number of tribal communities and all have their own culture, customs, traditions and practices in connection with their life style and livelihood (Mishra *et al.*, 2012). The plant genetic repository of the region is of great significance in the global context, and notably Koraput is one of the centres of diversity for many food crops and forest species (Mishra and Chaudhury, 2012).Recently Koraput has been declared by Food and Agriculture Organisation (FAO) as the global agricultural heritage site. Historically, tribal and rural people identified and collected plants for food and medicine from forests and developed a range of processing methods in accordance with their needs. With modernization and settled agriculture, this knowledge is becoming lost, a trend that may lead to decreased diversity of indigenous diets and poorer nutrition. The last five decades have witnessed a rapid decline in the genetic diversity of various crops and a depletion of forest species in the region (Nayar *et al.*, 2009).Now we can highlight some of the under-utilized crops resources of Koraput and their livelihood implications for improving health and nutritional security (Table 1.2).

Table 1.2. List of Plant Biodiversity for Improving Tribal Livelihood of Koraput

Sl.No.	*Purpose*	*Plant species*
1.	Food	*Amaraphophallus, Borassus, Cajanus, Caralluma, Cinnamommum, Citrus, Curcuma, Discoria, Glycine, Lablab, Limonia, Luffa, Mangifera, Momocardia, Musa, Oryza, Phylanthus, Phoenix, Piper, Rumex, Rubus, Sesamum, Setaria, Solanum, Sorghum, Syzygium, Eleusine Zinziber, Dioscorea etc.*
2.	Timber	*Boswellia, Chloroxylon, Gmelina, Hardwichia,Pittisporum, Polyalthia, Shorea, Toona, Zanthozylum etc.*
3.	Horticultural and ornamental	*Argyria, Bauhinia, Clematis, Cyanotis, Cycas, Dysophylla, Habenaria, Hardwickia, Oleo etc.*
4.	Medicinal purpose	*Aegle, Caesalpinia, Costrus, Cissus, Embellica, Gymnea, Hypericum, Piper,Pogostemon, Psychortia, Rouvolfia. Sarcostomma, Solanum, Strychnos, Terminalia,Tinospora, Tylophora, Vanda etc.*
5.	Fibre	*Crotolaria* and *Decaschistia*

Mishra (2009) reported 340 landraces of paddy, 8 species of minor millets, species of pulses, 5 species of oil seeds, 3 species of fibrous plants and 7 species of vegetables enabling tribal families to derive economic benefit in this region. Arunachalam *et al.* (2006) recorded some of the indigenous rice landraces with better cooking, possess good quality grains (Machhakanta, Haladichudi, Para Dhan, Muktabali, Sapuri and Umuriachudi) and Roy *et al.* (2016) reported different aromatic rice landraces such as Kalajeera, Gangabali, Kuyerkuling, Deulabhoga, Lactimachi, Sapuri, Dudhamani, Muktabali and Nadiarasa of this region and claimed to have superior aroma along with grain quality traits. Some of the medicinal rice landraces of Koraput was also reported by Edwina *et al.* (2014), theses are mostly pigmented rice landraces (Bhatamalli, Kandulakath, Haladiganthi, Kalachudi, Mallimakada, Maligoindi, Beda Gurumukhi). Different tribal communities cultivate various millet species importantly Mandia (*Eleusine coracana* (L.) Gaertn.), black and white Suan (*Panicum sumatrense* Roth. ex Roem. et Schult.) and black, white and red Kangu (*Setaria italica* (L.) P. Beauv.) (Mishra and Chaudhury, 2012). Grains and flour of small millets are important components of traditional diets and bioavailability of indigenous millets of Koraput to promote its utilization and bioprospecting.

A number of wild plants used by tribal populations contributed significantly to their livelihood and food security. Utilization of locally available resources in their diet among the tribal community is distinctly reflected in their traditional food habits. The tribal people possess rich knowledge and wisdom regarding wild edible plants including their usage (Swain *et al.*, 2013). Their distribution, conservation, mode of harvest by locals and optimal use require region-specific assessment in order to integrate them into developmental interventions. Recently, Padhan and Panda (2015) took up documentation of the wild edible plants and their importance towards food security of rural poor tribes of Koraput District by collecting data from 20 tribal villages belonging to six primitive tribal groups like Paroja, Bhumia, Gadaba, Bhatra, Koya and Kondha. A total of 156 plant species belonging to 105 genera and 71 families have been documented as wild edible plants in the study area. Of which, the most utilized part is green leaves rank first 51 species, tubers or underground parts, fruits and flowers are 38, 60 and 7 species respectively. Padhan *et al.* (2020) reported eight wild yam species (*Dioscorea oppositifolia* L., *D. hamiltonii* Hook.f., *D. bulbifera* L., *D. pubera* Blume., *D. pentaphylla* L., *D. wallichii* Hook.f., *D. glabra* Roxb. and *D. hispida* Dennst) are very popular and have been used as food by the tribal people of Koraput. Popularisation of these valuable resources is expected to lead to a path of improving tribal livelihood.

CONCLUSION

Biological diversity is fundamental to the fulfilment of human needs. An environment rich in biological diversity offers broadest array of options for sustainable economic activity. While challenges of Climate Change are looming large on our natural ecosystems, along with biodiversity assessments, there is the necessity of systematic planning on the part of conservationist for conservation of habitats and its pristine fauna and flora. Therefore, conservation and sustainable use of biodiversity is fundamental to ecologically sustainable development. Strategic approaches that integrate biodiversity conservation and livelihood goals have been part of the conservation toolbox for decades. While enough awareness and actions are required to mitigate climate change related vulnerabilities, the potential economic value of biodiversity should be translated into economic benefits for the communities. It is now required that our rich biodiversity need to be conserved for protection of life and livelihood of people as biodiversity is not an optional bonus in human affairs, but the very foundation of our existence.

REFERENCES

Abrams, J., Davis, E.J., Moseley, C. and Nowell, B. (2017). Building practical authority for community forestry in and through networks – The role of community based organisations in the U.S. West. *Environmental Policy and Governance*, 27 (4): 285-297.

Adams W.M., Aveling, R., Brockington, D., Dickson, B., Elliott, J., Hutton, J., Roe D. *et al.* (2004). Biodiversity conservation and the eradication of poverty. *Science*, 306(5699): 1146-1149.

Adams, H. D., Guardiola-Claramonte, M., Barron-Gafford, G.A., Camilo Villegas, J., Breshears, D.D., Zou, C.B., Troch, P.A. and Huxman, T.E. (2009). Temperature sensitivity of drought-induced tree mortality portends increased regional die-off under global-change-type drought. *Proceedings of the National Academy of Sciences,* 106(17): 7063–7066.

Altieri, M.A. (1990). Agroecology and rural development in Latin America. In: Altieri, M.A. and Hecht, S.B. (eds.) *Agroecology and Small Farm Development,* CRC Press, Florida, pp.113–118.

Amacher, G.S., Hyde, W.F., and Kanel, K.R. (1999). Nepali Fuelwood Production andConsumption: Regional and Household Distinctions, Substitution and Successful Intervention. *The Journal of Development Studies,* 35, (4): 138–163.

Andam, K.S., Ferraro, P.J., Pfaff, A., Sanchez-Azofeifa, G.A. and Robalino, J.A. (2008). Measuring the effectiveness of protected area networks in reducing deforestation. *Proceedings of the National Academy of Sciences,* 105(42): 16089-16094.

Anup, K.C. (2014). Opportunity Cost, Willingness to Pay and Cost Benefit Analysis of a Community Forest of Nepal. *International Journal of Environment,* 3: 108-124.

Arnold, M. and Townson, I. (1998). Assessing the Potential of Forest Product Activities to Contribute to Rural Incomes in Africa, *Natural Resource Perspectives,* 37, (November), 1-10.

Arunachalam, V., Chaudhury, S.S., Sarangi, S.K., Ray, T., Mohanty, B. P., Nambi, V. A. and Mishra, S. (2006). Rising on rice: The story of Jeypore. *Chennai: MS Swaminathan Research Foundation, 1:* 39.

Bellard, C, Bertelsmeier, C., Leadley, P., Thuiller, W., and Courchamp, F. (2012). Impacts of climate change on the future of biodiversity. *Ecology Letters,* 15: 365–377.

Bhattacharya, P. and Hayat, S.F. (2004). Sustainable NTFP management for rural development: a case from Madhya Pradesh, India. *International Forestry Review,* 6: 161-168.

Bouri, T. and Mukherjee, A. (2013). Documentation of traditional knowledge and Indigenous use of non- timber forest products in Durgapur forest range of Burdwan district, West Bengal. Paper presented in National Seminar on Ecology, Environment and Development 25 - 27 January, 2013, organized by Department of Environmental Sciences, Sambalpur University, Sambalpur.

Brauman, K.A., Daily, G.C., Duarte, T.K. and Mooney, H.A. (2007). The Nature and Value of Ecosystem Services: An Overview Highlighting Hydrologic Services. *The Annual Review of Environment and Resources,* 32: 6.1–6.32.

Bruijnzeel, L.A. 2004. Hydrological functions of tropical forests: not seeing the soil for the trees? *Agriculture, Ecosystems and Environment,* 104: 185–228.

Burrows, M.T., Schoeman, D.S., Buckley, L.B., Moore, P., Poloczanska, E.S., Brander, K.M., Brown, C., Bruno, J.F., Duarte, C.M., Halpern, B.S., Holding, J., Kappel, C.V., Kiessling, W., O'Connor, M.I., Pandolfi, J.M., Parmesan, C., Schwing,

F.B., Sydeman, W.J., and Richardson, A.J. (2011). The pace of shifting climate in marine and terrestrial ecosystems. *Science,* 334: 652-655.

Cernea, M.M. and Schmidt-Soltau. K. (2006). Poverty risks and national parks: policy issues in conservation and resettlement. *World Development,* 34(10): 1808-1830.

Chadha, S. and Kar, C.S. (1999). *Bhitarkanika: Myth and Reality.* Natraj Publishers, Dehradun. P. 388.

Chambers, R., and Conway, G. (1991). Sustainable Rural Livelihoods: Practical Concepts for the 21st Century. Retrieved February 3, 2010. http: //www. smallstock.info/reference/IDS/dp296.pdf

Chechinaa, M., Neveuxa, Y., Parkinsb, J.R. and Hamanna, A. (2018). Balancing Conservation and Livelihoods: A Study of Forest-dependent Communities in the Philippines. *Conservation and Society,* 16(4): 420-430.

Chen, I.C., Hill, J.K., Ohlemuller, R., Roy, D.B., and Thomas, C.D. (2011). Rapid range shifts of species associated with high levels of climate warming. *Science,* 333: 1024-1026.

Chileshe, R.A. (2005). Land Tenure and Rural Livelihoods in Zambia: Case Studies of Kamena and St. Joseph, PhD Thesis, University of Western Cape.

Cunningham, A.B., Sherr, S.J., McNeely, J.A. (2002). Matrix Matters: Biodivesrity Research for Rural Landscape Mosaics. http: //www.cifor.cgiar.org/ publications/pdf_files/research/environment/Matrix_Matters.pdf

Dash, S. and Misra, M. (2001). Studies on hill agro-ecosystems of three tribal villages on the Eastern Ghats of Odisha, India. *Agriculture Ecosystems and Environment,* 86(3): 287-302.

Dawson, T.P. (2011). Beyond predictions: Biodiversity conservation in a changing climate (vol 332, p. 53, 2011). *Science,* 332: 664-664.

Doak, D.F. and Morris, W.F. (2010). Demographic compensation and tipping points in climate-induced range shifts. *Nature,* 467: 959-962.

Doney, S.C., Ruckelshaus, M., Duffy, J.E., Barry, J.P., Chan, F., English, C.A., Galindo, H.M., Grebmeier, J.M., Hollowed, A.B., Knowlton, N., Polovina, J., Rabalais, N.N., Sydeman, W.J., and Talley, L.D. (2012). Climate change impacts on marine ecosystems. *Annual Reviews in Marine Science,* 4: 4.1–4.27.

Dudley, N., and Stolton, S. (2003). *Running pure: the importance of forest protected areas to drinking water.* World Bank/WWF Alliance for Forest Conservation and Sustainable Use.

Edwina, P., Ajitha, C., Eganathan, P., Gayathri, S., Saranya, J., Arivudai Nambi, V., Smita, M. and Susanta Sekhar, C. (2014). Biochemical and antioxidant activities of pigmented landraces of *Oryza sativa* - Koraput District, Odisha, India. *International Food Research Journal,* 1(5): 1941-1949.

Elmqvist, T., Maltby, E., Baker, T. *et al.* (2010). Chapter 2: Biodiversity, ecosystems and ecosystem services. *The Economics of Ecosystems and Biodiversity,* 1-96.

Ewel, J. (1986). Designing agricultural systems for the humid tropics. *Annual Review of Ecology and Systematics,* 17: 245–271.

FAO (1999) accessed from http: //www.fao.org/docrep/x2450e/x2450e0d.htm on 05.03.2013

FAO (2006). Global Forest Resources Assessment 2005: Progress Towards Sustainable Forest Management. Food and Agriculture Organization of the United Nations. Rome, Italy.

FAO (2008). Non Wood Forest Products, Rome, Italy.

FAO (Food and Agriculture Organization) (1999). Sustaining agricultural biodiversity and agro-ecosystem functions: Opportunities, incentives and approaches for the conservation and sustainable use of agricultural biodiversity in agro-ecosystems and production systems. Food and Agriculture Organization of the United Nations. Rome, Italy.

Ferraro, P.J., M.M. Hanauer, and K.R.E. Sims. 2011. Conditions associated with protected area success in conservation and poverty reduction. *Proceedings of the National Academy of Sciences,*108(34): 13913-13918.

Fischer, J., Brosi, B., Daily, G.C., Ehrlich, P.R., Goldman, R., Goldstein, J., Lindenmayer, D.B., Manning, A.D., Mooney, H.A., Pejchar, L., Ranganathan, J. and Tallis, H. (2008). Should agricultural policies encourage land sparing or wildlife-friendly farming? *Frontiers in Ecology and the Environment,*6(7): 380–385.

Fisher, M. (2004). Household Welfare and Forest Dependence in Southern Malawi. *Environment and Development Economics,* 9 (2): 135–154.

Geyer, J., Kiefer, I., Kreft, S., Chavez, V., Salafsky, N., Jeltsch, F., and Ibisch, P.L. (2011). Classification of Climate-Change-induced stresses on biological diversity. *Conservation Biology,* 25: 708-715.

Godoy, R. *et al.* (1998) Strategies of Rain-forest Dwellers Against Misfortunes: The Tsimane'Indians of Bolivia. *Ethnology,* 37 (1): 55– 69.

GoI – Government of India (2011). Agriculture. http://india.gov.in/sectors/agriculture/index.php

Gómez, A. (2017). **Integrating Livelihood and Conservation Goals-** A Retrospective Analysis of World Bank Projects. USAID Technical Brief. Environmental Incentives, LLC, Foundations of Success, and ICF. United States Agency for International Development (USAID)

Gopalakrishnan, R., Jayaraman, M., Malaviya, S., Rao, A.S., Sagadevan, A., Munsi, M., Indu, K., Tiwari, M.R., Ravindranath, N.H. (2011). Impact of climate change on Indian Forests. *Current Science,* no. 3.

Guiteras, R. (2007). The Impact of Climate Change on Indian Agriculture. http: //www.colgate.edu/portaldata/imagegallerywww/2050/ImageGallery/Guiteras Paper.pdf

Hajjar, R., Jarvis, D.I. and Gemmill-Herren, B. (2008). The utility of crop genetic diversity in maintaining ecosystem services. *Agriculture, Ecosystems and Environment* 123: 261–270.

Hansen, M.C., Potapov, P.V., Moore, R., Hancher, M., Turubanova, S.A., Tyukavina, A., Thau, D. *et al.* (2013). High-resolution global maps of 21st-century forest cover change. *Science,* 342 (6160): 850-853.

Hanski, I. (2005). Landscape fragmentation, biodiversity loss and the societal response. *EMBO Rep,* 6(5): 388–392.

Harlan, J. R. (1975). *Crops and man.* American Society of Agronomy, Madison Wisconsin, 295 p.

Heltberg, R., Arndt, T. C., and Sekhar, N. U. (2000). Fuelwood Consumption and ForestDegradation: A Household Model for Domestic Energy Consumption in Rural India. *Land Economics,* 76 (2): 213–232.

Heywood, V.H. (1999). *Use and potential of wild plants in farm households.* Food and Agriculture Organization, Rome. http: //agroforestry.net/overstory/ overstory55.html

Huq, S., Reid, H., Konate, M. *et al.* (2004). Mainstreaming adaptation to climate change in Least Developed Countries (LDCs). Climate Policy 4(1): 25-43

IFPRI – International Food Policy Institute (2009). Climate Change Impact on Agriculture and Costs of Adaptation. http: //www.undp-adaptation.org/ undpcc/files/docs/publications/pr21.pdf

IPCC (2018). Global Warming of 1.5 ?C. Intergovernmental Panel on Climate Change. IPCC, Switzerland.

Islam, M.A. and Quli, S.M.S. (2017). The Role of Non-Timber Forest Products (NTFPs) in Tribal Economy of Jharkhand, India. *International Journal of Current Microbiology and Applied Sciences,* 6(10): 2184-2195.

Joshi, J., Schmid, B., Caldeira, M.C., Dimitrakopoulos, P.G., Good, J., Harris, R., Hector, A., Huss-Danell, K., Jumpponen, A., Minns, A., Mulder, C.P.H., Pereira, J.S., Prinz, A., Scherer-Lorenzen, M., Siamantziouras, A.-S.D., Terry, A.C., Troumbis, A.Y. and Lawton, J.H. (2001). Local adaptation enhances performance of common plant species. *Ecology Letters* 4(6): 536–544.

Kandari, L.S., Gharai, A.K., Negi, T. and Phondani,.PC. (2012) Ethno botanical Knowledge of Medicinal Plants among Tribal Communities in Odisha, India. *Journal Forest Research Open Access,* 1: 104.

Kanvinde, H. (2003). Bhitarkanika. In: Bioresources Status in Select Coastal Locations. NBDB, New Delhi and MSSRF, Chennai. pp. 177-200.

Killman, W.(2003). "Non-wood News", n.10, March 2003, p.1.

Koehler, P., Chave, J., Ri´era, B. *et al.* (2003) Simulating the long-term response of tropical wet foreststo fragmentation. *Ecosystems,* 6(2): 114–128.

Kummer D.M. (1992). Upland agriculture, the land frontier and forest decline in the Philippines. *Agroforestry Systems,* 18(1): 31-46.

Kummer D.M. and Turner, B.L.I. (1994). The human causes of deforestation in Southeast Asia. *Bioscience,* 44(5): 323-328.

Laurance, W., Salicrup, P., Delamônica, P., Fearnside, P., D'Angelo, S., Jerozolinksi, A., Pohl, L. and Lovejoy, T. (2001). Rain forest fragmentation and the structure of Amazonian liana communities. *Ecology*, 82: 105–116.

Levang, P., Dounias, E., and Sitorus, S. (2005). Out of the forest, out of poverty? *Forests, Trees and Livelihoods*, 15(2): 211-235.

Linde-Rahr, M. (2003). Property Rights and Deforestation: The Choice of Fuelwood Source in Rural Vietnam. *Land Economics*, 79 (2): 217–234.

MA (2005). Millennium Ecosystem Assessment. *Ecosystems and Human Well-being: Synthesis.* Island Press, Washington, D.C.

Malhotra, K.C. and Bhattacharya, P. (2010). Forest and Livelihood. CESS Publisher, Hyderabad, India, Pages: 246.

McElwee, P.D. (2010). Resource use among rural agricultural households near protected areas in Vietnam: the social costs of conservation and implications for enforcement. *Environmental Management*, 45(1): 113-131.

Miller-Rushing AJ, Hoye TT, Inouye DW, and Post E. 2010. The effects of phenological mismatches on demography. *Philosophical Transactions of the Royal Society B-Biological Sciences*, 365: 3177-3186.

Mishra, S. (2009). Farming system in Jeypore Tract of Odisha, India. *Asian Agri History*, 13(4): 271-292.

Mishra, S. and Chaudhury, S. S. (2012). Ethno-botanical flora used by four major tribes of Koraput, Odisha, India. *Genetic Resources and Crop Evolution*, 59(5): 793–804.

Mishra, S., Chaudhury, S. S. and Nambi, V. A. (2012). Strengthening of traditional seed selection practices with improved knowledge and skills of tribal farm families in Koraput District. *Indian Journal of Traditional Knowledge*, 11(3): 461-470.

MoEF(2004). *India's Initial National Communications to the United Nations Framework Convention on Climate Change*, Ministry of Environment and Forests, New Delhi.

Molnar, A., Liddle, M., Bracer, C., Khare, A., White, A. and Bull, J. (2007). Community-based forest enterprises their status and potential in tropical countries. ITTO technical series. Rome: Food and Agriculture Organization of the United Nations.

Montoya, J.M., and Raffaelli, D. (2010). Climate change, biotic interactions and ecosystem services. *Philosophical Transactions of the Royal Society B-Biological Sciences*, 365: 2013-2018.

Nagendra H., Paul, S., Pareeth, S. and Dutt, S. (2009). Landscapes of protection: Forest change and fragmentation in northern West Bengal, India. *Environmental Management*, 44: 853-864.

Naik, D. (1998). Tribal culture in the context of Similipal. *Workshop Journal on Different Problematic Spects of Similipal Protection and Its Solutary Action Programmes*, pp. 63-64.

Naughton-Treves, L., Alix-Garcia, J. and Chapman, C.A. (2011). Lessons about parks and poverty from a decade of forest loss and economic growth around Kibale National Park, Uganda. *Proceedings of the National Academy of Sciences,* 108(34): 13919-13924.

Nayar, M. P., Singh, A. K. and Nair, K. N. (2009). Agrobiodiversity hotspots in India. *Conservation and Benefit Sharing*, vol 2. PPV and FR Authority, New Delhi. p 217.

Nye, J. A., Link, J. S., Hare, J. A. and Overholtz, W. J. (2009). Changing spatial distribution of fish stocks in relation to climate and population size on the Northeast United States continental shelf. *Marine Ecology-Progress Series,* 393: 111-129.

OECD – Organisation for Economic Co-operation and Development (2002). Climate change: India's perceptions, positions, policies and possibilities.http: //www.oecd.org/dataoecd/22/16/1934784.pdf

OECD–DAC. (2001). The OECD/DAC guidelines on poverty reduction. Available at: http: //www.oecd.org/dataoecd/47/14/2672735.pdf

Padhan, B., Biswas, M. and Panda D. (2020).Nutritional, anti-nutritional and physico-functional properties of wild edible yam (*Dioscorea* spp.) tubers from Koraput, India. *Food Bioscience*, 34, 100527.

Padhan, B., Panda D. (2015). Wild Edible Plant Diversity and its Ethno-medicinal use by Indigenous Tribes of Koraput, Odisha, India. *Research Journal of Agriculture and Forestry Sciences*, 3(9): 1-10.

Panigrahi, S., Meher, D. and Siri, P. (2019). Reliance and livelihood significance of non-timber forest products available in Odisha: A review. *Journal of Pharmacognosy and Phytochemistry*, 8(1): 2428-2432.

Parmesan, C. (2006). Ecological and evolutionary responses to recent climate change. *Annual Review of Ecology Evolution and Systematics,* 37: 637-669.

Pattanaik, B.K. (1997). Non-Timber Forest Produce in Similipal Forest. A Natural Habitat of Unique Biodiversity. *Orissa Environmental Society, Bhubaneswar,* 146-155.

Porter-Bolland L., Ellis, E.A., Guariguata, M.R., Ruiz-Mallén, I., Negrete-Yankelevich, S. and Reyes-García, V. (2012). Community managed forests and forest protected areas: an assessment of their conservation effectiveness across the tropics. *Forest Ecology and Management,* 268: 6-17.

Ravindranath, N. H. and Sathaye, J. (2002). *Climate Change and Developing Countries,* Kluwer Academic Publishers, Dordrecht, Netherlands.

Rockström, J., Steffen, W., Noone, K., Persson, Å., Chapin, F.S., Lambin, E., Lenton, T.M., Scheffer, M., Folke, C., Schellnhuber, H.J., Nykvist, B., de Wit, C.A., Hughes, T., van der Leeuw, S., Rodhe, H., Sörlin, S., Snyder, P.K., Costanza, R., Svedin, U., Falkenmark, M., Karlberg, L., Corell, R.W., Fabry, V.J., Hansen, J., Walker, B., Liverman, D., Richardson, K., Crutzen, P. and Foley, J. (2009). Planetary Boundaries: Exploring the Safe Operating Space for Humanity. *Ecology and Society,* 14(2): 32. (online).

Ross, A.B., Jones, J.M., Kubacki, M.L. and Bridgeman, T. (2008). Classification of macroalgae as fuel and its thermochemical behaviour. *Bioresource Technology*, 99: 6494–6504.

Rout, S.D. (2004). *Medicinal Plants of Similipal Biosphere Reserve*, Ph. D. Thesis. TM Bhagalpur University, Bhagalpur.

Rout, S.D., Panda, S.K., Mishra, N. and Panda, T. (2010). Role of Tribals in Collection of Commercial Non-TimberForest Products in Mayurbhanj District, Odisha. *Stud Tribes Tribals*, 8(1): 21-25.

Roy, P. Patnaik, S. A., Rao, G. J. N., Patnaik, S. S. C., Chaudhury, S. S. and Sharma, S. G. (2016). Participatory and molecular marker assisted pure line selection for refinement of three premium rice landraces of Koraput, India. *Agroecologyand Sustainable Food systems*, 41(2): 167-185.

Saastamoinen, O. (1996). Change and continuity in the Philippine forest policy. In: *Sustainable forestry challenges for developing countries* (eds. Palo M. and G. Mery). Pp.121-139. Dordrecht, the Netherlands: Springer.

Salganik, M.J. and Heckathorn, D.D. (2004). Sampling and estimation in hidden populations using respondent-driven sampling. *Sociological Methodology*, 34(1): 193-240.

Sathaye, J., Shukla, P. R. and Ravindranath, N. H. (2006) Climate change, sustainable development and India: Global and national concerns. *Current Science*, 9 (3): 314-325.

Saxena, H.O., Brahmam, M., Dutta, P.K. (1988). Ethno botanical studies in Similipal Forests of Mayurbhanj Distict (Odisha). *Bull Bot Surv India*, 10(1-4): 83-89.

Shukla, P. R., Sharma, S. K., Ravindranath, N. H., Bhattacharya, S. and Garg, A. (eds) (2003). *Climate Change and India: Vulnerability Assessment and Adaptation*, Universities Press, Hyderabad,

Swain, S. and Mohapatra, G. C. (2013). Multiple usages of forest trees by the tribes of Kalahandi District, Odisha, India. *International Journal of Biodiversity and Conservation*, 5(6): 333-341.

UN (2009). *World Population Prospects: The 2008 Revision*. UN Population Division Policy Brief No. 2009/1, United Nations, New York. http: //www.un.org/esa/population/unpop.htm

UNFCCC – United Nations Framework Convention on Climate Change (2009). Climate change: Impacts, vulnerabilities and adaptation in developing countries. http: //unfccc.int/resource/docs/publications/impacts.pdf

Williams, J. and Haq, N. (2002). Global research on underutilized crops: an assessment of current activities and proposals for enhanced cooperation. ICUC, Southampton, UK.

Wilshusen, P.R., Brechin, S.R., Fortwangler, C.L. and West, P.C. (2002). Reinventing a square wheel: critique of a resurgent "Protection paradigm" in international biodiversity conservation. *Society and Natural Resources*, 15(1): 17-40.

World Bank Report (1995). The Environmental Sector in Viet Nam.Washington, DC: World Bank.

Yadav, M., Dugaya, D. (2013). Non-timber forest products certification in India: opportunities and challenges. *Environment, Development and Sustainability,* 15: 567–586.

Zhu, Y.Y., Chen, H., Fan, J., Wang, Y., Li, Y., Chen, J., Fan, J.X., Yang, S., Hu, L., Leung, H., Mew, T.W., Teng, P.S., Wang, Z. and Mundt, C.C. (2000). Genetic diversity and disease control in rice. *Nature,* 406: 718–722.

Zimmerer, K.S., Galt, R.E. and Buck, M.V. (2004). Globalization and multi-spatial trends in the coverage of protected-area conservation (1980–2000). *AMBIO: A Journal of the Human Environment,* 33(8): 520-529.

Chapter 2

Vegetation Diversity and Litterfall Production of Similipal Biosphere Reserve, Odisha, India

Rabindra Kumar Mishra

Department of Wildlife and Biodiversity Conservation,
North Orissa University, Takatpur, Baripada – 757 003, Mayurbhanj, Odisha
e-mail: rabikumishra@gmail.com

Litterfall has a strong influence on biodiversity and the Physico-chemical characteristics of the soil. Its production can be quite variable over time and space, and influenced by both natural and anthropogenic factors. The study evaluated litterfall production and its relationship with species richness, and some phytosociological parameters of the upper storey vegetation. Hundred litter traps were constructed with 0.25 m^2 nylon meshes (1.0 mm) and randomly installed within 40000 m^2 area of the Similipal Biosphere Reserve (SBR) to quantify the above ground litter production. Litter samples were collected monthly from October 2012 to September 2013, and the collected material was classified, dried, and weighted. Phytosociological parameters of tree species (Upper storey vegetation) like species richness, density, basal area, bio volume, and canopy area was analyzed at each study site of the reserve through laid down of 20 m × 20 m plots surrounding each of the litter traps. The litterfall accumulation rate was 352 g/m^2/Yr, similar to values from other seasonally dry tropical forests. Litterfall production was continuous, with the greatest accumulation at the beginning of the dry season and the least during the rainy season. The different fractions of materials demonstrated distinct accumulation rates, with fresh litter being the principal category. Litterfall production was found to be related to tree density, basal area, canopy area, and bio-volume but no link was found to species richness. However, the relationship between species richness of herbs, shrubs and climbers with litterfall production was perfectly negative. Furthermore, the observed spatial and temporal heterogeneity of litterfall production demonstrated a strong link between soil Physico-chemical characters and the litter turnover dynamics of SBR.

Keywords: *Biodiversity, Litter production, Litter turnover, Phytosociology, Spatial variability, Temporal variability, SBR.*

INTRODUCTION

In deciduous forest, tree species have the characteristics of shedding their leaves, twigs, flower and fruits annually which is termed as the litterfall. It plays an important role in forest ecosystems in the way of the return of organic matter and nutrients from aerial parts of the plant community to the soil surface and enhances the soil fertility through the decomposition of the forest floor litter (Pragasan and Parthasarathy, 2005; Knoepp *et al.*, 2005; Menezes *et al.*, 2020). Not only it is an essential part of nutrient cycling but also acts as a protective layer by buffering changes in soil water content, temperature and hindering erosion, leaching and soil compaction (Sayer, 2006). Such activities performed by the forest floor are responsible for maintaining the integrity of an ecosystem through the transfer of matter and nutrients (Rajendraprasad *et al.*, 2000). In tropical ecosystems, maintenance of soil organic pool is achieved by the high and rapid circulation of nutrients through the fall and decomposition of litter. Standing crop of litter (total forest floor material) acts as an input-output system of nutrients and the rates at which forest litter falls and subsequently decays, regulates energy flow, primary productivity and nutrient cycling in forest ecosystems (Sunderpandian and Swamy, 1999).

Plant diversity is influenced by litter depth (Grime, 1979) and the heterogeneous nature of litter cover on the forest floor promotes species coexistence by facilitating seedling emergence and influencing seedling survival throughout the forest (Facelli and Pickett, 1991 a; Tesfay, 2020). The quantity of litterfall varies greatly over a range of spatial and temporal scales and is determined mainly by climate, seasonality, topography, and soil parent materials and species distribution. Climate change may affect forest floor litter biomass as a change in rainfall patterns and mean annual temperatures can also affect tree phenology and tree species distribution (Condit *et al.*, 1996; Mishra *et al.*, 2011) and increase in productivity (Zak *et al.*, 2003). The forest floor litter layer acts as an interface between the soil surface and the atmosphere and provides a degree of protection to the soil surface by intercepting rain (Bankobi *et al.*, 1993) and solar radiation (Ogee and Brunet, 2002), and buffering the soil surface against fluctuations in temperature (Ponge *et al.*, 1993) and water content (Ginter *et al.*, 1979). Litter is also the major source of soil organic matter which strongly influences the structure of the soil and increases its stability (Marshall *et al.*, 1996). Soil organic matter often plays a role in determining the pH of the soil surface horizons, as pH is regulated by humic acids to a large extent in some soils (Wilke *et al.*, 1993). Litter and soil organic matter affect soil porosity and aeration indirectly, as they constitute the major food supply to the soil biota (Gonzalez and Zou, 1999).

Forest floor litter biomass directly or indirectly influence plant growth and survival. Changes to the soil water balance and the nutrient cycle constitute direct influences on plant growth as they modify the availability of resources. Soil compaction and other changes to the soil structure indirectly affect tree growth by diminishing the available pore space, thus inhibiting root expansion (Sayer, 2006). An increase or decrease in organic matter content can cause changes in pH, base saturation and the water retention capacity of the soil (Marshall *et al.*, 1996), all of which may influence nutrient uptake, and therefore potentially growth and survival of the vegetation. Furthermore, forest floor litter can influence seed germination and seedling establishment and survival both positively and negatively, changing competitive outcomes between tree seedlings and forbs (Barritt and Facelli, 2001) and thus contributing to the species composition and structure of the forest (Facelli and Pickett, 1991b). Naturally occurring patchiness of litter and differing litter depths and structure facilitate the establishment of seedlings of some species while inhibiting others and thus contribute to small scale heterogeneity of the vegetation and coexistence of species (Dzwonko and Gawronski, 2002 a, b). Apart from this canopy cover and composition of tree species accounts forest floor litter biomass and show considerable variation. This spatial heterogeneity allows the coexistence of species and contributes to niche differentiation for seedlings (Sayer, 2006). In light of the importance of litterfall to the dynamics of tropical ecosystems, especially tropical regions, the present study sought to evaluate litterfall production rates in SBR and examine their relationships with vegetation structure (phytosociological attributes) and soil Physico-chemical characteristics.

Study Area

Location

The present study was conducted for a period of one year *i.e.* from Oct. 2012 to Sept. 2013 at five different sites of SBR (21° 16′- 22° 08′ N, 86° 04′ – 86° 37′ E), located in the district of Mayurbhanj, covers an area of 5569 sq. km. Sites were randomly selected for in-depth analysis of phytosociological attributes of upper storey vegetation, edaphic characters and litter dynamics. The sites are located in buffer and transitional areas of the reserve at different altitudes and aspects (Table 2.1).

Table 2.1. Characteristic Features of Study Sites

Name of the Study Site	*Elevation (m)*	*Aspect*	*Longitude*	*Latitude*
Lulung (S-1)	120	East	21°55′-22° 00′	86° 30′-86° 35′
Handipuhan (S-2)	280	North	21°55′-22° 00′	86° 05′-86° 15′
Podadiha (S-3)	80	South-East	21°00′-21° 35′	86° 20′-86° 25′
Gurguria (S-4)	614	North-West	21°45′-21° 55′	86° 10′-86° 25′
Kalikaprasad (S-5)	468	North-West	21°50′-21° 55′	86° 10′-86° 25′

Climate

The climate of Similipal is tropical, humid, warm and monsoonal. The temperature and humidity are moderate throughout the year. The average annual rainfall ranges from 28.11 to 423.96 mm. Most of the precipitation occurs from June to August. The rest of the rainfall occurs in summer. Summer is not unbearable, as the maximum temperature rarely goes above 40 °C. The average dry period in a year lies between 7 and 8 months (Figure 2.1).

Vegetation

The vegetation of this region is described as tropical moist mixed deciduous type (Champion and Seth, 1968). All the study sites exhibit an average tree height

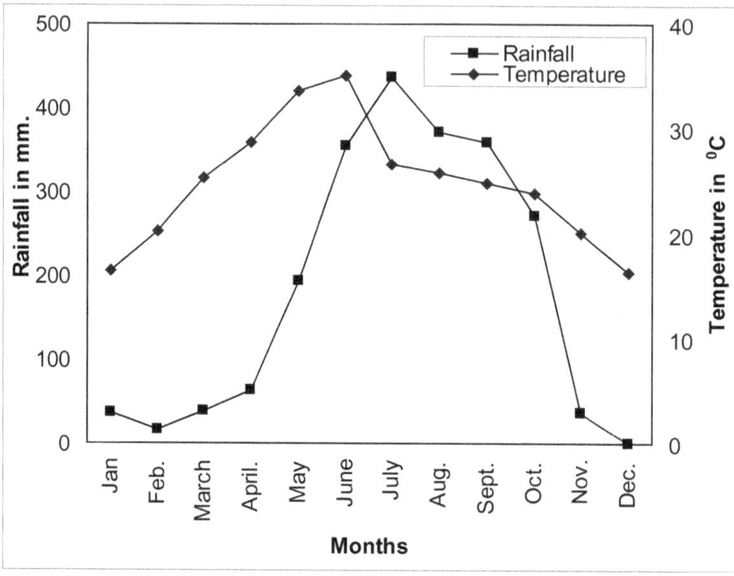

Figure 2.1. Monthly Variation in Temperature (°C) and Rainfall (mm) of the Similipal Biosphere Reserve.

of 8 to 10m, and also they are three-layered forests with a fair representation of trees in the middle and upper stories. *Shorea robusta* is the dominant plant species at all the study sites both in the upper and middle storey of the reserve. The associates of *Shorea robusta* in the upper storey are *Terminalia alata, Anogeissus latifolia, Dillenia pentagyna, Syzygium cumini, Adina cordifolia, Syzygium cerasoides, etc.* while the associates in the middle storey are *Bauhinia purpurea, Gardenia gummifera, Securinega virosa, Nyctanthes arbor-tristis, etc.*

Methodology

Vegetational Analysis

Ten study sites were selected for the study. Within each study site twenty quadrats of size 20 x 20m were laid and all the vegetational forms were counted and listed. The circumference at breast height (1.37m from the ground) was measured for each individual (except for seedling and herbaceous annuals). Frequency (per cent), density (Plants ha^{-1}) basal area (m^2 ha^{-1}), bio-volume (m^3 ha^{-1}) and canopy area (m^2 ha^{-1}) were calculated following Misra (1968) and Korhonen (2006). Species diversity and concentration of dominance were calculated following Magurran (1988). The crown area of each tree was viewed from below and the longest axis was identified. The two radii in the long axis were measured followed by the two radii in the axis perpendicular to the first axis. The four radii were then averaged. The averaged radius was then used to estimate crown area assuming a circular crown (Larson and Zaman, 1986). The bio-volume of trees was calculated using formula, following Pascal (1984).

Litter Dynamics

Forest floor litter was quantified by a stone block lined denuded quadrat technique (Pragasan and parthasarathy, 2005) in 20 replicates of 50 cm x 50 cm area placed regularly at 20 m intervals at each site. Forest floor litter was collected at monthly intervals and separated into fresh and partly decomposed groups following Pandey and Singh (1981). Forest floor litters were collected carefully to avoid soil contamination. Collected materials were taken into the laboratory in polyethylene bags, cleaned of soil particles and weighed after oven drying at 60°C for 48h to attain constant weight.

Turnover of Litter

The turnover rate (k) of litter was calculated following Olson (1963) and Pandey and Singh (1981) as K= A/(A+ F), where "A" is the annual increment of litter (*i.e.* annual litterfall) and "F" is the biomass of the litter at steady state (*i.e.* partly decomposed). Turnover time (t, years) is the reciprocal of turnover rate and is expressed as t = 1/k. In the present study, "F" was the partly decomposed litter of the forest floor and "A" included the annual tree litter fall plus shrub litter and herb litter.

Analysis of Soil

Soil samples were collected in the same period from each study site at a depth of 15cm (5 samples from each site) using a soil core having diameter 5cm. Bulk density, moisture content and soil porosity were determined following Misra (1968). Soil organic matter content was determined colorimetrically following Dutta *et al.* (1962). Available phosphorus of soil suspension was measured by Olsen's method (Olsen *et al.*, 1954) and nitrate-nitrogen by PDSA (Phosphoenol Disulphonic Acid) method (Jackson, 1973).

Statistical Analysis

Randomised block three-way analysis of variance was conducted with season, site and litter category of the SBR. F-ratio obtained was checked for the significance level. Linear regression analysis was done on the total forest floor litter biomass of each site. Multiple regression analysis was done on total forest floor taking litter biomass of each site with climatic factors like temperature, rainfall, and humidity.

Results and Discussion

Forest Floor Biomass

Monthly litter fall values at different sites are given in Figure 2.2. At all the sites, litter biomass started increasing from October and attain maximum value in March. However, at Lulung, Handipuhan and Podadiha the maximum litter biomass was observed in December. Low rates of litterfall were consistently recorded in August-September (Figure 2.2). This trend indicates that the litterfall rate may vary according to age, structure and type of vegetation, especially of the upper canopy even though generally high amounts of litter input occur during the dry season and low amounts in the wet season (Pragasan and Parthasarathy, 2005). The maximum litterfall biomass was accumulated at Gurguria followed by Kalikaprasad, Lulung, Podadiha and Handipuhan (Figure 2.2). The mean annual forest floor litter biomass obtained from different study sites of SBR varied from 304 to 584 g/m^2/yr, lie within the range reported for dry deciduous tropical forests of India (Rai and Srivastava, 1982; Rajendra Prasad *et al.*, 2000). The seasonal accumulation rate of total litter biomass across all study sites was in the order of winter > summer > rainy.

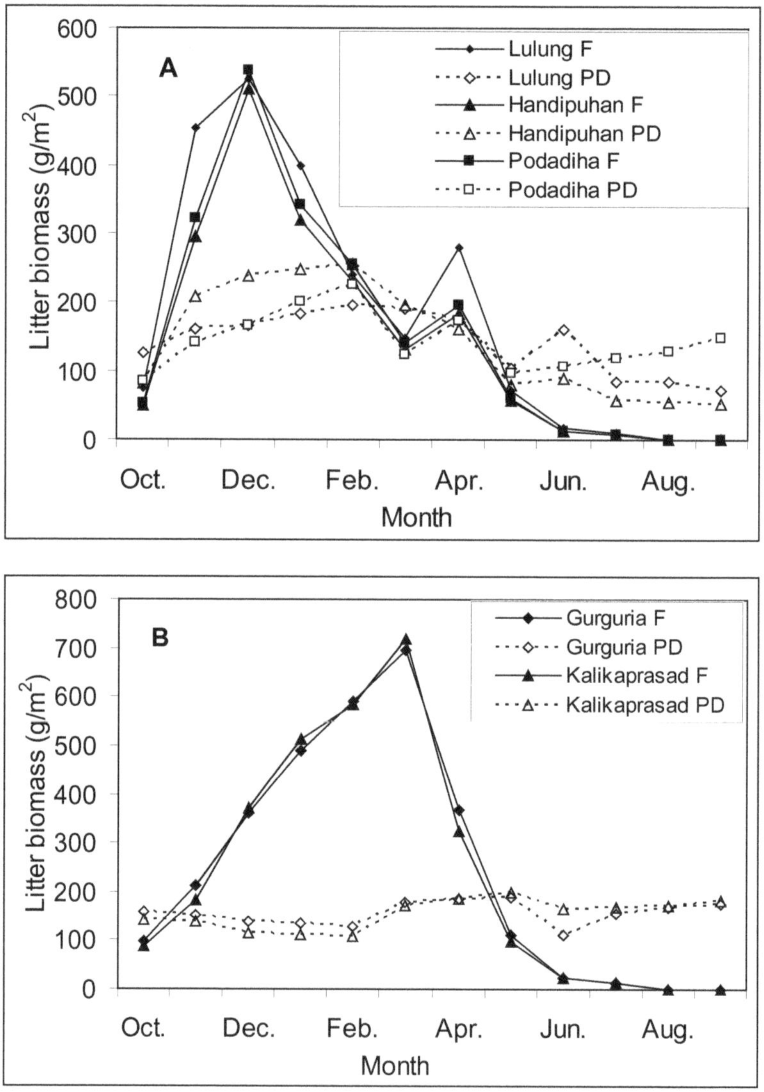

Figure 2.2. Variation in Monthly Forest Floor Litter Biomass (F = fresh and PD = partly decomposed) at Transition (A) and Buffer (B) Areas of the Similipal Biosphere Reserve.

Forest Floor Biomass Components

Forest floor litter was categorized into two components *i.e.* fresh and partly decomposed litter. The seasonal accumulation trend of fresh litter biomass follows the same trend as that of total forest floor biomass. While in case of partly decomposed biomass seasonal accumulation trend was in three different orders as:

1. Winter > summer > rainy at Lulung and Handipuhan and Podadiha;
2. Summer > rainy > winter at Kalikaprasad; and
3. Rainy > summer > winter at Gurguria (Figure 2.2).

Most of the study sites recorded lesser quantities of partly decomposed litter biomass in the rainy season compared to summer and winter seasons. The sites located in transition areas also received less litter fall than the buffer areas of the reserve. Two peaks of fresh litter input were marked in all the study sites of the reserve. In the transition areas, the first peak was observed in December and the second one in April while in buffer areas of the reserve first peak was observed in March and the second one was in May. However maximum fresh litter input in transition areas was recorded in December whereas maximum litter input occurred in March in buffer areas of the reserve (Figure 2.2). Analysis of variance indicated a significant difference in forest floor litter biomass between the sites, seasons and category. All these have also a significant interactive effect on forest floor litter biomass (Table 2.2). At all the study sites the maximum litterfall biomass was recorded in the drier part of the year and the lowest during the wetter part of the year (Figure 2.2). This behavior indicates that most of the upper canopy tree species are dry deciduous type and very few of them are evergreen and semi-evergreen type (Rajendraprasad *et al.*, 2000; Mishra *et al.*, 2006).

Table 2.2. Analysis of Variance (ANOVA) of Litter Production (Litter biomass), Turnover Rate (k), and Turnover Time (t) of Similipal Biosphere Reserve

Source		*Degree of Freedom*	*F- ratio*	*Significance Level*
Total litter biomass	Treatment	30	7.44	P < 0.001
	Season	2	44.28	P < 0.001
	Site	4	3.72	P < 0.001
	Category	02	25.21	P < 0.001
	Season × Site × Category	16	7.11	P < 0.001
	Error	209	-	
Turnover rate (k)	Treatment	29	4.5	P < 0.001
	Season	2	5.75	P < 0.01
	Site	4	15.1	P < 0.001
	Season × site	8	7.0	P < 0.001
	Error	60	-	
Turnover time (t)	Treatment	29	7.449	P < 0.001
	Season	2	10.761	P < 0.001
	Site	4	7.393	P < 0.001
	Season × site	8	7.111	P < 0.001
	Error	60	-	

Rate and Time of Forest Floor Biomass

Litter breakdown and mineralization are also governed by soil and forest floor micro and macro-organisms and its rate is regulated by the interaction between these organisms and the litter resources (Anderson and Swift, 1983; Upadhyay and Singh, 1989; Upadhyay *et al.*, 1989; Rajendraprasad, 1995). In the present study, a low turnover coefficient was observed during the monsoon period (July to October, the wet months in Odisha) followed by an increase in summer and winter seasons (Table 2.3). A site to site variation in the litter turnover was also marked in different seasons of a year (Table 2.4).

Table 2.3. Seasonal Variation in the Turnover Coefficient (k) at different Study Sites of the Similipal Biosphere Reserve

Site	*Winter*	*Summer*	*Rainy*
Lulung	0.83± 0.14	0.61 ± 0.13	0.41 ± 0.02
Handipuhan	0.78 ± 0.11	0.52 ± 0.09	0.44 ± 0.02
Podadiha	0.72 ± 0.12	0.51 ± 0.11	0.46 ± 0.025
Gurguria	0.89 ± 0.16	0.61 ± 0.14	0.44 ± 0.019
Kalikaprasad	0.86 ± 0.13	0.63 ± 0.15	0.42 ± 0.01

Table 2.4. Seasonal Variation in Turnover Time (t, years) at different Study Sites of the Similipal Biosphere Reserve

Site	*Winter*	*Summer*	*Rainy*
Lulung	1.21± 0.33	1.64 ± 0.28	2.44 ± 0.61
Handipuhan	1.28 ± 0.25	1.92 ± 0.51	2.27 ± 0.83
Podadiha	1.39 ± 0.28	1.96 ± 0.64	2.17 ± 1.14
Gurguria	1.12 ± 0.30	1.64 ± 0.58	2.27 ± 0.73
Kalikaprasad	1.16 ± 0.27	1.59 ± 0.47	2.38 ± 1.12

The annual value of the turnover coefficient was maximum at kalikaprasad (0.85) followed by Gurguria, Lulung, Handipuhan and Podadiha (Table 2.5). The turnover time for litter ranged from 1.18 to 1.45 years for complete decomposition of litter biomass on the forest floor of SBR (Table 2.5). Among the sites, maximum turnover time was required at Podadiha and minimum at Kalikaprasad (Table 2.5).

Table 2.5. Variation in the Litter Turnover Coefficient (k) and Litter Turnover Time (t) of the Study Sites of the Similipal Biosphere Reserve

Site	*Turnover Coefficient*	*Turnover Time (t)*
Lulung	0.82	1.22
Handipuhan	0.71	1.41
Podadiha	0.69	1.45
Gurguria	0.83	1.21
Kalikaprasad	0.85	1.18

The variation observed in the case of turnover rate and turnover time was significant between the sites and seasons. The site and season both have a significant interactive effect on the turnover rate and turnover time of the forest floor litter biomass of Similipal (Table 2.3). In the tropical climate of Similipal temperature and humidity are moderate throughout the year. Rainfall is abundant with rains for nearly 5 to 6 months in a year. But the major precipitation occurs between July and October and other rainy days are distributed over the rest of the months (Mishra *et al.*, 2006). According to Xiong and Nilsson (1997), such conditions are favorable for microbial activities except during certain months. In the present investigation during the heavy rain period, the decomposition rate was low and this could be explained in that, due to the continuous heavy rain, the soil becomes supersaturated with water accompanied by low oxygen content and low light intensity, resulting in retarded activity of the microflora and microbes. Altering wet and dry periods could also be important since Polunin (1984) suggested that the higher the frequency of change between wetting and drying, the greater the potential release of nutrients from the litter. Rajendraprasad *et al.* (2000) suggested that the aerobic digestion of cellulose is inefficient during the active rainy season but becomes important during the dry season due to the enhanced activity of microorganisms. These reasons may, therefore, account for the low turnover rate of litter during the rainy season and the high turnover rate of litter during the dry season and at the onset of periodical and interrupted summer rains during the months of April-May in SBR.

Effect of Abiotic Factors on Forest Floor Litter Decomposition and Turnover

The dry matter dynamics of the forest floor litter biomass is strongly determined by rain fall, temperature and humidity of the forest. The abiotic environmental factors like rainfall, temperature and humidity had significant influence on litter production ($Y = 1404 - 1.32X_1 + 0.09X_2 + 8.81X_3$, $r = 0.927$, $P < 0.05$, where Y = forest floor litter biomass, X_1 = rainfall, X_2 = temperature and X_3 = humidity) of SBR (Table 2.6).

Table 2.6. Analysis of Variance (ANOVA) Table for Testing the Significance of Multiple Correlation Coefficient (r) of Litter Biomass with Rainfall, Temperature and Humidity of SBR

Source	*Df.*	*SS*	*MSS*	*F-ratio (df.)*
Regression SS	3	719106.22	239702.1	16.312 (3,8)
Error	8	117559.31	14694.9	
D.f. (Difference) at P < 0.005 = 9.60 (Tabulated "F") Calculated "F" = 16.312 So, it is significant at P < 0.05				

Observations by Rajendraprasad *et al.* (2000) in the dry tropical sacred grooves of Kerala and Reich and Berchert (1984) in the tropical dry forest of Costa Rica also have established strong positive influence of abiotic parameters on litter biomass. Facelli and Pickett (1991) reported that in the upland ecosystems, climatic conditions are the major determinant of the litter dynamics. The production and decomposition of dry matter in an ecosystem are also strongly related to light availability during the growing season (Jordan, 1971) and temperature and precipitation (Vogt *et al.*, 1986). Site-to-site variation of litter production in SBR can be explained as the effect of biotic and abiotic factors of the ecosystems, which in turn influence the chemical properties of the soil. This may also have a selective influence on other biotas on the litter (Xiong and Nilsson, 1997). The tropical environmental conditions such as soil and atmospheric temperature and humidity influence the moisture status of the system which in turn influences the litter dynamics.

Effects of Vegetation on Forest Floor Biomass

Vegetation composition, profile and canopy cover of forest ecosystems influencing the quantification of forest floor litter biomass. Not only this but also a seasonal variation in litter production is influenced by plant physiognomic groups (Pragasan and Parthasarathy, 2005; Mishra *et al.*, 2011). From the present investigation, it was noticed that maximum forest floor litter biomass was collected during the dry seasons of the year in comparison to the wet season (Table 2.7).

Table 2.7. Seasonal Variation in Total Litter Biomass (g/m²) at different Study Sites of the Simlipal Biosphere Reserve

Site	*Summer*	*Winter*	*Rainy*
Lulung	291±66	556±107	110±39
Handipuhan	229±47	579±120	74±37
Podadiha	229±42	549±110	134±17
Gurguria	472±122	576±155	182±43
Kalikaprasad	473±111	547±165	186±26

Table 2.8. Floristic Wealth and Diversity of different Study Sites of Simlipal Biosphere Reserve

Name of the Study Site	*No of Tree Species*	*No. of Shrub Species*	*No. of Liana Species*	*No of Herb Species*	*The Density of Tree sp. (Plant/ha)*	*Basal Area of Tree sp. (m²/ha)*	*Bio-Volume (m³/ha)*	*Canopy Area (m²/ha)*
Lulung (S-1)	29	31	6	40	725	60	223	902
Handipuhan (S-2)	20	33	4	48	715	49	196	789
Podadiha (S-3)	29	29	4	45	680	52	218	864
Gurguria (S-4)	22	26	5	34	740	63	338	1026
Kalika-prasad (S-5)	23	23	3	36	750	79	333	1060

This is only due to the seasonality of litterfall. Inside the reserve, most of the upper canopy trees are deciduous and they shed their leaves towards the end of the winter and beginning of the summer (Mishra *et al.*, 2006). However, the difference in forest floor litter biomass among the study sites distributed in two different zones *i.e.* buffer and transition of the reserve may be due to the difference in above-ground phytomass, species diversity, tree density and structure of vegetation (Table 2.8).

Rajendraprasad *et al.* (2000) and, Pragasan and Parthasarathy, (2005) also reported the same type of phenomenon in sacred groves of Kerala and Coromandel coast of South India, respectively. Linear regression analysis taking litterfall as a function of tree density, basal area, bio-volume, canopy and species diversity showed significant correlation while significant negative correlation showed with the richness of herbs, shrubs and climbers (Table 2.9).

Table 2.9. Coefficient of Correlation (r) and Regression Parameters of Litter Biomass with Biotic (Density, basal area, bio-volume) and Abiotic Components of SBR

Parameters	*Correlation Co-efficient 'r'*	*Intercept*	*Slope*	*Probability*
Tree dens.Vs FFB	0.8	459.45	0.729	P<0.01
Basal area Vs FFB	0.91	12.5	0.13	P<0.001
Bio-volume Vs FFB	0.92	- 77.92	0.93	P<0.001
Tree species richness Vs FFB	0.4	19.23	0.002	NS
Shrub species rich ness Vs FFB	-0.96	46.47	-0.004	P<0.001
Climber species richness Vs FFB	-0.65	6.71	-0.0006	P<0.05
Herb species richness Vs FFB	-0.95	61.95	-0.005	P<0.001
Canopy cover Vs FFB	0.91	-125.55	0.257	P<0.001

Dens.: Density, FFB: Forest Floor Biomass, Bulk dens.: Bulk density, NS: Not significant.

A similar relationship between tree density, basal area and litter production has been reported for tropical forests (Martinez and Sarukhan, 1990; Rajendraprasad *et al.*, 2000).

Forest Floor Biomass and Soil Physical Characters

Soil surface covered with litter and living vegetation protects the soil from the direct impact of raindrops and through-fall (Sayer, 2006). Litter cover also decreases or prevents the destruction aggregates and separation of fine particles by raindrop impact, prevents soil surface compaction and sealing and consequently helps to prevent run-off and subsequent erosion during rainfall events (Marshall *et al.*, 1996). Besides such activities the forest floor litter also responsible for changing the soil's physical properties such as bulk density, pore space, *etc.* From the present investigation, it was found that bulk density of soil in two different zones of SBR is in the order of transition > buffer and also reflecting higher bulk density at low forest floor litter biomass sites in comparison to the higher forest floor litter biomass sites. However, the reverse trend was noticed for soil pore space and moisture content (Table 2.10).

Table 2.10. Physical Properties of Soil Across the Study Sites in Similipal Biosphere Reserve (Mean +SEM)

Site	*Bulk Density (g/m^3)*	*Per cent of Pore Space*	*Per cent of Moisture Content*
Lulung	1.279 ± 0.092	54.31 ± 1.177	9.872 ± 0.449
Handipuhan	1.103 ± 0.139	58.82 ± 2.217	5.978 ± 0.093
Podadiha	1.452 ± 0.064	58.36 ± 1.274	5.856 ± 0.695
Gurguria	1.589 ± 0.038	47.727 ± 0.984	6.362 ± 0.469
Kalikaprasad	0.949 ± 0.037	70.297 ± 2.594	7.257 ± 0.854

Linear regression analysis between forest floor biomass with physical properties of soil showed a significant positive correlation between forest floor biomass with soil moisture content while negative between forest floor biomass with bulk density (Table 2.11). A similar type of observation is marked by Sayer (2006) in different forest ecosystems of central Europe.

Table 2.11. Coefficient of Correlation (r) and Regression Parameters of Litter Biomass with Abiotic Components of the Similipal Biosphere Reserve

Parameter	*Correlation Co-efficient 'r'*	*Intercept*	*Slope*	*Probability*
Bulk dens. Vs FFB	- 0.789	9433.59	3965.95	P<0.05
Moisture Vs FFB	0.873	3162.34	104.69	P<0.001
Pore space Vs FFB	0.451	944.64	67.52	NS

FFB: Forest Floor Biomass, Bulk dens.: Bulk density, NS: Not significant.

Forest Floor Biomass and Soil Chemical Characters

Forest floor litter biomass forms an inherent part of the soil nutrient status, which directly or indirectly regulates the growth and development of plants (Killian, 1998). In tropical forests, litter removal is more frequent in comparison to any other forest. Thus in tropical forests, deciduous trees are usually badly adapted to low nutrient conditions (Givinish, 2002). Dzwonko and Gawronski (2002a) stated that forest floor litter biomass and their decay affects the chemical properties of soil and their removal from the forest floor reduces the soil nutrient status and also regulating the leaching of nutrients into different depths of soil. Results of the present findings envisaged that low forest floor biomass in transition than buffer areas of the reserve leads to a minimum value of soil organic carbon, nitrate nitrogen and available phosphorus in the transition in comparison to buffer areas of the reserve (Table 2.12).

Significant positive correlation of forest floor biomass with soil organic carbon, nitrate nitrogen and available phosphorus indicates that sites having higher forest floor biomass had highest value of soil nutrient (Table 2.13). Similar type of relationship between forest floor biomass and soil chemical nutrients is also marked by Mo *et al.* (2003) in tropical forests of China and Sayer (2006) in the forests of central Europe.

Table 2.12. Chemical Properties of Soil Across the Study Sites in Similipal Biosphere Reserve (Mean +SEM)

Site	*Organic Carbon (per cent)*	*Nitrate Nitrogen (Kg/ha)*	*Available Phosphorus (Kg/ha)*
Lulung	0.482 ± 0.011	4.072± 0.471	4.528± 0.229
Handipuhan	0.453± 0.010	5.956± 0.327	19.710± 0.549
Podadiha	0.524± 0.012	5.839±0.826	3.414± 0.199
Gurguria	0.618± 0.010	7.823± 0.737	8.782± 0.349
Kalikaprasad	0.579± 0.013	7.538± 0.519	11.38±0.354

Table 2.13. Coefficient of Correlation (r) and Regression Parameters of Litter Biomass with Chemical Properties of Soil of Similipal Biosphere Reserve

Parameter	*Correlation Co-efficient 'r'*	*Intercept*	*Slope*	*Probability*
FFB with soil organic carbon	0.792	0.079	0.0001	P<0.01
FFB with nitrate nitrogen	0.83	-15.49	0.006	P<0.01
FFB with available phosphorus	0.61	-4.99	0.004	P<0.05

FFB: Forest Floor Biomass.

CONCLUSION

Forest floor litter biomass plays two principal roles in forest ecosystems; firstly it forms an inherent part of the nutrient cycle of the forest and is a major source of soil organic matter. Secondly, the presence of forest floor biomass acts as protection against microclimatic fluctuations. Low forest floor biomass leads to a decline in fresh decaying organic matter in the soil surface layers and low nutrient status of soil in the transitional areas of the reserve. As Similipal biosphere reserve is a northern tropical moist mixed deciduous forest and the tropical deciduous plants are badly adapted to low nutrient conditions of soil, litter addition treatments in the transition and up to some extent in buffer areas of the reserve may help in maintaining the soil nutrient status which can be achieved through an increase in vegetation cover in the degraded habitats of the reserve.

ACKNOWLEDGEMENTS

The author is thankful to DST (Department of Science and Technology), Govt. of India, New Delhi for financial support and the Director Similipal Biosphere Reserve for providing permission to carry out the research work inside the reserve.

REFERENCES

Anderson, J.M. and Swift, M.J. (1983). Decomposition in tropical forests. PP. 287-309. In S. L. Sutton, A. C. Chadwick and T. C. Whitmore (Eds.). *Tropical Rain Forest: Ecology and Management*. Blackwell Scientific Publications, Oxford.

Barritt, A.R. and Facelli, J.M. (2001). Effects of *Casuarina pauper* litter and grove soil on emergence and growth of understorey species in arid lands of South Australia. *Journal of Arid Environments* 49: 569-579.

Benkobi, L., Trilica, M.J. and Smith, J.L. (1993). Soil loss as affected by different combinations of surface litter and rock. *Journal of Environmental Quality*. 22: 657-661.

Condit, R., Hubbell, S.P., and Foster, R.B. (1996). Assessing the response of plant functional types to climatic change in tropical forests. *Journal of Vegetation Science* 7: 405-416.

Datta, N.P., Khera, M. S. and Saini, T.R. (1962). A rapid colorimetric procedure for the determination of the organic carbon in soils. *Journal of Indian Society of Soil Science* 10: 67-74.

Dzwonko, Z. and Gawronski, S. (2002a). Effect of litter removal on species richness and acidification of a mixed oak pine woodland. *Biological Conservation* 106: 389-398.

Dzwonko, Z. and Gawronski, S. (2002b). Influence of litter and weather on seedling recruitment in a mixed oak pine woodland. Annals of Botany 90; 2454-251.

Facelli, J.M. and Pickett, S.T.A. (1991 b). Plant litter: Its dynamics and effects on plant community structure. *The Botanical Review* 57, 1-32.

Facelli, J.M. and Pickett, S.T.A. (1991). Plant litter its dynamics and effects on plant community structure. *The Botanical Review* 57: 1-32.

Ginter, D. l., McLeod, K. W. and Sherrod, C. (1979). Water stress in longleaf pine induced by litter removal. Forest Ecology and management 2: 13-20.

Givinish, T.J. (2002). Adaptive significance of evergreen vs. deciduous leaves: solving the triple paradox. *Silva Femina* 36: 703-743.

Grime, J.P. (1979). Plant Strategies and Vegetation Processes. John Wiley and Sons, Chichester.

Jackson, M.L. (1973). Soil Chemical Analysis. Prentice hall of India Pvt. Ltd., New Delhi.

Knoepp Jennifer, D., C., Reynolds Barbara, D. A. and Swank Wayne, T. (2005). Long-term changes in forest floor processes in southern Appalachian forests. *Forest Ecology and Management* 220: 300-312.

Larson, B.C. and Zaman, M.N. (1986). Thinning guidelines for teak (*Tectona grandis*). *Malay Forester* 48(4): 288-297.

Magurran, A.E. (1988). *Ecological diversity and its measurement*. New Jersey, Princeton University Press.

Marshall, T.J., Holmes, J.W. and Rose, C.W. (1996). Soil Physics. Cambridge University press, Cambridge.

Martinez-Yrizar, A. and Sarukhan, J. (1990). Litter fall patterns in a tropical deciduous forest in Mexico over a five year period. *Journal of Tropical Ecology* 6: 433-444.

Menezes, L. F.T., Souza, R.C., Pereira, M.G., Pires, F.R., Fanticelle, B.S. and Araujo-Filho, P.B. (2020). Different patterns of nutrient cycling In contiguous phytophysiognomies of Atlantic forest, Brazil. Floresta e Ambiente, 27(1): 1-12.

Mishra, R.K., Upadhyay, V.P., Bal, S., Mohapatra, P.K. and Mohanty, R.C. (2006). Phenology of species of moist deciduous forest sites of Similipal biosphere reserve. *Lyonia* 11(1): 5-17.

Mishra, R.K., Upadhyay, V.P., Mishra, P.K. and Mohanty, R.C. (2011). Ecological problems of tree species in two protected ecosystems of Eastern Ghats, India. *Journal of Environmental Biology*, 32(1): 115-119.

Misra, R. (1968). *Ecological Workbook*. Oxford and IBH Publishing Company, Calcutta.

Mo, J., Brown, S., Peng, S. and Kong, G. (2003). Nitrogen availability in disturbed, rehabilitated and mature forests of tropical China. *Forest Ecology and management* 175: 573-583.

Ogee,J. and Brunet, Y. (2002). A forest floor model for heat and moisture including a litter layer. *Journal of Hydrology*. 255: 212-233.

Olsen, S.R., Cole, C.V. Watanabe, F.S. and Dean, L.A. (1954). Estimation of available phosphorus in soils by extraction with sodium bicarbonate. Circ. U.S. Dept. Agric.

Olson, J. S. (1963). Energy storage and the balance of producers and decomposers in ecological systems. *Ecology* 44: 322-331.

Pandey, U. and Singh, J.S. (1981). A quantitative study of the forest floor, litter fall and nutrient return in an oak-conifer forest in Himalaya. I. Composition and dynamics of forest floor. *Acta Oecologia* 2: 49-61.

Pascal, J. P. (1984). *Forest humids sempervirentes des Ghats occidentaux de inde*. Institute Francals d' Pondicherry.

Polunin, N.V.C. (1984). The decomposition of emergent macrophytes in fresh water. *Advances in Ecological Research* 14: 115-166.

Ponge, J.F., Arpin, P. and Vannier, G. (1993). Collembolan response to experimental perturbations of litter supply in a temperate forest ecosystem. *European Journal of Soil Science* 29: 141-153.

Pragasan, L.A. and Parthasarathy, N. (2005). Litter production in tropical dry evergreen forests of south India in relation to season, plant life forms and physiognomic groups. *Current Science* 88(8): 1255-1263.

Rai, B. and Srivastava, A.K. (1982). Litter production in a tropical dry mixed forest stand. *Acta Oecol. Oecol. Appl.* 3: 169-176.

Rajendraprasad, M. (1995). *The floristic, structural and analysis of Sacred Groves of Kerala*. Ph. D. Thesis, Kerala University, Thiruvanthapuram.

Rajendraprasad, M., Krishnan, P.N. and Pushpangadan, P. (2000). Vegetational characterization and litter dynamics of the sacred grooves of Kerala, southwest India. *Journal of Tropical Forest Science* 12(2): 320-325.

Reich, P.B. and Borchert, R. (1984). Water stress and tree phenology in a tropical dry forest in the low lands of Costa Rica. *Journal of Ecology* 72: 61-74.

Sayer, E. J. (2006). Using experimental manipulation to assess the roles of leaf litter in the functioning of forest ecosystems. *Biological Review* 81: 1-31

Sunderpandian, S. M. and Swamy, P. S. (1999). Litter production and leaf litter decomposition of selected tree species in tropical forests at Kodayar in the Western Ghats, India. *Forest Ecology and Management* 123: 231-244.

Tesfay, F., Kibret, K., Gebrekirstos, A. and Hadgu, K. M. (2020). Litterfall production and associated carbon and nitrogen flux along exclosure chronosequence at Kewet district, central lowland of Ethiopia. *Environmental Sytem Research,* 9 (11): 1-12.

Upadhyay, V.P. and Singh, J.S. (1989). Patterns of nutrient immobilization and release in decomposing forest litter in Central Himalaya, India. *Journal of Ecology* 77: 127-146.

Upadhyay, V.P., Singh, J.S. and Meentemeyer. (1989). Dynamics and weight loss of leaf litter in central Himalayan forest: abiotic versus litter quality influences. *Journal of Ecology* 77: 127-146.

Vogt, K.A., Grier, C.C. and Vogt, D.J. (1986). Production, turnover and nutrient dynamics of above ground and below-ground detritus of world forests. *Advances in Ecological Research* 15: 303-377.

Wilke, B., Bogenrieder, A. and Wilmanns, O. (1993). Differenzierete Streuverteilung in Walde, ihre Ursachen und Folgen. *Phytocoenologia* 23: 129-155.

Xiong, S. and Nilsson, C. (1997). Dynamics of leaf litter accumulation and its effects on riparian vegetation: a review. *The Botanical Review* 63: 240-264.

Zak, D.R., Holmes, W.E., Finzi, A.C., Norby, R.J. and Schlesinger, W.H. (2003). Soil nitrogen cycling under elevated carbon dioxide: a synthesis of forest FACE experiments. *Ecological Applications* 13: 1508-1514.

Chapter 3

Diversity and Distribution of Invasive Alien Plant Species of Odisha, India

Shanta Sagarika Dash[1], *Shubhosmita Mohapatra*[2] *and Anil Kumar Biswal*[1] *

[1]*Department of Botany*, [2]*Department of Zoology, North Orissa University, Takatpur, Baripada – 757 003, Odisha*
**e-mail: anilkbiswal@yahoo.com*

Invasive species are animals or plants that are non-native to the region that were mainly introduced by human intervention. The invasive species quickly adapted to new environment and efficiently grow and propagate. Invasive alien plants become a major threat to many native taxa, biodiversity, ecosystem stability and impose a risk to agriculture, the economy as well as public health. Here we have summarized the distribution of invasive species of Odisha in relationship to their ecology and survival strategy.

INTRODUCTION

Plant species that move from one geographical region to another (either accidentally or intentionally), establish and proliferate there and threaten native ecosystems, habitats and species are known as invasive alien plants (Richardsond *et al.*, 2000). Across the globe, exotic alien plants are responsible for biodiversity crisis coupled with environmental changes.

Due to its diverse climatic and environmental conditions, the Indian subcontinent is highly vulnerable to biotic invasion. (Nayar, 1977) reported that around 18 per cent of Indian flora are adventive aliens, out of which 55 per cent are from America, 30 per cent from Asia, and 15 per cent from Europe and central Asia. Based on systematic review and analyses of studies (Khuroo *et al.*, 2011) compiled an inventory of the alien flora of India. The inventory comprises of 1,599 species belonging to 842 genera under 161 families, which constitutes 8.5 per cent of the told Indian flora. Biogeographically, more than one third (35 per cent) of the alien flora of the country has nativity in South America, followed by Asia (21 per cent), Africa (20 per cent), Europe (11 per cent), Australia (8 per cent) and North America (4 per cent) and cryptogenic (1 per cent).

Impact on Ecosystem

Invasive species cause loss of biodiversity, changes in structure and function of an ecosystem. For example, *Ageratum conyzoides, Chromolaena odorata, Mikania micrantha, Lantana camara* and *Parthenium hysterophorus* have undesirable impacts on terrestrial ecosystem whereas the species such as *Eichhornia crassipes, Salvinia cucullata* and *Ipoemia carnea* have become a nuisance for aquatic ecosystems (Raghubanshi *et al.*, 2005). *Lantana camara* has been rated as one of the worst invasive species identified by the Global Invasive species Database and is enlisted in ten top 100 invasive species of the world (GISD, 2010). Among the trees, *Prosopis juliflora* and *Leucaena leucocephala* predominate as invasive plants in India. The majority of the invasive plants belong to the family Asteraceae (Kohli *et al.*, 2012).

Ecology

The ecology of biological invasion is a new and challenging area of study especially in respect of conservation of biodiversity. The problem is to be addressed based on specific taxa, their mode of invasion, impacts on the ecosystem. In recent times, several workers have contributed immensely in this direction. Under the aegis of North Orissa University, a humble attempt has been made to undertake a study on the impact of alien species in Chandaka Wildlife Sanctuary, Khordha and Similipal Biosphere Reserve, Mayurbhanj, Odisha.

The invasion of exotic species, coupled with climate change has caused significant changes in the structure and composition of vegetation leading to a homogenization of flora (Cushrman and Gaffney, 2010). In a study, Shapma and Raghubanshi (2009) reported that invasion of *Lantana camara* in the dry deciduous forests is changing forest structure, accelerating the spread of the taxon. This taxon is commonly observed in forest cleaning and degraded forest of Odisha:

1. *Parthenium hysterophorus* (Family Asteraceae) an obnoxious invasive weed from tropical America has entered India accidentally in 1810 (Paul, 2010). It possesses many survival strategies like fast growth rate, adaptability to diverse habitat, high reproductive potential and unpalatability, *etc.* Upon invading any terrain it out-competes for the local vegetation and creates its territory forming impenetrable monoculture stands affecting not only biodiversity but also crop production, animal husbandry and human health (Batish *et al.*, 2012). By disturbing the food chain it also causes an ecological imbalance in an ecosystem. It prevents the growth of native plants leading to loss of species diversity.
2. *Chromolaena odorata* (Family Asteraceae) is commonly observed along roadsides; degraded forests open areas and grasslands. It has also invaded crop fields in many parts of India. Thus, it affects the vegetation, crop

yield and human settlements. *Ageratum conyzoides*, a native of tropical America is now observed throughout the world (Xuan *et al.*, 2004). It has been reported as an invasive weed of crops, plantations, grasslands, forest clearings and also along roadsides.

3. *Mikania micrantha* (Asteraceae) grows gregariously in agricultural lands, plantations, forest clearings and in gardens. It covers the entire photosynthetic surface of host trees and shrubs which affect seriously the survival need of the native vegetation. It has wide adaptability in diverse habitats.
4. *Eichhornia crassipes* (Pontederiaceae) is found abundantly in water bodies like ponds, tanks, slow-flowing rivers and other wetlands. The taxon multiplies very fast form a dense covering on the surface of the water. By the process, it eliminates the native aquatic flora. *Eichhornia crassipes* not only affect the aquatic ecosystem but also alters the herbaceous flora in the adjacent terrestrial ecosystem. It blocks the irrigation and navigational channels and reduces fish production.

The eight worst invasive plants of the state are *Lanterna camara, Chromolaena odorata, Eichhornia crassipes, Ipomea carnea, Parthenium hysterophorus, Mikania micrantha, Hyptis suaveolens* and *Senna alata*. The alien invasive species are equipped with better survival strategies such as fast growth rate, adaptability to diverse environmental conditions, high reproduction, early maturity and efficient seed dispersal mechanism *etc.* Moreover, may invasive plants possess prickles, spines and thorns which protect them from browsing animals.

Table 3.1. Invasive Alien Species of Odisha

Sl.No.	*Species*	*Family*	*Habit*	*Nativity*
1	*Acacia farnesiana* (L.) Wild	Mimosaceae	T	SAM
2	*Acanthospermum hispidum* DC.	Asteraceae	H	Brazil
3	*Aerva javanica* (Burm. f.) Juss.	Amaranthaceae	H	TAM
4	*Aeschynomene americana* L.	Fabaceae	H	TAM
5	*Ageratum conyzoides* L.	Asteraceae	H	TAM
6	*Alternanthera paronychioides* St. Hill.	Amaranthaceae	H	TAM
7	*Alternanthera philoxeroides* (Mart.) Griseb.	Amaranthaceae	H	TAM
8	*Alternanthera pungens* Kunth	Amaranthaceae	H	TAM
9	*Alternanthera tenella* Colla	Amaranthaceae	H	TAM
10	*Antigonum leptopus* Hook. and Arn.	Polygonaceae	M	TAM
11	*Argenone mexicana* L.	Papaveraceae	H	SAM
12	*Asphodelus ternuifolius* Cav.	Liliaceae	H	TAM
13	*Azolla pinnata R. Br.*	Azollaceae	H	TAF
14	*Bidens pilosa* L.	Asteraceae	H	TAM
15	*Blainvillea acmella* (L.) Philipson	Asteraceae	H	TAM
16	*Blumea eriantha* DC.	Asteraceae	H	TAM

Sl.No.	Species	Family	Habit	Nativity
17	*Blumea lacer* (Burm. f.) DC.	Asteraceae	H	TAM
18	*Blumea obliqua* (L.) Druce	Asteraceae	H	TAM
19	*Borassus flabellifer* L.	Asteraceae	H	TAM
20	*Calotropis gigantea* R. Br.	Asclepiadaceae	T	TAM
21	*Calotropis procera* (Ait.) R. Br.	Asclepiadaceae	S	TAM
22	*Cardamine trichocarpa* Hochst. Ex A. Rich.	Brassicaceae	S	TAM
23	*Cardiospermum halicacabum* L.	Sapindaceae	H	TAM
24	*Catharanthus pusillus* (Murr.) G. Don	Apocynaceae	C	TAM
25	*Celosia argentea* L.	Amaranthaceae	H	TAM
26	*Chamaecrista absus* (L.) Irwin and Barneby	Caesalpiniaceae	H	TAM
27	*Chamaecrista pumila* (Lam.) K. Larsen	Caesalpiniaceae	H	TAM
28	*Chloris barbata* Sw.	Poaceae	H	TAM
29	*Chromolaena odorata* (L.) King and Robins	Asteraceae	H	TAM
30	*Chrozophora rottleri* (Geis.) Spreng.	Euphorbiaceae	S	TAM
31	*Cleome gynandra* L.	Cleomaceae	H	TAF
32	*Cleome monophylla* L.	Cleomaceae	H	TAM
33	*Cleome rutidosperma* DC.	Cleomaceae	H	TAM
34	*Cleome viscosa* L.	Cleomaceae	H	TAM
35	*Conyzad bipinnatifida* Wall	Asteraceae	H	TAM
36	*Corchorus aestuans* L.	Tiliaceae	H	TAM
37	*Corchorus tridens* L.	Tiliaceae	H	TAF
38	*Corchorus trilocularis* L.	Tiliaceae	H	TAF
39	*Crassocephalum crepidioides* (Benth.) Moore	Asteraceae	H	TAM
40	*Crotalaria pallida* Ait.	Fabaceae	H	TAM
41	*Crotalaria retusa* L.	Fabaceae	H	TAM
42	*Croton bonplandianus* Baill.	Euphorbiaceae	H	SAM
43	*Cryptocoryne retrospiralis* (Roxb.) Kurth.	Cyperaceae	H	AS
44	*Cryptostegia grandiflora* R. Br.	Asclepiadaceae	H	MG
45	*Cuscuta chinensis* Lam.	Cuscutaceae	H	MR
46	*Cuscuta reflexa* Roxb.	Cuscutaceae	H	MR
47	*Cyperus difformis* L.	Cyperaceae	H	TAM
48	*Cyperus iria* L.	Cyperaceae	H	TAM
49	*Cyperus alopecuroides Roxb.*	Cyperaceae	H	TAF
50	*Datura innoxia* Mill	Solanaceae	S	TAM
51	*Datura metel* L.	Solanaceae	S	TAM
52	*Digera muricata* (L.) Mart.	Amaranthaceae	H	NAM
53	*Dinebra retroflexa* (Vahl) Panz.	Poaceae	H	TAM
54	*Echinochloa colona* (L.) Link	Poaceae	H	SAM
55	*Echinochloa crusgalli* (L.) Beauv	Poaceae	H	SAM
56	*Euphorbia hirta* (L.)	Euphorbiaceae	H	TAM

Sl.No.	*Species*	*Family*	*Habit*	*Nativity*
57	*Euphorbia indica* (Lam.) Croizat	Euphorbiaceae	H	TAM
58	*Echinops echinatus* Roxb.	Asteraceae	H	AF
59	*Eclipta prostrate* (L.)	Asteraceae	H	TAM
60	*Eichhornia crassipes* (Mart.) Solms -1	Pontederiaceae	H	TAM
61	*Emilia sonchifolia* (L.) DC.	Asteraceae	H	TAM
62	*Euphorbia cyathophora* Murr.	Euphorbiaceae	H	TAM
63	*Euphorbia heterophylla* (L.)	Euphorbiaceae	H	TAM
64	*Evolvulus nummularius* (L.) L.	Convolvulaceae	H	TAM
65	*Flaveria trinervia* (Spreng.) C. Mohr.	Asteraceae	H	TAM
66	*Fuirena ciliaris* L. Roxb.	Cyperaceae	H	TAM
67	*Glossocardia bosvallea* (L. f.) Dc.	Asteraceae	H	TAM
68	*Galinosoga parviflora* Cav.	Asteraceae	H	TAM
69	*Gnaphalium pesylvanicum* Willd.	Asteraceae	H	TAM
70	*Gnaphalium polycaulon* Pers.	Asteraceae	H	TAM
71	*Gomphrena celosoides* Mart.	Amaranthaceae	H	TAM
72	*Grangea maderaspatana* (L.) Poir.	Laminaceae	H	SAM
73	*Hyptis suaveolens* (L.) Poir.	Laminaceae	H	TAM
74	*Impatiens balsamina* L.	Balsaminaceae	H	TAF
75	*Imperata cylindrica* (L.) Raeusch.	Poaceae	H	TAF
76	*Indigofera linnae* Ali	Fabaceae	H	TAF
77	*Inigofera linifolia* (L. f.) Retz.	Fabaceae	H	TAF
78	*Ipomoea carnea* Jacq.	Convolvulaceae	S	TAF
79	*Ipomoea eriocarpa* R. Br.	Convolvulaceae	H	TAF
80	*Ipomoea hederifolia* L.	Convolvulaceae	H	TAM
81	*Ipomoea obscura* (L.) Ker-Gawl.	Convolvulaceae	H	TAF
82	*Ipomoea* pes-tigridis L.	Convolvulaceae	H	TAF
83	*Ipomoea quamoclit* L.	Convolvulaceae	H	TAM
84	*Lagascea mollis* Cav	Asteraceae	H	TAM
85	*Lantana camara* L. var. aculeata (L.) Mold	Verbenaceae	H	TAM
86	*Leonotis nepetiifolia* (L.) R. Br.	Laminaceae	H	TAF
87	*Leucaena leucocephala* (Lam.) de Wit	Mimosaceae	T	TAM
88	*Ludwigia adscendes* (L.) Hara	Onagraceae	H	TAM
89	*Ludwigia octovalvis* (Jacq.) Raven	Onagraceae	H	TAM
90	*Ludwigia perennis* L.	Onagraceae	H	TAM
91	*Malachra capitata* (L.) L.	Malvaceae	H	TAM
92	*Malvastrum coromandelianum* (L.) Garcke	Malvaceae	H	TAM
93	*Martynia annua* L.	Martyniaceae	C	TAM
94	*Mecardonia procumbens* (Mill.) Small	Scrophulariaceae	H	TAM
95	*Melilotus alba* Medik.	Fabaceae	H	Europe
96	*Melochia corchorifolia* L.	Sterculiaceae	H	TAM

Sl.No.	Species	Family	Habit	Nativity
97	*Mikania micrantha* Kunth	Asteraceae	C	TAM
98	*Mimosa pudica* L.	Mimosaceae	H	Brazil
99	*Mirabilis jalapa* L.	Nyctaginaceae	H	Peru
100	*Monochoria vaginalis* (Burm. f.) Presl	Pontederiaceae	H	TAM
101	*Nicotiana plumbaginifolia* Viv.	Solanaceae	H	TAM
102	*Ocimum americanum* L.	Lamiaceae	H	TAM
103	*Opuntia stricta* (Haw.) Haw.	Cactaceae	H	TAM
104	*Oxalis corniculata* L.	Oxalidaceae	H	Europe
105	*Parthenium hysterophorus* L.	Asteraceae	H	TAM
106	*Passiflora foetida* L.	Passifloraceae	H	SAM
107	*Pedalium murex* L.	Pedaliaceae	H	TAM
108	*Peperomia pellucid* (L.) Kunth	Piperaceae	H	TAM
109	*Peristrophe paniculata* (Forssk.) Brummitt	Acantaceae	H	TAM
110	*Physallis minima* L.	Solanaceae	H	SAM
111	*Pilea microphylla* (L.) Liebm.	Urticaceae	H	SAM
112	*Pistia stratiotes* L.	Araceae	H	TAM
113	*Portulaca oleracea* L.	Portulacaceae	H	SAM
114	*Portulaca quadrifida* L.	Portulacaceae	H	TAM
115	*Prosopis juliflora* (Sw.) DC.	Mimosaceae	H	TAM
116	*Ricinus communis* L.	Asteraceae	H	TAF
117	*Ruellia tuberosa* L.	Acantaceae	S	TAM
118	*Saccharum spontaneum* L.	Poaceae	H	ML
119	*Salvinia molesta* Mitch.	Salviniaceae	H	Brazil
120	*Scoparia dulcis* L.	Scrophulariaceae	H	TAM
121	*Sesbania bispinosa* (Jacq.) Wight	Fabaceae	H	TAM
122	*Senna alata* (L.) Roxb.	Caesalpiniaceae	S	WI
123	*Senna hirsuta* (L.) Irwin and Barneby	Caesalpiniaceae	S	TAM
124	*Senna obtusifolia* (L.) Irwin and Barneby	Caesalpiniaceae	S	TAM
125	*Senna occidentalis* (L.) Link	Caesalpiniaceae	H	SAM
126	*Senna tora* (L.) roxb.	Caesalpiniaceae	H	SAM
127	*Sida acuta* Burm. f.	Malvaceae	H	TAM
128	*Solanum torvum* Sw.	Solanaceae	H	WI
129	*Solanum viarum* Dunal	Solanaceae	S	TAM
130	*Solanum nigrum* L.	Solanaceae	H	TAM
131	*Solanum sisymbriifolium* Lam.	Solanaceae	H	SAM
132	*Sonchus asper* (L.) Hill	Asteraceae	C	MR
133	*Sonchus oleraceus* L.	Asteraceae	H	MR
134	*Spermacoce hispida* L.	Rubiaceae	H	TAM
135	*Stachytarpheta jamaicensis* (L.) Vahl	Verbenaceae	H	TAM
136	*Synadenium grantii* Hook f.	Euphorbiaceae	H	TAM

Sl.No.	*Species*	*Family*	*Habit*	*Nativity*
137	*Synedrella nodiflora* (L.) Gaertn.	Asteraceae	S	WI
138	*Tribulus terrestris* L.	Zygophyllaceae	H	TAM
139	*Tridax procumbens* L.	Asteraceae	H	TAM
140	*Triumfetta rhomboidea* Jacq.	Tiliaceae	H	TAM
141	*Turnera ulmifolia* L.	Turneraceae	H	TAM
142	*Typha angustata* Bory. and Choub.	Typhaceae	H	TAM
143	*Urena* lobata L.	Malvaceae	S	TAF
144	*Waltheria indica* L.	Storculiaceae	H	TAM
145	*Xanthium indicum* Koenig	Asteraceae	H	TAM
146	*Youngia japonica* (L.) Dc.	Asteraceae	H	SAM

Abbreviations Used: Life form: H-herb, S-Shrub, C-climber, T-Tree, AU-Australia, As-Asia, NAM-North Amerlca, SAM-South Amerlca, TAM-Troplcal Ablca, AF-Afghanlstan, WI-West Indles, MR-Medlterranean, MG-Madagascar, ML-Malaysia

REFERENCES

Babu, C.R. (1977). Herbaceous Flora of Dehradun, CSIR Publications, New Delhi.

Batish. D.R. Kohli, R.K., Singh, H.P. and Kaur. G. (2012). Biology, Ecology and spread of the Invasive weed *Parthenium hysterophorus* in India. In J.R. Bhatt *et al.* (eds.) Invasive Alien plants: An ecological approach for the Indian sub-continent. CAB international.

Biswal, A.K. and Choudhury, B.P. (1994). New Records of plants from Odisha, *J. Bomb. Nat. Hist. Soc.* 91(2): 351-352.

Cushman, J.H. and Gaffney, K.A. (2010). Community level consequences of invasion, impacts of exotic clonal plants on riparian vegetation. Biological invasion (8): 1073-1089 GISD (2010) *Global Invasive species Database.* http: www.issg.org/ database.

Gupta, O.p. (2001). Weedy Aquatic plants: Their utility, Menace and Management, Agrobios, Jodhpur, India. 273 pp.

Khuroo, A.A., Reshi, Z.A., Malik, A.H., Weber, E., Rashid, I and Dar, G.H. (2011) Alien flora of India: taxonomic composition, invasion status and biogeographic affiliations. *Biological invasion* 14(1): 99-119.

Kohli, R.K., Batish, D.R., Singh J.S., Singh, H.P. and Bhatt, J.R. (2012). in J.R. Bhatt *et al.* (eds.) Invasive Alien Plants: An ecological Appraisal for the Indian subcontinent. *CAB, International.*

Maheswari, J.K. and Paul, S.R. (1975). The alien flora of Ranchi, *J. Bomb. Nat. Hist. Soc.* 72: 158-188.

Nayak, S. K. and Satapathy, K. B. (2015). Diversity, uses and Origin of Invasive Alien Plants in Denkanal district of Odisha, India. *Int. Res. J. Biol. Sci,* 4(2): 21-27.

Nayar, M.P. (1977) Changing patterns of the Indian flora. *Bull. of the Bot. Survey of India.* 19: 145-154.

Negi, P.S. and Hajra, P.K. (2007). Allien flora of Doon Valley, North West Himalaya, *Curr. Sc.* (92) 698-978.

Panda, P. C., Choudhury, B. P. and Patnaik, S. N. (1985). *Cleome rutidosperma* DC. C (apparaceae) from Odisha. A new distribution record. *Bull. Pure and Applied Sci.* 4B(1): 59-60.

Paul, T.K. (2010) THE earliest record of *Parthenium hysterophorus* L. (Asteraceae) in India. *Curr. Scien.* 98: 1272.

Raghubanshi, A.S., Rai, L.C., Gaur, J.P. and Singh, J.S. (2005) Invasive alien species and biodiversity in India. *Cur. Sc.* 88: 539-540.

Reddy, C.S. (2008) Catalogue of invasive alien flora of India. *Life science Journal* 5 (2): 84-89.

Richardson, D.M., Pysek, P., Rejmanek, M., Barbour, M. G., Panetta, F.D. and West, C.J. (2000) Naturalization and invasion of alien plants: Concepts and definitions. *Diversity and Distributions* 6: 93-107.

Shah, A.M. and Reshi, A.Z. (2012). Invasion by Alien Macrophytes in Freshwater Ecosystem of India. In: J.R. Bhatt *et al.* (eds). Invasive Alien Plants: An ecological Appraisal for the Indian Subcontinent. *CAB International.*

Sharma, B.D. and Pandey. D.S. (1984) Exotic Flora of Allahbad District, BSI, Howrah.

Sharma, G.P. and Raghubanshi, A.S. (2009) How Lantana invades a dry deciduous forest in India: a case study from Vindhyan highlands. *Tropical Ecology* (51): 305-316.

Singh, K.P., Shukla, A.N. and Singh, J.S. (2010). State- level inventory of invasive alien plants, their source regions and use potential. *Curr. Sc.* 99 (1): 107-114.

Xuan, D., Shinkichi, T., Hong, N.H., Khanh, T.D. and Min, C. (2004). Assessment of phytotoxic action of *Ageratum conyzoides* L. (billy goat weed) on weeds. *Crop Protection* 23: 915-922.

Chapter 4

Medicinal Plants Conservation Areas in Mayurbhanj Districts, Odisha

Srustidhar Rout

Department of Wildlife and Biodiversity Conservation, North Orissa University, Takatpur, Baripada – 757 003, Odisha, India
e-mail: srusti_d_rout@rediffmail.com.

The study area is located in the eastern ghat region of Indian subcontinent and possess a great floristic diversity. The local people resides here are mostly dependent upon natural products for their day to day use and medicinal uses as well. An extensive survey was carried out in two medicinal plants conservation areas like Hatikote RF and Satkoshia RF of Mayurbhanj district to highlight some of the major medicinal plants and their uses through an appraisal approach with the local peoples. About 98 plants species belonging to 50 families have been found to be used frequently by the native people against different diseases. Therefore, due to various anthropogenic activities and unsustainable harvesting from wild, many medicinal plants species in India have become endangered and have been included in Red Data Book of threatened species. Thus, there is an urgent need to have measures taken to protect/conserve the natural habitats of these medicinal plants wealth of the Country

Keywords: *Medicinal plants, Similipal, Biodiversity conservation, Threatened species, Anthropologic activity.*

INTRODUCTION

India is one of the hot spots of world's biodiversity having wealth of medicinal plants. Forests are the main habitat of Medicinal Plants and a large number of species are also endemic that gives India a unique position in the world. Medicinal plants form the major resource base of our traditional systems of medicine or AYUSH systems of medicines and folklore medicines. Even allopathic medicines also used plant based raw material for the preparation of medicine. However, due to various anthropogenic activities and unsustainable harvesting from wild, many medicinal plants species in India have become endangered and have been included in Red Data Book of threatened species. Thus, there is an urgent need to have measures taken to protect/conserve the natural habitats of these medicinal plants wealth of the Country (Biswas *et al.*, 2017). India has a rich resource base of medicinal plants, plush with about 8,000 different species. According to the Government of India (GoI), traditional medicines are the sole means of health care for about 65 per cent of the population.

The expanding domestic and global demand of herbal products has put the native medicinal plant resources under significant stress. Wild populations of many medicinal plant species, those growing inside the forests as well as outside forests, have depleted to such an extent that their very survival has become a cause of concern. Whereas several of these species are facing threat of extinction on account of their narrow distribution or endemicity and degradation of their specific habitat, many of these have come to this pass on account of indiscriminate and un-sustainable harvesting. It is a pity that most of these species have never been the focus of any dedicated management intervention. Whereas no management protocols are in existence for the species found outside forests, management focus in respect of the forest species has been primarily on the timber species of high commercial value with little priority accorded to the management of medicinal plants. This lack of management focus has been a cause, as well as a consequence, of inadequacy of data related to the diversity of medicinal plant species being harvested and the quantum of such harvests.

There are 108 Medicinal Plant Conservation Areas (MPCA's) established in India, Since 1993 through the Projects coordinated by FRLHT. Medicinal plants play a vital role in the life of local community by providing health security and a source of income. The conservation and sustainable utilization of Med. Plants is one of the thrust areas of GoI-UNDP Environment Programme Support. Support envisages the need for both *in situ* and *ex situ* conservation approaches involving the local communities. MPCPs serve as community conservation Education Centres and repositories of the Medicinal plant resources and local health knowledge of the region. It also serves as a training centre for rural households, Kaviraj, local herbal doctors, tribal heads, schools, colleges and Government departments on the conservation and sustainable use of medicinal plants. Urgent conservation of the full spectrum of medicinal plants occurring in our wilderness areas is now the need (Rout *et al.*, 2005)

Study Area

Mayurbhanj district being the largest district in Odisha, possess 38 per cent of its area under forest cover and has a great diversity of plant resources especially in medicinal plants. Baripada forest division and Karanjia forest division are the important forest division in Mayurbhanj. once considered to be the maximum revenue earning division of the state.

Mayurbhanj a predominantly tribal populated district of Eastern India in the state of Odisha, has 4,525,04 km^2 of forestland, covering 43 per cent of its total geographic area (21°172 –22°342 N,85°402 –87°102 E), bordering the districts of Balasore in the east and Keonjhar in the south and west; to the north is Singhbhum in Jharkhand State and Medinapur in West Bengal State. The district is famous for its rich flora and fauna. Similipal Biosphere Reserve extends over 5569 km^2 and has over3000 plant species, including 200 high-value medicinal plants (Rout 2005). Uncontrolled and ruthless collection of the forest-based medicinal wealth from Similipal Biosphere Reserve has threatened the very survival of many critically endangered medicinal plants. The challenge of conservation has led to scientific domestication and cultivation of these plants, and an assured market and favourable agroclimatic conditions has encouraged the cultivation of medicinal plants by local people in Mayurbhanj. For the present exploration work on Medicinal Plants Conservation Areas in Mayurbhanj two reserve forest such as Hatikote Reserve Forest (Hatikote RF) under, Baripada Forest division and Satkoshia Reserve Forest (Satkoshia RF) have been selected.

Hatikote Reserve Forest (Hatikote RF), bordering Similipal Biosphere Reserve, abounds in highvalue medicinal plants and has been protected by local people since 1983. It covers an area of about 5.85 km^2 and has a north–south orientation and elongated shape. Nuadihi village, near Hatikote, on the banks of the Budhabalanga River,is the lifeline for the Mayurbhanj district. The soil is mainly lateritic and about 10 per cent of the area issandy loam. The forest was denuded and depleted through over-exploitation, but the local peoplehave formed the VSS to protect it. Now Hatikote RF is rich in Sal (*Shorea robusta*) coppice crops with Mahula (*Madhuca longifolia* var *latifolia*), Chara (*Buchanania lanzan*), Kendu (*Diospyros melanoxylon*), Dhaura (*Anogeissus latifolia*), Asan (*Terminalia alata*), Kaima (*Haldinia cordifolia*) and many species of medicinal plants. Because of the abundance of medicinal plants in this forest, Orissa State Medicinal Plant Board has inventoried 200 ha of this RF for medicinal plant species to ensure their protection and sustainable harvests and to develop it as a model site for other VSS to follow.

The Satkoshia Reserve Forest (Satkoshia RF) is comes under Karanjia forest division which is one of the important forest division in Mayurbhanj comprising 3077.05 km^2 area. It spreads from 85°40′12″ E to 86°21′50″ E and longitudinally 21°15′58″ N to 22°05′57″ N and has an elevation of 365m MSL. It is surrounded by Rairangpur Forest Division in north, Keonjhar Wildlife Division in south, Similipal Forest in east and Keonjhar Division in west. There are two major types of forests found in this Division *i.e.* dry deciduous sal forest and moist deciduous sal forest with few patches of semi evergreen forests inside Similipal. Due to heavy biotic interference and fire hazard, most of the forests change from moist deciduous to dry deciduous with its associates.

Materials and Methods

For Hatikote RF the survey of medicinal plants was carried out by plot sampling. For this purpose, the entire area of 200 ha was divided into 20 segments. In each segment, a sampling area of 100 m^2 (14.14 m ×7.07 m) were chosen to obtain a rectangular plot of1: 2 (Brower *et al.*, 1990). The girth at breast height of all plants above 3-m tall was recorded. For bushes, shrubs (< 3 m), a sampling area of 10

m^2(2.24 m × 4.47 m) was nested inside the aforementioned100-m^2 plot. A sampling plot of 1 m^2(0.71 m × 1.41 m) was also nested inside the 10-m^2plot to inventory all herbaceous vegetation. The medicinal plants were identified with the help of the *Flora of Orissa* (Saxena and Brahmam 1994–96).Computation of plant communities was done for density, frequency, coverage and relative values and importance (IV) following Misra (1968) and Browar *et al.* (1990).

For Satkosia RF an extensive survey under taken during three months forest work experience in 2016. The local people residing in the nearby areas were interviewed for the first hand information on ethnomedicinal uses of the plants. Further interaction session were held at different VanaSuraksha Samities (VSS) named Godabhangi, Hiliriposi, Dhatkidiha, and Colony sahi. The plant species were identified with the help of standard flora of Odisha (Haines 1921-24) and Saxena and Brahmam (1994-96).

Results and Discussion

In Hatikote RF the study reveals that there are 21 medicinal species in the over wood, 35 species in the middle storey, 17 species in the ground flora and 16 species of climbers, representing 38 families of vascular plants.

The dominant species are *Shorea robusta* in the overwood, *Croton roxburghii* in the middle storey and *Curculigo orchioides* in the ground flora. *Shorea robusta* is the dominant tree species in this study plot, having maximum density (0.169) and frequency (0.17). In terms of density/hectare, this species is followed by *Buchanania lanzan, Terminalia alata, Madhuca longifolia* and *Diospyros melanoxylon*. The coverage of *Buchanania lanzan* is maximum, followed by *Syzygium cumini, Ixora undulata, Madhuca longifolia* and *Careya arborea*. It is evident from the ratio of coverage to frequency that the distribution of tree species is random.

The importance value (IV) of *Shorea robusta* is highest (1.017), followed by *Madhuca longifolia* (0.339), *Syzygium cumini* (0.250) and *Buchanania lanzan* (0.184). *Albizia chinensis, Careya arborea* and *Strycnos nux-vomica* recorded the lowest IV (0.008)and (0.009), respectively. In the middle storey, *Hollarhena pubescens* was the most common species, followed by *Croton roxburghii* and *Casearia tomentosa*. Among the herbs, *Curculigo orchioides* was followed by *Chlorophytum arundinaceum* and *Ichnocarpus frutescens*.

Similarly, in Satkoshia RF the study provides information on 30 plant species belonging to 22 families. The plants parts used for medicinal preparation were roots, barks, shoots, seeds, leaves, flowers, fruits and whole plants. Most frequently utilized plant parts were leaves (16), followed by root (08) and bark (04). This study gives a brief account of the various ethno-medicinal plant parts against different diseases and infections. The study area is rich in natural medicinal plant resources. An attempt was made to collect information on traditional medicinal knowledge of the local peoples residing in villages on the periphery of the area. More depth information maybe explored from these villagers and tribes which will give new impetus to the traditional system of healthcare.

Constraints for Conservation

The IUCN Red Data book lists 34,000 plants with endangered status. The Botanical Garden Conservation International (BGCI) 2000 database indicates that there are about 1846 botanic gardens. In-order to put efforts for ex situ conservation; these botanical gardens have to cultivate several hundreds of endangered, rare and vulnerable plant species, which requires elaborate facilities and extraordinary efforts. Therefore, biologists feel that the *ex situ* conservation should be considered as a complimentary measure of *in situ* conservation for holistic strengthening of conservation.

Conservation Measures

For the Conservation of medicinal plants the following short and long term measures should be taken. Banning export of medicinal plant species, Important to develop scientific harvesting technologies,Urgently agro technologies need to be developed, Agro technology developed in experimental fields needs to be transferred from Lab to land and ultimately to the market place, State Forest Dept. and local communities can jointly contribute for *in situ* conservation, Industry, research institutes, NGO's and local communities (women group) can play a very effective role in long term conservation, It provides scope for income, employment and empowerment of primary users of medicinal plants.

CONCLUSION

During the course of the study it has been seen that the local people of the Baripada and Karanjia Forest Division are having a great idea on utilization of natural products in their daily life especially against diseases. The study area is a good resource of medicinal plants, but due to lack of attention some of the plants may lost in future. So steps should be taken to conserve these plants and make valuable utilization of them.

India, considered as a store house of medicinal plants, is losing its medicinal plant diversity at a very high rate due to indiscriminate and destructive collection of plant materials, population growth and habitat destruction. It is feared that unless immediate actions are taken most of the important medicinal plants will be lost or endangered. To prevent extinction of important medicinal plants of India, stride by scientist and government has been made but the situation is still out of control. The various conservation efforts need to be geared up to combat the existing challenges. Biotechnological techniques of plant conservation need to be harnessed to its full potential. Large scale production of medicinal plants through tissue culture can reduce the existing ever-growing burden on wild population of medicinal plants and thus will improve the conservation status of important medicinal plants.

ACKNOWLEDGEMENTS

The author thanks the Member Secretary, Orissa State Medicinal Plant Board, Forests and Environment Department, Government of Odisha for financial assistance. The assistance of the staff of Baripada and Karanjia Forest Division in carrying out the fieldwork is highly appreciated.

REFERENCES

Biswas, S., Rawat, M. S., Tantray, F. A., and Sharma, S. (2017). Medicinal plants conservation and development areas (MPCDAs) – An initiative towards conservation of medicinal plants. *Medicinal Plants – International Journal of Phytomedicines and Related Industries, 9*, 143–149.

Brower J., Zar, J. and Ende, C. (1990). *Field and Laboratory Methods for General Ecology*. Third Edition. USA: Wm. C. Brown Publishers.

Haines, H.H. (1921-24). *The Botany of Bihar and Orissa*, Vol I-IV, Government of Bihar and Orissa.

Misra, K.C. (1968). *Ecology work book*. Calcutta: Oxford and IBH Publishing Co.

Rout, S.D. Kumar, J. and Swain, D. (2005). Conservation strategy for medicinal plants in Odisha, India. *International Journal of Biodiversity Science and Management* 1: 205–211

Saxena, H.O. and Brahmam, M. (1994-96). *The Flora of Orissa,* Vol I-IV, Regional Research Laboratory, Bhubaneswar.

Chapter 5

A Checklist of Orchids in Similipal Biosphere Reserve and its Conservation Approaches

Debashish Behera[1]*, *Sulagna Subhasmita Jena*[2] *and Debaraj Parida*[3]

[1]*Botanical Survey of India, Kolkata – 700 064, West Bengal*
[2]*Regional Plant Resource Center, Nayapalli, Bhubaneswar – 751 015, Odisha*
[3]*Dhenkanal Autonoumous College, Dhenkanal – 759 001, Odisha*
**e-mail: debashish.nou@gmail.com*

The orchids were intensively searched and recorded inside Similipal Biosphere Reserve. Field observation was recorded in field note book. Various herbaria like Central National Herbarium, Botanical Survey of India, Kolkata, West Bengal and Regional Plant resource Centre, Bhubaneswar, Odisha were consulted for reference, verification and identification of orchid flora. Total number of orchid species along with its habitat and most suitable phorophytes were documented in all the epiphytic orchids; however, the terrestrial orchids were documented either from field observation or literature. All together 99 orchid species along with ten most suitable phorophytic species (*Shorea robusta, Terminalia alata, Mangifera indica, Terminalia bellirica, Madhuca indica, Diospyros malabarica, Buchanania lanzan, Schleichera oleosa, Careya arborea, Terminalia chebula*) were documented. Among which 55 species are epiphytic 41 species are terrestrial, 2 species are grown in the entire three habitat *i.e.* terrestrial, lithophytic and epiphytic habitat. Only one species (*Epipogium roseum*) is having saprophytic habitat reported in 2017. The most orchid rich patches in SBR are Chahala, Upper Barakamuda, Bhanjabasa, Meghasai and Kabatghai area.

Keywords: *Orchids, Habitat, Flowering, Conservation.*

INTRODUCTION

Orchids are one of the most beautiful and self-life flowers among all the flowering plants. These beautiful flowered orchids are creating interest for several taxonomical people along with fascinating peoples to do research on it. The ancient Greeks were the first to take note of these strange plants. The word orchid is derived from the Greek term "orchis", which means testicle, because of the similar appearance of the root tubers of some orchidaceous species with testicle. It was the famous Greek philosopher Theophrastus, (370-285 BC), who in his *Enquiry into Plants* (a set of ten books) first referred the group of plants having paired, testiculate tubers as *Orchis*. As the Greeks believed in the *Doctrine of Signature,* such plants were considered good for human vitality. Later, Dioscorides (1st century AD) in his Materia Medica adopted the name, where description of two orchids was given along with several species of medicinal plants. Indian Vedic scriptures also mentioned about these plants under the name "*Vanda*" which has been adopted as the generic name of one of the most important orchids. Orchids have been attracting botanists, naturalists, and ecologists since time immemorial due to their incredible range of diversity in shape, size and colour of flowers. They have been very closely associated with human beings since ancient times. The Chinese term "Lan" means 'beautiful orchid', appeared in the oldest book on Chinese philosophy, *Eki-kyo*, of which Confucius (551-479 BC) was a co-author. In 280 BC, "Emperor Shen Nung (Father of Chinese medicine) in his Materia Medica described Bletilla hyacinthine along with Dendrobium species". Orchids are monocotyledonous angiospermic herbaceous plants that are grouped in the family, Orchidaceae, which includes approximately 750 genera and 17, 000 species (Misra, 2007). Moreover, thousands of hybrid orchids are added in the Royal Horticultural Society, UK, for last few years. Orchids are well known for their floricultural value long before the arrival of Europeans in Mexico. The Aztecs (people of Central Mexico in the 15th century AD) discovered the orchid genus *Vanilla* and used it extensively as flowering species.

Taxonomy of Orchids

Orchid species are grouped into larger taxonomic units *i.e.* 'genera' based on some similar characteristics such as habitat, shape and form of the plant, leaves, flowers or some other morphological features. Just how a particular species is placed under a genus is not a rigid science and as a result, a species is sometimes found listed under different genera depending on which classification is being followed. Standardization of the orchid classification system is an ongoing issue. In view of global attempts at updating the catalogue of seed plants (Govaerts, 2003), a large number of redundant names, cryptic synonyms and poorly resolved aggregates, especially in orchids, there is an urgent need to carry out the detail taxonomic studies on the orchid taxa from various regions.

Distribution of Orchids

World Distribution

Orchids are found in all parts of the world except in the extremely dry cold (the artic), hot regions (the deserts) and the aquatic and marine ecosystems. According to the mode of distribution, orchids could broadly be grouped as tropical and temperate. They are distributed throughout the world. The majorities of orchids are however, natives of tropical climates and occur in their greatest profusion in humid tropical forest of Central and South America. Orchids are also found in Australia, Afghanistan, Bangladesh, Colombia, Finland, France, Island, India, Italy, Iran, Indonesia, Japan, Kenya, Kuwait, Korea, Myanmar, Mexico, Madagascar, Mauritius, Malaysia, Nepal, Peru, Poland, Portugal, Pakistan, Panama, Philippines, Srilanka, Singapore, Spain, Sudan, Siberia, Thailand, Turkey, Vietnam, and many other countries. The Orchidaceae have about 28,000 currently accepted numbers of species, distributed in about 763 genera in the world.

Orchid Flora of India

Altitudinal range plays a major role in floristic distribution. In India, this ranges from the sea level to about 8000m above sea level along with a wide variation in temperature and rainfall. That is why Indian sub-continent is bestowed with widely diverse climatic conditions. This provides opportunity for wide diversity of orchid flora, in this subcontinent. Although in Hooker's flora, 1600 species were reported from undivided British India, only 1150 orchids species are reported by Kumar and Manilal (1994) and 1331 species were reported by Misra (2007) in India.

The Indian subcontinent can geomorphologically be divided into:

1. The Himalayan region
2. The Peninsular region
3. The Andaman and Nicobar Islands

1. The Himalayan Region

The Himalayan region is the meeting point of several countries and so their flora. There are many Malaysian, Chinese and Indian elements found in this region. Several genera like *Dendrobium, Eria, Flickingeria, Habenaria, Oberonia, Phalaenopsis* and *Vanda* are found in the tropical forests. In the sub-tropical forests are found genera like *Anoectochilus, Eulophia* and *Galeola* are found. The temperate region harbours genera like *Corybas, Pleions, Satyrium* and *Spiranthes* while genera like *Cypripedium, Didiciea, Epipactis, Herminium etc.* are met in the alpine area. A total of 1023 species in 167 genera have been recorded from this Himalayan region. Approximately 868 taxa, belonging to 79 genera are found exclusively in the Himalayan region. Nearly 107 species of this region are common with the orchid flora of the peninsula, while 20 species are found in the Andaman and Nicobar Islands.

In terms of numbers of species and genera, the Himalayan region has the richest orchid flora in India. Arunachal Pradesh has the largest number of wild orchid in India followed by Sikkim. However, there are only 195 endemic species in this region, which is about 22.3 per cent of the total endemic orchids of India (Misra, 2007).

2. The Peninsular Region

The northern boundary of the peninsular India is an imaginary horizontal line from the Kutch district in Gujarat state in the west to the delta of the river Ganga in West Bengal. The chief geographical features of this region are the mountain systems of the Western Ghats and the Eastern Ghats. Vegetation of the peninsula is considerably diverse owing to the variation in the latitude and climate of the land. The Western Ghats is almost parallel to the west coast of India for an unbroken length of about 1600 km from the Tapti River in Gujarat to Kanyakumari in Tamil Nadu. Two hundred and sixty seven species of orchids have been reported (Satish Kumar, 1991) from this region. The eastern coastal area is much broader than the western coast. In the Peninsula, orchid flora consists of 371 species in 86 genera.

3. The Andaman and Nicobar Islands

This is a group of 572 Islands. The orchid flora is not very well known from this region. Only 121 taxa in 60 genera have been reported, from this 17 taxa are endemic to these islands.

Orchid Flora of Odisha

Odisha is known to be the meeting ground of the flora of the Northern and the Southern India (Misra, 1997). That is why it enjoys a wide range of phyto-climatic conditions due to its geographic diversity. This has influenced the composition and distribution of the orchid flora in the state. The orchid flora of Odisha comprises of 139 species. Out of these 139 species, 5 species are endemic to Odisha. This consists of 3 epiphytic species of the sympodial form, *Cirrhopetalum*, *Eria*, *Liparis*, and two terrestrial taxa of *Habenaria*.

Importance of Orchid

Orchids have been closely associated with human beings since ancient time. Indian Vedic scriptures also mentioned about these plants under the name *Vanda*. This name has been adapted as a generic name for one of the most beautiful monopodial group of orchids now commercially grown in Malaysia, Singapore, Thailand and Taiwan.

Ornamental Importance and Importance in Cut-flower Industry

India's annual flower production stands at around 1000 tons and its floriculture industry have a miniscule 0.01 per cent share in the international market. Even though since the last few years, orchids have made their presence felt in the Indian cut-flower trade, orchid cultivation and commercialization in India is still at a primary stage (Chugh *et al.*, 2009). Varied agro-climatic conditions, zones, and ever-growing high-end consumer markets, make it a highly profitable proposition to grow orchids in India (Singh *et al.*, 2008).

Orchids are marketed both as cut flowers and as potted plants. The largest exporters of potted orchids are Taiwan, Thailand, UK, Italy, Japan, New Zealand and Brazil while the largest importer of potted orchids is the United States. Today, orchids such as *Cymbidium*, *Dendrobium*, *Oncidium* and *Phalaenopsis* are marketed globally and the orchid industry has contributed substantially to the economy of many Association of South East Asian Nation (ASEAN) countries (Hew, 1994; Laws, 1995).

Use in Traditional Medicine

In many countries like India, China and in some parts of Europe and America, Australia and Africa, orchids have been used as traditional drugs for a very long time. The ethno botanical values of several orchids are also discussed in *'Charaka Samhita'* a classic Indian medicinal treatise written by Charaka in Sanskrit in the past. The people of middle age even believed that orchid plants came from the drops of semen which fell to earth in meadows where animals came together to breed. Thus to the European eye, the orchid is a symbol of sex. The usefulness of different drugs of these plants is also indicated in the treatment of stomach worms and in allaying pains of the joints. Recently, it has been reported that orchid molecules are important in reducing fever, increasing the white blood cell count, curing eye diseases, treating fatigue and headache, and most importantly, functioning as anti-cancer agent (Bulpitt, 2005). In India, the orchids are employed for a variety of therapeutic use in different systems of traditional medicine. In the Ayurvedic system of medicine, a group of eight drugs, known as Astavarga, is employed in the preparation of a number of rejuvenating formulations and tonics. Astavarga is an important ingredient of various classical Ayurvedic formulations (Singh and Duggal, 2009) like Chyawanprash.

Mycorrhizal Association of Orchids

All orchids have an obligate relationship with mycorrhizal symbionts during seed germination and being similar to *Rhizoctonia*-like fungi (Arditti, 1992). Understanding the mycorrhizal symbiosis is a great importance, because availability of the fungal symbionts may play a key role in determining orchid distribution and diversity. Furthermore, most studies of orchid mycorrhizal interactions have concentrated on terrestrial orchids from temperate regions, whereas the majority of orchid species are epiphytes in tropical regions.

Study Area

The study was conducted in the area of Similipal Biosphere Reserve (SBR), which is located in the district of Mayurbhanj in the state of Odisha. SBR, for its location, geography and climate has a rich flora and fauna.

Location

The study area Similipal Biosphere Reserve (SBR) is located between 21° 28′ to 22° 08′ North longitude and 86° 4′ to 86° 37′ East latitude. The SBR is divided into mainly three zones. The core zone of Similipal forest consists of 845 Km^2, whereas the buffer zone is approximately 2174 Km^2. The transitional zone that encircles buffer area by 10 Km is around 2559 Km^2. Therefore, the whole forest is approximately 5578 Km^2 in area. This forest consists of tropical to subtropical forests. The ground elevations vary from a minimum of 500-600 meter, from Mean Sea Level (MSL) of the outer area to 1000-1100 meter along the central area. The highest elevation is 1178m at Khairiburu followed by 1175m at Meghasani.

Topography

The geological formation in Similipal hills holds three huge impervious quartzite bowls concentrically placed and their interspaces filled with pervious volcanic rocks. This is in a way of a unique feature. The water retention capacity of these hills is very high. Similipal Biosphere Reserve occupies approximately two-third part of Mayurbhanj district. Udala in south, Jashipur in the north, Bisoi in the east and Thakurmunda in the west are around SBR. Baripada, the heart of the Mayurbhanj district is situated in the southeast portion of SBR. The Biosphere Reserve is dissected by several systems of drainage.

Climate

The climate of SBR is tropical with warm and humid summer. Only three distinct seasons are felt during the year. Winter creeps in gradually from mid-October and becomes severe in December. The spring is pleasant. The southern and western aspects are cooler and northeastern aspects are warmer in Similipal. The rainy season starts from middle of June till October with rainfall of about 1250mm in plateau. Southwest monsoon (June-July) and retreating northeast monsoon (August) are the main cause of rainy season. Low pressure is another reason for rain in the forest of Similipal Biosphere Reserve. Its topography, water-holding capacity with moderate high rainfall has given birth to numerous streams and rivers. It is virtually a land of perennial waters.

Temperature

The temperature of Similipal Biosphere Reserve varies from season to season. The maximum temperature in summer goes up to 45 °C. The nights are cool and pleasant. In winter, the maximum temperature usually comes down to 4 °C, but sometimes it may come down as low as 0 °C.

Humidity

Humidity is also not constant in the different regions of SBR. It varies between 70-100 per cent in core region due to the dense forest. Most of orchids thrive well and flower when the atmospheric moisture is not less than 30 per cent at night and more than 80 per cent during the daytime.

Materials and Methods

Field Survey

Field visits were conducted for many times for last one decade (2008-2018) inside SBR, to locate the natural habitats of orchids. Mostly in three seasons (winter, summer and monsoon. Though it was difficult to go inside the forest during monsoon, the author took sufficient precautions and hired local people for assistance during field study. Permission was obtained from the Field Director, Similipal Tiger Reserve and the sites were visited in the presence of forest guards.

Inventory of Wild Orchids

During the present work, orchids were located in their natural habitats.

Photographs were taken for both vegetative and reproductive parts, for documentation without disturbing their natural habitats. Only some flowers were plucked whenever necessary, for identification and photography in the field. The species were identified in the spot by the help of Orchid Flora of Orissa (Misra, 2004), Orchids of India (Misra, 2007) and later it was confirmed by the help of Flora of Orissa (Saxena and Brhamam, 1995). In addition, the authors were take information from different literature and books during their field visit. In addition, Forest Department of Government of Odisha constructed an Orchidarium at Gudgudia range, which sheltered more than 60 native orchids, were examined inside that orchidarium.

Results and Discussion

In the present investigation, 99 species under 43 genera were documented from Similipal Biosphere Reserve. There was a lot of variation of their habitat, *i.e.* epiphytic, lithophytic, saprophytic and terrestrial orchids (Table 5.). The most dominant species were *Vanda tessellata* followed by *Acampe praemorsa*. Although ten phorophytic plant species (*Shorea robusta, Terminalia alata, Mangifera indica, Terminalia bellirica, Madhuca indica, Diospyros malabarica, Buchanania lanzan, Schleichera oleosa, Careya arborea, Terminalia chebula*) were documented from field the most common host plants were *Mangifera indica, Shorea robusta, Madhuka indica and Careya arborea*. Among all the 98 orchid species, 55 species grown as epiphytic in habitat, 41 species are terrestrial, 01 species are saprophytic in nature and finally 02 species have diverse habitat for grown as epiphytic, lithophytic and terrestrial also (Figure 5.1). One species Vanda *tessellata* is flowering two times in a year. Among these 99 orchid species, rarest orchids among all were *Eria meghasaniensis, Odisha cleistantha, Tainia penangiana, Nervilia infundibulifolia, Cymbidium bicolor* and *Cirrhopetalum panigrahianum*. The highest number of species was found in *Dendrobium* genus (11 species) followed by *Habenaria* genus (8 species). A detail of information on botanical name, habitat and its flowering calendar is mentioned in Table 5.1.

Table 5.1. List of Orchids of Similipal Biosphere Reserve

Sl.No.	*Botanical Name*	*Habitat*	*Flowering Calendar*
1	*Acampe carinata* (Griff.) Panigr.	E	October – January
2	*Acampe ochracea* (Lindl.) Hochr.	E	December – February
3	*Acampe praemorsa* (Roxb.) Blatt. and McC.	E	June – July
4	*Acampe rigida* (Buch.-Ham. ex J.E. Sm.) Hunt.	E	August – October
5	*Acanthephippium bicolor* Lindl.	T	May
6	*Acanthephippium sylhetense* Lindl.	T	May – June
7	*Aerides multiflora* Roxb.	E	June – July
8	*Aerides odorata* Lour.	E	June-July
9	*Bulbophyllum cariniflorum* Rchb. f.	E	August
10	*Bulbophyllum crassipes* Hook.f.	E	October
11	*Bulbophyllum macraei* (Lindl.) Rchb. f.	E	July – August
12	*Bulbophyllum polyrhizum* Lindl.	E	March

Sl.No.	Botanical Name	Habitat	Flowering Calendar
13	*Bulbophyllum triste* Rchb. f.	E	February – March
14	*Bulbophyllum umbellatum* Lindl.	E	April – May
15	*Calanthe triplicata* (Willem.) Ames	T	
16	*Chiloschista parishii* Seidenf.	E	March – July
17	*Cirrhopetalum panigrahianum* (S.Misra) S. Misra	E	June
18	*Cleisostoma appendiculatum* (Lindl.) Benth. and Hook. f. ex Seidenf.	E	August – October
19	*Crepidium purpureum* (Lindl.) Szlach.	T	June – July
20	*Cymbidium aloifolium* (L.) Sw.	E	April – June
21	*Cymbidium bicolor* Lindl.	E	February – March
22	*Dendrobium aphyllum* (Roxb.) Fisch.	E	February – April
23	*Dendrobium bicameratum* Lindl.	E	August
24	*Dendrobium cathcartii* Hook. f.	E	March – April
25	*Dendrobium crepidatum* Lindl. and Paxt.	E	February – April
26	*Dendrobium formosum* Roxb. ex Lindl.	E	May – July
27	*Dendrobium herbaceum* Lindl.	E	February – April
28	*Dendrobium macrostachyum* Lindl.	E	April – July
29	*Dendrobium moschatum* (Buch.-Ham.) Sw.	E	May – June
30	*Dendrobium peguanum* Lindl.	E	December – February
31	*Dendrobium regium* Prain	E	April – June
32	*Dendrobium transparens* Wall. ex Lindl.	E	April – June
33	*Dienia ophrydis* (Koen.) Ormer. and Seidenf.	T	June – August
34	*Diploprora championi* (Lindl.) Hook. f.	E	June – July
35	*Disperis neigherrensis* Wight	T	June – July
36	*Epipogium roseum* (D. Don) Lindl.	S	May – June
37	*Eria bambusifolia* Lindl.	E	December – January
38	*Eria meghasaniensis* (S. Misra) S. Misra	E	September- October
39	*Eulophia explanata* Lindl.	T	April – May
40	*Eulophia graminea* Lindl.	T	April – May
41	*Eulophia mackinnonii* Duthie	T	June – July
42	*Eulophia spectabilis* (Dennst.) Suresh	T	May – June
43	*Flickingeria macraei* (Lindl.) Seidenf.	E	July and September
44	*Geodorum densiflorum* (Lamk.) Schltr.	T	July – August
45	*Geodorum recurvum* (Roxb.) Alston	T	May – June
46	*Goodyera fumata* Thw.	T	February – March
47	*Goodyera hispida* Lindl.	T	August
48	*Goodyera procera* (Ker-Gawl.) Hook.	T	March – May
49	*Goodyera thailandica* Seidenf.	T	February – March
50	*Habenaria commelinifolia* (Roxb.)Wall. ex Lindl.	T	August – October
51	*Habenaria digitata* Lindl.	T	July - August

Sl.No.	*Botanical Name*	*Habitat*	*Flowering Calendar*
52	*Habenaria furcifera* Lindl.	T	July – September
53	*Habenaria gibsoni* var. *foetida* Blatt. and McC.	T	July – August
54	*Habenaria plantaginea* Lindl.	T	August – November
55	*Habenaria reniformis* (D. Don) Hook. f.	T	July – September
56	*Habenaria roxburghii* R. Br.	T	June August
57	*Habenaria stenopetala* Lindl.	T	August – November
58	*Kingidium deliciosum* (Rchb. f.)Sweet	E	May – September
59	*Liparis elliptica* Wight	E	November – January
60	*Liparis nervosa* (Thunb.) Lindl.	T/L	June-July
61	*Liparis odorata* (Willd.) Lindl.	T	July – August
62	*Liparis viridiflora* (Bl.) Lindl.	E	November –December
63	*Luisia brachystachys* (Lindl.) Bl.	E	April
64	*Luisia trichorhiza* (Lindl.) Bl.	E	March – July
65	*Lusia tristis*	E	February – March
66	*Luisiopsis inconspicuua*	E	June – August
67	*Micropera pallida* (Roxb.) Lindl.	E	June – July
68	*Nervilia aragoana* Gaud.	T	May – July
69	*Nervilia crociformis* (Zoll. and Mor.) Seidenf.	T	May – July
70	*Nervilia infundibulifolia* Blatt. and McC.	T	June
71	*Nervilia plicata* (Andr.) Schltr.	T	May – June
72	*Oberonia ensiformis* (J. E. Sm.) Lindl.	E	October – November
73	*Oberonia falconeri* Hook. f.	E	September – October
74	*Oberonia gammiei* King and Pantl.	E	August
75	*Oberonia mucronata* (D. Don) Ormer and Seidenf.	E	September –November
76	*Oberonia proudlockii* King and Pantl.	E	February – March
77	*Oberonia pyrulifera* Lindl.	E	October – November
78	*Odisha cleistantha* S. Misra	T	October
79	*Pecteilis gigantea* (J. E. Sm.) Rafin.	T	August – October
80	*Pelatantheria insectifera* (Rchb. f.) Ridl.	E	October – December
81	*Peristylus constrictus* (Lindl.) Lindl.	T	June – September
82	*Peristylus goodyeroides* (D. Don) Lindl.	T	July – August
83	*Peristylus lawii* Wight	T	June – August
84	*Peristylus parishii* Rchb. f.	T	June July
85	*Peristylus plantagineus* (Lindl.) Lindl.	T	August – September
86	*Pholidota imbricata* Lindl.	E	July – September
87	*Rhynchostylis retusa* (L.) Bl.	E	May – June
88	*Seidenfia versicolor* (Lindl.) Marg. and Szlach.	T/L/E	July – August
89	*Smitinandia micrantha* (Lindl.) Holtt.	E	April – June
90	*Spiranthes sinensis* (Pers.) Ames	T	January – March
91	*Staurochilus ramosus* (Lindl.) Seidenf.	E	April – May

Sl.No.	Botanical Name	Habitat	Flowering Calendar
92	*Tainia penangiana* Hook.f..	T	January – March
93	*Thunia bracteata* (Roxb.) Schltr.	E	August – September
94	*Tropidia angulosa* (Lindl.) Bl.	E/T	August – September
95	*Tropidia pedunculata* Bl.	T	May – July
96	*Vanda tessellata* (Roxb.) Hook. f. ex G. Don	E	March – May and September –November
97	*Vanda testacea* (Lindl.) Rchb. f.	E	April – May
98	*Zeuxine affinis* (Lindl.) Benth. ex Hook. f.	T	January —-February
99	*Zeuxine nervosa* (Wall. ex Lindl.) Benth. ex Clarke	T	February – March

*E: Epiphyte; T: Terrestrial; L: Lithophyte; S: Saprophyte.

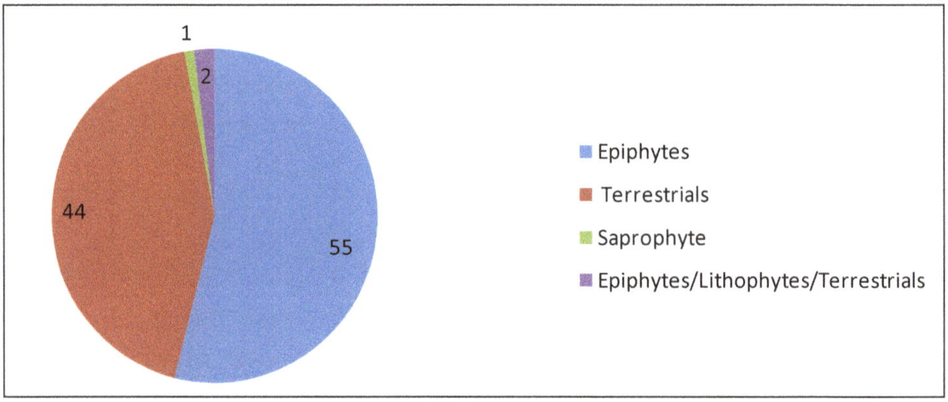

Figure 5.1. Showing the Orchid Species Grown in different Habitat.

Conservation Approaches

Orchid plant bears the most dazzling and self-life flowers/inflorescence among all the flowering plants. In the current scenario, the climate change, deforestation and development of urbanization decline the flora and fauna in the nature. Along with the above said cause, the ethno-medicinal uses of some species of orchids make these species extinct. Further, slow growth and less germination capacity (less than 0.1 per cent) of seeds due to lack of myccorizal association in orchids are made these plants become more rare and endangered. Moreover, the lack of suitable phorophytic species at the nearby area of mother orchids, create the key role for rarity day by day. Mass propagation through tissue culture and reintroduction of the wild species to their natural habitat may sustain this rare group of plants in nature. In addition, awareness among people and conservation of the orchid habitat may sustain these rare plants for future research and preservation. Although, one orchiderarium is present within SBR, the condition and maintenance of that is need to improve and one-person trend in these orchids for proper maintain it.

Plate 5.1: Orchids of SBR.

(a) ***Dendrobium regium*** Prain; (b) ***Dendrobium formosum*** Roxb. ex Lindl.; (c) ***Vanda tessellate*** (Roxb.) Hook. f. ex G. Don.; (d) ***Cymbidium bicolor*** Lindl.; (e) ***Geodorum recurvum*** (Roxb.) Alston; (f) ***Odisha cleistantha*** S. Misra.

ACKNOWLEDGEMENTS

The Authors are grateful to the Ministry of Environment and Forest, Govt. of India and Director, Botanical Survey of India for Financial support and Facilities for carryout the research. The authors also acknowledged the PCCF and Director Similipal Tiger Reserve, Govt. of Odisha for carry out the research inside SBR.

REFERENCES

Arditti, J. (1992). Fundamentals of Orchid Biology. *John Wiley and Sons, New York.*

Bonnardeaux, Y., Brundrett, M., Batty, A., Dixon, K., Kock, J. and Sivasithamparam, K. (2007). Diversity of mycorrhizal fungi of terrestrial orchid: Compatibility webs, brief encounters, lasting relationships and alien invasion. *Mycological Research*, 11: 51-61.

Bose, T.K., Bhattacharjee, S.K., Das, P. and Basak, U.C. (1999). Orchids of India. Calcutta, 1-487.

Bulpitt, C. J. (2005). The uses and misuses of orchids in medicine. *QJM*, 98(9): 625-631.

Chugh, S., Guha, S. and Usha Rao, I. (2009). Micropropagation of orchids: A review on the potential of different explants. *Scientia Horticulturae*, 122: 507-520.

Govaerts, R. (2003). World checklist of monocotyledons database in ACCESS: 1-71827. *The Board of Trustees of the Royal Botanic Gardens, Kew.*

Hew, C.S. (1994). Orchid cut-flower production in ASEAN countries. *Arditti, J. (Ed.), Orchid Biology: Reviews and Perspectives*, 6: 363-401.

Kar, T., Mohan, M. and Mandal, K. K. (2017). Disperis and Epipogium (Orchidaceae): Two new generic record for the floraof Odisha. *Nelumbo*, 59(2): 159-163.

Laws, S. N. (1995). Cut orchids in the world market. *Flora Culture International*, 5: 12-15.

Misra, S. (1997). Orchids of the Similipal forest. *P.C. Tripathy and S.N. Patra (Eds.), Similipal: A natural habitat of unique biodiversity, Orissa Environmental Society*, 73-91.

Misra, S. (2004). Orchids of Orissa. *Bishen Singh and Mahendra Pal Singh, Dehra Dun*, 774.

Misra, S. (2007). Orchids of India. *Bishen Singh and Mahendra Pal Singh, Dehra Dun*, 402.

Mishra, S. and Kar, T. (2017). Extended distribution of *Eulophia macckinnonii* (Orchidaceae) in eastern India. *Nelumbo*, 59(1): 61-65.

Satish, K.C. (1991). Endemic orchids of Western Ghats: A struggle for survival. *Proceedings of the seminar on rare, endangered and endemic plants of Western Ghats, Trivandrum*, 209-214.

Saxena, H.O. and Brahamam, M. (1995). The Flora of Orissa. *Orissa Forest Department Corporation, Bhubaneswar*, (I-IV), 2918.

Singh, A. and Duggal, S. (2009). Medical orchids: An overview. *Ethno-botanical leaflets*, 13: 351-363.

Singh, K., Gutgutia, A. and Gutgutia, S.K. (2008). Prospects of Commercial Orchid Cultivation in India. *National Conference on Orchids: Science and Society, Bangalore, India*, 97.

Chapter 6

Floristic Diversity in a Dry Deciduous Forest of Similipal Biosphere Reserve, Odisha, India

*Manas Ranjan Mohanta, Amrita Manjari and Sudam Charan Sahu**

Department of Botany, North Orissa University,
Baripada – 757 003, Odisha, India
**e-mail: sudamsahu.bdk@gmail.com*

Similipal Biosphere Reserve (SBR) is known for its rich biodiversity, medicinal plants and migratory path for many Himalayan and south Indian biomes. The present study aims to study the floristic diversity of a dry deciduous forest in Lulung, SBR. Random sampling were carried out by laying 10 quadrats of size 50 m × 20 m each for tree species. Besides this 1 ha sampling, floristic surveys were also conducted to document the herbs, shrubs and climbers. A total of 166 plant species were reported, belonging to 147 genera and 85 families. Out of 166 species, 70 species were herbs (42.17 per cent) followed by 56 species of trees (33.73 per cent), 28 species of shrubs (16.86 per cent) and 13 species of climbers (7.23 per cent). Fabaceae, Acanthaceae and Rubiaceae were the most specious families in case of tree, herb and shrub category, respectively. Similarly, Liliaceae, Combretaceae, Fabaceae and Asclepiadaceae were the most specious families in climber category. The average number of species per family was calculated as 3.5, 4.5, 2.75 and 1 for tree, herb, shrub and climber, respectively. The plants having high medicinal values were *Asparagus racemosus* Willd., *Aegle marmelos* (L.) Correa, *Bacopa monnieri* (L.) Wettst., *Andrographis paniculata* (Burm. f.) Nees *etc.* with healthy distribution in the study area. The information in this study will help forest managers and administrators to plan effective management policies towards the conservation of floral diversity of SBR.

Keywords: *Floristic diversity, Medicinal plants, Conservation, Similipal Biosphere Reserve.*

INTRODUCTION

India has a total geographical area of about 329 million hectares with a coastline of over 7500 km. The ecological ecosystem diversity of the country is enormous, ranging from sea level to highest mountainous ranges in the world; hot and arid conditions in northwest to cold arid conditions in the trans Himalayan regions; tropical weight ever green forest in north east India and the western Ghats; mangroves of Sunderbans and fresh water aquatic to marine ecosystem (Sharma and Singh, 2000). It has twelve bio-geographical provinces, five biomes and three bio-regions domains (Cox and Moore, 1993). The country support a diverse array of habitants or ecosystems such as forest, grass lands, wet lands, coastal, marine and dessert, each with rich and unique floristic diversity. These biological attributes are further enhanced by the geography location of the country at the confluence of three major global bio geographic realms, *viz.* Indo-Maleasian, Eurasian and Afro-Tropical, thus allowing intermingling of floristic region as well and making it one of the 17 mega diversity countries in the world, recognised by the world's conservation monitoring centre in 2000. The majorly concentrated in the four biodiversity hotspots, namely Eastern Himalayans, Western Ghats, Indo-Burma region and Sundaland, out of 34 biodiversity hotspots in the world. The floristically significant areas exhibit exceptional concentration of endemic species and also experiencing loss of habitat with higher occurrences of threatened plant species.

Forests are very rich in "flora and fauna". It plays a major and indispensable role in improving the food security of rural and tribal people, enhancing livelihood requirements for rural community and in maintaining ecological balance. Over 53 million tribal people in India belong to 550 communities of 227 ethnic groups and about 60 per cent of the rural communities directly rely on forest for their day-to-day requirement. The rate of forest loss due to deforestation as reported by Food and Agriculture Organization (2001) is 15.2 million hectare per year from 1990 to 2000. Forest ecosystem is one of the fundamental goals of ecological research. It is essential for providing information of plant diversity on ecosystem function and stability (world conservation monitoring centre, [WCMC] 1992, Tilman 2000, Townsend *et al.*, 2008). It has attracted attention of ecologist because of the growing awareness of its importance on the massive depression on the other living organisms (Singh, 2002). Indigenous people have good knowledge of the floristic wealth of their surroundings and having close association with biodiversity rich areas. Thus, they store maximum information by way of diversity of species and life forms. Forest at different latitudes form distinctly different eco-zones: boreal forests near the poles tend to consist of evergreens, while tropical forests near the equator tend to be distinct from the temperate forests at mid-latitude. The amount of precipitation and the elevation of the forest also affect forest composition. Human society and forests influence each other in both positive and negative ways. Forest provides ecosystem services to human and serves as tourist attractions. Forest can also impose costs, affect people's health, and interfere with tourist enjoyment. Human activities, including harvesting forest resources, can negatively affect forest ecosystem.

The ecological conditions which favour the forest growth are availability of moisture throughout the year. Tropical forests are highly productive, structurally complex, genetically rich renewable resources (Roy *et al.*, 2002). The tropical deforestation contributes to increases the CO_2 atmospheric layer affecting the climate and biodiversity. Though such type of forest occupies less than 7 per cent of land surface and they have the higher distinction of harbouring 50 per cent of all plants and animal species (Mayer, 1992).

Forest also helps in generating employment for those tribal people and is a reliable source of their income (Harris and Mohammed, 2003; Hegde and Enters, 2000; Kutty and Kothari, 2001; Poffenberger and McGean, 1996). Tropical and subtropical forests are the house to many living organisms and harbour maximum diversity of plant species found on the earth (WCMC, 1992). It is a precious resource provided by nature. The organisms living in forests are interdependent on each other. Life in forests is governed by factors like air, water and sunlight. There are varieties of plants available in most forests: herbs, shrubs and trees depending upon the climate of the region. These forests are enriched with medicinal and economically important plants. Plants make their own food by the process of photosynthesis and animals depend on plants and other animals for their food. Sometimes plants also depend on animals for processes like pollination and seed dispersal. There are many forests spread over large areas across the globe. Forest can be classified as: tropical, evergreen, partly evergreen, deciduous and dry forests based on the climatic conditions and types of trees present. Forests also comprise of non-living components such as lakes, ponds, soil, rocks, *etc.* A forest is defined as an area forming an ecosystem. Forests are of great importance to mankind. However, increase in human population and unmindful activities of human beings like increase in industrial areas, mining, construction of dams, canals, highways, encroachment of forest areas for agricultural purposes, *etc.* has led to forest degradation, loss of biodiversity and forest cover (Anitha *et al.*, 2003; Arjunan *et al.*, 2005; Davidar *et al.*, 2007; Gadgil and Guha 1992; Karanth *et al.*, 2006). Mostly tropical forests are exposed to rapid land cover fluxes than natural restoration, tree species composition, distribution and diversity are particularly susceptible to change from continuous stress factors such as bushfires, logging and cultivation (Uhi, 1987; Lykke, 1998).Over-exploitation has resulted in a rapid loss of tropical forests which is considered as one of the most serious environmental and economic problems all over the world (Hare *et al.*, 1997).Therefore, tree component of vegetation must be constantly monitored and managed in order to direct successional processes towards maintaining species and habitat diversity (Pickett and White, 1985; Turner, 1987). The causes of forest loss and degradation need to be identified to formulate better management and policy decisions.

The Indian subcontinent is rich in biodiversity and acts as flourishing ground for many flora and fauna. So, India is considered as one of the mega-diversities of the world. But all the 17 mega-diversity centres of the world are facing problem in disappearance of tropical forests at an alarming rate due to anthropogenic activities and invasion of invasive species (Ramachandran *et al.*, 2016). About 37.34 per cent of the geographical area in Odisha is covered under forest area that is about 7.66 per cent of country's forest having very fast deforestation rate during 1935 to 2010 (Reddy *et al.*, 2013). Therefore, many studies and surveys are being conducted throughout the India to know the diversity of various species of plants and animals. Some of the floristic diversity studies are Sahu *et al.*, 2012 (Niyamgiri hill range), Mishra *et al.*, 2008 (Similipal Biosphere Reserve), Kumar *et al.*, 2005 (Garo Hills), Mishra *et al.*, 2018 (Tropical forest of Balasore), Gandhi and Sundarpandian, 2014 (Tiruvannamalai forest), *etc.* However, no floristic diversity studies were conducted in the Lulung region of Similipal till date. So, viewing of above fact the present study was aimed to study the floristic diversity of a dry deciduous forest of Lulung, Similipal Biosphere Reserve (SBR), Odisha. These data are hoped to be helpful in making various policies for the conservation and management of forest in the future.

Materials and Methods

Study Area

Similipal Biosphere Reserve is rich in diversity and cultural significance due to which it regarded as a prominent ecological hot-spot and is one of the seventeen biosphere reserves of India. It is also included in World Network of Biosphere Reserve (WNBR) since 6th May, 2009. It stretches between 20°17′ to 22°34′N latitude and 85°40′ to 87°10′ E longitude, located in the central part of Mayurbhanj district, Odisha (Figure 6.1). It extends over an area of 5569.00 km^2 with a core area of *ca.* 1194.75 km^2, buffer area *ca* 1335.88 km^2 and transition area *ca.* 3038.39 km^2. Similipal has huge influence over the climate of the state and its neighbour-hood and is a treasure house of floristic diversity of Odisha. Therefore, it is considered as Himalayas of Odisha.

The place Lulung, is noted for its natural beauty being surrounded on three sides by hills of the Similipal ranges. It is a place in Baripada, subdivision is one of the entry points to Similipal National Park and is located on its eastern fringers. Lulung is situated on the banks of the Tributary river, Palpala. This river has a perennial water flow. The dense forest is an idyllic picnic spot for many nature lovers and their crystal clear water with pebbles of different sizes on its river bed.

Lulung is a good example of tropical dry deciduous forest which is predominantly composed of taxa namely *Shorea robusta, Terminalia alata, Haldinia cordifolia, Anogeissus latifolia, Schleichera oleosa, Pterocarpus marsupium, Syzygium cumini, Dillenia pentagyna, Terminalia bellirica, Kydia calycina, Mitragyna parvifolia, Ardisia solanacea, etc.* The shrubby plants are mostly evergreen and commonly present near waterfalls, streams, *etc.*, where the climate is somewhat moist. *Bauhinia vahlii, Butea superba and Milletia extensa* are most predominant climbers.

Field Sampling

Field surveys were carried out in different areas of Lulung, Similipal Biosphere Reserve during December 2017- February 2018. During the botanical survey, sites with specific species were demarcated/identified with hand held GPS. Ten plots of size 50 x 20 m each were laid for tree species in dry deciduous forest of Lulung, SBR. Besides this 1 ha sampling, floristic surveys were also conducted to document climbers, shrubs and herbs. All the available literature sources documenting the presence of flowering plants have been analyzed for the identification of plants

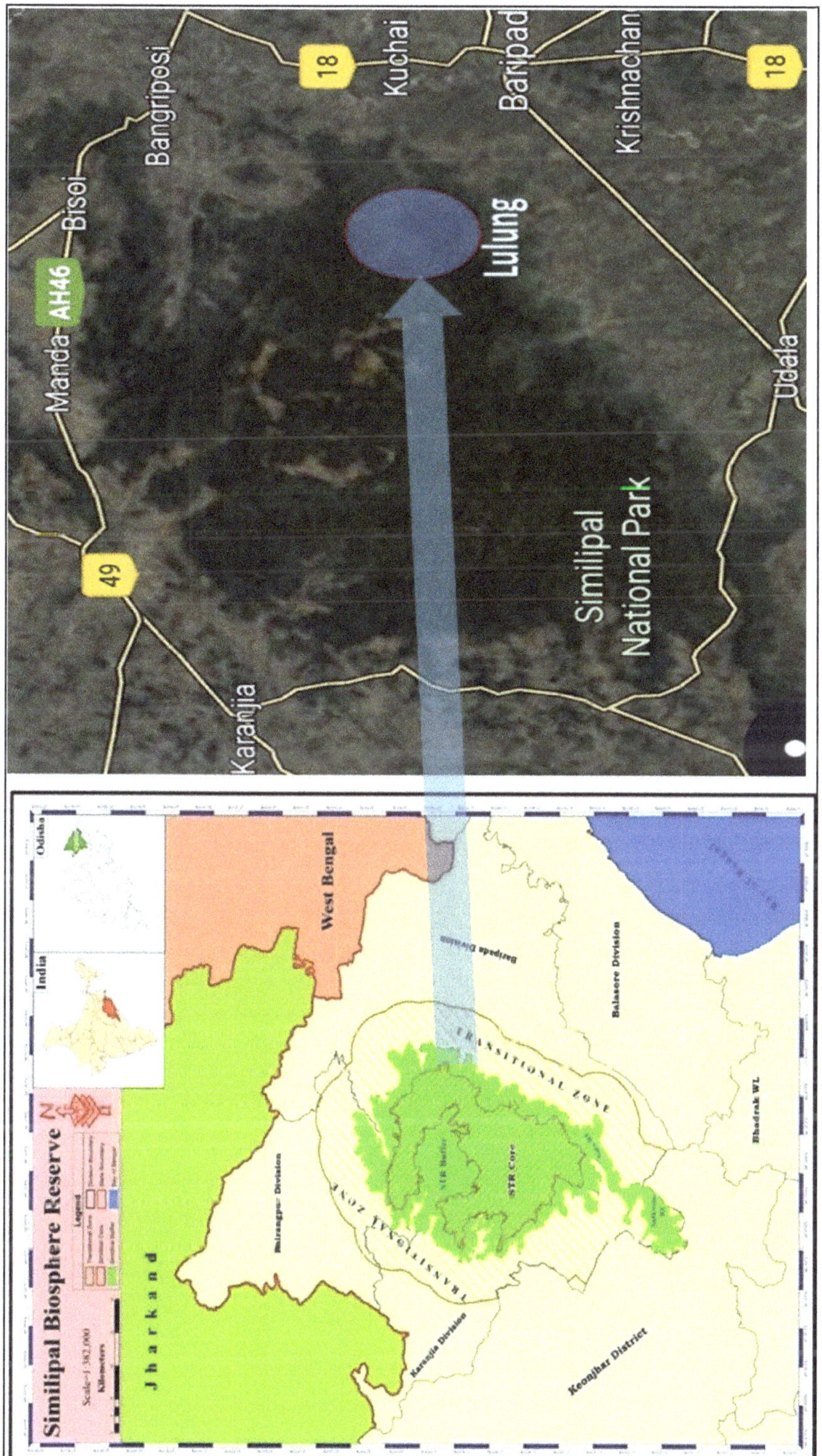

Figure 6.1. Google Earth image of Similipal Biosphere Reserve indicating Study Area.

(Flora of Presidency of Madras (Gamble, 1915-1935), Botany of Bihar and Orissa (Haines, 1921-1925), Supplement to the Botany of Bihar and Orissa (Mooney, 1950), Flora of Orissa (Saxena and Brahmam, 1994-1996), Flora of Bilaspur district (Murty and Panigrahi, 1989-1999), and Flora of Bihar (Singh *et al.*, 2001).

Results

A total of 166 species under 68 families were recorded from the study area, out of which, 56 were tree species (under 50 genera and 28 families), 12 climber (under 12 genera and 9 families), 28 shrub (under 26 genera and 17 families) and 70 herb species (under 60 genera and 34 families) (Tables 6.1–6.4). The most common family were Rubiaceae and Fabaceae which is represented by 15 and 14 species respectively; followed by Acanthaceae (9 species), Asteraceae (7 species), Malvaceae, Combretaceae, Zingiberaceae, Euphorbiaceae, Cyperaceae (6 species each), Phyllanthaceae, Lamiaceae, Apocynaceae, Commelinaceae (4 species each); Ebenaceae, Anacardiaceae, Annonaceae, Moraceae and Rutaceae (3 species each) *etc.* In case of trees most common families were Fabaceae and Rubiaceae (6 species) followed by Combretaceae and Phyllanthaceae (4 species each). The family with highest number of species found in herbs was Acanthaceae (8 species) which was followed by Asteraceae, Cyperaceae (6 Species each) and Fabaceae, Rubiaceae, Commelinaceae (5 species). The most common family of shrubs was Rubiaceae (5 Species) which is followed by Fabaceae (3 Species) and Euphorbiaceae, Rutaceae, Verbenaceae, Moraceae, Vitaceae (2 species each). The most common families in case of climber were Liliaceae, Combretaceae, Asclepiadaceae (2 Species each). So, the average number of species per family in case of tree was 2, herbs were 2.05, shrubs was 1.64 and climber was 1.33. The percentage of tree species was 33.73 per cent, herb species was 42.17 per cent, Shrub species was 16.86 per cent and climber species was 7.23 per cent found in Lulung (SBR) (Figure 6.2).

The plants with highest economic value includes species are *Andrographis paniculata, Asparagus rucemosus, Basella alba, Bauhinia vahili, Curcuma aeromatica, Dioscorea bulbifera, Diospyros menanoxylon, Madhuca indica, Terminalia chebula, etc.* were found abundantly in the area taken for study in Lulung (SBR).

The plants with high medicinal value are widely distributed throughout Lulung area of SBR. Some of them are *Aegle marmelos, Andrographis paniculate, Aspargus racemosus, Celastrus paniculatus, Curcuma aromatica, Dillerea pentagyna, Dioscorea bulbifera, Lantana camara, Madhuca indica etc.*

Species like *Aegle marmelos, Commelina benghalensis, Curcuma aromatica, Dioscorea bulbifera, Lantana camara, Madhuca indica, Terminalia chebula, etc.* is edible and is a source of food for many tribal communities. They are abundantly present in Lulung area of Similipal. Plants like *Bauhinia vahili* and *Lantana camara* is generally used for fire wood purposes. Generally, tree species like *Shorea robusta, Madhuca indica, etc.* used for timber purpose.

Table 6.1. List of Tree Species with their Local Names in Lulung, SBR

Sl.No.	*Species Name*	*Family*	*Local Name*
1	*Aegle marmelos* (L.) Correa	Rutaceae	Bela (O)
2	*Alangium salviifolium* (L.f.) Wangerin	Cornaceae	Ankal (O)
3	*Alstonia scholaris* (L.) R. Br.	Apocyanaceae	Chhatian (O)
4	*Anogeissus latifolia* (Roxb. Ex DC) Wall. Ex Bedd.	Combretaceae	Aakli, dhau (O)
5	*Antidesma acidum* Retz.	Phyllanthaceae	Marmuri (O)
6	*Bombax ceiba* L.	Malvaceae	Simuli (O)
7	*Bridelia retusa* (L.) A.Juss.	Phyllanthaceae	Kasi (O)
8	*Buchanania lanzan* (Lour.) M.R. Almeida	Anacardiaceae	Chara (O)
9	*Careya arborea* Roxb.	Lecythidaceae	Kumbhi (O)
10	*Casearia graveolens* Dalzell.	Salicaceae	Kokra (O)
11	*Cassia fistula* L.	Fabaceae	Sunari (O)
12	*Catunaregam spinosa* (Thunb.) Tirveng.	Rubiaceae	Patua, Salara (O)
13	*Cleistanthus collinus* (Roxb.) Benth. exHook.f.	Phyllanthaceae	Karada (O)
14	*Croton persimilis* Müll.Arg.	Euphorbiaceae	Pathudi, Masundi (O)
15	*Dalbergia latifolia* Roxb.	Fabaceae	Pahadi sisoo (O)
16	*Dalbergia paniculata* Roxb.	Fabaceae	Sisoo (O)
17	*Desmodium oojeinense* (Roxb.) H. Ohashi	Fabaceae	Bandano (O)
18	*Dillenia pentagyna* Roxb.	Dilleniaceae	Rai (O)
19	*Diospyros melanoxylon* Roxb.	Ebenaceae	Kendu (O)
20	*Diospyros Montana* Roxb.	Ebenaceae	Kansa, Bhadrika (O)
21	*Diospyros sylvatica* Roxb.	Ebenaceae	Kalicha (O)
22	*Ehretia laevis* Roxb.	Boraginaceae	Masania (O)
23	*Garcinia xanthochymus* Hook.f. ex T.Anderson	Clusiaceae	Sitmba, Rajamba (O)
24	*Haldina cordifolia* (Roxb.) Ridsdale	Rubiaceae	Koima (O)
25	*Holarrhena pubescens* Wall. ex G.Don	Apocyanaceae	Kuduchi (O)
26	*Kydia calycina* Roxb.	Malvaceae	Baranga, Kapasia (O)
27	*Lagerstroemia parviflora* Roxb.	Lythraceae	Sidha (O)
28	*Lannea coromandelica* (Houtt.) Merr.	Anacardiaceae	Mahi, Jia (O)
29	*Macaranga peltata* (Roxb.) Müll.Arg.	Euphorbiaceae	Jinka (O)
30	*Madhuca latifolia* var. latifolia (Roxb.) A.Chev.	Sapotaceae	Mahula (O)
31	*Mallotus philippensis* (Lam.) Müll.Arg.	Euphorbiaceae	Sinduri, Kapilagundi (O)
32	*Meyna spinosa* Roxb. ex Link	Rubiaceae	Nangal Kanta (O)
33	*Miliusa tomentosa* (Roxb.) J.Sinclair	Annonaceae	Gandhasalia (O)
34	*Miliusa velutina* (A.DC.) Hook.f. and Thomson	Annonaceae	Damsal, Gandhasala (O)
35	*Mitragyna parvifolia* (Roxb.) Korth.	Rubiaceae	Gudi Kaima (O)

Sl.No.	Species Name	Family	Local Name
36	*Morinda pubescens* Sm.	Rubiaceae	Achhu (O)
37	*Phyllanthus emblica* L.	Phyllanthaceae	Aonla (O)
38	*Polyalthia cerasoides* (Roxb.) Bedd.	Annonaceae	Potamasu, Panjon (O)
39	*Protium serratum* (Wall. ex Colebr.) Engl.	Burseraceae	Rimili (O)
40	*Psydrax dicoccos* Gaertn.	Rubiaceae	Dalsingh, Karuna (O)
41	*Pterocarpus marsupium* Roxb.	Fabaceae	Piasal, Bija (O)
42	*Pterospermum xylocarpum* (Gaertn.) Oken	Sterculariaceae	Muchukundu (O)
43	*Schleichera oleosa* (Lour.) Merr.	Sapindaceae	Kusuma (O)
44	*Semecarpus anacardium* L.f.	Anacardiaceae	Bhalia (O)
45	*Shorea robusta* Gaertn.	Dipterocarpaceae	Sal, Sala (O)
46	*Sterculia villiosa* Roxb.	Malvaceae	Kadala (O)
47	*Streblus asper* Lour.	Moraceae	Sahada (O)
48	*Strychnos potatorum* L.f.	Loganiaceae	Nirmali (O)
49	*Syzygium cumuni* (L.) Skeels	Myrtaceae	Jamu (O)
50	*Tectona grandis* L.f.	Lamiaceae	Sagwan (O)
51	*Terminalia bellirica* (Gaertn.) Roxb.	Combretaceae	Bahada (O)
52	*Terminalia chebula* Retz.	Combretaceae	Harida (O)
53	*Terminalia tomentosa* Wight and Arn.	Combretaceae	Asana (O)
54	*Wrightia tinctoria* R.Br.	Apocyanaceae	PitaKaruan, Kryo (O)
55	*Xylia xylocarpa* (Roxb.) Taub.	Fabaceae	Tangan (O)
56	*Ziziphus glabrata* B.Heyne ex Roth	Rhamnaceae	Karattai (O)

O: Oriya.

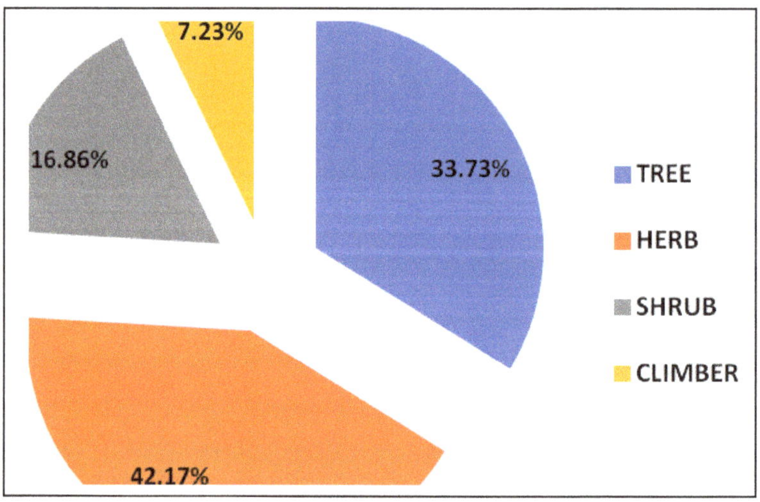

Figure 6.2. Pie Chart Showing Per cent of Trees, Herbs, Shrubs and Climbers in SBR.

Table 6.2. List of Herb Species with their Local Names in Lulung, SBR

Sl.No.	*Botanical Name of the Plant*	*Family*	*Local Name*
1	*Abutilon indicum* (L.) Sweet	Malvaceae	Atihala (O)
2	*Artemisia japonica* Thunb.	Asteraceae	Black Yarrow (E)
3	*Ageratum conyzoides* (L.) L.	Asteraceae	White weed (E)
4	*Allium sativum* L.	Amaryllidaceae	Garlic (E)
5	*Alpinia malaccensis* (Burm.f.) Roscoe	Zingiberacea	Ornamental Ginger (E)
6	*Anchomanes difformis* (Blume) Engl.	Araceae	Morton (E)
7	*Andrographis paniculata* (Burm. f.) Nees	Acanthaceae	Bhuin Nimba (O)
8	*Aquilegia vulgaris* L.	Rununculaceae	Columbine (E)
9	*Artemisia vulgaris* L.	Asteraceae	Mugwort (E)
10	*Bacopa monnieri* (L.) Wettst.	Scrophulariaceae	Brahmi (O,H)
11	*Barleria strigosa* Willd.	Acanthaceae	Blue bristly (E)
12	*Basella alba* L.	Basellaceae	Spinach (E)
13	*Beta vulgaris* L.	Amaranthaceae	Beet (E)
14	*Brassica juncea* (L.) Czern.	Brassicaceae	Mustard plant (E)
15	*Capcicum frutescens* L.	Solanaceae	Bird's eye chilli (E)
16	*Celosia argentea* L.	Amaranthaceae	Silver crocks comb. (E)
17	*Centrosema pubescens* Benth.	Fabaceae	Butterfly pea (E)
18	*Cheilocostus speciosus* (J.König) C.Specht	Zingiberaceae	Crepe-ginger (E)
19	*Chromolaena odorata* (L.) R.M.King and H.Rob.	Asteraceae	Siam weed (E)
20	*Cinchona officinalis* L.	Rubiaceae	Coffee beans (E)
21	*Cleome viscosa* L.	Capparaceae	Bagra (O)
22	*Commelina benghalensis* L.	Commelinaceae	Kana (O)
23	*Commelina maculata* Edgew.	Commelinaceae	Spotted Dayflower (E)
24	*Commelina paludosa* Blume	Commelinaceae	Himalayan monsoon flower (E)
25	*Curculigo orchioides* Gaertn	Hypoxidaceae	Kali tulsi (O)
26	*Curcuma amada* Roxb.	Zingiberaceae	Turmoric (E)
27	*Curcuma longa* L.	Zingiberaceae	Turmeric (E)
28	*Curcuma aromatica* Salisb.	Zingiberaceae	Wild Turmeric (E)
29	*Cyperus corymbosus* Rottb.	Cyperaceae	Gola methi (B)
30	*Cyperus involucratus* Rottb.	Cyperaceae	Papyrus (E)
31	*Cyperus pilosus* Vahl.	Cyperaceae	Fuzzy flat sedge (E)
32	*Cyperus rotundus* L.	Cyperaceae	Nut grass (E)
33	*Cyperus triceps* Endl.	Cyperaceae	White Water Sedge (E)
34	*Desmodium triflorum* (L.) DC.	Fabaceae	Creeping tick trefoil (E)

Sl.No.	*Botanical Name of the Plant*	*Family*	*Local Name*
35	*Dicliptera bupleuroides* Nees	Acanthaceae	Thorowax Foldwing (E)
36	*Echinacea purpurea* (L.)	Asteraceae	Purple cone flower (E)
37	*Eragrostis ciliata* (Roxb.) Nees	Poaceae	Pea (E)
38	*Eranthemum purpurascens* Wight ex Nees	Acanthceae	Gulsham (O)
39	*Eulophia spectabilis* (Dennst.) Suresh.	Orchidaceae	Amer kand (O)
40	*Evolvulus alsinoides* (L.) L.	Convolvulaceae	Vishnu kranta (O)
41	*Evolvulus nummularius* (L.) L.	Convolvulaceae	Nimulwel (E)
42	*Fimbristylis aestivalis* Vahl	Cyperaceae	Summer Fimbry (E)
43	*Floscopa scandens* Lour.	Commelinaceae	Konda-amadikada (T)
44	*Gaura lindheimeri* Engelm. and A.Gray	Onagraceae	Variegated gaura (E)
45	*Ginkgo biloba* L.	Ginkgoaceae	Maiden hair (E)
46	*Gonostegia pentandra* (Roxb.) Miq..	Urticaceae	Telpuri (O)
47	*Knoxia sumatrensis* (Retz.) DC.	Rubiaceae	*Knoxia* (E)
48	*Lepidagathis fasciculata* (Retz.) Nees	Acanthaceae	Rasna (O,H)
49	*Lobelia alsinoides* Lam.	Lobeliaceae	Chick weed (E)
50	*Marantochloa leucantha* (K.Schum.) Milne-Redh.	Marantaceae	Yoruba soft cane (E)
51	*Melochia corchorifolia* L.	Sterculiaceae	chocolate weed (E)
52	*Mimosa pudica* L.	Mimosaceae	Lajakuli (O)
53	*Oldenlandia verticillata* L.	Rubiaceae	Oldenlandia (E)
54	*Olyra latifolia* L.	Poaceae	Papda (O)
55	*Orthosiphon rubicundus* (D.Don) Benth.	Lamiaceae	Red jare tea (E)
56	*Pancratium* triflorum Roxb.	Amaryllidaceae	Pankusum (O)
57	*Perilepta auriculata* (Nees) Bremek.	Acanthaceae	Kapur Mingar (H)
58	*Peristrophe speciosa* Nees	Acanthaceae	Showy Fold wing (E)
59	*Phoenix acaulis* Roxb.	Arecaceae	Khajuri (O)
60	*Phyllanthus fraternus* G.L.Webster	Euphorbiaceae	Bhunin anal (O)
61	*Ruellia simplex* C.Wright	Acanthaceae	Mexican petunia (E)
62	*Rumex acetosa* L.	Polygonaceae	Garden sorrel (E)
63	*Rungia pectinata* (L.) Nees	Acanthaceae	Rungia (O)
64	*Salvia leucantha* Cav.	Lamiaceae	Mexican sage (E)
65	*Sida cordifolia* L.	Malvaceae	Bajromali (O)
66	*Spermacoce pusilla* Wall.	Rubiaceae	Safed Pholli (H)
67	*Symphyotrichum cordifolium* (L.) G.L.Nesom	Asteraceae	Blue wood aster (E)
68	*Uraria picta* (Jacq.) DC.	Fabaceae	Dabra (O)
69	*Zingiber officinale* Roscoe	Zingiberacea	Ginger (E)
70	*Zonia gibbosa* Span.	Fabaceae	Birmoch (S)

O: Oriya; E: English; T: Telugu; H: Hindi.

Table 6.3. List of Shrub Species with their Local Names in Lulung, SBR

Sl.No.	Botanical Name of the Plant	Family	Local Name
1	*Aidia genipiflora* (DC.) Dandy	Rubiaceae	*Aidia* (E)
2	*Antidesma ghaesembilla* Gaertn.	Euphorbiaceae	Black Currant (E)
3	*Ardisia solanacea* (Poir.) Roxb.	Myrsinaceae	Bisi (O)
4	*Bixa orellana* L.	Violaceae	Latkan(O)
5	*Catunaregam spinosa* (Thunb.) Tirveng.	Rubiaceae	Gurol (O)
6	*Cipadessa baccifera* (Roth) Miq.	Meliaceae	Nalbila (O)
7	*Citrus medica* L.	Rutaceae	Lembu (O)
8	*Clausena excavata* Burm.f.	Rutaceae	Agnijal (O)
9	*Croton caudatus* Geiseler	Euphorbiaceae	Sabarjala (O)
10	*Ficus asperifolia* Miq.	Moraceae	Nyanja (O)
11	*Ficus exasperata* Vahl	Moraceae	Brahma's Banyan (E)
12	*Flemingia chappar* Benth.	Fabaceae	Bhandari (O)
13	*Gardenia gummifera* L.f.	Rubiaceae	Dekamali (O)
14	*Hypericum gaitii* Haines	Hypericaceae	*Hypericum* (E)
15	*Hyptianthera stricta* (Roxb. ex Schult.) Wight and Arn.	Rubiaceae	Bana dadri (O)
16	*Indigofera cassioides* DC.	Fabaceae	Saknya (O)
17	*Jasminum arborescens* Roxb.	Oleaceae	Bara Kunda (O)
18	*Lantana camara* L.	Verbenaceae	Raimuniya (O)
19	*Leea asiatica* (L.) Ridsdale	Vitaceae	Banchalita (O)
20	*Leea indica* (Burm. f.) Merr.	Vitaceae	Kalad Chana (O)
21	*Melastoma malabathricum* L.	Melastomataceae	Shapti (O)
22	*Pavetta tomentosa* Roxb. ex Sm.	Rubiaceae	Yellow Orchid (E)
23	*Pogostemon benghalensis* (Burm. f.) Kuntze	Lamiaceae	Pangala (O)
24	*Rotheca serrata* (L.) Steane and Mabb.	Verbenaceae	Baharangi (O)
25	*Sesbania bispinosa* (Jacq.) W.Wight	Fabaceae	Tentua (O)
26	*Urena lobata* L.	Malvaceae	Rakta Phrni (O)
27	*Vernonia anthelmintica* (L.) Willd.	Asteraceae	Kalijiri (O)
28	*Woodfordia fruticosa* (L.) Kurz	Lythraceae	Dhataki (O)

O: Oriya; E: English.

Discussion

The forest ecosystems of Similipal Biosphere Reserve are experiencing disturbances of various magnitudes. Practices for removal of forest biomass in the form of grazing, lopping, surface burning and litter removal are the continuous disturbance affecting the stability of the ecosystem and retarding the successional processes in the area. These areas can be clearly demarcated on the basis of different phytosociological parameters of the vegetation.

Table 6.4. List of Climber Species with their Local Names in Lulung, SBR

Sl.No.	Botanical Name of the Plant	Family	Local Name
1	*Aganosma heynei* (Spreng.) ined.	Apocynaceae	Malati (O)
2	*Asparagus racemosus* Willd.	Liliaceae	Satavari (O)
3	*Bauhinia vahlii* Wight and Arn.	Caesalpinaceae	Maloo (O)
4	*Combretum album* Pers.	Combretaceae	White *Combretum* (E)
5	*Dioscorea bulbifera* L.	Dioscoreaceae	Ratallo
6	*Entada rheedii* Spreng.	Mimosaceae	Gila (O)
7	*Getonia floribunda* Roxb.	Combretaceae	Ukshi (O)
8	*Hemidesmus indicus* (L.) R. Br. ex Schult.	Asclepiadaceae	Anant Mul (O)
9	*Lygodium flexuosum* (L.) Sw.	Lygodiaceae	Kala Mahajal (O)
10	*Millettia extensa* (Benth.) Baker	Fabaceae	Marda, Gerul (O)
11	*Pergularia daemia* (Forssk.) Chiov.	Asclepiadaceae	Hunturi, Iturhi (O)
12	*Smilax ovalifolia* Roxb. ex D.Don	Liliaceae	Kumarika (O)

O: Oriya; E: English.

A total of 203 species were recorded from SBR by Mishra *et al.* (2008), out of which 98 were tree species, 29 shrub and 76 herb species, where as in our case a total of 166 species belonging to 68 families were recorded from the study out of which 56 were tree species, 12 climber, 28 shrub and 70 herb species which is less than the above presented values. In some other studies like Reddy *et al.*, 2008 total of 137 tree species were recorded, representing 98 genera in 44 families which is lower than readings found in Lulung (SBR). Odum (1971) stated that under natural conditions, a clumped distribution of plants is normal. A higher percentage of random and regular distribution reflects the greater magnitude of disturbance' such as grazing and lopping in natural forest stands. All the vegetation layers showed generally clumped type of distribution in the present study. It is important for understanding regeneration and magnitude of disturbances and future stability of tree species population in forest communities (Upreti, 1982). The distribution pattern of other vegetational layers did not show any distinct difference among the study sites. The trees at Podadiha, Bangirposi, Handipuhan and Ghodabindha are faced with more biotic disturbance; Kendumundi and Kalika Prasad with moderate biotic disturbance; and Joranda, Chahala, UBK and Jenabil with no biotic disturbance.

The distribution of climbers on the trees of the reserve was considerably low, being nearly equal to 7.23 per cent. This may be due to high canopy coverage, thereby allowing low light to reach the forest floor and not triggering vigorous growth of climbers (Babweteera *et al.*, 2001). The impact of climbers on the vitality of trees is negative (Toledo- Aceves and Swaine, 2008) causing loss of foliage and thereby reducing the surface area available for metabolic processes and reproductive potential as well as impeding or obstructing forest succession (Toledo-Aceves and Swaine, 2008). Notwithstanding their negative impacts, climbers form bridges between the forest canopies, thereby facilitating the movement of arboreal animals across the forest. They also protect weaker trees from strong winds (Schnitzer and Bongers, 2002).

CONCLUSION

Overall analysis indicates that species rich communities of the dry deciduous tropical forests are not only being reduced in area but they are also becoming species poor and less diverse due to rapid deforestation and forest fragmentation. The community is also changing in response to increased anthropogenic disturbance. The study has shown that SBR is highly rich in plant diversity and is one of the treasure houses of good ecological wealth of Eastern Ghats, India. The long history of timber exploitation prior to its conversion into a biosphere reserve has resulted in the alteration of structure of the forest whereby most of the tree species were affected by very few individuals. The ecological importance of most of the tree species was also low, which reflected rarity of most of the species. However, the abundance of small trees coupled with the colossal sum of saplings and seedlings reflects a high regeneration potential of the forest. The forest management issues of SBR could be addressed by collection and analysis of long term ecological data which requires scientific baseline studies. We have covered extensively structural parameter analysis which is helpful to know the present state of ecological health of the ecosystem. But due to the various forms of anthropogenic pressure the habitat is destroyed with for logging, illegal hunting, and other challenges (mining in periphery, *etc.*). The conservation efforts have not so far yielded desired result. With continued biotic pressure and consequent change in structure and function of ecosystem, the management methodology also needs to be modified developing a Long Term Research Network. Similipal is a globally recognized ecosystem covered under UNESCO's Biosphere Reserve housing wide range of flora and fauna. We need to carry out Research and education activities to create an institutional platform to academicians, researchers and scientists. This ecosystem is under pressure. Continued destruction of old-growth and pristine forests of Similipal with high biodiversity will have a regional impact on social and ecological sustainability.

The over exploitation of natural resources in tropical world for meeting the basic needs of food, fodder and shelter of local population has disturbed the landscapes causing rapid depletion of biodiversity. Our research results may be helpful to develop conservation and management of biodiversity of SBR. Lack of data base and structural and functional characters of the ecosystem at regular intervals will not help to develop a long term strategy for sustainable development. Thus continuous collection of data as per long term action plan on successional status of species level up external and local pressures on the ecosystem, soil fertility management and linkage between social and ecological processes is needed. The community participation and use of traditional technologies as tools for natural resource management should be integrated to achieve sustainable resource management and ecological rehabilitation.

REFERENCES

Anitha, V., Muraleetharan P.K. and Binilkumar A.S. (2003). Natural resource depletion in protected areas: Socio economic linkages. *Indian Journal of Social Development*, 103(3): 44-59.

Arjunan, M., Puyravaud J.P. and Davidar, P. (2005). The impact of resource collection by local communities on the dry forests of the Kalakad-Mundanthurai Tiger Reserve, India. *Tropical Ecology*, 46: 135-143.

Babweteera, F., Plumptre, A.J. and Obua, J. (2001). Effect of gap size and age on climber abundance and diversity in Budongo Forest Reserve, Uganda. *African Journal of Ecology*, 38(3): 230 – 237.

Cox, C.B. and Moore, P.D. (1993). Biogeography: an ecological and evolutionary approach (5th edn). *Oxford: Blackwell Scientific Publications.*

Davidar, P., M. Arjunan, P. C. Mammen, J. P. Garrigues and J.-Ph. Puyravaud. (2007). Forest degradation in the Western Ghats biodiversity hotspot: resource collection, livelihood concerns and sustainability. *Current Science*. 93: 1573-1578.

Gadgil, M. and Guha, R. (1992). This Fissured Land – An Ecological History of India. Oxford University Press. New Delhi.

Gamble, J.S. (1915-1935). Flora of Presidency of Madras. London; reprinted 1967. Culcutta.

Gandhi, D.S., Sundarapandian, S. (2014). Inventory of trees in tropical dry deciduous forests of Tiruvannamalai district, Tamil Nadu, India. *Biodiversitas* 15(2): 159-169.

Haines, H.H. (1921-1925). The Botany of Bihar and Orissa. Adlard and Son Ltd., London.

Hare, M.A., Lantagne, D.O., Murphy, P.G. and Chero, H. (1997). Structure and tree species composition in a subtropical dry forest in the Dominican Republic: Comparision with a dry forest in Puerto Rico. *Tropical Ecology* 38: 1-17.

Harris, F. and S. Mohammed. (2003). Relying on nature: Wild foods in Northern Nigeria. *Ambio* 32: 24-29.

Hegde, R. and Enters, T. (2000). Forest products and household economy: A case study from Mudumalai Wildlife sanctuary, Southern India. *Environmental Conservation* 27: 110.

Karanth, K.K., Curran, L.M. and Reuning-Scherer, J.D. (2006). Village size and forest disturbance in Bhadra Wildlife Sanctuary, Western Ghats. India. *Biological Conservation* 128: 147-157.

Kumar, R. and Shahabuddin, G. (2005). Effects of biomass extraction on vegetation structure, diversity and composition of forests in Sariska Tiger Reserve, India. *Environmental Conservation* 32**:** 1-12.

Kutty, R. and Kothari, A. (2001). Protected Areas in India: A Profile. Kalpavriksh Publications, New Delhi.

Lykke, M.A. (1998). Assessment of species composition change in savanna vegetation by means of woody plants' size class distributions and local information. *Biodiversity and Conservation*, **7:** 1261-1275.

Mayer, H. (1992). Waldbau 4., neu, bearbeiteteth edn. Grundlage. Gustav Fischer, Stuttgart, New York.

Mishra, R.K., Upadhyay, V.P. and Mohanty, R.C. (2008). Vegetation ecology of the Similipal biosphere reserve, Orissa, India. *Applied Ecology and Environmental Research*, 6(2): 89-99.

Mishra, R.K., Parhi. S. and Biswal, A.K. (2018). Diversity of over storey plant communities of tropical forest covers of Balasore district, Odisha, India. *Adv Plants Agric Res*. 8(1): 20-26.

Mooney, H.F. (1950). Supplement to the Botany of Bihar and Orissa, Catholic Press, Ranchi.

Murty, S.K. and Panigrahi, G. (1989-1999). Flora of Bilaspur district, M.P. *BSI Publication*, Calcutta.

Odum, E.P. (1971). Fundamental of Ecology. 3rd Edition, *W.B. Saunders*, Philadelphia.

Pickett S.T.A. and White, P.S. (1985). The theory of natural disturbance and patch dynamics. *Academic Press Inc.*, New York, NY.

Poffenberger, M. and Mc Gean, B. (1996). Village Voices, Forest Choices: Joint Forest Management in India. *Oxford University Press*, Delhi.

Ramachandran, A., Radhapriya, P., Jayakumar, S., Dhanya, P. and Geetha, R. (2016). Critical analysis of forest degradation in the southern eastern ghats of India: comparison of satellite imagery and soil quality index. *Plos One.* **11**(1): 1-19.

Reddy, C.S., Babar, S., Giriraj, A., Reddy, K.N. and Rao K.T. (2008). Structure and Floristic Composition of Tree Diversity in Tropical Dry Deciduous Forest of Eastern Ghats, Southern Andhra Pradesh, India **1** (1): 57-64.

Reddy, C.S., Dutta, K. and Jha, C.S. (2013). Analysing the gross and deforestation rates in India. *Current Science* 105(11): 1492-1500.

Roy, P.S., Dutt, C.B.S. and Joshi, P.K. (2002). Tropical forest resource assessment and monitoring. *Journal of the Indian Society of Remote Sensing*, 43(1): 21-37.

Sahu, S.C., Dhal, N.K. and Mohanty, R.C. (2012). Tree species diversity and soil nutrient status in a tropical sacred forest ecosystem on Niyamgiri hill range, Eastern Ghats,

Saxena, H.O. and Brahmam, M. (1994-1996). The Flora of Orissa. Vol. I-IV. *Orissa Forest Development Corporation Ltd*, Bhubaneswar, India.

Schnitzer, S., Bongers, F. (2002). The ecology of lianas and their role in forests. *Trends in Ecology and Evolution*, 17(5): 223–230.

Sharma, J.R. and Singh, D.K. (2000) Status of plant diversity in India: an overview. In. P.S. Roy, Sarnam Singh and A.G. Toxopeus, Proceedings of Biodiversity and Environment. Remote sensing and Geographical information system prospective. *Indian institute of Remote sensing*. NRSSA, Dehradun, 69-105.

Singh, S. (2002). Multi-national corporations and agricultural development: a study of contract farming in the Indian Punjab. *Journal of International Development*, 14(2): 181–194.

Singh, N.P., Mudgal, V., Khanna, K.K., Srivastava, S.C., Shaoo, A.K., Bandopadhyay, S., Aziz, Das M., Bhattacharya, R.P. and Hajara, P.K. (2001). Flora of Bihar. Analysis. *BSI publication,* Culcutta.

Tilman, D.C. (2000), consequences and ethics of biodiversity. *Nature,* 405: 208–211.

Toledo-Aceves, T. and Swaine, M.D. (2008). Biomass allocation and photosynthetic responses of lianas and pioneer tree seedlings to light. *Acta Oecologica,* 34(1): 38–49.

Townsend, A.J., de Lange, P.J., Duffy, C.A.J., Miskelly, C.M., Molloy, J., Norton, D. (2008). New Zealand threat classification manual. Wellington, Dept. of Conservation. http: //www.doc.govt.nz/upload/documents/science-andtechnical/sap244.pdf

Turner, M.G. (1987). Landscape heterogeneity and disturbance. *Springer-Verlag,* New York.

Uhi, C. (1987). Factors controlling succession following slash-and-burn agriculture in Amazonia. *Journal of Ecology,* 75: 377-407.

Upreti, N., Tewari, J.C. and Singh, S.P. (1985). The Oak Forests of the Kumaun Himalaya (India): Composition, Diversity, and Regeneration. *Mountain Research and Development,* 5(2): 163-174.

World Conservation Monitoring Centre (WCMC). (1992). Global Biodiversity: *Status of the Earths Living Resources.* Chapman and Hall, London.

Chapter 7

Biodiversity Conservation and Sustainable Use of Some Medicinal Plants in Vindhyan Region of Uttar Pradesh

Somesh Gupta and Satya Prakash Yadav*

Uttar Pradesh State Biodiversity Board,
Lucknow – 226 010, Uttar Pradesh
**e-mail: ftisomesh@gmail.com*

India is a mega diverse country harboring nearly 7-8 per cent of all recorded species including over 45000 species of plants and 91,000 species of animals on only 2.4 per cent of the world's land area. It has a long history of working for conservation of its unique biodiversity with multi-stakeholder participation. More than 70 per cent of India's biodiversity is confined within forests. Uttar Pradesh is the most populous state in the country, having geographical area of two, 40,928 Km^2, which constitutes 7.33 per cent of the total area of the country (3,287,263Km^2). The total forest/tree cover of Uttar Pradesh is 22,121 Km^2, which represents 9.18 per cent of total geographical area of the state (FSI Report 2017). The forests of the state are of typical northern dry deciduous type and can be broadly divided into drydeciduous forests, mixed forests, Sal forests, scrub forests, grasslands and aquatic vegetation. The diversity of flora and fauna displayed here due to vast area, big and small rivers, varieties of climatic conditions, and different kinds of soil are hard to find elsewhere. In recent years, and the global demand of herbs has led to a quantum jump in volume of medicinal plants traded within and across the countries. It is estimated that 300 plants are of medicinal use and more than hundreds are being utilized in preparing Ayurveda formulations as they contain medicinal properties. The medicinal plants have been identified as one of the most important plant diversities for the rural development. The forest areas in U.P. are very rich in variety of medicinal plant species particularly in the Vindhyan region where various medicinal plants grow naturally and have great scope for their conservation and cultivation.

This paper communicates the efforts carried out to document some important medicinal plants of Vindhyan region, which would not only conserve local biological diversity, but will also help to sustain the bio-cultural diversity of Vindhyan region of Uttar Pradesh.

Keywords: *Biodiversity, Conservation, Sustainable use, Vindhyan region and Medicinal plants.*

INTRODUCTION

India is a mega diverse country harbouring nearly 7-8 per cent of all recorded species including over 45000 species of plants and 91,000 species of animals on only 2.4 per cent of the world's land area. It has a long history of working for conservation of its unique biodiversity with multi-stakeholder participation. More than 70 per cent of India's biodiversity is confined within forests. These are five hundred tribal and aboriginal communities in India living in close proximity to forests since time immemorial.

Due to the close and long association with forests, they have acquired tremendous knowledge of plants, plant produce and their uses in their daily needs and health care. (Ambasta, 1986; 1986; Dhawan *et al.*, 1977; Singh *et al.*, 1994; Singh and Prakash 1996; Jain1975, 1991; Jain and Puri 1984; Prakash and Singh 2000). India ranks sixth under world's twelve mega bio-diversity zones. Out of these, two of them exist in our country. India possesses tremendous ecological biodiversity. It contains 5 per cent of the world's bio-diversity on 2 per cent of the earth's surface. It is estimated that approximately 45000 plants species *i.e.* 12 per cent of global wealth of flowering plants (Jain, 1987) exist, out of which nearly 33 per cent are endemic (Gupta *et.al.*, 2002). In India's scenario, floristic surveys of many wetlands were carried out Rao (1994) listed 300 aquatic plant species belonging to 69 families, out of which 42 of these are designated as endemic plants.

In recent years, the global demand of herbs has led to a quantum jump in volume of medicinal plants traded within and across the countries. The medicinal plants have been identified as one of the most important plant diversities for the rural development. The ethno botanical research on medicinal plants is getting much importance and it has now been considered as thrust area in the field of biological research.

Uttar Pradesh is the most populous state in the country, having geographical area of 2,40,928 Km^2 which constitutes 7.33 per cent of the total area of the country (3,287,263Km^2). The total forest/tree cover of Uttar Pradesh is 22,121 sq km., which represents 9.18 per cent of total geographical area of the state (FSI Report 2017). The forests of the state is of typical northern dry deciduous type and can be broadly divided into dry deciduous forests, mixed forests, Sal forests, scrub forests, grasslands and aquatic vegetation. The diversity of flora and fauna displayed here due to vast area, big and small rivers, varieties of climatic conditions, and different kinds of soil are hard to find elsewhere.

The studies in the floristic status of the present political Uttar Pradesh state reveals that it comprises 2711 species belonging to 1088 genera and 185 families, which shows their occurrence in Terai, Bundelkhand and Vindhyan region (Srivastava 2004). Khanna (2001) has enumerated 119 taxa of flowering plants endemic to Uttar Pradesh including Uttrakhand. Of these, 10 taxa are endemic to present Uttar Pradesh belonging to eight genera and six families. It is estimated that nearly 300 plants are of medicinal use and more than hundreds are being utilized in preparing ayurvedic formulations as they contain medicinal properties (Srivastava 2004).

In Uttar Pradesh, Vindhyan region is enriched with dense forests, but in recent years depletion of this resource has started due to increase in developmental activities, agricultural land expansion and many other human activities like felling, encroachment *etc.* Factors like abiotic and edaphic, have also played a key role in determining the growth of different species in this area. Vindhyan region being native of many important medicinal plants, have a great potential for their conservation and cultivation.

This paper deals with the efforts carried out to document some important information of medicinal plants of this region, which would not only conserve local biological diversity, but will also help to sustain the biocultural diversity of Vindhyan region of Uttar Pradesh.

Materials and Methods

Area for Conservation

The forest areas in U.P. are very rich in variety of medicinal plant species particularly in the Vindhyan region where various medicinal plants grow naturally. According to Planning Commission Report (2001) the primary goal of biodiversity conservation as envisaged in World Conservation Strategy may be summarized as below:

1. Maintaining of essential ecological process and its life support system.
2. Preservation of genetic diversity
3. Sustainable Management.

Topography and Climate

The Vindhyan region consists of the Vindhyan plateau and the hills in the southern part of the state and lies between 23°52′- 24°21′ N Latitude to 82°42′-88 °24′E Longitude. The soil of this region varies considerably but most of the area has red lateritic soils with often-pronounced nodules locally called "Murram". The depth of the soil also varies and much of the area has a soil depth of just a few cm. Deep and fertile soils are often found in the valleys between folds of hills where the soil washed from hill slopes has accumulated.

The climate of the area is characterized by a long and intensely hot summer, low rainfall and a short mild winter. The hot weather usually begins from middle of March and extends up to the break of the monsoon till June. The summer temperature goes up to 48°C in May and June. However, the average temperature varies between 8° to 34°C. The rainfall varies from 750-mm. to about 1200 mm.

Most of the rainfall occurs in the month of June, July and August. There is a little winter rain also which occur generally in January and February and are sometimes substantial, regular as compared with other part of the state (Dubey *et al.*, 2001).

Figure 7.1. Status of the Vegetation.

The information on uses of threatened medicinal plants was collected during the several field visits and with the help of local people/tribes, traditional practitioners and experience of local forest officials and through the literature survey.

Figure 7.2. Collection of Samples during Field Visits.

Extensive surveys were conducted to collect the information; a questionnaire in this context was also formulated and widely distributed to gather the information of flowering and fruiting of different species. Interactions with the farmers, local vaidhs, local tribes and Hakims were carried out to document the part used and uses of the selected species.

Results

As the outcome of the various survey/field visits, the potential uses of some important medicinal plants as claimed by the local peoples/tribes were documented, which are described in Table 7.1. The table bears the detailed information about the phenology, propagation, part used and uses of some important medicinal plant species. Some figures of some important medicinal plants of Vindhyan region have been presented in Plate 7.1. The paper reveals to educate the local people of Vindhyan region regarding the importance of the biodiversity and its conservation for their future generations.

Figure 7.3. Interaction with Local Peoples.

Dicussion

The Vindhyan region of Uttar Pradesh is enriched with diversity of species, but in recent years, depletion of this resource has started due to increase in developmental activities, agricultural land expansion and, of course, spurt in forest offences, like- illicit felling, encroachment, poaching *etc.* Factors, like- abiotic and edaphic, have also played a key role in determining and also restraining the growth of different species in this area. This paper communicates the efforts carried out to document some important medicinal plants of this region, which would not only conserve local biological diversity but will also help to sustain the biocultural diversity of Vindhyan region of Uttar Pradesh.

During the regular surveys, it was observed that due to overexploitation and biotic pressure of some important medicinal plants species prevailing in the Vindhyan range. These medicinal plants are either threatened or they are at the

Table 7.1. List of some Important Useful Medicinal Plants of Vindhyan Region, Uttar Pradesh

Sl.No.	*Local Name*	*Botanical Name*	*Family*	*Category*	*Phenology*	*Propagation*	*Part Used*	*Uses*
1	Aloe vera	*Aloe barbadensis*	Liliaceae	Herb	Fl: Mar.-Apr. Fr: - Jul-Aug.	Root suckers and rhizome cuttings	Leaves	Skin diseases and Blood purifier
2	Anantmool	*Hemidesmus indicus*	Asclepiadaceae	Creeper	Fl: Oct. Fr: - Jan.	Stem cutting	Roots	Blood purifier, leprosy, asthma, bronchitis, skin disease and fever
3	Ashwagandha	*Withania somnifera*	Solanaceae	Herb	Fl: Sept.-Oct. Fr: - Mar.-Apr.	Seeds	Roots, leaves and Seeds	Tonic, arthritis, blood pressure and skin disease
4	Brahmi	*Bacoppa monerii*	Scrophulariaceae	Creeper	Fl: Sept.-Oct. Fr: - Sept.-Oct.	Cuttings	Whole plant	Asthma, bronchitis, diarrhea and snake bite
5	Chirayta	*Sweretia chirayta*	Gentianaceae	Herb	Fl: Jul. – Sept. Fr: -Jul. – Sept.	Seeds	Whole plant	Ulcer, diabetes and skin disease
6	Dhatura	*Datura somnifera*	Solanaceae	Shrub	Fl: Jun.–Jul. Fr: -Aug.	Seeds	Whole plant	Rheumatism, bronchitis, asthma and narcotics
7	Ghumchi	*Abrus precatorius*	Fabaceae	Climber	Fl: Aug.-Oct. Fr: - Dec.-Jan.	Seeds and rhizome	Seeds, leaves and roots	Diuretic, ulcer, leucoderma and arthritis.
8	Giloe	*Tinospora cardifolia*	Menispermaceae	Climber	Fl: Jan.– Mar. Fr: -Jan.– Mar.	Seeds	Stem	Malaria fever
9	Gudmar	*Gymnema sylvestre*	Asclepiadaceae	Herb	Fl: Apr.– Jun. Fr: -Jul.– Sept.	Seeds and cuttings	Leaves and roots	Diabetes and urinary troubles
10	Guggal	*Commiphora wightii*	Burseraceae	Shrub	Fl: Mar.–Apr. Fr: -Mar–Apr.	Seeds and cuttings	Stem	Typhoid, fever and bronchitis

Sl.No.	*Local Name*	*Botanical Name*	*Family*	*Category*	*Phenology*	*Propagation*	*Part Used*	*Uses*
11	Kalmegh	*Andrographis panniculata*	Acanthaceae	Herb	Fl: upto Dec. Fr: -upto Dec.	Seeds	Whole plant	Wounds, diarrhea, dysentery and fever
12	Mandookparni	*Centella asiatica*	Apiaceae	Creeper	Fl: Apr.–Oct. Fr: -Apr–Oct	Seeds and cuttings	Leaves	Nerve tonic
13	Marorphali	*Helicteres isora*	Tiliaceae	Shrub	Fl: Oct.–Nov. Fr: -Dec–Jan	Seeds	Stem, bark and roots	Skin diseases, griping bowels or stomachache
14	Mulethi	*Glycyrrhiza glabra*	Fabaceae	Herb	Fl: Mar. Fr: -Aug.	Root cutting	Rhizome	Cough, cold and conjunctivitis
15	Nagarmotha	*Cyprus rotendus*	Cyperaceae	Herb	Fl: Jul. Fr: -Dec.	Seeds	Tuber and leaves	Tonic, stomachache, syphilis and weakness
16	Sankhpuspi	*Evolvulus alsinoides*	Convolvulaceae	Creeper	Fl: Jul.–Aug. Fr: -Jul.–Aug.	Seeds	Whole plant	Urinogenital disorder, eye disorder, diabetes and memory power.
17	Sarpgandha	*Rauwolfia serpentina*	Apocynaceae	Herb	Fl: Oct.–Nov. Fr: -Sept.–Dec.	Seeds and cuttings	Roots	Blood pressure and snake bite
18	Satawar	*Asparagus racemosus*	Liliaceae	Shrub/ climber	Fl: Dec.–Jan. Fr: -Dec–Jan.	Seeds and rhizome	Tuber	Aphrodisiac, blood purifier and kidney disorders
19	Vach	*Acorus calamus*	Araceae	Herb with creeping rhizome	Fl: Apr.–Aug. Fr: -Apr.–Aug.	Rhizome	Rhizome	Snakebite, cough and cold

Plate 7.1. Some Important Useful Medicinal Plants of Vindhyan Region, Uttar Pradesh.

Aloe barbadensis **Aloe Vera**

Hemidesmus indicus **Anantmool**

Withania somnifera **Ashwagandha**

Bacoppa monerii **Brahmi**

Sweretia chirayta **Chiraita**

Datura somnifera **Dhatura**

Abrus precatorius **Ghumchi**

Tinospora cardifolia **Giloe**

Gymnema sylvestre **Gudmar**

Commiphora wightii **Guggal**

Andrographis panniculata **Kalmegh**

Centella asiatica **Mandookparni**

Helicteres isora **Marorphali**

Glycyrrhiza glabra **Mulethi**

Cyprus rotendus **Nagarmotha**

Evolvulus alsinoides **Sankhpuspi**

Rauwolfia serpentina **Sarpgandha**

Asparagus racemosus **Satawar**

Acorus calamus **Vach**

verge of extinction. The major causes of biodiversity loss of the Vindhyan region are as discussed below:

Habitat Loss, Deterioration and Fragmentation

A number of species have lost their habitat (place where a species lives) because of increased human interference like construction of dams, roads, railway tracts or bridges across natural ecosystems, mining activity or industries. Habitat fragmentation due to deforestation or any other biotic stress has caused much harm to precious biodiversity.

Deforestation rate is increasing due to burgeoning population and this has destroyed the natural homes of species. In fresh water ecosystems, construction of dams has destroyed large habitat of hundreds of aquatic flora and fauna. Likewise, in marine ecosystems, the coastal development has affected the communities, particularly species rich ecosystems like coral reefs.

Exotic Species

Exotic species are those which have been introduced either purposely or entered accidentally in some environmentally distinct zone from the other geographically different areas. These have caused much harm to native plant communities than expected and enhanced the extinction (loss of species) rate. The phenomenon is more common on the islands or isolated ecosystems. For example, *Lantana camara* introduced as an ornamental hedge in India has now become a serious invader of forests. The example of accidental entry is the Congress grass (*Parthenium hysterophorus*) that is now a major problematic weed in India, Australia and other parts of the world. A number of reasons such as fast growth, rapid colonization, better and efficient resource utilization, wider adaptability and the absence of natural enemies in the invaded areas.

Overexploitation

Some of the species have become threatened and even gone extinct due to their over exploitation by the human beings for their food, feed or some other beneficial purpose. In fact, a number of ecosystems have been damaged both in terms of number and in terms of variety of species that these are beyond repair in the near future.

Environmental Pollution

Air, water, and soil pollution is one of the major reasons for biodiversity loss in the modern era. Modernization, urbanization, increasing population, and changing life style are some of the prime reasons for the increased levels of pollution in the natural environment. It has not only affected the number of existing species but has also caused the loss of a variety of species that could not tolerate the increased pollutant levels in the atmosphere and have gone extinct. Excessive use of synthetic chemicals as herbicides, pesticides and insecticides has polluted the soil and water environment and greatly threatened the diversity and richness of the species. Due to increased industrialization, the intensity of the acid rain has increased and greatly affected the natural vegetation and the forests in different regions of the world.

Global Warming

Climate change due to global warming is another reason for the depletion of natural biodiversity in the recent times. The increase in the earth's temperature has already shown its impact in terms of depletion of forest resources in many parts of world. Additionally, the temperature rise also alters the tolerance limit of the species and greatly changes the structure and ecosystem function. The increased global temperature would result in melting of ice caps and rising of sea levels. It would result in submerging of many islands and inundation of coastal areas thereby greatly affecting the biodiversity of such regions.

Threats of Biodiversity Conservation

Because of depletion, fragmentation and degradation of natural habitats various floral and faunal species in the state have been listed under threatened category. On one hand conservation of forests, wetlands, riverine and other ecosystems and biodiversity therein is important as they provide multiple ecosystem benefits. On the other hand, these ecosystems also provide direct economic benefits and livelihood opportunities to the poor people. Balancing the conservation goals and meeting the local community needs is a difficult task. Biodiversity resources are illegally exploited due to high demand in the urban and international markets, with contributes to the depletion of biological wealth. However, concerted efforts are required to protect the ecosystems and the biodiversity resources.

Need for Conservation

In recent years, the habitat loss due to developmental programmes, overgrazing, animal husbandry and tourism has resulted in the loss of biodiversity. Natural causes such as floods, earthquakes and landslides also add to this tragedy. Many species are extinct or on the verge of extinction before they are known for their uses. It is estimated that about 25,000 plant species are on the verge of extinction throughout the world (Jain and Sastry 1980). In order to categorize and update the list of threatened species, the International Union for Conservation of Nature and Natural Resources (IUCN) has recognized the categories based on geographical range, populations and fragmentations of populations. The earlier categories, Extinct, Endangered, Vulnerable and Rare, have been superseded by the categories as Extinct (EX.), Extinct in the Wild (EW), Critically Endangered (CR), Endangered (EN), Vulnerable (VU), Conservation Dependent (CD), Data Deficient (DD), Low Risk (LR), Not Evaluated (NE) (Nayar 1996). Jain and Sastry (1980) reported 135 species as being rare and threatened in Indian flora. The native communities, with their unique traditional cultures and indigenous knowledge are also depleting gradually due to modernization, urbanization, industrialization, and other socio-economic developmental programmes for the upliftment of life and economy, worldwide. The indigenous knowledge (IK) or local knowledge is empirical knowledge of a community relating to all activities ranging from land use and natural resources use, to management and other activities associated with the life sustainable process. The ethno medicinal plant wisdom as a part of indigenous knowledge plays a vital role in the primary healthcare of the native communities. It has the potential for isolation of safe and effective drugs and for sustainable utilization of ethnomedicinal plant genetic resources (PGR) and their conservation is highly needed (Prakash 2011).

CONCLUSION

The loss of biodiversity resulting in extinction of many useful species has also posed a serious concern to the global medical scientists and researchers. In the present scenario, medicinal plants have become one of the main concerns of the world conservation organizations. Systematically more attempts to study and document the medicinal plants are seriously needed. There is also a need to establish herbal drug centers for collecting, processing and preparation of ethno-medicine and to develop cultivation, farming and documentation of potential and promising ethno medicinal plants for improving the life and economy of the local tribal and rural peoples. There is also a need to conserve the valuable biodiversity in existing protected areas including hot spots *vis-à-vis* to establish more protected areas for protection and conservation of our natural resources. Further, documentation of useful threatened plants is required on priority basis for their *in situ* and *ex situ* conservation and sustainable utilization in health care and human welfare.

ACKNOWLEDGEMENTS

The author is thankful to the Divisional Forest Officers of Mirzapur, Sonbhadra and Renukoot Divisions for providing necessary help and facilities during regular survey/field visits. The author is also thankful to the Gram Pradhans, Range Forest Officer and local officials of the Forests departments for their help and assistance during the visits in the forests and tribal villages.

REFERENCES

Annual Report (2001). Planning Commission, Government of India, New Delhi.

Ambasta, S.S. (1986). The Useful Plants of India. *Publications and Information Directorate CSIR*, New Delhi, India.

Dhawan, B.N., Patnaik, G.K., Rastogi, R.P., Singh, K.K. and Tandon J.S. (1977). Screening of Indian plants for biological activity. *Indian J. Exp. Biol.*, 15: 208-19.

Dubey, P., Mishra, P., Srivastava, R.J., Burfal, B.S. and Dixit, R.K. (2001). Technology to Conserve Biodiversity of Vindhyan Region. *Proc. Int. Symposium Tropical Forestry Research:* Challenges in the New Millennium, pp.59-61.

Gupta, S., Saxena, V. and Mishra, R. (2002). Conserving Biodiversity of Bundelkhand Region of U.P. through People's Participation. *72nd Session Annual Session of NASI, Shillong*, pp. 62.

Jain, S.K. (1975). Medicinal Plants. *National Book Trust*, New Delhi.

Jain, S.K. and Sastry, A.R.K. (1980). Threatened Plants of India State of Art Report. Botanical Survey of India, Howrah.

Jain, S.P. and Puri, H.S. (1984). Ethno-medicinal Plants of Jaunsar-Bawar hills Uttar Pradesh, India. J. of *Ethnopharmacol*, 12: 213-222.

Jain. S.K. (1987). The problem of endangered species: its study and solution. Presidential address delivered in Biological Sciences, *57th Academy of Sciences*, Allahabad.

Jain, S.K. (1991). Dictionary of Indian Folk medicine and Ethno-botany, *Deep publications*, New Delhi

Khanna, K.K. (2001). Endemic Plants of Uttar Pradesh (Angiosperms). Economic Botany Information Service, NBRI, Lucknow. *Phytotaxonomy*, 1: 71-72.

Nayar, M.P. (1996). Hot Spots of endemic plants of India, Nepal and Bhutan. *Tropical Botanical Garden and Research Institute*, Thiruvananthpuram.

Prakash, A., and Singh, K.K. (2000). Observations on some high valued ethno medicinal plants among the tribal's of Uttar Pradesh. *J. Med. and Arom. Plants*, 22: 148-163.

Prakash, A. (2011). Uses of some threatened and potential ethno medicinal plants among the tribals of Uttar Pradesh and Uttrakhand in India. *National Conference on Forest Biodiversity: Earth's Living Treasure*, 22: 93-99.

Rao, R.R. (1994). Biodiversity in India, Bishen Singh Mahendra Pal Singh, Dehra Dun.

Singh, K.K. and Anand, P. (1994). Indigenous Phototherapy among the Gond tribe of Uttar Pradesh, India. *Ethnobotany*, 6 (1 and 2): 37-41.

Singh, K.K. and Anand, P. (1996). Observations on ethno botany of Kol tribe of Varanasi district Uttar Pradesh, India. *J. Econ. Tax. Bot.*, 12: 133-137.

Srivastava, S.K. (2004). Floristic diversity in Uttar Pradesh - An overview. *J. Eco. Taxon. Bot.*, 28(2): 292-334.

State of Forest Report (2017). India State of Forest Report (ISFR) published by *Forest Survey of India*, Dehradun.

Chapter 8

Plant Derived Anticancer Agents and Future Perspectives

*Shubhasmita Mohapatra and Priya Ranjan Debata**

Department of Zoology, North Orissa University, Takatpur, Baripada – 757 003, Mayurbhanj, Odisha
**e-mail: prdebata@gmail.com*

Cancer is one of the leading causes of death and the incidence of cancer is steadily increasing across the globe. Although various types of treatment options are available none of them gives optimum results. The major problem in cancer chemotherapy is the toxicity of the established drugs and the development of chemoresistance. Several chemotherapy drugs used for treating cancer are derived from plants. Besides some phytocompounds are also under investigation. Several plant-derived products have been shown effective in vitro condition and some of them are under clinical trial. Characterization of plants for anticancer properties is important for the identification of new therapeutic agents. Understanding the molecular mechanism, safety, and efficacy of phytocompounds will play an important role in the management of cancer in the future.

Keywords: Cancer, Medicinal plants, Chemotherapy, Potential drugs.

INTRODUCTION

Cancer is a general term applied to series of malignant diseases characterized by a rapid and uncontrolled formation of abnormal cells, which may mass together to form a growth or tumor, or metastasize to newer places in the body and initiating abnormal growth. The development of cancer is thought to be multifactorial. Several lines of evidence suggest that cancer development is due to alteration in genetic, epigenetic, and metabolic pathways of the cell. In a general sense, cancer arises due to specific effects of environmental factors such as physical carcinogens (UV rays, X-rays, Gamma rays), chemical carcinogens (Aflatoxins, Asbestos, Benzene,

Cadmium, Coal tar, *etc.*) and biological agents (Oncogenic virus, Bacteria) (Parsa, 2012). The cancer cells are characterized by six hallmarks which include (1) Growth signal autonomy: which means the cell grows even without a growth signal, (2) Overrides the stop signals: the cancer cells don't respond to antigrowth factors in the cell cycle, (3) Evades apoptosis: which means the cancer cells adopt various strategies to override apoptosis including amplifying the anti-apoptotic machinery, down-regulating the pro-apoptotic program, or both, (4) Sustained angiogenesis: where new blood vessels are formed from pre-existing vessels to support the growth, (5) Unlimited replicative potential: a cell keeps replicating no matter how many times it has already done, (6) Invasion and metastasis: cancer cells move into areas away from their origin(Hanahan and Weinberg, 2011). Most of the anticancer agents used today are very expensive and the patient develops enormous side effects. Besides, the cancer cell also develops chemoresistance and formation of cancer stem cells which hinder the cancer treatment. Cancer chemoprevention is another area of research. Various epidemiological studies suggest that natural phytochemicals and dietary compounds have chemo preventive properties which protect body from development of cancer by modulating various signaling pathways involved in cell proliferation and apoptosiss (Surh, 2003). Thus, prevention of cancer remains evidently an essential part of the contest against cancer in the world. Natural products play an important role in the treatment of cancer as a significant number of compounds used for the treatment of cancer are plant derived products.

India is one of the 12-major biodiversity centers having about 10 per cent of the world's biodiversity wealth, which is scattered across 16 agro-climatic zones (Chitale *et al.*, 2014). Around 20,000 medicinal plant species have been recorded in our country (Pandey *et al.*, 2013). There is considerable interest in the screening of plant and other natural product extracts in modern drug discovery programs, since structurally novel chemotypes with potent and selective biological activity may be obtained (Atanasov *et al.*, 2015). In recent years the characterization of anticancer drugs has been taken momentum due to the development of new technologies which helps large scale screening and characterization. India has the greatest diversity of plant species in the world and systematic screening for plant species with anticancer properties has not done well. The screening and evaluation of plants have tremendous potential for the discovery of new molecules for cancer treatments and chemoprevention.

Conventional Cancer Treatments and Therapy

Chemotherapy, surgery, and radiotherapy are the most common types of treatment options for cancer patients (Arruebo *et al.*, 2011). Cancer can be cured if the diagnosis becomes early however cancer detected late has a poor prognosis.

Surgery

It is used for the excision of a tumor. It is the earliest therapy established for cancer and the most widely used. Its disadvantages include the possible damage of healthy tissues or organs (such as lymph nodes) and the inability to remove metastasized cancer cells or tumors not visible to the surgeons.

Radiation

X-rays and gamma rays of a cancerous tumor causing cancer cell death or apoptosis preserves the anatomical structures surrounding the tumor and destroy nonvisible Cancer cells. However, they cannot kill metastasized cancer cells.

Chemotherapy

Itis based on the systemic administration of anticancer drugs that travel throughout the body via the blood circulatory system. In essence, chemotherapy aims to wipe out all cancerous colonies within the patient's body; including metastasized cancer cells. This kind of treatment also has many side effects, such as nausea, anemia, weakening of the immune system, diarrhea, vomiting, and hair loss

Molecular Mechanism of Conventional Chemotherapy

Many different compounds are currently used. Chemotherapy is the most rapidly developing field of cancer treatment, with new drugs being constantly tested and screened. These include also plant metabolites. Chemotherapeutic drugs are classified in general groups:

1. Anti-metabolites act as nonfunctional analogs of essential metabolites in the cell, thus blocking the physiological functions of the tumor.
2. Alkylating agents chemically bond with DNA through alkyl groups, thus disrupting gene structure and function, or with proteins, thus inhibiting enzymes.
3. Topoisomerase inhibitors inhibit DNA replication in rapidly dividing cells, as in the case of tumors.
4. Plant alkaloids also inhibit tumor cell division by blocking microtubule depolymerization, which is an essential step for chromosome detachment during mitosis.
5. Antibiotics are derived from diverse groups of microorganisms or synthesized and block DNA replication and protein synthesis.
6. Anthracyclines are a subgroup of antibiotics, associated with considerable toxic side effects on the heart and bone marrow.
7. Enzymes, in particular proteolytic and fibrinolytic ones, as well as tyrosinase inhibitors, such as *Gleevec*, a new cytotoxic drug used for treating chronic myeloid leukemia.
8. Hormones are substances interfering with other chemotherapeutic agents by regulating the endocrine system. They find specific applications against carcinomas of breast, prostate, and endometrium.
9. Immunomodulators act by inhibiting tumor proliferation through the stimulation of the host's immune system.

Table 8.1. Some of the Plant-derived Compounds with Potential Anticancer Properties

Sl.No.	*Name of Compound*	*Name of Plant*	*Type of Studies*
1	3,5-Dimethylbenzenemethanol, Alpha-Bisabolol, Alpha-pinene, Beta-pinene, and Baccatin III	Ferula hermonis	Laboratory research
2	Baicalein, Baicalin, Oroxylin A	Scutellaria baicalensis	Clinical Studies
3	Cannabidiol	Cannabis sativa	Laboratory research
4	Colchicine	Gloriosa superba	Clinical trial
5	Curcumin	*Curcuma longa*	Clinical trial
6	Etoposide, Teniposide	*Podophyllum peltatum*	Clinical trial
7	Flavonoids, and vitamins C and E	*Triticum aestivum*	Clinical trial
8	Gallotannins	*Caesalpinia spinosa*	Laboratory research
9	Ginkgolide B	*Ginkgo biloba*	Clinical use
10	Glucomoringin, glucosinolate	*Moringa oleifera*	Laboratory research
11	Mistletoe	*Viscum album*	Clinical trial
12	Myricetin	*Bryonia dioica*	Clinical trial
13	Oenothein B	*Epilobium angustifolium*	Laboratory research
14	Oleandrin	*Nerium oleander*	Clinical investigations
15	Paclitaxel	*Taxus brevifolia*	Clinical use
16	Piperamides	*Piper nigrum*	Laboratory research
17	Punicalagin, Pedunculagin, punicallin, Ellagic acid	*Punica granatum*	Clinical trial
18	Quinic acid, cinnamic acid, rhoifolin, and malic acid	*Artemisia absinthium*	Laboratory research
19	Resveratrol	*Vitis vinifera*	Clinical trial
20	S-allylcysteine	*Allium sativum*	Laboratory research
21	Silymarin	*Silybum marianum*	Clinical trial
22	Vinblastine, Vincristine, Vinorelbine	*Vinca rosea*	Clinical use
23	Withanolides, withanolide D, withaferin A and withaniamides	*Withania somnifera*	Laboratory research
24	β-Elemene	*Curcuma rhizoma*	Clinical trial

Plant-Derived Anti-Cancer Agents

Vinca Alkaloids

The first agents introduced in clinical use were vinca alkaloids, vinblastine (VLB), and vincristine (VCR), isolated from the *Catharanthus roseus* (Apocynaceae). These drugs were discovered during an investigation for oral hypoglycemic agents. Further extraction and fractionation led to the isolation of two active alkaloids namely vincristine and vinblastine. The plant was originally endemic to Madagascar, but the samples used in the discovery of vincristine and vinblastin were collected in the Philippines and the Jamaican. Recently semi-synthetic analogs of vinca alkaloids are vinorelbine (VRLB) and vindesine (VDS). These are primarily using alone or in combination with other chemotherapeutic drugs to cure a variety of cancers. VLB is using for the treatment of lymphomas, leukemias, breast cancer, testicular cancer, lung cancers, and Kaposi's sarcoma. VCR had also shown efficacy against leukemia, particularly acute lymphocytic leukemia in childhood (Moudi *et al.*, 2013)

Podophyllotoxin Derivatives

The species of Podophyllaceae family such as *Podophyllum peltatum* Linn. *P. emodii* Wallich have been reported with a long history of therapeutic use, including the treatment of skin cancers and warts. The Native Americans for the treatment of "cancer" have used *Podophyllum peltatum.* Extensive research studies at Sandoz Laboratories in Switzerland in the 1960s and 1970s led to the development of etoposide and teniposide as clinical agents, which are being used in the treatment of lymphomas and bronchial and testicular cancers (Pang *et al.*, 2018)..

Taxanes

More recent advancement in the development of plant-derived chemotherapeutic agents in the development of a class of molecules called taxanes. Paclitaxel also names as taxol was first isolated from the bark of *Taxus brevifolia* Nutt. (Taxaceae). Paclitaxel is used in the treatment of a wide variety of cancers including breast, ovarian, and non-small-cell lung cancer, and has shown efficacy against Kaposi sarcoma (Schiff *et al.*, 1979).

Campothecin Derivatives

Another advancement that was made in the anti-cancer drug is the class of clinically active agents derived from camptothecin. Camptothecin was first isolated from the Chinese ornamental tree, *Camptotheca acuminate* Decne (Nyssaceae), and known in China as the tree of joy. Camptothecin derivatives, Topotecan (Hycamtin), and Irinotecan are now in clinical use. Topotecan is used for the treatment of ovarian and small-cell lung cancers, while Irinotecan is used for the treatment of colorectal cancers (Rahier *et al.*, 2005).

Homoharringtonine

Other plant-derived agents, which are in clinical use, are homoharringtonine. Homoharringtonine was originally isolated from the Chinese tree *Cephalotaxus harringtonia* var. *drupaceous* (Cephalotaxaceae). A racemic mixture of harringtonine and homoharringtonine (HHT) is being used successfully to combat acute myelogenous leukemia and chronic myelogenous leukemia in China (Efferth *et al.*, 2007).

Potential Plants for Cancer Drug Development

Plants are used medicinally in different countries and are the source of potential and powerful drugs. In India, extracts of medicinal plants are used both directly as folk medicines in different indigenous systems of medicine like Siddha, Ayurveda, and Unani. Plant-derived natural products with documented anticancer/ antitumor properties are classified into the following chemical groups: Aldehydes, alkaloids, annonaceous acetogenins, flavonoids, glycosides, lignans, lipids, nucleic acids, phenols and derivatives, polysaccharides, proteins, terpenoids, unidentified compounds (Kintzios and Barberaki, 2004). Medicinal herbs are also a significant source of synthetic and herbal drugs. Phytoconstituents derived from the different herbs have been used in various formulations to enhance the activity of immune cells of the body that promotes the production of cytokines including interleukin, interferon, tumor necrosis factor and colony-stimulating factor (Sakarkar and Deshmukh, 2011). These formulations help the body to fight cancer more effectively and reduce the toxic side effects of chemotherapy and radiotherapy stages of cancer. The plants are discussed below:

Curcuma longa

Curcumin is the principal curcuminoid of turmeric found in the rhizome of *Curcuma longa*. Curcumin has been shown to have antioxidant, antineoplastic, antiviral, anti-inflammatory, antibacterial, antifungal, antidiabetic, anticoagulant, antifertility, cardiovascular protective, hepatoprotective, and immunostimulant activity in animals(Kumar and Kumar Sakhya, 2013).

Paronychia Bauer

The bark of the Australian plant *Acronychia Baueri* Schott (*Bauerella australiana* Borzi), has resulted in the isolation of the triterpene lupeol and the alkaloids melicopine, acronycine, and normelicopidine. The plant experiment proves the anti-tumor activity associated with the crude alkaloidal mixture obtained from the ether extract is attributable to acronycine. The present study leads in the search for agents effective in the chemotherapeutic management of human neoplasms(Svoboda *et al.*, 1966).

Allium sativum

It is the oldest of all cultivated plants. It has been used as a medicinal agent for thousands of years. It is a remarkable plant, which has multiple beneficial effects such as antimicrobial, antithrombotic, hypolipidemic, antiarthritic, hypoglycemic and antitumor activity. The present studies have demonstrated the chemopreventive activity of garlic by using different garlic preparations including fresh garlic extract, aged garlic, garlic oil, and some organosulfur compounds derived from garlic. In addition to this, some organosulfur compounds derived from garlic is Sallylcysteine, which is used to retard the growth of chemically induced and transplantable tumors in several animal models. Therefore, the consumption of garlic may provide some kind of protection from cancer development (Bayan *et al.*, 2014).

Aloe vera

Aloe vera contains aloe emodin, which activates the macrophages to fight cancer. *Aloe vera* also contains acemannan, which enhances activity of the immune cells against cancer (Majumder *et al.*, 2019) (Hamman, 2008).

Echinacea angustifolia

Echinacea angustifolia contains arabinogalactan, which protects the body from cancer by activating the macrophages. *Echinacea angustifolia* is used to treat metastatic carcinoma of the esophagus and the colon (Chicca *et al.*, 2007).

Zingiber officinalis

Zingiber officinalis (ginger) rhizomes offer a rich package of gingerols-phenolic antioxidants that possess pronounced anti-inflammatory activity-that inhibit various cancers. The chemo preventive effects of ginger and its components have been reported from the laboratory studies in a wide range of experimental models(Shukla and Singh, 2007).

Withania somnifera

Withania somnifera contains withanolides, which possess immuno-modulatory activity are known to inhibit the growth of cancer. An *in vitro* study shows withanolides from *Withaniasomnifera*inhibited growth in human breast, central nervous system, lung, and colon cancer cell lines comparable to doxorubicin (Palliyaguru *et al.*, 2016).

Ginkgo biloba

Ginkgo biloba contains Ginkgolide-B, which protects the body against cancer. It inhibits growth of cancer by regulating activity of the platelet-activating factor (El Mesallamy *et al.*, 2011).

Glycine max

Glycine max (Soya bean) is rich in zinc, selenium, vitamins (A, B1, B2, B12, C, D, E and K), amino acids, isoflavones, protease inhibitors, saponins and phytosterols. The studies have shown that these compounds convert the cancerous cells to normal cells by inducing cell-differentiation. It inhibits growth of hormone dependent cancers of the breast and the prostate(Van Die *et al.*, 2014).

Camellia sinensis

Camellia sinensis contains polyphenolics, which are known to possess antimutagenic and anticancer activity. Some evidence suggests that tea has a protective effect against stomach and colon cancers (Dreosti, 2009).Green tea polyphenol has chemoprotective role.(Sur and Panda, 2017).

Betula utilis

Betula utilis contains betulin that can be easily converted into betulinic acid. Studies have revealed that betulinic acid inhibits growth of malignant melanoma (Surowiak *et al.*, 2009) and other cancers (Fulda, 2008).

Colchicum luteum

Colchicum luteum, C. autumnale contains tropolone groups of alkaloid colchicines. Colchicine shows antimitotic activity and used in cancer for the dispersal of tumors and the treatment of various neoplastic diseases (S. *et al.*, 2013).

Combretum caffrum

Combretum caffrum contains combretastatin, which has been isolated recently. Camptothecin is a well-known anticancer agent. Derivatives of camptothecin such as 18-OH-camptothecin, 11-OH-camptothecin, and 10-OH-camptothecin have been found to possess a strong antileukemic activity (Pettit *et al.*, 1989).

Arctium lappa

Arctium lappa contains potent anticancer factors that prevent mutations in the oncogenes. It has been used in the treatment of malignant melanoma, lymphoma, and cancers of the pancreas, breast, ovary, esophagus, bladder, bile duct, and bone. This study reveals that it reduces the size of the tumor, relieves the pain, and prolongs the survival period(Susanti *et al.*, 2012).

Annona species

Annona species contain acetogenins, which possess significant cytotoxic activity against leukemia and sarcoma. Acetogenins are found to be effective in the treatment of nasopharyngeal carcinoma (Pieme *et al.*, 2014).

Angelica sinensis

Angelica sinensis is a Chinese medicinal herb used for treating gynecological and gastrointestinal disorders, and also in conjunction with cancer chemotherapy (Kan *et al.*, 2008).

Ananas comosus

Ananas comosus contains bromelain, which is a mixture of proteases and some other enzymes. It inhibits the growth of cancer cells (Béez *et al.*, 2007)

CONCLUSION

The incidence and death due to cancer is steadily increasing and cancer is considered as a global health problem. Although current treatments work in controlling the cancer progression, but produce severe side effects. Scientific and research interest is drawing its attention towards naturally-derived compounds as they are considered to have less toxic side effects. Medicinal plants have been used for thousands of years in folk medicines. A number of phytocompounds are produced by plant kingdom to fight against attacking pathogens. These compounds have a tremendous capability to inhibit cancer cell proliferation. With the advancement of different molecular tools and screening techniques, the discovery of anticancer compounds from plants has become significantly simpler. Thus, identification and characterization of phytocompounds for anticancer properties holds tremendous potential for the development of new therapeutic regimens to treat cancer in future

REFERENCES

Arruebo, M., Vilaboa, N., Sáez-Gutierrez, B., Lambea, J., Tres, A., Valladares, M., and González-Fernández, Á. (2011). Assessment of the evolution of cancer treatment therapies. *Cancers*. 3(3): 3279–3330.

Atanasov, A. G., Waltenberger, B., Pferschy-Wenzig, E. M., Linder, T., Wawrosch, C., Uhrin, P., Temml, V., Wang, L., Schwaiger, S., Heiss, E. H., Rollinger, J. M., Schuster, D., Breuss, J. M., Bochkov, V., Mihovilovic, M. D., Kopp, B., Bauer, R., Dirsch, V. M., and Stuppner, H. (2015). Discovery and resupply of pharmacologically active plant-derived natural products: A review. *Biotechnology Advances*, 33(8): 1582–1614.

Bayan, L., Koulivand, P. H., and Gorji, A. (2014). Garlic: a review of potential therapeutic effects. *Avicenna Journal of Phytomedicine*, 4(1): 1–14.

Béez, R., Lopes, M. T. P., Salas, C. E., and Hernández, M. (2007). In vivo antitumoral activity of stem pineapple (*Ananas comosus*) bromelain. *Planta Medica*, 73(13): 1377–1383.

Chicca, A., Adinolfi, B., Martinotti, E., Fogli, S., Breschi, M. C., Pellati, F., Benvenuti, S., and Nieri, P. (2007). Cytotoxic effects of Echinacea root hexanic extracts on human cancer cell lines. *Journal of Ethnopharmacology*, 110(1): 148–153.

Chitale, V. S., Behera, M. D., and Roy, P. S. (2014). Future of endemic flora of biodiversity hotspots in India. *PLoS ONE*, 9(12): e115264.

Dreosti, I. E. (2009). Bioactive Ingredients: Antioxidants and Polyphenols in Tea. *Nutrition Reviews*, 54(11): S51–S58.

Efferth, T., Li, P. C. H., Konkimalla, V. S. B., and Kaina, B. (2007). From traditional Chinese medicine to rational cancer therapy. *Trends in Molecular Medicine*, 13(8): 353-61.

El Mesallamy, H. O., Metwally, N. S., Soliman, M. S., Ahmed, K. A., and Abdel Moaty, M. M. (2011). The chemopreventive effect of *Ginkgo biloba* and *Silybum marianum* extracts on hepatocarcinogenesis in rats. *Cancer Cell International*, 11(1): 38.

Fulda, S. (2008). Betulinic acid for cancer treatment and prevention. *International Journal of Molecular Sciences*, 9(6): 1096–1107.

Hamman, J. H. (2008). Composition and applications of Aloe vera leaf gel. *Molecules*, 13(8): 1599–1616.

Hanahan, D., and Weinberg, R. A. (2011). Hallmarks of cancer: The next generation. *Cell*, 144(5): 646–674.

Kan, W. L. T., Cho, C. H., Rudd, J. A., and Lin, G. (2008). Study of the anti-proliferative effects and synergy of phthalides from *Angelica sinensis* on colon cancer cells. *Journal of Ethnopharmacology*, 120(1): 36–43.

Kintzios, S., and Barberaki, M. (2004). Plants and Cancer. In Plants that Fight Cancer.

Kumar, N., and Kumar Sakhya, S. (2013). Ethnopharmacological Properties of *Curcuma longa*: A Review. *International Journal of Pharmaceutical Sciences and Research*, 4(1): 103–112.

Majumder, R., Das, C. K., and Mandal, M. (2019). Lead bioactive compounds of Aloe vera as potential anticancer agent. *Pharmacological Research*, 148: 104416.

Moudi, M., Go, R., Yien, C. Y. S., and Nazre, M. (2013). Vinca alkaloids. *International Journal of Preventive Medicine*, 3(3): 3279–3330.

Palliyaguru, D. L., Singh, S. V., and Kensler, T. W. (2016). Withania somnifera: From prevention to treatment of cancer. *Molecular Nutrition and Food Research*, 60(6): 1342–1353.

Pandey, M. M., Rastogi, S., and Rawat, A. K. S. (2013). Indian traditional ayurvedic system of medicine and nutritional supplementation. In Evidence-based Complementary and Alternative Medicine.DOI: 10.1155/2013/376327

Pang, X. C., Zhang, L., and Du, G. H. (2018). Podophyllotoxin. In Natural Small Molecule Drugs from Plants (pp. 545–550).

Parsa, N. (2012). Environmental factors inducing human cancers. *Iranian Journal of Public Health*, 41(11): 1–9.

Pettit, G. R., Singh, S. B., Hamel, E., Lin, C. M., Alberts, D. S., and Garcia-Kendal, D. (1989). Isolation and structure of the strong cell growth and tubulin inhibitor combretastatin A-4. *Experientia*, 45(2): 209–211.

Pieme, A. A., Kumar, G. G., Dongmo, S. S., Moukette, M. M., Boyoum, F. F., Ngogang, Y. Y., and Saxena, K. K. (2014). Antiproliferative activity and induction of apoptosis by *Annona muricata* (Annonaceae) extract on human cancer cells. BMC Complementary and Alternative Medicine, 14(1): 516.

Rahier, N. J., Thomas, C. J., and Hecht, S. M. (2005). Camptothecin and its analogs. In Anticancer Agents from Natural Products, (pp. 5-22). CRC Press.

S., S., Y., B., Nandani, S., S., G., S., K., N., A., B., and K.L., D. (2013). Colchicine and its various physico-chemical and biological aspects. *Medicinal Chemistry Research*, 22(2): 531–547.

Sakarkar, D. M., and Deshmukh, V. N. (2011). Ethnopharmacological review of traditional medicinal plants for anticancer activity. *International Journal of PharmTech Research*, 217: 63–82.

Schiff, P. B., Fant, J., and Horwitz, S. B. (1979). Promotion of microtubule assembly in vitro by taxol [19]. *Nature*, 77(5698): 665–667.

Shukla, Y., and Singh, M. (2007). Cancer preventive properties of ginger: A brief review. *Food and Chemical Toxicology*, 45(5): 683–690.

Sur, S., and Panda, C. K. (2017). Molecular aspects of cancer chemopreventive and therapeutic efficacies of tea and tea polyphenols. *Nutrition*, 43–44: 8–15.

Surh, Y. J. (2003). Cancer chemoprevention with dietary phytochemicals. *In Nature Reviews Cancer* (Vol. 3, Issue 10: pp. 768–780).

Surowiak, P., Drag, M., Materna, V., Dietel, M., and Lage, H. (2009). Betulinic acid exhibits stronger cytotoxic activity on the normal melanocyte NHEM-neo cell line than on drug-resistant and drug-sensitive MeWo melanoma cell lines. *Molecular Medicine Reports*, 2(4): 543–548.

Susanti, S., Iwasaki, H., Itokazu, Y., Nago, M., Taira, N., Saitoh, S., and Oku, H. (2012). Tumor specific cytotoxicity of arctigenin isolated from herbal plant *Arctium lappa* L. *Journal of Natural Medicines*, 66(4): 614–621.

Svoboda, G. H., Poore, G. A., Simpson, P. J., and Boder, G. B. (1966). Alkaloids of Acronychia Baueri Schott I: Isolation of the alkaloids and a study of the antitumor and other biological properties of acronycine. *Journal of Pharmaceutical Sciences*, 55(8): 758–768.

Van Die, M. D., Bone, K. M., Williams, S. G., and Pirotta, M. V. (2014). Soy and soy isoflavones in prostate cancer: *A systematic review and meta-analysis of randomized controlled trials. BJU International*, 113(5b): E119-30.

Chapter 9

Protection and Conservation of Tiger Habitats in Odisha: An Overview

*Rajkishore Mohanta**

O/o the PCCF (WL) and CWW, Prakruti Bhawan,
Plot No.1459, Saheed Nagar, Bhubaneswar – 751 007, Odisha
**e-mail: brownbear143@gmail.com*

Tiger is considered as a flagship/umbrella species. A viable population of tiger represents a healthy ecosystem. Apart from the two tiger reserves in Odisha, namely Similipal and Satkosia tiger reserves in north and central parts, the state harbors sporadic tiger populations in many of its forest patches. While tiger population inside tiger reserve is facing several threats from poaching and excessive anthropogenic activities, the habitats of tigers outside the reserves are not adequately protected. Once distributed across the whole landscape of the state, the tiger habitats have shrunk to patches due to large scale habitat fragmentation. Also many of the tigers were poached for heavy demand of their body parts in international markets. Systematic planning to connect the fragmented habitats and a master plan for protecting tigers in landscape level approach will definitely help in conserving these big cats. A viable tiger population warrants inviolate areas, good prey base, habitat connectivity, trained staff strength, protection from poaching and most importantly a good Government policy. It needs urgent attention of the forest department authorities to save them before they get completely wiped out from the state. Tiger conservation also needs dedicated support of non-Government organizations, local communities and media. So a holistic approach will definitely lead to success in saving this charismatic species in Odisha.

Keywords: *Tiger, Conservation, Habitat, Odisha.*

INTRODUCTION

Tigers are a large cat species that inhabit varied habitat types in Asia and the Russian far-east. They are easily identifiable by the distinct dark stripes on a reddish-orange background. They are apex predators and act as both flagship and umbrella species. Conservation of tigers thus ensures the protection of their habitats and the various components of biodiversity present along-with the maintenance of associated ecosystem services (Wildlife Institute of India, 2018). Tiger is locally Known as "Bagha" in Odisha. Tiger is the most charismatic animal in India, the majestic Tiger has an orange coat patterned with broad black stripes. It has black ears, each with a winking white spot on the back, powerful forepaws, and a long-banded tail. The Indian Tiger is one of six living races or subspecies distributed through Asia. Today, shrinking Tiger numbers and habitat are major conservation issues (Menon, 2003).

Tigers are members of the mammalian order Carnivora with well-developed olfactory lobes in the brain that is reciprocated by a cranial structure for keen olfactory sense. Further, the highly developed carnassials (incisors) enable their predatory mode of life (Nowak and Nowak, 2005). Similar to other felids they rely on their highly developed vision and hearing for their nocturnal predatory mode of life. Their skull morphology that is rounded and shortened at the fore-end allows them to deliver powerful bites that enable them to hunt large prey (Mazák, 1981). Similar to other members of the sub-family Pantherinae the imperfectly ossified hyoid bone provides greater flexibility to the larynx thus enabling them to roar (Pocock, 1917; Wozencraft, 2005). Eight subspecies of tiger (*Panthera tigris*), are traditionally recognized, of which three recently became extinct, are commonly recognized based on geographic isolation and morphological characteristics (Mazák, 1981). Luo *et al.* (2004) suggested the segregation of extant tigers into six subspecies taking number of all tiger species to nine. The classification scheme at the sub-species level for tigers as proposed by Luo *et al.* (2004) has been accepted by the IUCN Redlist of threatened species (Goodrich *et al.*, 2015) and the same is used here. A comparison of the traditionally recognized eight and the presently accepted six subspecies are presented in Table 9.1. The tigers of Sunderban have been identified as a distinct Evolutionary Significant Unit with isolation from other Bengal tiger populations in the last 2000 years (Singh *et al.*, 2015). The general taxonomy and biology attributes of the tiger are given in Table 9.2.

In this connection, the information generated by the earlier three cycles of tiger status evaluation exercises resulted in major changes in the policy and management of tiger populations. The major outcomes were 1) Tiger Landscape Conservation Plans, 2) designation and creation of inviolate critical core and buffer areas of Tiger Reserves, 3) identification and declaration of new Tiger Reserves, 4) recognition of tiger landscapes and the identification of important habitat corridors, 5) integrating tiger conservation with developmental activities using reliable information in a Geographic Information System, and 6) planning reintroduction and supplementation strategies for tigers in the future to prioritize conservation investments and conserve unique gene pools (Jhala *et al.*, 2019).

So, prey species either directly or indirectly influence the population dynamics of their predators. Therefore, to conserve and manage endangered or threatened species, it is essential to understand prey densities in their natural habitat (Larue *et al.*, 2007). The estimation of population size and status assessment of prey density in forested habitat is important for wildlife management. Although ungulate (prey)

Table 9.1. Comparison of Sub-species Level Classification and Distribution of different Tiger Subspecies

Distribution Range	*Sub-species*		*Present Status of Populations*
	Mark (1982)	*Luo et al. (2004)*	
Indian sub-continent	*Panthera tigris*	*P. t. tigris*	Extant populations
Russian Far East and northeastern China	*P. t. altaica*	*P. t. altaica*	
South China	*P. t. amoyensis*	*P. t. amoyensis*	
Sumatra	*P. t. sumatrae*	*P. t. sumatrae*	
Indochina north of the Malayan Peninsula	*P. t. corbetti*	*P. t. corbetti I*	
Malayan peninsula		*P. t. corbetti II* (*P. t. jacksoni*)	
Bali	*P. t. balica*		Extinct populations (Nowell and Jackson, 1996)
Eastern Anatolia, Iran, Caucasus region and Central Asia	*P. t. virgata*		
Java	*P. t. sondaica*		

Table 9.2. Taxonomy, Biological Attributes and Behaviour of Bengal Tiger

Taxonomy	*Biology Attributes*	*Males*	*Females*	*Behaviour*
Kingdom: Animalia	Body length	270-310 cm	240-265 cm	The tiger prefers to hunt large deer, especially Sambar where available. Quite adaptable, it can survive on smaller prey or even fish in mangroves. It may attack humans if pray is unavailable or if it is old or injured. This behaviour has earned it the fearsome tittle of man eater, although other mammals such as the leopard or the elephant take for more human lives.
Phylum: Chordata	Body weight	175-260 kg	100-160 kg	
Class: Mammalia	Age at first reproduction (free ranging)	4.8 years (mean)	3.8 years (mean)	
Order: Carnivora	Age at first reproduction (in captivity)	2.5 years	2.5 years	
Family: Felidae	Longevity	12-15 years in wild; 20 years or more in captivity		
Sub-family: Pantherinae	Gestation period	106 -112 days		
Genus: Panthera	Litter size	Ranging from 1 –7 cubs (mostly 2-4)		
Species: tigris	Social unit	Solitary		
Sub-species: tigris	Stratum	Terrestrial		
	IUCN Status/WPA	Endangered/I		

census may be relatively easy in open grassland or meadow areas, it is much harder in forest habitats (Lowe, 1969). However, in the recent past, continued depletion of prey population and fragmentation of natural habitats apart from poaching is a serious issue acknowledged by conservationists as well as managers. These two major factors led to the present dilemma of tigers in the wild and will determine its survival in the future (Karanth and Stith, 1999; Sunquist *et al.*, 1999).

Wildlife Biodiversity Conservation in Odisha

The state of Odisha has a geographical area of 1,55,707 sq. Km. The recorded Forest Cover in the State area is 51,619 sq.km. which is 33.15 per cent of the State's geographical area as per the State of Forest Report (ISFR-2019) of Forest Survey of India (India State of Forest Report, 2019). The State is quite rich in wildlife. The important wildlife found in the state are Elephants, Tigers, Melanistic Tiger, Leopards, Black Panther, Hyena, Wolf, Fishing Cats, Leopard Cats, Jungle Cats, Gaur, Sambar, Spotted Deer, Wild Boar, Giant squirrels, Peafowl, Hill Myna, Marsh Crocodile, Gharial, Saltwater Crocodile, King Cobra, Sea Turtles, and Monitor Lizard, *etc.* To protect wildlife and their habitat, a network of protected areas has been constituted. Special programs have also been initiated for special care of endangered and threatened species.

The State has 19 Sanctuaries including Gahirmatha Marine Wildlife Sanctuary, one National Park (Bhitarkanika), one proposed National Park (Similipal) which have been notified under the provisions of Indian Wildlife (Protection) Act, 1972 for *in situ* conservation. These Protected Areas constitute 10.37 per cent of the total forest area and 5.36 per cent of the total geographical area of the State. One large Zoological Park (Nandankanan), three Small Zoos, and seven Mini Zoos have been established for *ex situ* conservation and management of wildlife outside protected areas. The State has the singular distinction of having three stretches of mass nesting beaches of endangered Olive Ridley Sea Turtles (*Lepidochelys olivacea*) including the world's largest nesting ground at Gahirmatha. It has many natural wetlands. Chilika is the largest wetland of Asia and another one Bhitarkanika has been declared as "Ramsar Sites". Seven Eco-sensitive Zones are finally notified *i.e.* Kapilash, Chandaka, Balukhanda, Debrigarh, Kuldiha, Bhitarkanika, and Sunabeda. There is one Biosphere Reserve and three elephant reserve. Besides this, two Tiger Reserve i.e Similipal and Satkosia, Sunabeda Tiger Reserve is the third one which is under the process of notification. One proposal for declaring Tiger Reserve *i.e.* Debrigarh which is under progress (Wildlife Odisha, 2018). A list of tiger reserves in Odisha is shown in Figures 9.1 and 9.2.

Wildlife Species Diversity

A total of >800 species of fishes, 29 species of amphibians,129 species of reptiles including three crocodilian species, 537 species of birds and 114 species of mammals have so far been recorded in the state which includes 54 species of threatened animals (17 species of reptiles,15 species of birds and 22 species of mammals) as per the definition of IUCN Red Data Book. The state is blessed with an extremely rich and unique assemblage of wildlife, inhabiting their respective habitats spread over two different bio-geographic zones (Decan Peninsula and Coasts) and three provinces (Eastern Plateau, Chhota- Nagpur and Eastern Coast Life) in the state (Wildlife Odisha, 2018; Wildlife Odisha, 2019). Various developmental projects and mining *etc.* are creating disturbances for wildlife and their habitat leading to man-animal conflicts. Hence, habitat fragmentation and the ensuring man-animal conflicts are some of the major challenges for the state of Odisha.

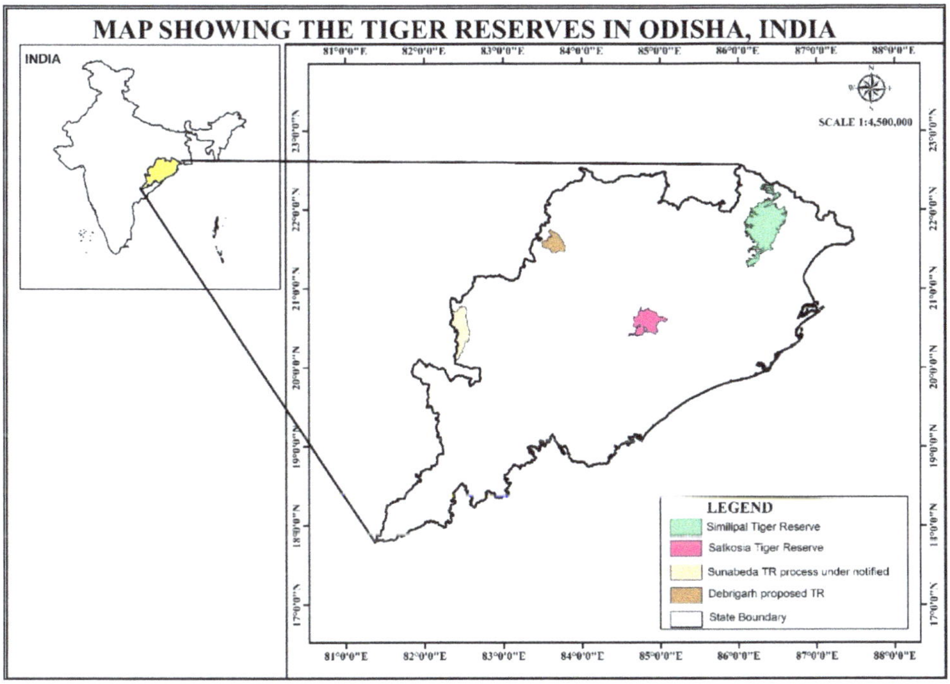

Figure 9.1. Map Showing Tiger Reserves in Odisha.

Methods of Estimation of Population and Monitor the Conservation

The government initiated several measures to estimate their population and monitor the conservation and protection of the species. Different scientific organisations and research institutions are employed by specialists to study mammals for various methods of research activity. So, documentation, research, and monitoring are a recognized mandate of the wildlife for the conservation of species. Its approaches are some of the few areas generating a lot of data that are used in the planning and execution of wildlife conservation programs. Population estimation of various wildlife species has been done regularly for a better understanding of their habitat which forms a vital input for better management.

Tiger Conservation in Odisha

Late Saroj Raj Choudhury was credited with several initiatives in the field of wildlife conservation. He introduced the "pugmark methodology" for the census of tigers in India, a technique he introduced while heading the first tiger census in India in 1972 which later became a popular method used all over India (Abe Books Profile, 2015). Pugmark technique was involved for collection of pugmark tracking and plaster casts from the field and analysis of these separately for individual male, female and cub Tiger and Leopard and their diagnostic track dimensions and study of the spatial distribution of tigers.

Royal Bengal Tiger in Similipal Tiger Reserve (Source: Wildlife Odisha-2018)

Melanistic Tiger in Similipal Tiger Reserve (Source: Wildlife Odisha-2018)

White Tiger in Nandankanan Zoological Park (Source: www.nandan.org)

Khairi Tigress with S.R. Choudhury with Others (Source: incredibleorissa.com)

Figure 9.2. Sighting of Tigers in Odisha Landscape.

Several aspects of the pugmark technique have been improved by (Singh, 2000). Pieces of evidence of Melanistic tiger of the Similipal Tiger reserve were reported by (Swain and Beheura, 2013). Similarly, Monitoring the Status of Tiger Report, 2018 by (Wildlife Odisha, 2019; Nayak, 2020) also important Guidelines, Protocols, and Order related to Tiger issued by Wildlife Odisha (2019).

In the year 2004-05, the Project Tiger authorities and the wildlife Institute of India together devised a monitoring method. In this method, there were two approaches to collect field data, (i). information on habitat condition and different signs for the presence of tiger as evidence to be collected and (ii). by using camera traps Tiger photographs were to be obtained from all Tiger areas. The first national-level assessment of tiger status along with that of Co-predators and their pray was undertaken in 2006. All Indian Tiger estimation was conducted by the National Tiger Conservation Authority (NTCA) and Wildlife Institute of India (WII) by Camera Trap methods in the year 2018. Different software tools were used such as SPACECAP/DENSITY for estimating tiger population size, DISTANCE for estimating densities of prey, and used the MSTrIPES app to assist in wildlife protection and also helps upload geo-tagged pictures into the central GIS database (Wildlife Odisha, 2016; Jhala *et al.*, 2019; Wildlife Odisha, 2019).

Project Tiger in Odisha

Similipal

Similipal National Park is a tiger reserve in the Mayurbhanj district in the Indian state of Odisha covering 2,750 sq. km. Similipal holds most of the tiger population of the state of Odisha. Less than thirty tigers were surviving here in the year 1975. Sustained protection measures and management initiatives have resurrected the dwindling population of tigers. Recent census of tigers conducted by the use of both pugmark and camera trap techniques in the year 2018 shows presence of 28 tigers. Similipal is the only place in the world that houses the source population of melanistic tigers. Most of the tigers from North Similipal are moving to south Similipal that remains a better habitat, and the leopards, more adaptable than the tiger, are appearing in most of the habitats vacated by the tiger. These observations suggest that Similipal habitat may be reaching the maximum of its carrying capacity for the tiger. The future seems dependent on adjoining forests of Hadgarh and Kuldiha that are better known as elephant habitats.

Odisha is the only state to have melanistic tigers, which are rarely found in the world. The melanistic tiger was first reported in Similipal in the year 1993. The first photographic evidence about the presence of the melanistic tiger was captured in 2007.

Assessment of Tiger Population

'Project Tiger' and Conservation Practices of Tigers

Tigers across their distribution range are threatened by extensive poaching, habitat destruction as a consequence of development of linear infrastructure. The development of linear infrastructure is also responsible for fragmentation of populations leading to reduced gene-flow between populations.

The royal Bengal tiger, *Panthera tigris tigris*, has always been an integral part of the life and legend of India. At the beginning of the 1900s, the Indian tiger population was estimated at 40,000 animals. The first official estimate, done in 1972, recorded only about 1800 Tigers. This led to the establishment of a task force under the Indian Board of Wildlife, and based on their recommendations, 'Project Tiger' was launched on 1 April 1973 with the following objectives: i. to maintain a viable population of tigers in India for scientific, economic, aesthetic, cultural, and ecological values; and ii. to preserve, for all times, areas of biological importance as a national heritage for the benefit, education, and enjoyment of the people. At the beginning of the project, 9 Tiger Reserves were created. Currently, there are 50 Tiger Reserves in India (Warndeep, 2004; Goodrich *et al.*, 2015), which are governed by Project Tiger which is administrated by the National Tiger Conservation Authority (NTCA).

The fourth cycle of the assessment was undertaken in 2018 and 2019 using the best available science, technology and analytical tools. In this cycle, recording of primary field data digitally through mobile phone application like M-STrIPES (Monitoring System for Tigers - Intensive Protection and Ecological Status), that uses GPS to geotag photo-evidences, and survey information made this exercise more accurate, with smaller margins of human error. The unique feature of this cycle of assessment, in keeping up with "Digital India", is the development and use of innovative technological tools in the collection and processing of data to reduce human errors (Jhala *et al.*, 2019). The fourth cycle of the National tiger status assessment of 2018-19 is the most accurate survey conducted. The survey covered 381,400 km of forested habitats in 20 tigers occupied states of India. A foot survey of 522,996 km was done for carnivore signs and prey abundance estimation. In these forests, 317,958 habitat plots were sampled for vegetation, human impacts and prey dung. Camera traps were deployed at 26,838 locations. These cameras resulted in 34,858,623 photographs of the wildlife of which 76,651 were of tigers and 51,777 were of leopards. The total area sampled by camera traps was 121,337 km^2. The total effort invested in the survey was 593,882 man-days. We believe that this is the world's largest effort invested in any wildlife survey till date, on all of the above criteria. A total of 2,461 individual tigers (>1 year of age) were photo-captured. The overall tiger population in India was estimated at 2,967 (SE range 2,603 to 3,346) (Jhala *et al.*, 2019).

In Odisha, Tiger estimation was carried out by the Pugmark technique during 2004 and 2016. A total of 192 tigers were estimated to be found during the year 2004 and 40 in 2016 by pugmark technique (Table 9.3). Out of 51 divisions Tiger estimation was conducted in 47 divisions (Except City Forest, Bhadrak Wildlife, MFD Rajnagar and Chilika Wildlife Division) in the year 2018. According to the National Tiger Conservation Authority and Wildlife Institute of India, there were estimated only 28 tigers in existence in Odisha (Table 9.4) (Jhala *et al.*, 2019; Wildlife Odisha, 2019), which is different from the earlier process of sampling through pugmark survey. Hence, they are not comparable.

Satkosia

Satkosia Tiger Reserve is located in the Central Zone of Odisha and spreads over Angul, Cuttack, Nayagarh and Boudh districts covering an area 1136.70sq. kms and comprises of Moist Deciduous Forests, Dry Mixed Deciduous Forest, Bamboo Brakes, Dry Tropical Riparian Forest and Teak Plantations. It encompasses the Satkosia Gorge Sanctuary constituted in May 1976 and the Baisipalli Sanctuary constituted in November 1981. The Tiger Reserve is rich in large cats and their prey along with wetland fauna (Gharial, Mugger, Freshwater Turtles, *etc.*) in the Mahanadi and rich biodiversity in the terrestrial ecosystem.

Besides Tigers, Satkosia Gorge is the home of the most endangered Gharial, the Mugger crocodiles and a variety of freshwater turtles. Other faunal elements include Tiger, Leopard, Wild dog, Gaur, Nilgai, Sambar, Spotted Deer, Barking Deer, Bear, Ratel, Jackal, Porcupine, Indian Hare, and Giant Squirrel *etc.*

Table 9.3. Odisha Tiger Estimations during the Year 2004 and 2006

Name of the Circle	Year 2004				Year 2006			
	Male	Female	Cub	Total	Male	Female	Cub	Total
1. Bhawanipatna	12	13	12	**37**	02	03	00	**05**
2. Koraput	04	03	00	**07**	00	00	00	**00**
3. Angul	04	06	01	**11**	01	00	00	**02**
4. Sambalpur	03	04	00	**07**	00	00	00	**00**
5. Berhampur	03	05	13	**21**	00	00	00	**00**
6. Rourkela	00	00	00	**00**	01	03	00	**04**
7. Bhubaneswar	03	02	02	**07**	00	00	00	**00**
8. Baripada	28	42	32	**102**	09	17	03	**26**
Odisha Gr. Total	**57**	**75**	**60**	**192**	**13**	**24**	**03**	**40**

Source: Wildlife Odisha 2019.

Table 9.4. Showing all Odisha Tiger Estimation by Camera Trap Methods by NTCA

Name of the Tiger Reserve	Year 2006	Year 2010	Year 2014	Year 2018
1. Similipal Tiger Reserve	20 (17-23)	23(12-34)	Whole State	Whole State
2. Satkosia Tiger Reserve	6	8 (7-9)		
3. Sunabeda Tiger Reserve	9	Census not done		
4. Sporadic Occurrence	6			
Odisha Total	**45 (37-53)**	**32 (20-44)**	**28**	**28**

Source: Wildlife Odisha 2019.

Village Relocation from Tiger Reserve and Sanctuary of Odisha

The relocation of villages from critical protected areas is an important step for conserving flagship species like Tigers which require vast inviolate habitats. It not only makes the protected areas inviolate but also enables the villagers to assess various development schemes of the state. The following villagers were relocated from Tiger Reserve and Sanctuary.

1. Relocation of Raigoda village from Satkosia Tiger Reserve

i. Raiogada village to New Raigoda Village (Saruali): 78 families

2. Relocation of Forest villages in Similipal Tiger Reserve/Sanctuary

i. UBK and Bahagarh to Asankudar village: 73 families
ii. Kabatghai to Manada village: 47 families
iii. Kiajhari to Khandiadar village : 79 families
iv. Ramjodi to Sialnai village: 73 families
v. Sanbarakamuda (Matakocha) to Sialnai village: 43 families

3. Relocation of Forest villages in Debrigarh Sanctuary

i. Dibrugarh to Outside Sanctuary: 08families

ii. Jhagadabehera to Outside Sanctuary: 37 families

Awareness meetings were conducted to motivate the households as well as family members to relocate their village in Pallisabha/Gramasabha. The villagers agreed to relocate voluntarily as per the prevailing guideline of NTCA, Govt. of India, and additional incentives of State Government of Odisha.

Following facilities were provided to each eligible family, the supply of cooked food for 30 days, regular health check-up, social security schemes, provision of the temporary shed, common cattle, kitchen, and dining shed, approach road, solar light, drinking water facilities and pond and rehabilitation assistance @ 10.00 lakh and 10 decimal homestead land. NGOs have been requested to provide capacity building for livelihood improvement through Income Generation Activities (IGA) in various technical fields. After the relocation process was completed, concentrations of wild animals increased in the forests. Presently tiger movement inside the sanctuary has been noticed and captured in camera traps. The relocation of villages outside the sanctuary strengthens the wildlife habitat and wildlife population (Wildlife Odisha, 2018).

Relocation of Tiger in Satkosia Tiger Reserve (STR)

Satkosia Tiger Reserve comes within the jurisdiction of two divisions namely Satkosia Wildlife and Mahanadi Wildlife Division. STR covering two sanctuaries *i.e.* Satkosia Gorge Wildlife Sanctuary and Baisipalli Wildlife sanctuary covering an area of 693.87 sq. km. An area of 359.79 sq. km will be added to STR for the free and safe movement of the tiger. Various eco-development activities to augment the livelihood option of the villagers living in and around STR were also taken up by the Forest and Environment Department, Government of Odisha.

STR is one of the six proposed recovery sites of the tiger in India as identified by National Tiger Conservation Authority (NTCA) to establish viable, free-ranging tiger populations by supplementing wild tigers from other similar habitats and rebuilding the tiger populations by providing a secure and conducive environment.

Sunabeda

The Sunabeda Sanctuary area comes under the Administrative control of Sunabeda Wildlife Division with headquarters at Nuapada district of Odisha. Sunabeda Sanctuary covers hills and plateaus over 600 sq. km of area. The geographical location is Longitude 82° 20′ 06″E and 82° 34′ 42″E and latitude 20° 24′N and 20° 44′N. The Division constitutes three ranges *i.e.* Nuapada, Komna, and Sunabeda, having headquarters at Nuapada, Komna, and Sunabeda respectively. Sunabeda has received 'in-principle-approval' of the National Tiger Conservation Authority, proposals for its notification as per NTCA-guidelines is under preparation. The proposed tiger reserve extends over 956.17 sq. km. Sunabeda Tiger Reserve is home to varied engaged fauna like Royal Bengal Tiger, Leopard, Indian Bison, Hyena, Barking Deer, Chital, Sambar, Wild Pig, Ratel, Sloth Bear, Nilgai, Porcupine,

Hares, Jackals, Peafowl, Partridge, Red Jungle Fowl and many other colourful and migratory birds and several reptilian species also inhabit the sanctuary.

Tiger Conservation Programme is going on in Sunabeda since 1997. As part of this programme special emphasis has been laid upon tiger conservation in the forests of Sunabeda-Khariar in the Nuapara district of Odisha, which constitutes a vital tiger habitat. These forests harbour a reasonably large tiger gene pool, with long term conservation potential. Wildlife wing collaborates with District Collector to facilitate income-generating activities for villagers to create a conducive wildlife-human interface.

Debrigarh

Debrigarh Wildlife Sanctuary (83° 30′E - 83° 46′E, 21° 28′N - 21° 44′N) located in Bargarh district, Odisha, covering a total area of 346.91 sq. km. The sanctuary is bound on the east and north by the huge Hirakud reservoir. It is one of the few select sanctuaries in the state supporting both terrestrial and aquatic biodiversity. Debrigarh Wildlife Sanctuary, an important site for in situ conservation of wildlife and its habitat in the state of Odisha is home to an immense array of biodiversity, a large number of species of mammals, birds, reptiles, amphibians, fishes, odonates, butterflies and spiders are found here. The sanctuary is famous for its sylvan beauty and pristine wilderness. The dry deciduous forests with varied flora and fauna attract the nature-lovers of Odisha and neighbouring states every year. The combination of dry deciduous mixed forests with rich wildlife, Hirakud reservoir and attractive topographical features are the important features of Debrigarh wildlife sanctuary. It is said to be one of the vibrant wildlife sanctuaries of the state. It is the home to a wide variety of flora and fauna.

Herbivores such as Spotted Deer, Sambhar, Four-horned Antelope, Nilgai, and Carnivores such as Leopard and other animals such as Indian Gaur, Bear, Wild Boar, Civet, Porcupine, Hanuman Langur, *etc.* are noticed. Among birds Peafowls are seen in plenty. Other birds such as Eagle, Owl, Parrot, Flycatcher, Sunbird, Hornbill, Crow, Pigeon, Nightjar *etc.* are seen. Python, Cobra and other snakes are rarely seen.

Debrigarh Wildlife Sanctuary is an Eco-sensitive Zone from an ecological and environmental point of view and it is required to prohibit industries or class of industries and their operations and processes in the said Eco-sensitive Zone.

Nandankanan Zoological Park

Nandankanan Zoological Park is one among premier Large zoos of India located amidst beautiful natural surroundings and spreads over an area of 3.62sq. Km. The zoological park was dedicated to public on 29th December, 1960. The undulating topography with natural moist deciduous forest, water bodies and enchanting landscape provides the zoo inhabitants and an appropriate ambience to live in harmony with nature.

It has achieved the distinction of an outstanding Zoo through World Class Conservation, Education, Research and enriching visitor experiences by connecting people to wildlife and ex situ conservation. The Zoo provide environmental education, eco-system awareness, and nature interpretation. They facilitate research on wild animals maintained in captivity and help in maintaining the gene pool of the critically endangered species. They also serve as rescue centres for straying wild animals rescued from nearby human habitations (Nandankanan Zoological Park). The Nandankanan Zoological Park has observed several exciting events like i). Birth of white tiger from normal coloured parents in the year 1980 and Melanistic Tiger cubs in 2014 are unique in itself, ii). White Tiger Safari was established on 01.10.1991, It is the first zoo in India with a white tiger safari and iii). After successful mating of zoo bred tigress "Megha" with wild tiger "Nandan" in Nandankanan, it was the day to remember when tigress Megha gave birth to two cubs on 20.04.2016 as Nandankanan got wild tiger gene for the first time in its history.

CONCLUSION

Bengal tigers across their distribution range in the Indian subcontinent are threatened by extensive poaching and habitat destruction leading to its inclusion under Schedule I, Part I of the Wildlife Protection Act, 1972 and as endangered under IUCN Red data list. Tigers are charismatic animals that are a flagship species of the varied habitats they inhabit. Continued threats to tiger populations have necessitated intensive conservation efforts. Management of captive populations to ensure a genetically viable and demographically stable population is a part of these efforts.

Tigers across their distribution range are threatened by extensive poaching, habitat destruction as a consequence of development of linear infrastructure. The development of linear infrastructure is also responsible for fragmentation of populations leading to reduced gene-flow between populations. The present analysis was carried out to develop the breeding recommendations and determining the future fate of the current population lead to the conclusion that the current population is demographically stable and genetically viable. Management interventions should be focussed on managing the genetic diversity that is present in the population. This can be achieved by using the pairing recommendations suggested to ensure retention of desired levels of genetic diversity in the current population.

ACKNOWLEDGEMENTS

The author is thankful to the Principal CCF (Wildlife) and Chief Wildlife Warden, Odisha for rendering necessary help to carry out this piece of work. Thanks, are also due to field staff of Wildlife Wing for their help and co-operation.

REFERENCES

Abe Books profile. Abe Books. 2015. Retrieved 7 July 2015.

Goodrich, J., Lynam, A., Miquelle, D., Wibisono, H., Kawanishi, K., Pattanavibool, A., Htun, S., Tempa, T., Karki, J., Jhala, Y. and Karanth, U. (2015). Panthera tigris. The IUCN Red List of Threatened Species 2015: e.T15955A50659951. http://dx.doi.org/10.2305/IUCN.UK.2015- 2.RLTS.T15955A50659951.en

India State of Forest Report (2019) (Vol.1): Ministry of Environment Forest and Climate Change, Kaulagarh road, P.O. IPE Dehradun-248195, Uttarakhand, India.

Jhala, Y. V., Qureshi, Q. and Nayak, A. K. (eds). (2019). Status of tigers, co-predators and prey in India 2018. Summary Report. National Tiger Conservation Authority, Government of India, New Delhi and Wildlife Institute of India, Dehradun. TR No./2019/05.

Karanth, K.U., Stith, B.M. (1999). Prey depletion as a critical determinant of tiger population viability, Riding the Tiger: Tiger Conservation in Human Dominated Landscapes (Editors: J. Seidensticker, S. Christie and P. Jackson), Cambridge University Press, Cambridge, pp. 100-122.

Larue, M.A., Nielsen, C.K. and Grund, M.D. (2007). Using Distance Sampling to Estimate Densities of White-Tailed Deer in South-Central Minnesota, *The Prairie Naturalist*, 39: 57- 68.

Lowe, V.P.W. (1969). Population dynamics of red deer (*Cervus elaphus*, L.) on Rhum, *Journal of Animal Ecology*, 38: 425-457.

Luo, S.J., Kim, J.H., Johnson, W.E., Van Der Walt, J., Martenson, J., Yuhki, N., Miquelle, D.G., Uphyrkina, O., Goodrich, J.M., Quigley, H., Tilson, R., Brady, G., Martelli, P., Subramaniam, V., Mcdougal, C., Hean, S., Huang, S.Q., Pan, W., Karanth, U.K., Sunquist, M., Smith, J.L.D. and O'Brien, S.J. (2004). Phylogeography and genetic ancestry of tigers (*Panthera tigris*). *PLoS Biology*, 2: 2275-2293.

Mazák, V. (1981). Panthera tigris. Mammalian Species, 152,: 1-8.

Menon, V. (2003). A Field Guide to Indian Mammals, Wildlife Trust of India. Pp-200

Nayak, A. K. (2020). Compendium of Guidelines, Advisories and Orders by NTCA, MoEF and CC, Govt. of India Vol. II, January 2020 pp.-248.

Nowak, R. M. and Nowak, R. M. (2005). Walker's carnivores of the world. Baltimore: Johns Hopkins University Press.

Nowell, K. and Jackson, P. (1996). Wild cats: Status survey and conservation action plan. Gland, Switzerland: IUCN-World Conservation Union. 406 p.

Pocock, R. I. (1917). The Classification of existing Felidae. *The Annals and Magazine of Natural History, Series 8*, XX: 329–350.

Singh, L.A. K. (2000). Improved approach to tiger counting through pugmarks. Discussion based on power-point presentation at the Seminar: "Wildlife in the wake of the third millennium" Venue: Indian Museum Auditorium, Calcutta Date: 18 January 2000.

Singh, S.K., Mishra, S., Aspi, J., Kvist, L., Nigam, P. and Pandey, P. *et al.* (2015). Tigers of Sundarbans in India: Is the Population a Separate Conservation Unit?. *PLoS ONE*, 10(4): e0118846.

Sunquist, M.E., Karanth, K.U. and Sunquist, F. (1999). Ecology, Behaviour and Resilience of the Tiger and its Conservation Needs, Riding the Tiger: Tiger Conservation in Human Dominated Landscapes (Editors: J. Seidensticker, S. Christie and P. Jackson). Cambridge University Press, Cambridge, pp.5-18.

Swain, D. and Beheura, B. K. (2013). Melanistic Tigers of Similipal Tiger Reserve, *Tiger Paper*, 40 (1): 4.

Warndeep, H.S. (2004). Wildlife Conservation Strategies and Management in India: An Overview pp.12.

Wildlife Institute of India (2018). National Studbook of Bengal Tiger (*Panthera tigristigris*): IV Edition, Wildlife Institute of India, Dehradun and Central Zoo Authority, New Delhi. TR. No.-2018/34 Pages: 668.

Wildlife Odisha (2019). Wildlife Organisation, Forest and Environment Department, Govt. of Odisha, Bhubaneswar.

Wildlife Odisha(2018). Wildlife Organisation, Forest and Environment Department, Govt. of Odisha, Bhubaneswar.

Wozencraft, W. C. (2005). "Order Carnivora". In Wilson, D. E.; Reeder, D. M. Mammal Species of the World (3rd ed.). Johns Hopkins University Press. pp. 545–548.

Chapter 10

Sloth Bears in and Around Similipal Biosphere Reserve: Threats and Conservation

Prakash Mardaraj[1,2]* *and Janmejay Sethy*[2]

[1]*I.U.C.N/SSC- Bear Specialist Group, Po-Rajnilgiri, Balasore – 756 040, Odisha*

[2]*Amity Institute of Forestry and Wildlife, Amity University, Noida – 201 313, Uttar Pradesh*

**e-mail: pmardaraj@gmail.com*

Four species of bear, namely, sloth bear (*Melursus ursinus*), Asiatic Black Bear (*Ursus thibetanus*), Himalayan Brown Bear (*Ursus arctos*) and Malayan Sun Bear (*Ursus malayanus*), are found in India. Listed as Vulnerable by the IUCN, sloth bear is listed in Appendix I of CITES and is protected under Schedule I of the Indian Wildlife (Protection) Amendment Act, 2006. Sloth bear is endemic to the Indian Sub-continent and is found in India, Sri Lanka, Nepal and Bhutan. In India, it is distributed from the southern tip of the Western Ghats mountains to the foothills of the Himalayas. The quality of its habitat is determined by the availability and seasonal variations in food, shelter, and vegetation cover. Nearly 30 districts, in the state of Odisha, the distribution of sloth bear are recorded in 12. Out of 50 forest divisions in the state, bear presence has been recorded from 44 forest divisions (including wildlife and territorial divisions). Bear occurs in 14 out of 19 protected areas of the state. Sloth bear is a key species of the hills and mountains of the dry deciduous region of eastern India. The majority of sites where human sloth bear conflicts occurred were around the villages adjoining the Similipal. Human mauling cases were reported from 7 out of 11 fringe ranges of Similipal wildlife sanctuary. It is found that dense forest habitat, particularly forests along escarpments is more used by sloth bears in the study area and thus it is more crucial for bears; on the other hand the same habitat is used by humans for various purposes like MFP collection, cattle herding, local transport *etc.* This situation has been the major cause of bear human conflict in the study areas. Conservation threats to the sloth bear population such as loss, degradation and fragmentation of habitats and retaliation killing have discussed.

Keywors: *Sloth bears, Human animal conflict, Habitat fragmentation, Similipal, Conservation.*

INTRODUCTION

The bears are one of the most diverse groups of large mammals. The bear species currently exist in more than 60 countries on four continents (Servheen, 1990). Family Ursidae is a small group of mostly large mammals, with 8 species in 5 genera (Ursus, Tremarctos, *Melursus*, Helarctos, and Ailuropoda). Although Ursidae is not diverse, species in this family are widespread and culturally significant to human populations throughout their range. (Flynn *et al.*, 2005 and Vaughan, Ryan, and Czaplewski, 2000) The members of family Ursidae do not occur in Africa, Madagascar, Australia, various oceanic islands, and Antarctica, except for polar bears, inhabiting the Arctic region (Nowak and Paradiso, 1983). There are significantly more bears in the northern hemisphere than in the southern hemisphere. The spectacled bear of South America is the only bear found predominantly in the southern hemisphere.

The Sloth Bear (*Melursus ursinus*) also known as lip bear, as endemic to the Indian subcontinent, has entirely tropical and subtropical in its distribution (Yoganand *et al.*, 2006). Most tropical regions are impacted by expanding human population and accompanying socioeconomic inequality under development. The sloth bear range has shrunk in recent times, and the populations have become fragmented of the existing population and their degrading habitat is necessary for the long term conservation of the species. The Sloth bear is listed as "vulnerable" by the IUCN (1996), listed under appendix I of CITES protected under schedule I of the Indian Wildlife (Protection) Act of 1972, and under the fauna and flora protection Ordinance of Sri Lanka. Sloth bears are typically black, but sometimes have a blackish brown tinge at the coat. Rarely fully brown individuals have been observed (Pocock, 1941, Prater, 1980, Brander, 1982). A, V or U shaped whitish buff coloured breast patch is present, although lacking in rare cases as in some Sri Lankan individuals (Pocock, 1941). Pocock distinguished two races; the first *Melursus ursinus ursinus* occurring in central India and second the *Melursus ursinus inornatus*, found only in Sri Lanka. The Sri Lankan race has much shorter hair on the body, making it appear less shaggy. Also, the race *inornatus* is smaller in general dimensions, including the dimension of teeth (Erdbrink, 1953). Since Sloth bear is endemic to the Indian sub-continent, it is found in India, Nepal, and Sri Lanka in most low altitude, non-arid areas where forest cover still remains. In India it occupies a wide range of habitats on the mainland including wet or dry tropical forest, savannas, scrubland, and grasslands (Joshi *et al.*, 1995, Akhtar *et al.*, 2004). It ranges from the southern tip of the Western Ghats, mountain ranges to the foothills of the Himalaya. It`s western distribution is limited by the desert of the Rajasthan (Yoganand *et al.*, 2006). To the East the range is bounded by the wet forest of the Naga Hills of the north eastern India. The Himalayan foothill is ranges overlap slightly with the range of the Asiatic Black bear (*Ursus thibetanus*). In the North-Eastern portion of its ranges in the state of Assam and Mizoram, it may overlap with the range of Malayan sun bear (*Ursus malayanus*). Sloth bears subsist primarily on termites, ants and fruits. This is the only species of bear adapted specifically for myrmecophagy (ant and termite eating). Ward and Kynaston, 1995 and Middleton, 1996 said that the Sloth bears are found mainly in the warmer area. They use the "day den" after spending time for nocturnal foraging. An important factor affecting the behaviour of the wild animals living in human use landscapes is the extent of human activity in their habitats. The human animal interactions mostly manifest as the conflict rather than being compatible. Conflict resolution may influence the survival of the species and therefore has crucial conservation implications. Wildlife management itself is frequently is about managing these conflicts that occur at various levels. The study of animal behavioral ecology cannot be completed without dealing with the question. Now a day the conflict is the great threat to the wild animal.

Sloth bear are widely distributed in India. But now due to continuous habitat loss, fragmentation and degradation and poaching, Sloth bear population has drastically decreased in the wild. Sloth bear are facing the problem of its poaching and habitat degradation by the increasing human population (Kemf *et al.*, 1999; Garshelis *et al.*, 1999b; Servheen, 1990 and Cowan, 1972). The increased human population altered forests and harassed bears and thereby resulted in a decrease of their populations (Cowan, 1972 and Schoen, 1990). Sloth bear population outside the protected areas has declined significantly because of dramatic decrease in forest cover and poaching within the last 50 years (Garshelis *et al.*, 1999). (Santipillai and Santipillai, 1990), (Khan, 1982) and (Servheen, 1990) have reported that Sloth bear population is decreasing in India, Srilanka and Nepal, and they are facing precarious future. The Sloth bear has been reported to be disappeared from many forest areas of its former range (Krishnan, 1972). A serious threat to bears in Asia and America is the killing and capture of bears for use of their parts in primitive medicine, aphrodisiacs, or as pets. In a rough estimate the total world population of Sloth bears is in-between 10000-23000. Garshelis *et al.*, 1999, conducted a survey in India during 1993-94 and estimated the median density of 12 bears/100 sq.km. in 23 protected areas and showed an estimate of about 5000 Sloth bears within the protected areas.

Sloth bear population in Indian sub-continent is declining over a past century, primarily due to loss of habitat due to fragmentation and human encroachment. Besides this, conflict with human was considered a threat in over 40 per cent of all Sloth bear areas (Yoganand *et al.*, 2006). The conflicts become very high and frequent when the bear population is abundant in a smaller forest patch. Conflicts are also found to be more acute with the degree of use of the same habitats by human. The Quarries are extremely damaged the Sloth bear environment and play a vital role in influencing Human Sloth bear conflicts (Mardaraj and Dutta, 2011). In fact, that there is increasing biotic pressure, encroachment on the forest land and continuous habitat degradation, the sloth bear populations have become overabundant and showing an increase in conflicting trend. Likely that the sloth bears have now started straying out from these sanctuaries more and more into cultivation areas and human habitations, which results is serious human-sloth bear conflict in form of human casualties and agricultural crop raiding. Eventually over time, the status of the sloth bears is not only is critical in these areas, but it may also lead to more conflicting situation, which might be directly affecting the conservation efforts. The sloth bears come out in the evening or after sunset, to hunt for food all night, and retire in the morning (Bargali, 2004).

This kind of observation is also noticed in Nilgiri Range may be because of the continuous degradation and fragmentation of the habitat, available habitat is not suitable, and the food is also scarce for sloth bears in Nilgiri (Mardaraj and Chauhan, 2011). The change in the predatory behaviour of sloth bear appears to be due to the increasingly degraded and fragmented nature of the habitat, and a shortage of natural food for bears in the Nilgiri Range (Mardaraj *et al.*, 2012). It is found that dense forest habitat, particularly forests along escarpments is more used by sloth bears in the study area and thus it is more crucial for bear; on the other hand, the same habitat is used by humans for various purposes like MFP collection, cattle herding, local transport, *etc*. This situation has been the major cause of bear human conflict (Mardaraj and Nayak, 2014).

Figure 10.1. Grassland of Similipal Biosphere Reserve.

Similipal

From the hoary past till the present, throughout the annals of the history of Mayurbhanj, many a chronicle and document referred to Similipal extensively at its various forms. At the rise and fall of Mayurbhanj in its glories and glooms, Similipal always remained as epicenter of activities. It was witnessed as a mute spectator to the political development of Mayurbhanj; it shaped and groomed the economy and the culture of the people. Similipal at its fullest version is no less than an organic whole. For common man of Mayurbhanj, Similipal is a symbol of reverence and awe to them. It has the appeal of the religious sanctity, a coherent bend of cultural assimilation and a perennial source of livelihood. Hence, Similipal, is not only hills and valleys, grasslands and meadows (Figure 10.1) streams and rivers, ravines and waterfalls (Figure 10.2), but a bewildering panorama of many hundreds of millions of trees, depicting a huge and mammoth canopy bloomed in green veils; it is more than a worldly endeavor, as it represents the heartthrob and the emotions of the people of Mayurbhanj. The biodiversity of Similipal reserve forests with its typical ecosystems, is the benevolent gift of nature to living beings. The regulatory and functionary role of the massif in maintaining the ecological balance and regulating the ecosystems suffer from temporary setback and dis links caused by the biotic factors especially by fire, grazing, poaching, clear felling and frost. The nifty massif covered by forest growth stands as a barrier to the tropic of cancer on the north eastern edge of the Decan Plateau.

Figure 10.2. Riverine Habitat of Similipal.

Similipal Biosphere reserve (SBR) which is located in the Mayurbhanj district in the state of Odisha. SBR comprises the entire Similipal sanctuary and the proposed National park. The state is covered with 57,183 km^2 of forest area, out of which about 12.65 per cent fall under a network of two national parks and 18 sanctuaries. Similipal sanctuary falls under latitude 210.20′ to 10.40′ north and longitude 840.10′ to 840.30′ east. Sanctuary was notified in the year 1979, the length of periphery of Similipal Sanctuary is about 600 kms. Similipal contributes 38 per cent of the total area of the Protected Area (PA) network in Orissa (Singh, 1997). Similipal hill range forms a part of the northern plateau and occupies the central part of Mayurbhanj district in Orissa. The SLS has twelve ranges falling into three forest divisions, namely: Baripada, Karanjia and Rairangpur. Similipal is an ecosystem entirely with forest vegetation (mainly sal trees). The park covers total area of 2,200 sq km. The average elevation is 559.31 meters (1,835.0 ft). However, the entire Similipal area is undulating, rising from 600 meters (2,000 ft) to 1,500 meters (4,900 ft). The landscape of Similipal is comprises of numerous rolling hills covered with thick forest vegetation. It is the richest watershed in Odisha, At least 12 rivers cut across the plain area giving rise to perennial rivers such as the Budhabalanga, Khadkei, Khairi, Bhandan, West Deo, Salandi, East Deo, Sanjo and Palpala. These rivers are lifelines for the people of Mayurbhanj, Baleswar and Bhadrak districts. The high hills of Similipal are surrounding Meghasani. There are 64 villages inside the sanctuary and 61 villages in the buffer zone of the sanctuary. Kharias, a primitive tribe inside the sanctuary subsist on collection of non-timber forest products. Besides, people from outside also collect Although Gonds were living in few pockets in the dense forest of Similipal. Other common tribes are Bhumija, Bathudi, Kolha, Gonda, Santhal and Mankadia. The tribal population comprise of 73.44 per cent, the scheduled caste population 5.21 per cent and other castes 21.35 per cent. About 4.5 lakhs people constitute the 80,000 families who live within or close to Similipal, and about 50 per cent of them earn a part of or the complete livelihood from Similipal.

Sloth Bear Distribution

Sloth bears are patchily distributed in the protected and unprotected forests in SBR. The adjoining forests falling in the Kendujhar district and reportedly contain sloth bears and are contiguous with forests of Karanjia Division. The majority of sites where human sloth bear conflicts occurred were around the villages adjoining the Similipal Sanctuary. Human mauling cases were reported from 7 out of 11 fringe ranges of Similipal wildlife sanctuary. 11 Range areas of protected and non-protected forests in the district were surveyed to know the presence of bear (Mardaraj and Nayak, 2014).

Figure 10.3. Sloth Bear in its Natural Habitat.

The maximum entropy model for the potential distributions of Sloth bear in the Similipal Biosphere Reserve shows about 10 per cent of Similipal is suitable for sloth bear habitat. It has been found that the Sloth bear is very intensively distributed in the south-western part of Similipal biosphere reserve. The northern and the eastern part has an uninhabitable climate for the sloth bear and also found to be tigers core habitat region. Some what the core region provides some suitable habitat for the bears. Various environmental variables may be contributing to this spatial distribution of the Sloth bear. As the environmental variable that is the Bio 2, the mean diurnal temperature range has a great effect on the distribution pattern of sloth bear whereas the Bio 3, Isothermality has less effect on the distribution pattern. Any certain change in major contributing environmental parameters can leads to great fluctuation in habitat as well as in distribution pattern (Jena and Nandi 2017).

Threats

The human activities observed in the Similipal Biosphere Reserve were domestic livestock grazing, illegal cutting of trees, poaching, akhand sikar (annual mass hunting), collection of non-timber forest products, collection of medicinal plants, and tourism. People affect biodiversity in both direct and indirect ways. Agriculture and animal husbandry alter the biological diversity by destroying or modifying the native biota. It is estimated that since 1950, the area of forest and woodland worldwide has decreased by 15 per cent, primarily in Africa, Asia and Latin America (Bhatnagar and Rao, 1997).

Domestic Cattle

Grazing is prohibited in the core area of Similipal Sanctuary. But incidences of grazing have been noticed around the villages in the core area, as well as at the other villages located just adjacent to the core area. A livestock population survey conducted during August 1997 in the four villages revealed that the livestock population consisted of cows, buffaloes and goats (Figure 10.4). These population fluctuate seasonally, with the lowest number of goats and sheep found during

Figure 10.4. Domestic Cattle Grazing Inside the Protected Area.

the rainy season, as villagers sell these animals to earn their livelihood. Cows and buffaloes are less productive in respect of milk. Their utility is more for manure and as draught animals. Excessive exploitation of forests has seriously diminished our forest resources. The illicit felling of trees is mainly confined buffer areas are more prone to illicit felling and illegal removal of timber. The situation in certain pockets near the buffer zone is quite alarming. All out efforts to minimizing illicit felling are being made by the respective forest.

Akhand Sikar

The tribal hunt on the day following Pana Sankranti (during mid-April) in Mayurbhanj district is widely known as Akhanda Sikar which is in vogue since Darbar Rule and is observed for one day. But the Akanda Sikar now-a-days spreads over more, now this unrestricted hunting occurs throughout the year, barring a few months during the rainy season. This change of practice is due to the strong anti-hunting measures undertaken by the park management during April. With indiscriminate killings of all types of game without discrimination, followed by picnics and merry makings. Previously, they simply killed animals, cooked and ate them inside the forest. But at present, consumerism and commercialism prompt them to sell their kills in the markets of the foothills in contravention of traditional practice. Another significant deviation in their practice is that of carrying the prey animals back to villages (Rout, 2008). To resist and combat the mass killing operation of Adivasis.

Figure 10.5. Akhand Shikar Gathering of Villagers (Photo credit- © newindianexpress.com).

Plant Collections

The Similipal Biosphere Reserve is an ideal habitat for approximately 600 plant species, representing 74 families in 34 compartments, which are used in the treatment of different diseases (Rout, 2005). Tribals living inside and outside the forests possess the knowledge to some extent about these medicinal plants and how they are used in treatments. Various ethnomedicinal plant parts have been documented by the people of the Mayurbhanj district for use against diseases. The bark of the Paja tree (*Litsea monopetala*) is collected by the locals (Figure 10.6) and selling them to the professional smugglers. The illegal and unscientific collection by local people, and over-exploitation by village vaidyas and local kavirajs has led to a drastic reduction in the number of medicinal plant species (Pandey and Rout, 2002).

Figure 10.6. Collections of Bark of the Paja Tree (*Litsea monopetala*) by the Villagers.

Fuelwood Collection

However, often the human pressure on forests is not sustainable and can cause forest loss and degradation. Fuelwood is a major source of energy for rural households in India and other developing countries (Cecelski *et al.*, 1979). Wood harvesting is widespread because wood is abundant, inexpensive, and easily available in protected forests. As compared to other sources of energy such as electricity or LPG, its abundance makes it a reliable source of energy. Alternatives to fuelwood are difficult to put into practice. Complicated renewable energy devices are not easily accepted in the rural areas (Arjunan *et al.*, 2006). Villagers residing around the Similipal Biosphere Reserve depend on the forest for firewood as their principal source of energy, particularly in rural areas and among the below poverty level groups of people in the villages. Even if cooking gas is affordable by the people but they depend on forest as it is free of cost, firewood is collected mostly by unregulated commons and, as a consequence it is very sensitive to overexploitation as the population increases. Illegal felling of tree, cutting of bushes for the fuel wood for the nearby villagers play a main role in habitat loss.

Timber Wood Collection

Excessive exploitation of forests has seriously diminished our forest resources. In the core area of Similipal Tiger Reserve, the illicit felling of trees is mainly confined to two economic species, namely *Dalbergia latifolia* and *Pterocarpus marsupium*, and the felling intensity is not high except in the north-eastern fringe areas. But buffer areas are more prone to illicit felling and illegal removal of timber. The situation in certain pockets near the buffer zone is quite alarming. All out efforts to minimize illicit felling is being made by the respective

Collection of Non-timber Forest Products (NTFP)

The tribe in and around the sanctuary, subsist on the collection of non-timber forest products. Honey, gum, arrowroot and wild mushrooms are collected by them. The collection of the seeds and flowers of Mahua (Figure 10.7) and of sal seeds also cause disturbances in the area. Forests are deliberately burnt to clear the floor to collect these items. Besides timber and bamboo, non-timber forest products also play a major role in the day-to-day life of tribal in generating economic. However, in the core area, the intensity of disturbance is comparatively less than in the buffer areas. The percentage of income from this source increases with the poverty of a household (Patnaik, 1997). Although sloth bears have large and widely distributed food resources, availability of food items was low (Rajpurohi and Chauhan 1996). This was mainly because many important food items such as jamun, ber, tendu, bel, karvanda, mango and gunda; mahua flowers, shoots of bamboo, tubers and root and honey from the rock bee (*Apis dorsata*) and smaller forest bee (*Apis indica*) were also collected by the villagers as NTFP (Mardaraj and Udgata, 2019).

Figure 10.7. Burning of Dry Leaves for Collection of Mahua Flowers.

Sericulture

Sloth bear use fringe forest areas for foraging, but the areas are highly used by the villagers as plantation of white mulberry tree (*Morus alba*) for cocoon farming, agriculture, stone mining, *etc.* Thus, the sloth bear is forced to invade into human habitation in search of food thus resulting into conflicts. Tasar silk industry has directly and indirectly been providing gainful employment to family in rural and semi-urban areas. Mostly tribal people are engaged in the production of silk. Out of the total tribal population of about 38 million nearly 30 million (77 per cent) are located in tasar cultivating states. In recent times there has been fragmentation of forest landscapes. There is sporadic distribution of forests with tasar silkworm host plants like *Shorea robusta* and *Terminalia tomentosa*. The tropical tasar silkworm which has immense commercial value as it also provides bread and butter to the thousands of poor tribal families through its lucrative and invaluable silk (Achary *et al.*, 2015)

Figure 10.8: Sericulture in the Peripheral Villages.

Conservation Strategy

Factor leading to degradation and fragmentation of sloth bear habitats should be identified area-wise in and around the protected areas and strategies should be developed to remove these threats. Cattle grazing, illicit cutting and lopping of trees should be completely banned in and around the forest area. Local people venture into forest anytime of the day to collect non-timber forest produce, which may be of bear interest *i.e.* food plants. So, there should be restriction on collection of these forest products from the bear habitat. Villagers should be discouraged from collecting bear food items from the forest. This situation has been the major cause due to habitat loss. It is hereby suggested that such fringe habitats and forest patches should be identified and protected from humans' use. Further human interference should be limited up to some specific season and time of the day to reduce the bear human confrontation. These threats are the major risk not only to the bear population but also to the overall ecosystem leading to habitat loss and human infiltration in the whole Biosphere habitat. Restoration measures could include reducing the causes of degradation, facilitating regeneration of plants, and planting trees species including key food plants to augment cover and food for bears in those habitats. Establishment of strict protection in these ranges for regulating human infiltration and maintaining habitat may be a better step for the future of sloth bears. Unlike Sri Lanka, the cases of bear hunting and poaching are very less in Odisha but similarly a large population of tribal people and local villagers are depending on the collection of forest produces which can be regulated and managed.

REFERENCES

Akhtar, N. (2004). Habitat use, ranging pattern and management of sloth bear (*Melursus ursinus*) in North Bilaspur forest division, Madhya Pradesh. Ph.D. thesis, Wildlife Institute of India, Dehradun.

Achary, S.N., Nandi, D. and Kar. P. K. (2015). Status of Wild Tasar Silkworm Populations in Mayurbhanj Revealed Through GIS and Remote Sensing. Biotechnology for Sustainable Utilization of Bioresources. *Daya Publishing House* ISBN: 978-93-88982- 00-9 (HB) page 167-179.

Arjunan M., Holmes, C., Puyravaud, J.P.H. and Davidar, P. (2006). Do developmental initiatives influence local attitudes toward conservation? A case study from the Kalakad– Mundanthurai Tiger Reserve, India. *Journal of Environmental Management*. 79: 188– 197.

Bargali, H.S. (2004). The ecology of the problematic sloth bear (*Melurusus ursinus*) and mitigation of human bear conflict in Bilaspur forest division, Madhya Pradesh. Ph.D thesis, Wildlife Institute of India, Dehradun.

Bhatnagar, A. and Rao, R.J. 2004. Biodiversity Conservation in the proposed Amarkantak Biosphere Reserse-Central Asia. *Tiger paper* 31(4): 10-16

Brander, A.A.D. (1982). Wild animals in central India. Natraj Publishers, Dehradun, India. 296 pp.

Cecelski, B., Dunkerley, J. and Ramsay, W. (1979). *Household Energy and the Poor in the Third World. Resources for the Future*, Ramsey W. Mbi E. *FAO*, Rome (Italy). Washington DC.

Cowan, I. McT. (1972). The status and conservation of bears (Ursidae) of the world. International Conference on Bear Research and Management. 2: 343-367.

Erdbrink, D.P. (1953). A review of fossil and recent bears of the world. with remarks on their phylogeny based upon their dentition. Volume 1. Deventer-Drukkerij Jan De Lange. 1953. 2 vols. xii, 597 p.

Figure 3: https: //www.newindianexpress.com/states/odisha/2016/mar/19/Poaching-Foiled-in- Similipal-913451.html

Flynn, J., J. Finarelli, S. Zehr, J. Hsu, M. Nedbal. (2005). Molecular phylogeny of the Carnivora (Mammalia): Assessing the impact of increased sampling on resolving enigmatic relationships. *Systematic Biology*, 54: 317-337.

Garshelis, D.L., Joshi, A.R., Smith, J.L.D. and Rice, C.G. (1999b). Sloth Bear Conservation Action Plan. In Bears: Status Survey and Conservation Action Plan. Eds. Christopher Servheen, Herrero and Bernard Peyton. IUCN/SSC Bear Specialist Group, 225-240 pp.

Jena, P. S. and Nandi, D. (2017). Using maximum entropy modeling to delineate the distribution patterns of sloth bear (*Melursus ursinus*) in Similipal Boisphere reserve, Odisha, India. *International Journal of Current Advanced Research*, Volume 6; Issue 6; June 2017; Page No. 4226-4231

Joshi, A. R., Garshelis, D. L. and Smith, J. L. D. 1995: Home ranges of sloth bears in Nepal: Implications for conservation. *Journal of Wildlife Management* 59: 204–213.

Kemf, E., Wilson, A., and Servheen, C. (1999). Bears in the Wild. 1999. WWF Species Status Report. World Wide Fund For Nature, Gland, Switzerland, pp. 44.

Krishnan, M. (1972). An ecological survey of the larger mammals of peninsular India. *Journal of Bombay Natural History Society*. 69: 26-54.

Mardaraj, P and Chauhan, N.P.S. (2011). Factors influencing the human-sloth bear conflict in Nilagiri, Orissa, India. 20th International Conference on Bear Research and Management, July 17-23 in Ottawa, Canada.

Mardaraj, P and Dutta, S. K. (2011). Human-Sloth Bear Conflict in Balasore Forest Division, Eastern India, "A Case Study in Human Dominated Landscape", *LAP Lambert Academic Publishing GmbH and Co.* KG, 1-116 pp.

Mardaraj, P., Debata, S. and Swain, K.K. (2012). Sloth Bears Prey on Domestic Animals in Response to Habitat Degradation in Eastern India. *International Bear News*. 21(1): 18-19.

Mardaraj, P. and Nayak, A.K. (2014). Status and extent of human-Sloth bear (*Melursus ursinus*) conflict in and around Similipal wildlife sanctuary, Odisha, India. *Project Report*.

Similipal Tiger Reserve, Baripada, India. 42 pp.

Mardaraj, P. and Udgata, H. B. (2019). Assessment of Bear Human Conflict in Swarnachuda Reserve Forest, Balasore, Eastern India.

Nowak, R. and Paradiso, J. (1983). Walker's mammals of the world, Vol. II. John Hopkins Univ. Press, Baltimore, 1362 pp.

Pandey, A. K., Rout, S.D. and N. Pandit. (2002). Medicinal Plants of Similipal Biosphere Reserve - Perspectives of Plant Biodiversity. (ed. A.P. Das); Bishen Singh Mahendra Pal Singh, Debra Dun, 681-696.

Pocock, R.I. (1941). The black and brown bears of Europe and Asia. Part II. The sloth bear (*Melursus*) and the Asian black bear (*Selenarctos*) and the Malayan sun bear (*Helarctos*). *Journal of Bombay Natural History Society*. 36: 101-138.

Prater, S.H. (1980). The Book of Indian Animals. *Bombay Natural History Society*, Bombay.

Rajpurohit, K.S., and Chauhan, N.P.S. (1996). Study of animal damage problems in and around protected areas and managed forest in India phase I: Madhya Pradesh, Bihar and Orissa. Wildlife Institute of India, Dehradun, India.

Rout, S.D. (2005). Medicinal plants of Similipal Biosphere Reserve. Ph. D. Thesis, T.M. Bhagalpur University, Bhagalpur,

Rout. S. D. (2008). Anthropogenic threats and biodiversity conservation in S imilipal biosphere reserve, Orissa, India. *Tiger paper*. 35(3): 22-26.

Santiapillai, A and Santiapillai, C. (1990). Status, distribution and conservation of sloth bears (*Melursus* ursinus) in the Sri Lanka. *Tiger Paper* 13-15 PP.

Schoen, J.W. (1990). Forest management and bear conservation. Report of Division of Wildlife Conservation, Alaska 1-7.

Servheen, C. (1990). The Status and conservation of the bears of the world. *International Conference on Bear Research and Management Monograph*, 2. 32 pp.

Vaughan, T., Ryan, J. and Czaplewski, N. (2000). Mammalogy, 4th edition. New York: Saunders College Publishing.

Yoganand, K., Rice, C.G., Johnsing, A.J.T., and Seidensticker, J. (2006). Is the sloth bear in India secure? A preliminary report on distribution, threats and conservation requirements. Journal of the ***Bombay*** Natural ***History*** Society, 103 (2-3): 172-181.

Chapter 11

Migration, Movement and Human-Elephant Conflict in Peripheral Area of Similipal Biosphere Region, Using Remote Sensing and GIS

Debabrata Nandi, Sandeep Ojha, Tribeni Nayak and Snehalata Ghadei*

Department of Remote Sensing and Geographic Information System, North Orissa University, Takatpur, Baripada – 757 003, Mayurbhanj, Odisha
**e-mail: debabrata.gis@gmail.com*

India is home to 60 per cent of the Asian elephant population. According to Project Elephant, government initiative for elephant conservation, there were 28,000 elephants in the country in 2011. Man, animal conflict has been on the rise in the forest border areas with herds of wild s straying into human habitation. Till recently powerful searchlights, crackers, and drums to push the wild elephant herds back into the jungle were used. The Kumki (Rescue) elephants were used more in hill areas to chase the herds of wild elephants into the forests. Surveillance and tracking of these herds are difficult due to their size and nature of movement. The loss due to destruction in the farms is more. The elephants are also subject to attack by humans resulting in danger to the life of elephants. It is generally the borders of forests that are the focal points of conflict. Minimizing human elephant conflict to reduce the risk of life of both human beings and elephants is of utmost importance. Taking these problems in to consideration an early warning system to minimize the human elephant conflict in the forest border areas is proposed in this research work. The model developed helps mitigate such conflicts in two ways: By providing warning to people about the anticipated entry of elephants into human habitation. By providing advance information to the authorities to take to chase the pachyderms back to the forest. The objective of this research is to develop innovative solutions to the problems related to elephant monitoring in the forest border areas. Four methods are proposed in this work for predicting the presence of the intruding elephants in the human living areas using intelligent systems.

INTRODUCTION

The Asian elephants in India are distributed in four distinct geographical areas in five isolated populations. There are two populations in southern India, one each on either side of the Palghat gap. The other three populations inhabit the central, northeastern and northwestern regions of the country. The Asian Elephant (Elephas maximus), is categorized as Endangered on the IUCN Red List (Choudhury *et al.*, 2008. In India, the state of Orissa holds 70 per cent of the Elephant population in Eastern India (Patnaik *et al.*, 2008). The Mayurbhanj Elephant of Odisha includes the part of Similipal Tiger Reserve, Kuldiha Wildlife and Hadgarh Wildlife Sanctuary. ThisReserve also serve as corridor for interstate migration with adjoining Protected Areas of Jharkhand and West Bengal. Due to their migratory and feeding habit, which necessitates large continuous forest and large water bodies, it is rightly projected as vertical flagship species to manage large landscape. But illegal logging, mining, hunting is causing mass destruction of the pristine forests. Mining is the single largest threat to Elephants and their habitat within the state (Swain and Patnaik, 2002),

Peak depredation season were September to December and the most vulnerable time Morning and evening. Most human killing was by male elephants and depredation of crop by elephant herd due to trampling.

Elephant is one of the largest and wide ranging animals which migrate over long distances in search of food, water and use forest corridors for many hundreds of years. (Sukumar. R 1989) If these corridors are blocked, they enter human settlements. Elephants are also attracted to human settlements by crop or by domestic alcohol distilleries that they can smell from miles away. The resulting human-elephant conflict is a major threat to the survival of the species (Menon V 2003) Depredation of crops by wild elephants is common along the fringe villages of elephant habitats.

Human deaths due to attack by wild elephants have been reported by (Pradhan *et al.*, 2013., Veeramani *et al.*, 1996; Dey S.C.,1991; Swain, D., 2001) The present study aims to document the consequence of human-elephant conflict from Baripada division and Karanjia division of Mayurbhanj Elephant Reserve, which is one of the largest and important Elephant Reserve of the State due to regular arrival of migratory elephants besides the presence of large number of native elephants. The study had specific objectives to understand the elephant habitat composition and structure, movements and habitat utilization patterns, dietary spectrum and impact of elephant feeding on vegetation and to work out social organization of elephants in Baripada and Karanjia Divison of Mayurbhanj.

During the period of seven years from 2001-02 to 2007-08, a total of 365 human beings killed in elephant attack, which more or less shows an increasing trend with maximum deaths during 2006. Maximum of human deaths recorded during this period were mostly from Mayurbhanj district, followed by Sambalpur and Sundargarh. In our country, most wildlife species including Elephants are nearing extinction as expansion for agriculture and human settlements have taken over their habitats. All these pressures call for rational long-term solutions.

Study Area

The study area is conduct 2 division of the peripheral area of Simlipal biosphere region, which are Baripada Division, Karanjia Division.Baripada forest division

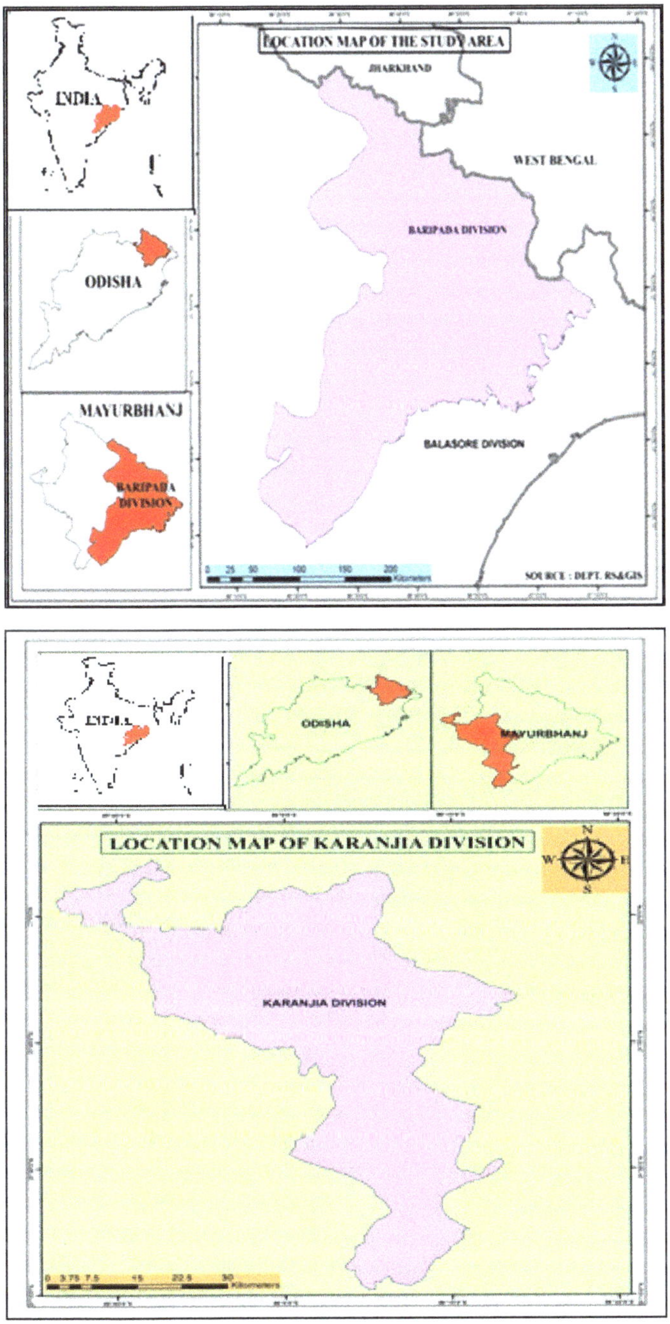

Figure 11.1. Location Map of Study Area.

is situated towards northern most part of Odisha between 220 33′ 45′ and 210 17 North latitude and 85045′ 30′ and 870 13′ 15′ east longitude and is bounded on the North by the Singhbhum Medinapur district of West Bengal on the south by Balasore district and Keonjhar district on the east. Total area of Baripada division is 4666.539753sqkm.As the division situated at the border of the state its division boundary touch the West Bengal (70km) and Jharkhand state (40km). Karanjia Division is located in the south- west part of Mayurbhanj district, the Karanjia forest division was mostly affected by HumanElephant Conflict. The highest number of crop depredations occurred in Gourigoda village (17 cases; 6.8 acres) but of the greatest area damaged was in Paliabeda (10 cases; 8.01 acres). This division consists of 4 ranges - Kaptipada,Udala, Pithabhata and Dukuda range.

Methodology

Field Investigation and Site Condition Monitoring

The field study was conducted between15th januawary2015 to 15th june2015, extensive field study was conducted in all the 3 division where the elephant and human conflict has more identify the interstate movement of elephants well as the intrastate movement during the study period. Public awareness among the local people plays a vital role during the study period. Also the access the crop damage, cause the house damage, behavior of the elephant, herd of the elephant composition also studied during the field survey.

Interviewing Local People

Local people of the affected villages including victims of conflict, forest staffs were interviewed through questionnaires to understand movement Pattern of elephant in the area, the way people were affected, their attitudes towards elephant and on different aspects of conflict.

Official Collection of Information

Data on human-elephant conflict 2008 to 2018 were collected from the local Divisional Forest Offices to make comparative study on human-elephant conflict in Baripada and Karanjia in the peripheral area of Simlipal biosphere reserve.Satellite images of the areas based on 1980, 1991, 1998 and 2004 were analyzed along with intense ground assessment to assess the current state of forest cover and loss of elephant habitat.

Landscape Element Types of Simlipal Biosphere Region

The landscape of Simlipal biosphere region was surveyed to classify the land cover. Different elements of the landscape were identified using differences in canopy structure, phonological characteristics, land use pattern and degree of disturbance in the landscape. Ground-truthing was carried out using a Global Positioning System (GPS) receiver. Multiple GPS points were collected from the different landscape elements (LSE) types. From each GPS location the following information were recorded: a) latitude and longitude, b) elevation, c) type of landscape element, d) details of terrain characteristics of the adjacent land.

Human-Elephant Conflict

Habitat loss, degradation, fragmentation, Conversion and resource exploitation due to human activities result in alterations of the extent and spatial configuration of habitats available for wild species. Wide-ranging animals, exemplified by species such as Asian elephants (Elephas maximus) and African elephants (Loxodonta africana) are threatened by many human activities. For Asian elephants, large scale conversion of forests to monoculture plantations, Croplands, and developed areas has drastically reduced and fragmented available habitats. This has resulted in compression of elephant herds in protected areas causing escalation of human elephant conflict in the adjoining human-dominated landscapes. Landscape variables such as spatial heterogeneity in vegetation cover, seasonal change in resource availability and human density, influence elephant distribution and use of natural and man-modified habitats, and patterns of human-elephant conflicts and encounters. Close and frequent encounters between humans and elephants in Landscape with settlements and croplands have detrimental effects on long-term conservation of elephants. Because of fragmentation and the ensuing interspersion of human habitations, Cultivation and natural habitats, conflicts between humans and elephants are now wide spreads across in Asia.

Conflict between man and elephant is not new.Rather it's as old as the human civilization itself. However, in today's world this issue has grabbed serious concern due to the fast decrease of elephant population worldwide. Each year human-elephant conflict results in about 300 human deaths and damage to 10,000-15,000houses and 810million hectares of crops while over 200 elephants death due to human related activities which include poaching for ivory or meat, poisoning, cattle-borne disease, electrocution and collision with train.

Results and Discussion

Human-Elephant Conflict in Baripada Division

About 36 villages were severely affected due to elephant in Baripada Division. Human death, elephant death, crop damage, house damage, harm to livestock, injury to human are the consequences of this conflict. Peak depredation seasons were October to December during this period migration of elephant occurred since last six years. Most human killings were by Solitary male elephants and depredation of Crop by Family groups. Baripada Division is one of the major places for migratory elephants from West Bengal and Jharkhand.

Description of HEC

From the year, 2008 to 2018 collected data from field level and from the Divisional forest office. Crop damage was the most common incident during the study period. Elephant damage more than different type of cultivation plants which Banana, Paddy, were most common. In Baripada forest Division Fuljhari, Saria, Nadapur are most affected and Vulnerable Villages due to interstate migration of elephant migrated from dalma (Jharkhand state).

Human Death and Human Injury

A total 21 human Death and 19 human injury occurred from 2008 to 2018.All of them are accidental as the victim came face to face with wild elephants. Human Death and Human injury also occurred adjoin to the Reserve forests where the elephant shelter during the daytime at the same time the local people entered into the forest to collect the Non-Timber forest products. Most victims are Woman, older ages and some person in drunken stage. In Betnoti Range elephants attacked in the search of the houses where people storage the paddy crops, Mahua and Handia.

Crop Damage and Compassionate Paid

From 2008 to 2018, total 3236.36 acre of crop Damage occurred in Baripada forest Division. Mostly affected ranges were Rasgovindapur,Betnoti and Deuli Range where the movement of Migratory elephants occurred from last five years. Similarly Total compassionate amount paid 21518400 during the study period and rate of amount paid gradually increase from the year 2009 where the incidents of migration of elephant frequently occurred.

Affected Villages of Baripada Division

About 36 villages were severely affected due to elephant in Baripada Division. Human death, elephant death, crop damage, house damage, harm to livestock, injury to human are the consequences of this conflict.

Table 11.1. Consequence of Human-Elephant Conflict in Baripada Division

Year	*Man Killed by Elephant*	*Human Injured by Elephant*	*Crop Damage by Elephant*	*Compassionate Paid*
2008	01	01	143.64	245640
2009	03	-	190.75	490750
2010	03	01	939.48	1241480
2011	03	-	647.83	947830
2012	01	-	328.85	528850
2013	01	01	283.53	3116020
2014	05	12	-	-
2015	3	19	-	-
2016	2	10	-	285674
2017	1	6	-	-
Total	**23**	**48**	**2534.08**	**6327394**

Causes of Conflict

Though there are different causes behind human- elephant conflict but the three main causes of conflict are discussed below.

Habitat Loss

Habitat loss is one of the major problems leading elephant movement towards human habitation. This problem is related to the increase in India's human population. Biotic pressures from such as a large population have led to the loss of forests there by loss of sufficient fodder for this large species. Irrespective of the suitability of the area for permanent agriculture, degradation of such lands into wastelands happens due to defective agricultural practices and absence of appropriate soil conservation measures. This has further increased the dependence of the local people and their cattle on the forests and is ultimately accelerated throughout the elephants' range in the country.

Alteration of Corridors

Over a few decades the vast elephants, ranges have been fragmented and precariously connected by narrow corridors due to expansion of human habitations and agriculture, indiscriminate growth of various developmental activities, severe biotic pressures *etc.* Forest corridors can be defined as the narrow strips of forests connecting two habitats that facilitate major functions such as exchange of genes between populations, dispersal, provide access to variety of seasonal foraging grounds and prevention of faunal collapse. Elephant corridors have received lot of attention in conservation and are widely used in devising conservation strategies especially in recent years, with the reduction of Contiguous habitats into islands. To address the issue and for long-term conservation of elephants, it is important to maintain viable populations with in viable habitats and this could be maintained by linking the fragmented habitats by protecting and strengthening the existing corridors. But these corridors are lost day by day due to increase anthropogenic pressure. In Baripada Forest Division one corridor, *i.e.* Deuli-Suliapada-West Bengal plays an important role of migration and movement of elephant from west Bengal to Odisha.

Poaching

According to estimation, Ivory poachers between 1980 and 1986 killed about 100 male elephants in the country every year. The poachers, which results in elephant death, targeted the tusker elephant of the migrated elephants. The levels of poaching in south India seem to have reached unacceptable level during the past few years. In the states of Karnataka, Kerala and Tamil Nadu about 100-150 male tuskers were shot annually during 1980-83 (Sukumar 1985).

Human-Elephant Conflict in Karanjia Division

About 45 villages were severely affected due to elephant Karanjia Division. Human death, elephant death, crop damage, house damage, harm to livestock, injury to human are the consequences of this conflict. Most human killing by depredation of Crop by Family groups. Karanjia Division is also one of the major places for migratory elephants from West Bengal and Jharkhand.

Description of HEC

From the year, 2008 to 2017 collected data from field level and from the Divisional forest office. House damage was the most common incident during the study period. Elephant damage more than different type of cultivation plants which

Banana, Paddy, were most common. In Karanjia forest Division Thakurmunda, Satkusia and Kendumundi Vulnerable Villages due to interstate migration of elephant migrated from Jharkhand state.

Human Death and Human Injury

A total 20 human Death and 20 human injury occurred from 2008 to 2017.All of them are accidental as the victim came face to face with wild elephants. Human Death and Human injury also occurred adjoin to the Reserve forests where the elephant shelter during the daytime at the same time the local people entered into the forest to collect the Non-Timber forest products. Most victims are Woman, older ages and some person in drunken stage. In Satkusia Range elephants attacked in the search of the houses where people storage the paddy crops, Mahua and handia.

Human Death and Human Injury

A total 20 human Death and 20 human injury occurred from 2008 to 2017.All of them are accidental as the victim came face to face with wild elephants. Human Death and Human injury also occurred adjoin to the Reserve forests where the elephant shelter during the daytime at the same time the local people entered into the forest to collect the Non-Timber forest products. Most victims are Woman, older ages and some person in drunken stage. In Satkusia Range elephants attacked in the search of the houses where people storage the paddy crops, Mahua and handia.

Table 11.2. Consequence of Human-Elephant Conflict in Karanjia Division

Year	*Man Killed by Elephant*	*Human Injured by Elephant*	*Crop Damage by Elephant*	*Compassionate Paid*
2013	01	01	98.31	3116020
2014	05	01	247.73	-
2015	02	07	145.78	-
2016	01	09	-	285674
2017	-	04	135.67	-
Total	9	22	627.49	3401694

Movement and Migration of Elephant

Movement of elephants is considered one of the most important ecological factors, which influence the distribution of other small herbivores. Elephants travel Long distances as part of their migration activities and at the same time, they stay within different forest habitats that are enriched with water and fodder.

Elephant Movement

An elephant is a sensitive and intelligent animal and likes to move in free and quite environment. Whenever the herd finds any obstacle on the migratory or movement track, it tries to avoid the route and sometimes even adopt a longer path to fulfill their need. They generally move very cautiously in the group to protect the younger elephants. They normally move in herd in a spreader pattern but remain in visible range. Baby elephants generally stay under the mother and whenever they come across any steep ditch or obstacle, the mother lifts the baby elephant with her trunk. In general, there is an intense bond and love between the elder elephants and the younger ones. Sometimes, group clashes among the elephants also results in death. The elephants consume 75-150 kg of food and 80 to 160 liter of water every day. Their food consists of mostly grass, tender shoots, twigs, barks, leaves and fruits.

Migration of Elephant

An elephant is a long-distance migratory animal. Elephants devote their three Fourth life in moving in search of water and fodder. In general, they stay in an area for a few days and migrate to other places. Elephants like to take bath daily in summers and that too in deep waters. It has been observed that during the Warm weather, groups of elephants try to reach water source in the noon or after -Noon. Since the availability of water and fodder changes seasonally in the park, the elephants migrate in the park to even shorter distances in search of water And fodder. The purpose of this migration is to get water, fodder and a mating partner. They normally move in herds in a spread pattern, but within visible range, often using torrents or natural depressions as their path of movement.

Movement and Migration of Elephant in Baripada Division

Normally in Baripada Division elephant of Jhrkhand (Dalma Wildlife santuary) use to migrate through the corridor that extends from West Bengal-Deuli- Suliapada. From the year 2005, there Is deviation noticed in their normal track, changing their route from Rasgobindpur (Baripada forest division) – Nilgiri of Balasore Wildlife Division through Morada- Chitrada (Rasgovindpur Range) to Asanbani – Jugal-Dalki of betnoti range. Dalma wildlife sanctuary is at a distance of 10 km south of Jamshedpur in Jharkhand state, having an area of 193 sq. km. It carries around 158 Elephant population (as per 2007 census), which is beyond the carrying capacity of that sanctuary. Since last ten years herd of more than 100 elephants entered to Baripada Division area through Deuli and Rasgobindpur range. They use to stay in Baripad Forest Division for at least 15 to 22 days through Deuli-Rasgobindpur-Betnoti ranges raiding standing paddy and other cash crops, damaging houses, injuring human beings during their onward journey to Nilgiri Range of Balasore Wildlife Division. Migratory route and forest used by elephant in Baripada division are Phuljhari forest, Dalki RF, Nadapur forest and Asanbani Forest. The most vulnerable area in Baripada division is Sarai under Betnoti Range and Baladia under Rasgobindhpur range. In Nilgiri area of Balasore WL Division, elephants usually raid mostly paddy crop fields and cause house damage in search of paddy, rice and liquor. The vulnerable villages are Routraipur, Pundal, Kathapal, Patina, Gopalpur, Ayodhya, Hatimunda, Ambajhar and adjoining villages.

Movement and Migration of Elephant in Karanjia Division

Karanjia Division is one of the major migration areas for elephants of Keonjhar Division. Elephants frequently migrate from Joda and Badbil of Keonjhar Division to Karanjia Division for several reasons such as developmental activities going on

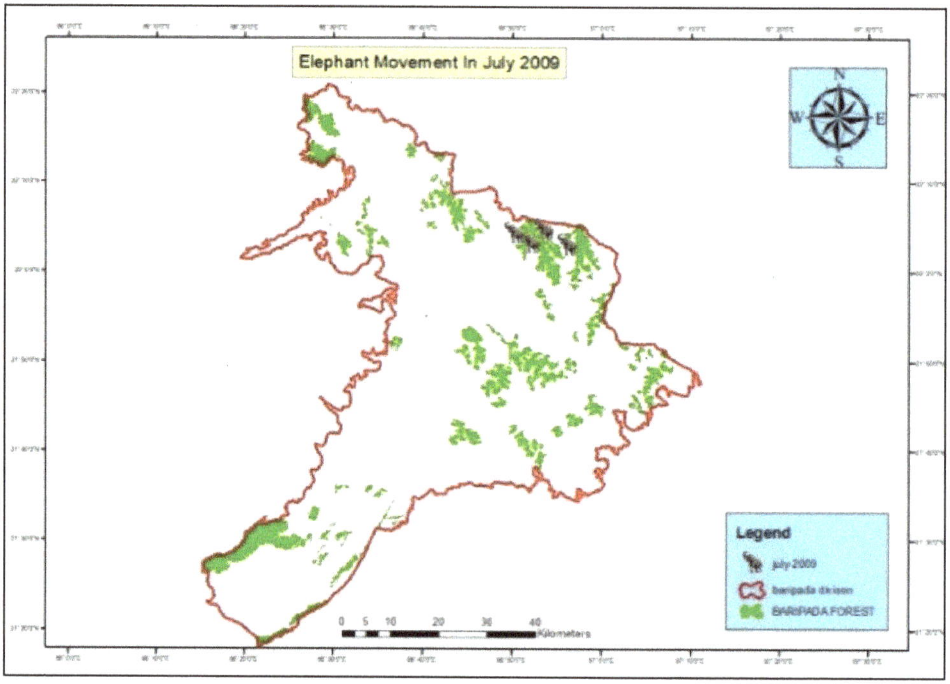

Figure 11.2. Elephant Migration and Movement of Baripada Division during August, 2008.

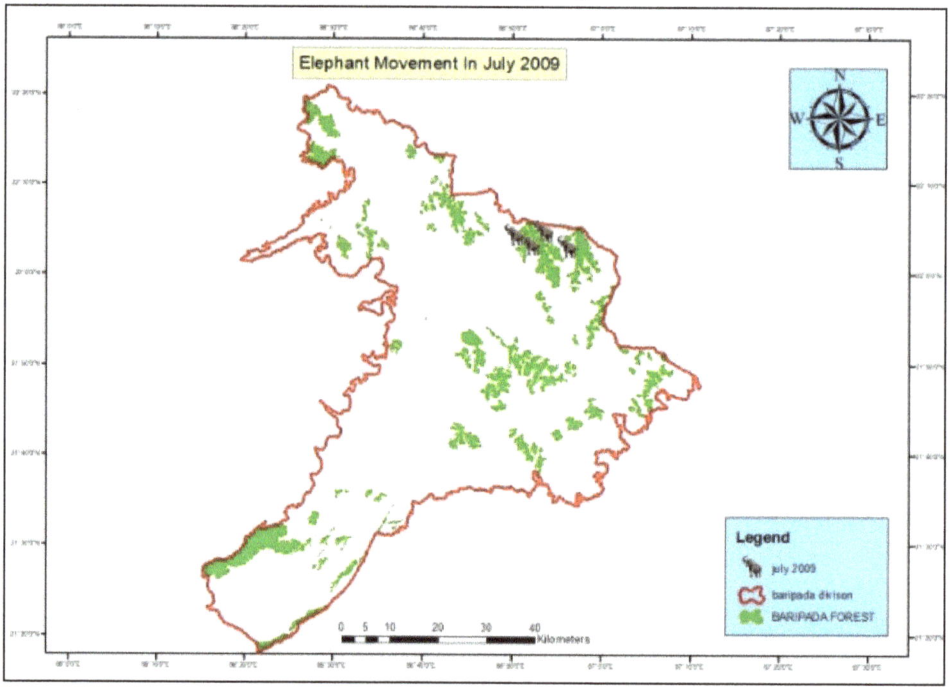

Figure 11.3. Elephant Migration and Movement of Baripada Division during July, 2009.

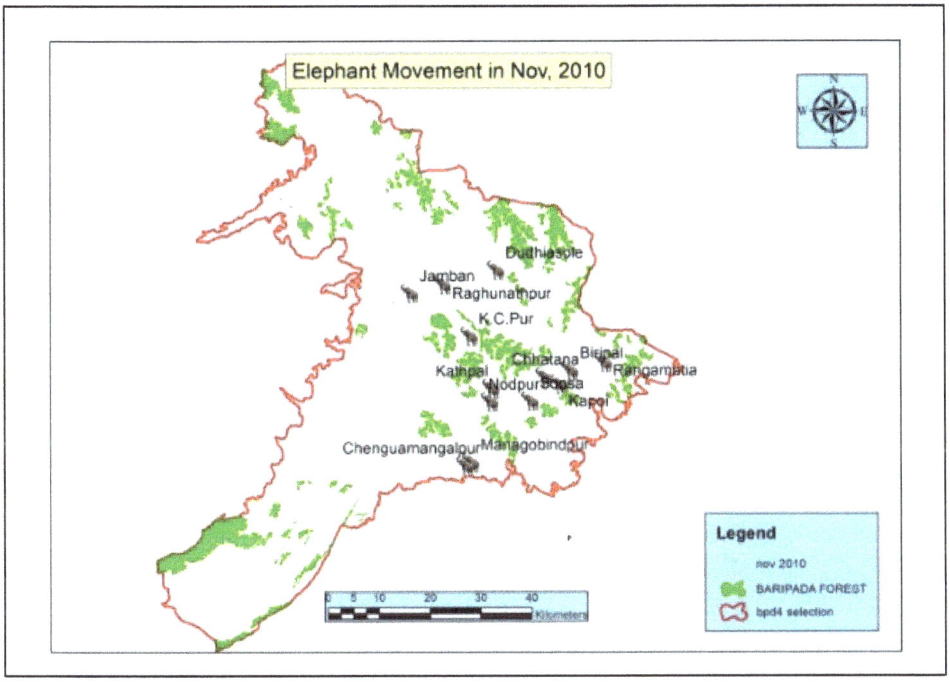

Figure 11.4. Elephant Migration and Movement of Baripada Division during Nonvember, 2010.

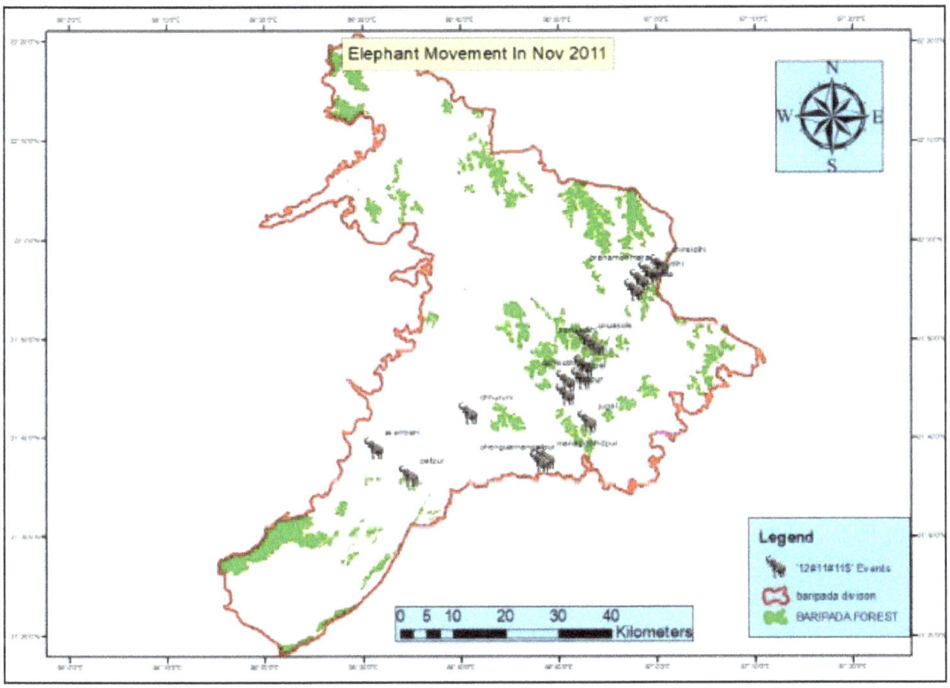

Figure 11.5. Elephant Migration and Movement of Baripada Division during November 2011.

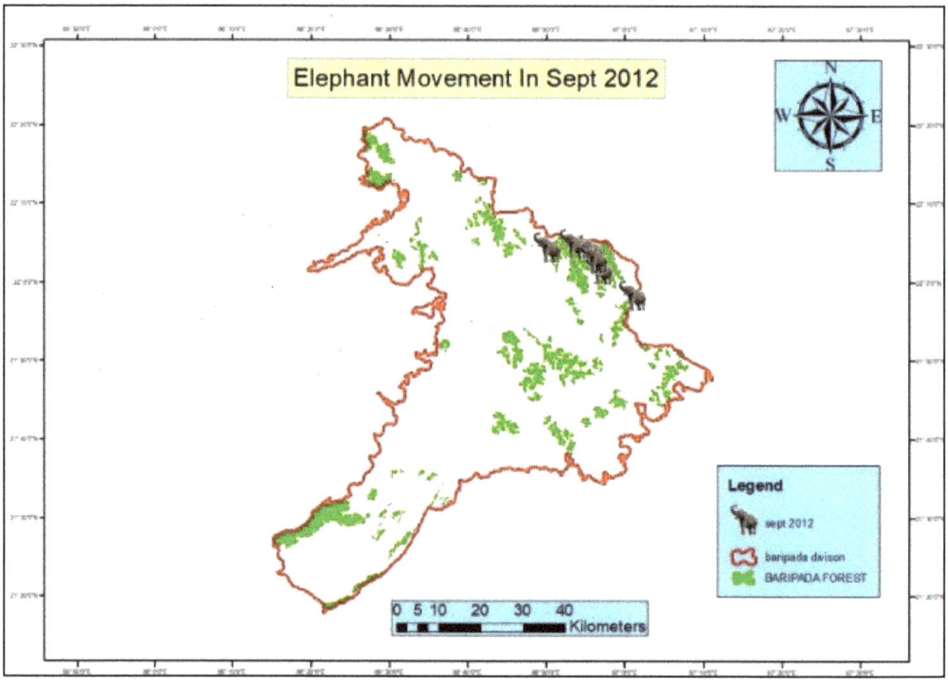

Figure 11.6. Elephant Migration and Movement of Baripada Division during September 2012.

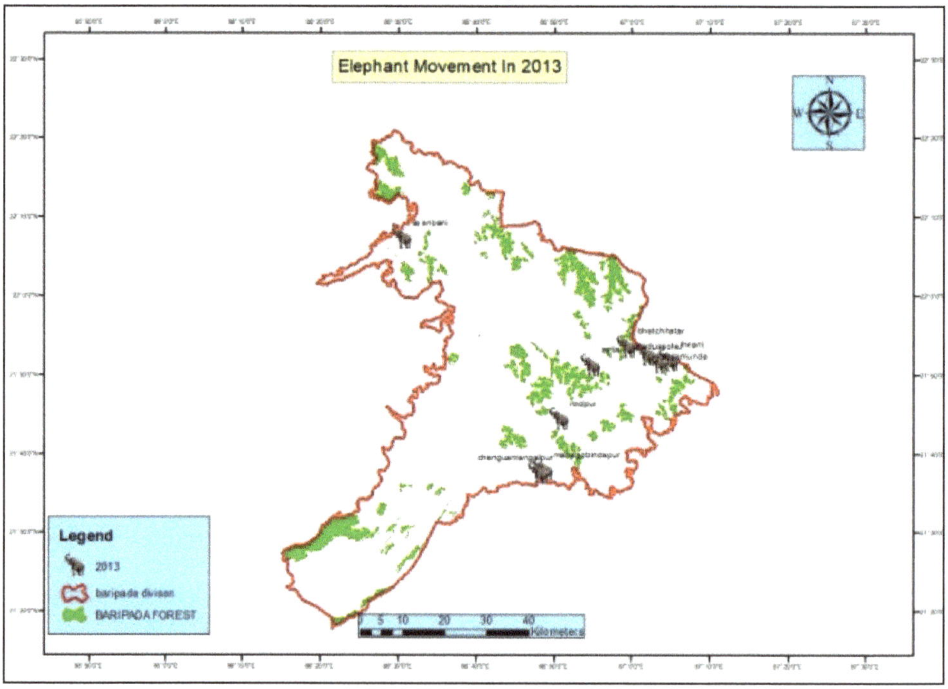

Figure 11.7. Elephant Migration and Movement of Baripada Division during Year 2013.

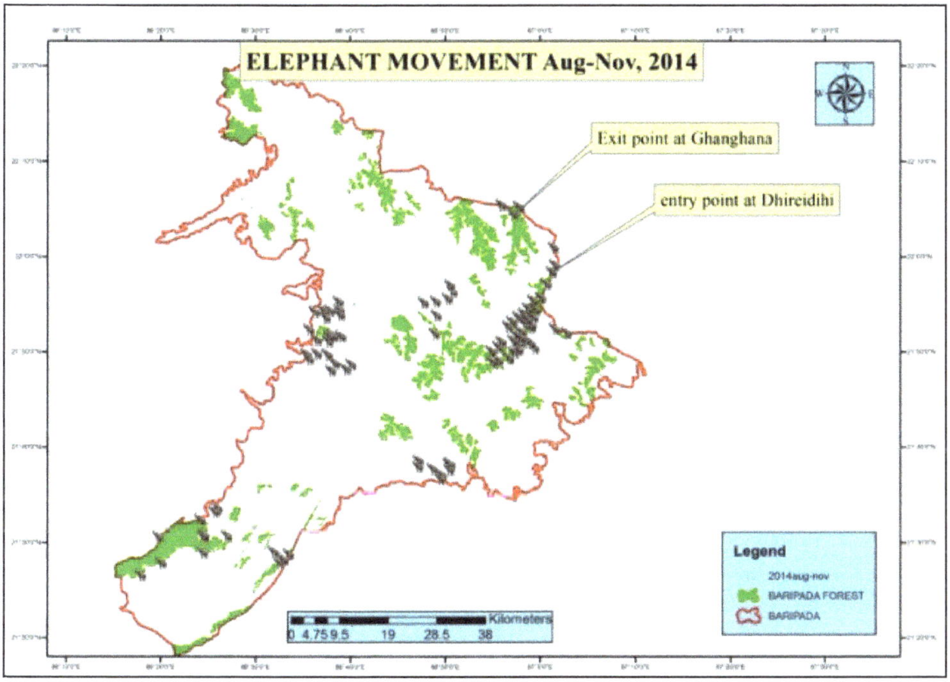

Figure 11.8. Elephant Migration and Movement of Baripada Division during Aug-Nov 2014.

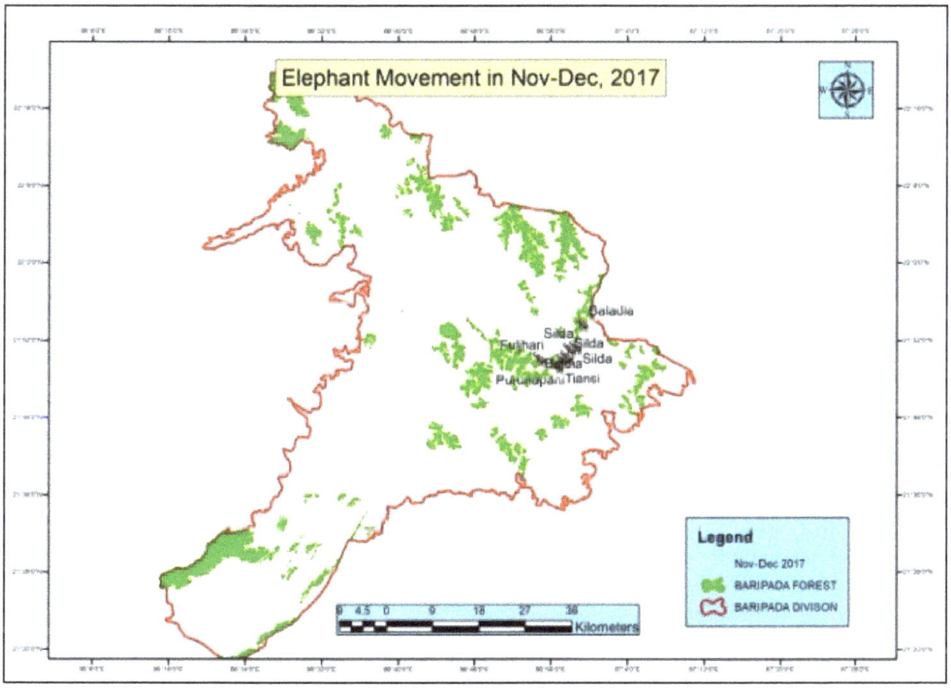

Figure 11.9. Elephant Migration and Movement of Baripada Division during Nov-Dec 2017.

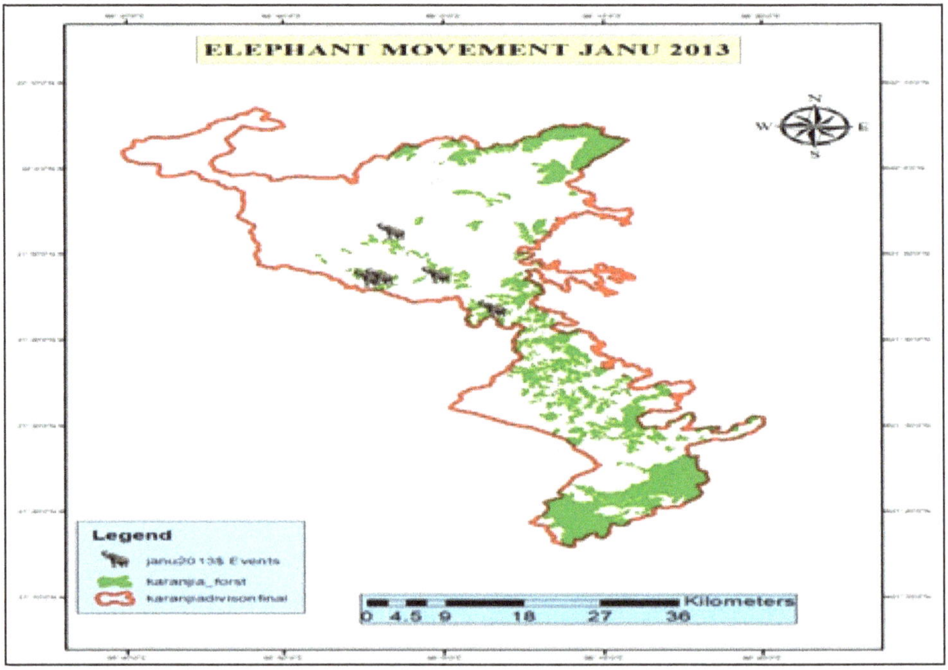

Figure 11.10. Elephant Migration and Movement of Karanjia Division during Jan, 2013.

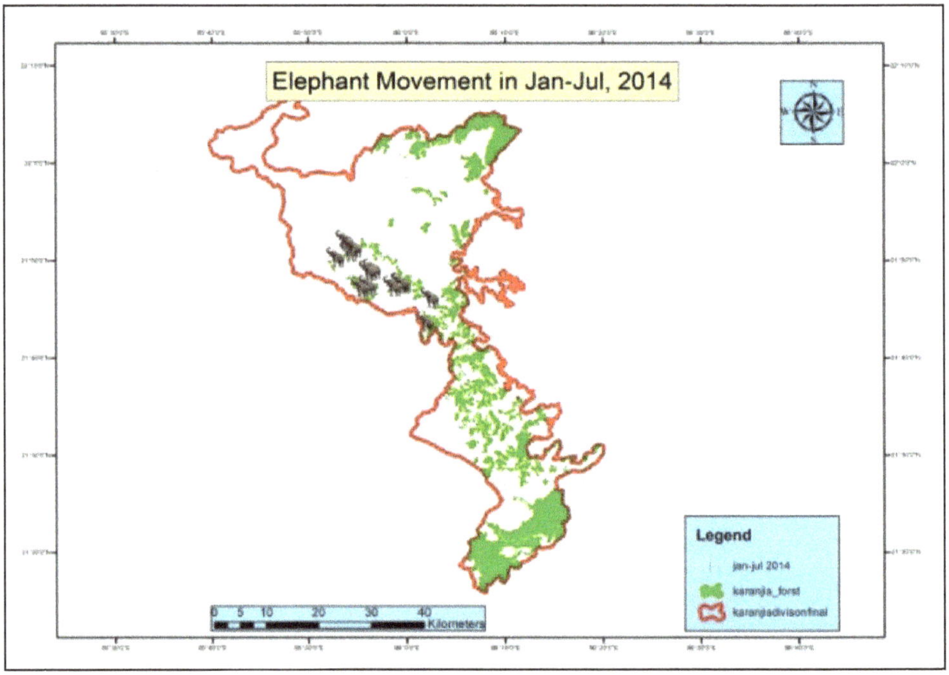

Figure 11.11. Elephant Migration and Movement of Karanjia Division during Jan-Jul, 2014.

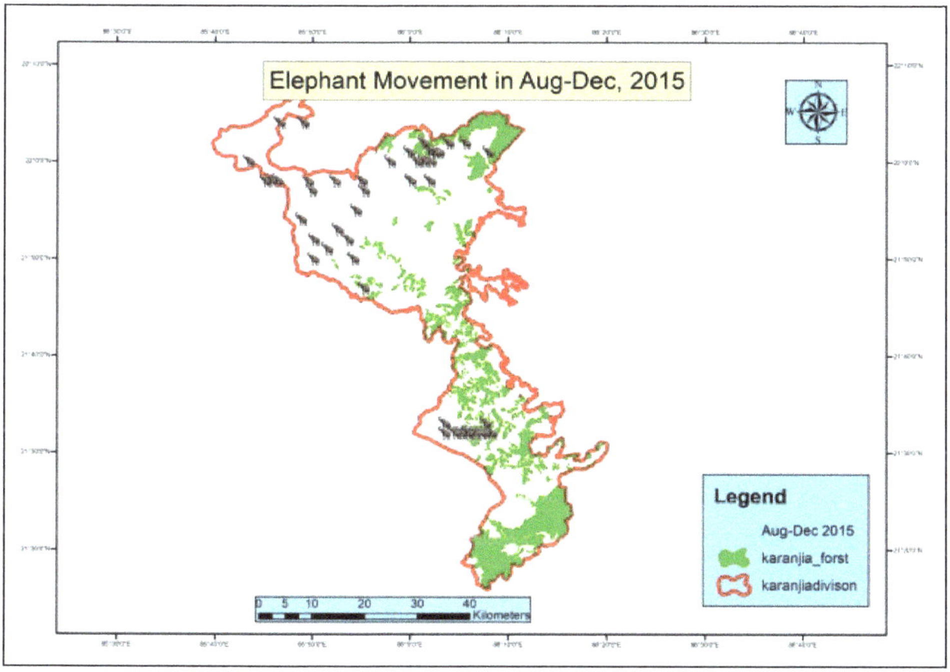

Figure 11.12. Elephant Migration and Movement of Karanjia Division during Aug-Dec, 2015.

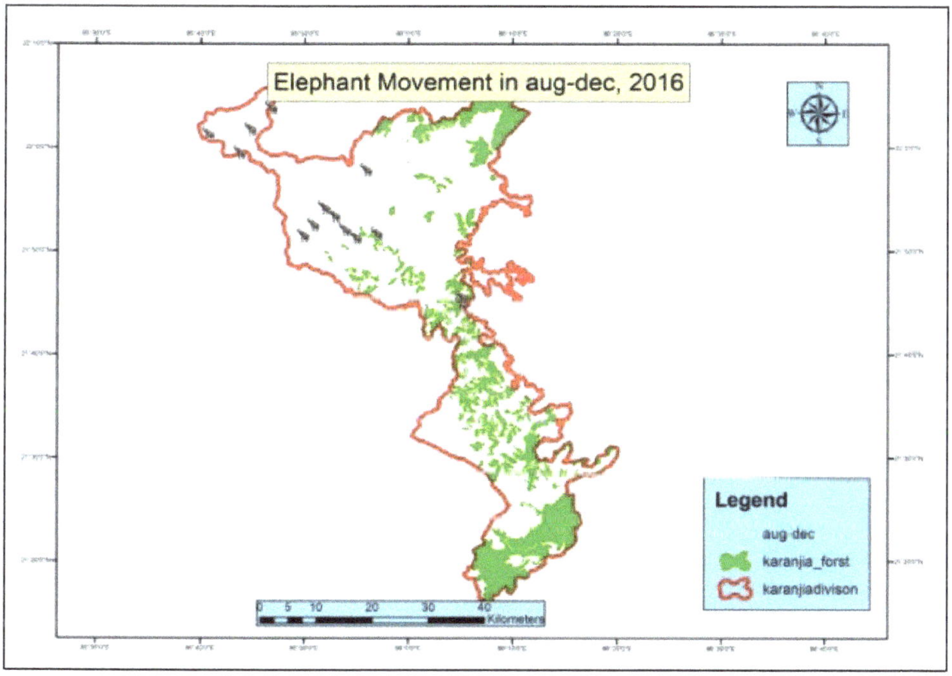

Figure 11.13. Elephant Migration and Movement of Karanjia Division during Aug-Dec, 2016.

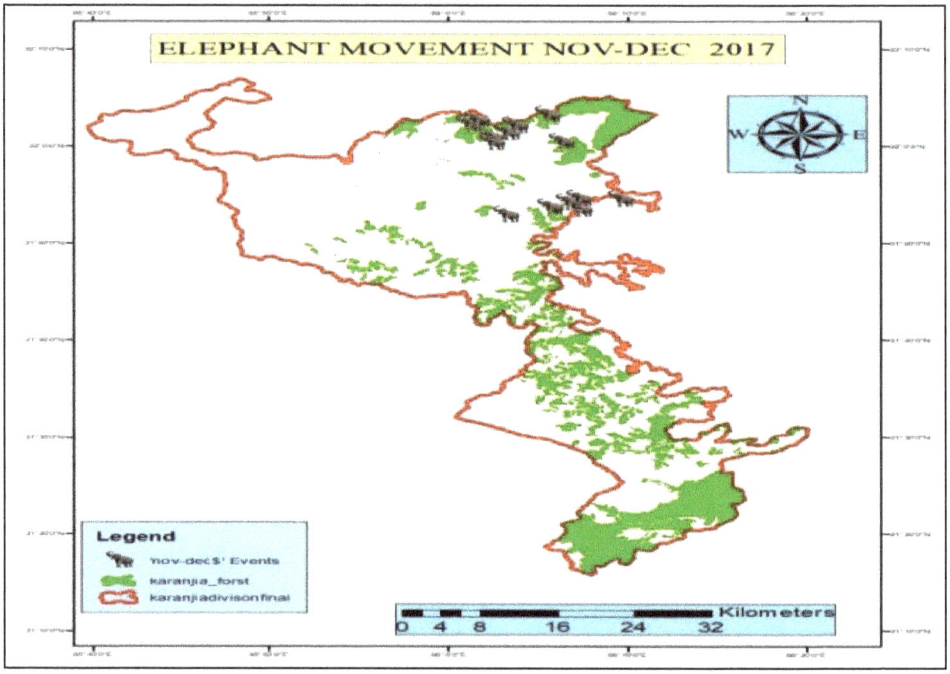

Figure 11.14. Elephant Migration and Movement of Karanjia Division during Nov-Dec, 2017.

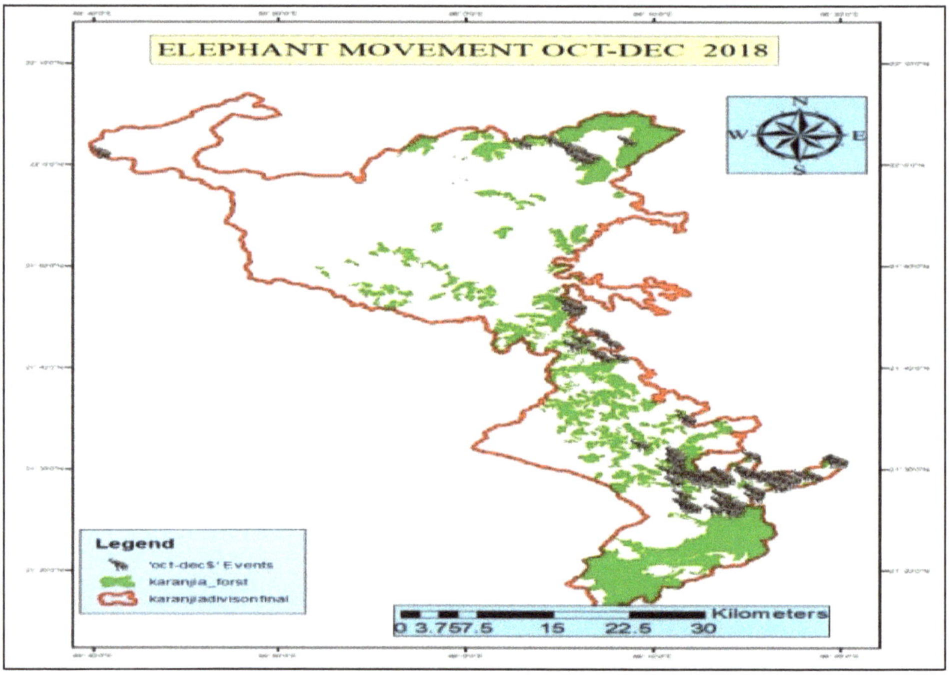

Figure 11.15. Elephant Migration and Movement of Karanjia Division during Oct-Dec, 2018.

in Joda and Badbil, mining, habitat destruction, and scarcity of food and water, which is easily available in Karanjia Division. Elephants prefer a variety of food plants in their diet. It was found that elephants also consume bark and small stems of a number of plants. At the end of winter there is leaf shedding and the grasses and sedges dry out, making the forest floor a perfect stage for catching fire. Fire burns out many of the elephant's food plants. At this time elephants depend on bark and small stems. During the rainy season and prior to winter, grasses and sedges constitute the bulk of elephant diet. Important species grazed by elephants are *Cyperus* sp.; *Eragrostis* sp., *Digitaria* sp., *Ischaemum* sp., *Themeda* sp., *Panicum repens*, *Cynodon dactylus* and *Brachiaria* sp. Elephants also like to eat the bark of Sal (Shorea robusta). The plant species which serve as the staple food of elephants are *Kydia calycina* (leaves), *Grewia tiliaefolia* (leaves and small stems), *Helicteres isora* (leaves and branches), *Mallotus philippinensis* (leaves and branches), *Cassia fistula* (leaves and small stems), *Madhuca indica, Dioscorea* sp. (tubers).

CONCLUSION

The impact of the unrest has completely devastated the infrastructure of the Simlipal biosphere reserve and information vacuum has further hampered the management of key lifeline that protects the sanctity of the biosphere reserve. After a series of negotiation, the political settlement addressing the problems faced by the tribal Communities has been evolved by the Government, which has offered a conducive environment in the area to initiate rigorous scientific research. Simlipal biosphere reserve is now in the process of revival. People who disregarded Simlipal biosphere reserve a few years back now have come forward to bring back the past glory of Simlipal again. The researchers, conservationists, nature lovers, politicians, teachers, lawyers, students and the local community have congregated altogether in a common platform with a strong believe to save and restore Simlipal biosphere reserve keeping hand in hand.

This project has already achieved the prime information's on Elephant corridor, land use land cover using GIS/RS, movement patterns and interaction between humans and elephants in the fringe villages. Initiatives for capacity building of the young researchers, local communities and forest officials to foster and encourage scientific research, sustainable livelihood and conservation have also undertaken which would be the foundation for the next phase of our research works in similar. The long-term conservation of the Asian elephant in Simlipal biosphere reserve lies in a systematic data generation for consecutive years along with people's participation in conservation initiatives would be helpful to prepare a future blue print of Asian elephant conservation in Simlipal biosphere reserve.

AKNOWLEDGEMENT

The authors thankful to department of Remote Sensing and GIS, North Orissa University and Baripada wildlife and forest division for providing necessary facility to carry out this research work

REFERENCES

Bist, S.S. (2002). An Overview of Elephant Conservation in India. *The Indian Forester*, 128: 121-136.

Choudhury, A, Lahiri Choudhury, D.K., Desai, A., Duckworth, A. and others (2008). Elephas maximus. *www. iucn redlist.org/apps/redlist/details/7140/0* (accessed 19 April 2011)

Dey, S.C. (1991). Depredation by Wildlife in the fringe areas of North Bengal Forests with special reference to elephant damage. *Indian forester*, 117(10): 901-907

Menon, V. (2003). Mammals of India, *DK Publication*, Delhi, India

Nagamani and Ramachandran (2003). Landuse/Land Cover in Pondicherry using Remote Sensing and GIS, Proceedings of the Third International Conference on Environment and Health, Chennai, India, 15-17 December, 2003. Chennai: Department of Geography, University of Madras and Faculty of Environmental Studies, York University, pp. 300 - 305.

Nobi, E. P.etal. (2009). Land Use and Land Cover Assessment along Pondicherry and its Surroundings Using Indian Remote Sensing Satellite and GIS. *American-Eurasian Journal of Scientific Research*, 4 (2): 54-58.

Patnaik, B.K., Singh, L.A.K., Kar S.K. and Kar C.S (2008). Wildlife Census in Orissa. Wildlife Headquarters, Forest Department, Govt. of the Orissa, Bhubaneswar.

Pradhan, R.N., Behera, P., Mohaptra, R.K. and Sahu, H.K. (2013). Human-Elephant Conflict in Dhenkanal Forest Division, Odisha, India. *Journal of Basic and Applied Sciences*, 5(4): 193-199.

Sukumar, R. (1990). Ecology of the Asian Elephant in southern India, Feeding habits and crop raiding pattern, *Trop. Eco.*, 33-53.

Swain, D. (2001). Man and Wild Elephant Conflict in Orissa, 127(10): 1134-1142

Swain, D. and Patnaik, S.K. (2002). Elephants of Orissa, Conservation Issues and Management Options. *Indian Foresters*, 128(2): 145-154.

Veeramani A., Jayson, E.A. and Easa P.S. (1996). Man-wildlife conflict: cattle lifting and human casualties in Kerala. *Indian Forester*, 122(10): 897-902

Chapter 12

A Review of Recent Mortality Trends of Asian Elephants in Simlipal Tiger Reserve, Odisha, India

Biswajit Mohanty

Wildlife Society of Orissa,
Shantikunj, Link Road, Cuttack – 753 012, Odisha
e-mail: kachhapa@gmail.com

Wild population of elephants have fallen throughout the globe. Asian elephant has been listed as endangered on the IUCN Red list. The major factors behind declining in elephant population is the habitat degradation, fragmentation and poaching. The present paper deals with mortality trends of Asian Elephants in Odisha, India with special reference to Similipal Tiger Reserve. The work reveals the importance of detection of the actual cause of death as soon as the carcass is detected and further emphasizes on various plausible threats due to targeted poaching of tuskers.

INTRODUCTION

According to the last all India elephant census carried out in 2012 by Project Elephant, India is home to nearly 29,391-30,711 Asiatic elephants (*Elephas maximus*). Elephants are distributed in southern, eastern and northeastern states and a few northern states of India though the highest concentration is found in the southern Indian states of Karnataka, Kerala and Tamil Nadu. The North East has also a sizeable population of about 8,592 elephants (2012 census). The 2015 elephant census figures revealed that Jharkhand, West Bengal and Chattisgarh together have about 1,105 elephants. Among these, Odisha has the largest population (64 per cent) of the eastern Indian elephants with reported population of 1,954 individuals in the state. In 2015, Simlipal Core Division recorded 337 individuals while Karanjia, Rairangpur and Baripada Divisions reported 172 elephants in total.

Several experts in both captive and wild populations have studied trends in elephant mortality. Detection of elephant mortality in the wild though should be easy due to the large size of the mammal. However, remote forest areas make detection difficult and it is believed that many go undetected. Moreover, there are also attempts not to record mortality by forest department staff in some areas as elephant death news usually attracts huge media attention as well as disciplinary action by government.

Poaching of elephants for ivory has been a serious conservation issue in African elephants (*Loxodonta africana*) as expressed by several researchers populations (Western, 1987; Douglas-Hamilton 1988; Poole and Thomsen, 1989; Dublin, Milliken and Barnes, 1995). Asian elephant (*Elephas maximus*) populations are also affected by poaching (Sukumar, 1989; Menon *et al.*, 1997). In most cases only the male is targeted for poaching in India because of their tusks. The proportion of male Asian elephants having tusks varies widely among populations, from over 90 per cent in southern India to under 10 per cent in Sri Lanka (Kurt, 1974; Sukumar, 1989). Selective removal of breeding males from a population has affected them. Sex ratios have become heavily female biased; genetic diversity has eroded and demographic changes are expected due to less female mating success. Studies have revealed that removal of tuskers due to poaching/capture and females mating with makhnas can result in lower inheritance of tusks in the population (Tiedemann and Kurt, 1995).

An audit of the state of elephant conservation in Asia revealed that in the southern states of India, severe ivory poaching has reduced the numbers of adult bulls and led to distorted sex ratios of between 1: 12 and 1: 100 (Jepson and Canney, 2003). A pragmatic study of elephant deaths and poaching during the period from 1994-1998 at Bandipur Tiger Reserve has revealed how elephants continued to be poached inside a well patrolled tiger reserve due to the high demand for ivory for both domestic and export use (Verma, 2000). Thus, conservation intentions of the government are difficult to translate into field action due to vast range of the population as well as understaffed protection units. Similar study on elephant mortality and analysis of compiled records of 291 elephant deaths over a 33-year period (1979–2011) in the Mudumalai Tiger Reserve and the reserved forests of Nilgiri North and South divisions of south India reveal the role of poaching in loss of adult males. Adult male elephant's median life span was recorded to be 22.45 years while it was 31.84 years for females. Poaching caused the majority of deaths (40 per cent). It was found that almost 82 per cent of the male elephants died by poaching (Wright *et al.*, 2015).

Ivory poaching in the seventies and eighties has destroyed African elephant family social structures and disrupted complex social behavior. It has also lead to reproductive behavioural changes, which in turn has resulted in lower breeding and rise in calf mortality Absence of breeding-aged healthy adult males is likely to lead to lowered conception rates, delaying the age of first conception apart from increasing the inter-calf gestation periods (Poole and Thomsen, 1989). Poaching of elephants is a severe problem in most Elephant Reserves of Odisha according to a study on conservation status carried out in 2004(Sar and Varma, 2004).

This work is a review of elephant poaching cases reported in Odisha with special reference to Similipal Tiger Reserve and Mayurbhanj Elephant Reserve.

Results and Discussion

During the last 26 years, approximately 1,269 dead Asiatic elephants have been recorded in Odisha by the forest department (Wildlife Wing of Odisha Forest Department). It is seen that death rate has increased over the decades from average of 33 per year (1990 to 2000) to average of 73 per year (2010 to 2016). Mayurbhanj district located on the eastern part of the state is an important elephant habitat with substantial population of elephants (> 500). Mayurbhanj has reported 76 dead elephants in six years, which is the highest for any district in Odisha. Analysis of death records has shown that the district has been leading in elephant mortality in 3 years out of the last 6 years. Simlipal is the fourth largest tiger reserve in India and was among the first batch to be notified as Tiger Reserves in 1973 by Government of India. The Park is a repository of bio diversity and has huge assemblages of flora and fauna some of which are unique. The melanistic black tigers of Simlipal have acquired world fame being endemic to this Reserve only. The Similipal Tiger Reserve (TR) is in central portion of the Mayurbhanj district stretching from 20° 17′ and 22° 34′ North Lat and 85° 40′ and 87° 10′ East Long. declared as 'Tiger Reserve' on 04-12-1973 under 'Project Tiger' of Government of India. The Critical Tiger Habitat (Core) extends over 1194.75 km^{2}and 1555.25 km^2 is the buffer zone.

The Park forms an ideal elephant habitat with huge patches of meadows and availability of water throughout the year. Fortunately, due to the absence of developmental structures like canals, railway lines, busy metaled roads, mines or industries elephants enjoy an undisturbed habitat.

Since the adjacent reserve and protected forests are contiguous, they form one huge elephant habitat that is used by the Simlipal TR elephants also. Hence, it is thought that no analysis will be complete without considering these deaths too.

A spree of elephant poaching by poisoning had occurred inside Simlipal TR in early 2010, which had led to discovery of more than 12 dead elephants. Many carcasses were discovered after 3-4 weeks of death leading too difficult in post mortem due to acute decomposition. From 1.4.2015 to 12.6.2016 forest officials found 39 highly decomposed bodies of elephants in Odisha. The age of the carcass and number of elephants found was visually estimated. The carcasses were categorized as 1- 4 weeks (n = 30), 1 month - 3 months (n= 6), 3 months – 6 months (n= 3). In the case of carcasses of more than 3 months, only scattered skeletal bones were found.

In Simlipal TR and adjacent forest, areas 7 highly decomposed bodies that were more than 1 to 2 weeks old were discovered. This indicates lack of adequate and regular patrolling in a highly protected area like a Tiger Reserve. How can the cause of death be determined in such cases?

It is important to identify the true cause of elephant death in order to ensure proper management interventions. In June, 2010, National Tiger Conservation Authority (NTCA) appointed a two member team in which the author was a member to enquire into the mass killings of elephants inside Simlipal as reported by a number of media agencies in both print and electronic media. A total of 84 elephant deaths were recorded in the Simlipal TR and adjacent areas of Mayurbhanj district during the period from 2010-11 to 2016-17 (Tables 12.1 and 12.2).

Table 12.1. Recorded Elephant Mortality in Mayurbhanj District

Total Deaths	*Simlipal National Park*	*Karanjia Div*	*Baripada Div*	*Rairangpur Div*
84	42	14	15	13

Core area of the Tiger Reserve is a proposed National Park under the Wildlife (Protection) Act, 1972and comprises about 1,192 sq.kms. The three divisions of Karanjia, Rairangpur and Baripada control the buffer zone and have recorded almost the same number of deaths, which reveals that the elephant abundance and mortality are distributed equally over the entire Tiger Reserve.

Table 12.2. Recorded Cause of Elephant Deaths in Mayurbhanj District

Year	*Natural*	*Unknown*	*Poaching (Weapons)*	*Poaching (Poisoning)*	*Total*
2010-11	N.A.	2	1	8	11
2011-12	2	N.A.	4	2	8
2012-13	3	1	5	3	12
2013-14	7	2	1	N.A.	10
2014-15	7	1	N.A.	N.A.	8
2015-16	14	5	2	N.A.	21
2016-17	7	7	N.A.	N.A.	14
Total	40	18	13	13	84

Note: Natural includes accidental fall, disease, infighting and old age.

Thus, about 48 per cent of the total deaths, which were reported in Mayurbhanj district, are attributed to natural causes. There is a significant recorded percentage of deaths caused by unknown cause (21 per cent) that is the consequence of late discovery of carcass. Poisoning seems to be controlled. However, many of the unknown cause cases could be actually due to poisoning as tissue samples were destroyed too much due to late discovery to yield viable tissue samples for toxicity tests.

Anthrax as a cause has been reported since 2014-15. From 1.12.2014 to 28.3.2017, the forest department claimed that 16 elephants died due to Anthrax. Many carcasses were buried after being covered with polythene sheets to prevent the spores from spreading. Curiously, not a single ungulate or domestic cattle death was reported though feeding areas are common in the study area.

16 tissue samples including 10 from Simlipal TR were sent to Animal Disease Research Institute, (ADRI), Phulnakhara run by the Fisheries and Animal Resources Department of Government of Odisha for tests 11 samples proved negative after

confirmatory test reports were received from National Test Labs. 3 samples from Simlipal TR proved positive while 2 samples from Baripada Forest Division were positive (RTI information dtd. 16.1.2017). Thus, Anthrax caused mortality should be declared only after tests confirm it as 69 per cent of cases proved negative even though local forest officials had claimed Anthrax. Thus, many of the deaths passed off as Anthrax were actually due to other causes. Late discovery of the carcass would have made ascertaining the cause extremely difficult.

Table 12.3. Poaching Death Rate (per cent) of Total Deaths of Simlipal vs. State

	2010-11	*2011-12*	*2012-13*	*2103-14*	*2014-15*	*2015-16*	*2016-17*
STR	73 per cent	75 per cent	67 per cent	10 per cent	0	10 per cent	0
State	19 per cent	12 per cent	8 per cent	16 per cent	15 per cent	15 per cent	7 per cent

The above table clearly illustrate the huge divergent trend of elephant mortality caused by poaching in Simlipal TR vis-a-vis the poaching trends of the whole state. Since protection measures in Simlipal Tiger Rreserve are not significantly more rigorous than the rest of the state, there is no reason that its poaching rate should be different from the rest of the state. Thus while 73 per cent of the total elephant mortality of Simlipal Tiger Reserve was ascribed to poaching in 2010-11 compared to 19 per cent for the whole state this poaching death percentage has fallen to as low as 0 per cent in 2016-17 which defies the natural trend; the state's poaching death rate has declined from 19 per cent in 2010-11 to 7 per cent in 2016-17.

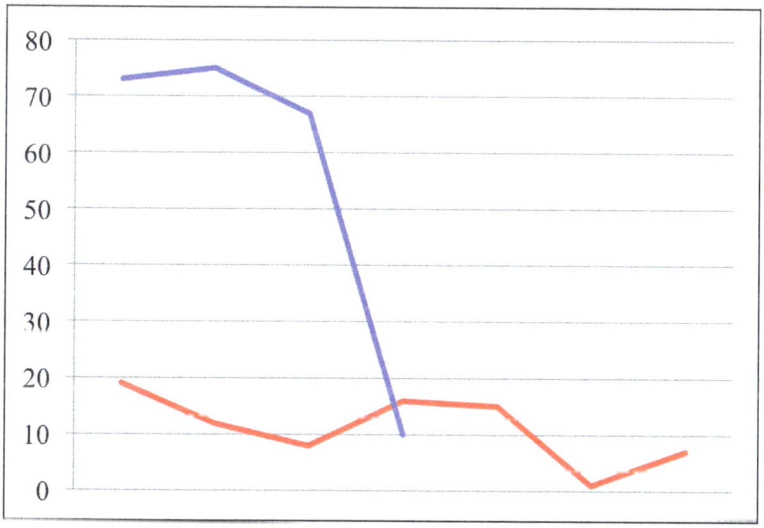

Figure 12.1. Poaching Death Rate of Str. vs State – 7 Years (1.4.2010 - 31.3.2017).

Poaching death trend (per cent of total elephant mortality) in Simlipal TR vis a vis State poaching trend as above:

(Bottom line= State Trend of poaching deaths; Top Line = Poaching deaths trend of Simlipal TR)

CONCLUSION

Organized elephant poaching gangs since the last two decades as is apparent by the thriving ivory trade in the district have targeted Simlipal. Gangs from Arunachal Pradesh and Manipur have been detected in the past in this area. The systematic poaching of adult male elephants has resulted in loss of vital breeding individuals that could have a long-term impact on the local population as has been pointed out by researchers in case of South India. At present, adult tuskers,> 30 years have become a rare sight in the Tiger Reserve. Most herds comprise of females, calves and young males with small tusks.

It is clear the true cause of mortality of most dead Asiatic elephants that are detected inside Simlipal TR and adjacent forests of Mayurbhanj district is not being ascertained. The chief reason is the late discovery of carcasses. Most carcasses are half eaten by wild boar and other scavenging carnivores like jackals and hynae making it impossible to carry out a proper post mortem to ascertain cause.

Anthrax has been claimed in most cases, which appears to be based on superficial post mortem analysis since out of 16 reported cases analyzed above, 11 were reported as negative by the test laboratory. The department should not claim Anthrax as the cause unless confirmatory test report arrives. In such cases where cause of death is not apparent from a pure visual and physical examination of the carcass, "unknown cause" can be cited rather than a cause, which is later on disproved. The author is not aware of what has happened to those cases recorded as "Anthrax" that were subsequently disproved by the test lab results.

Unless dead bodies of elephants are discovered within a few days of death cause of death would be extremely difficult to determine during the post mortem process. Regular patrolling is urgently required to ensure this. Otherwise, many elephant poachers might go scot free due to lack of registration of poaching cases based on post mortem report.

ACKNOWLEDGEMENTS

I would like to thank Odisha Forest Department for providing records on elephant mortality in Odisha. Elephant Family and WPSI for generous support for field visits. Prof. S.K. Dutta, former Head, Post Graduate. Department of Zoology, North Orissa University, Baripada for valuable inputs in preparing this paper.

REFERENCES

Western, D. (1987). African elephants and rhinos: Flagships in crisis. *Trends in Ecology and Evolution*, 2(11): 343-346.

Douglas, H. (1988). African elephants: population trends and their causes. *Oryx*, 21, 1124,

Dublin, H.T., Milliken, T.and Barnes, R.F.W. (1995). Four years after the CITES ban: illegal killing of elephants, ivory trade and stockpiles. Report of the IUCN/ SSC African Elephant Specialist Group. World Conservation Union. Gland, Switzerland.

Jepson, P. and Canney, S. (2003), The state of wild Asian elephant conservation in 2003. An independent audit for elephant family. *Elephant family and Conservation Direct*, London and Oxford.

Joyce, H.P. and Jorgen, B.T. (1989). "Elephants are not beetles implications of the ivory trade for the survival of the African elephant". *Oryx*, 23(4).

Kurt, F. (1974). Remarks on the social structure and ecology of the Ceylon elephant in the Yales National Park. The Behaviour of Ungulates and its Relationship to Management, (eds V. Geist and F. Walther), IUCN, Switzerland, 2, 618–634.

Menon, V., Sukumar, R. and Kumar, A. (1997). A God in Distress: Threats of Poaching and the Ivory Trade to the Asian Elephant in India. WPSI, New Delhi, India: 84 pp.

Sukumar, R. (1989). The Asian Elephant. Ecology and Management, Cambridge University Press.

Sukumar, R. Ramakrishnan, U. and Santosh, J.A. (1998). Impact of poaching on an Asian elephant population in Periyar, southern India: a model of demography and tusk harvest. *Animal Conservation*, 1: 281-291.

Sar, C.K. and Varma, S. (2004). The Asian elephants in Orissa, India - Population status, Conservation and Management of Asian elephants (*Elephas maximus*) in Elephant Reserves of Orissa, India. Occasional report of Asian Nature Conservation Foundation (ANCF) -Asian Elephant Research and Conservation Centre, Bangalore, India

Tiedemann, R. and Kurt, F. (1995). A stochastic simulation model for Asian elephant *Elephas maximus* populations and the inheritance of tusks. *Acta Theriologica*, Suppl., 3: 111-124.

Varma, S. (2000). Elephant Poaching in Bandipur Tiger Reserve, southern India; A Study on the Mortality of Asian Elephants (*Elephas maximus*) due to Poaching and Other Causes, Poachers and Anti-Poaching Strategies. Asian Elephant Research and Conservation Centre (AERCC- a Division of Asian Nature Conservation Foundation-ANCF), C/o Centre for Ecological Sciences, Indian Institute of Science, Bangalore.

Wright, B. (2015) Mortality records (1979–2011) shed light on threats to Asian Elephants *Elephas maximus* Linnaeus, 1758 (Mammalia: Proboscidea: Elephantidae) in Nilgiris, southern India". *Journal of Threatened Taxa*, 7(8): 7436–7442.

Other Sources

RTI reply dated 16.1.2017 regarding Anthrax tests on tissue samples of dead elephants from Simlipal TR obtained by author from the Animal Disease Research Institute, Phulnakhara run by Fisheries and Animal Resources Department, Government of Odisha.

Chapter 13

Similipal Biosphere Reserve and Elephant Conservation

Dillip Mishra

Department of Social Work, North Orissa University,
Takatpur, Baripada – 757 003, Odisha
e-mail: pdillipm@gmail.com

Similipal Biosphere Reserve is the natural home to Asiatic elephants. Due to anthropogenic activities elephants are under severe pressure. Developmental activities such as mining, deforestation, railway line and irrigation projects brings obstruction to the elephant migration and movement. This led to conflict with human for food and space. Besides diseases and poaching for ivory are important threat to Similipal elephants as well as the elephants of the state of Odisha

INTRODUCTION

Similipal Biosphere Reserve is located in the Mayurbhanj district of Eastern Indian state of Odisha. This is situated in the Latitude: 21°10′N – 22°12′N, Longitude: 85°58′E – 86°42′E and the Centre Point: 21°40′N – 86°19′E. It is also a part of the the Eastern Ghats range and is very rich in faunal diversity (Ramakrishna *et al.*, 2006). Similipal is named after abundance of the tree "Simili", which is the local name of the red silk cotton tree. Similipal has been declared as a Sanctuary, National Park and a Biosphere Reserve. It was formally designated a tiger reserve in 1956 and brought under Project Tiger in May 1973. The Similipal Biosphere Reserve has been notified by Government of India on the 22nd June, 1994.

Traditionally each biosphere reserve is designated as core area(s), buffer zones and transition areas. The "Core area(s)" is the highly protected area, with main focus is the conservation of biological diversity and the monitoring of little disturbed ecosystems. The core area also contributes to the maintenance of ecosystem services (provisioning, regulating and cultural services), for example; carbon capture, soil stabilization or the supply of drinking water, among others. Besides certain economic activities such as environmental education, research, environmental conservation and rehabilitation activities and also recreational and Eco touristic activities can be developed inside the areas. "Buffer zone", surrounds the core areas. This zone also reduces the impact of human activities on the core area and is essential for maintaining biological and cultural diversity. It also promotes biological connectivity that acts as a natural corridor between the core zone and the transition zone. The "Transition area" is fundamental for the management and sustainable development of natural resources. A much wider range of economic activities that benefit the socio-economic development of local populations can take place here. However, these activities must respect the specific goals of each reserve. Similipal Biosphere Reserve covers a total surface area (terrestrial): 556,900 ha where the Core area(s): 84,500 ha, Buffer area(s): 212,900 ha and Transition area(s): 259,500 ha. Similipal is home to a range of wild animals including 55 species of mammals, 304 species of birds, 20 species of amphibians, 62 species of reptiles and 37 species of fishes.

Odisha hosts the large number of central Indian population of elephants' where 7 per cent of the total population of elephants in India are in Odisha. Government of Odisha has identified three elephants' reserves. This includes (1) Mayurbhanj Elephant Reserve (Constituted: September 2001), (2) Mahanadi Elephant Reserve (Constituted: June, 2002) (3) Sambalpur Elephant Reserve(Constituted: March, 2002). The elephant census data of 2015 revealed the elephant population in Odisha is 1954, a little higher than previous census. With a comparison to different forest zones, Similipal forests are more protected than any other forest divisions of Odisha. Similipal Biosphere Reserve is a natural home to Asiatic elephant and has a unique elephant habitat. In 2010 total 59 elephants were estimated in four ranges of Baripada division (Mishra *et al.*, 2014). The herds of elephants majestically walking across the roads and rivulets could be a regular sight of these areas. The study focuses on the habitat structure in Similipal Biosphere Reserve for elephants and probable reasons for the declining in elephant population.

Methodology

Observation method and anthropological way of analysis were used to assess probable causes of the elephant's mortality and a way of their conservation. The research was conducted by taking direct interviews with community members, communities intimately related to Similipal areas, tribal leaders, members of local NGOs and local youths. Qualitative and quantitative research methods are used for data analysis. The volunteers of the fieldwork areas were interviewed in the spots from time to time for progressing the research study related to Elephant Conservation.

Results and Discussion

Similipal an Unique Habitat for Elephants

Similipal has unique habitats ranging into the Forests, Grasslands and Wetlands. The upper slope of the hills and their flat valleys are often covered with grasslands.

The Sal forest gradually encroaches upon the grassland. Among the tree species 1. *Syzygium crasoideum*, 2. *Symplocos ralemosa*, 3.*Dillenia pentagyna* are more common. The Similipal forest has thick cover due to the large varieties of plant species dominated by Sal. The Sal trees forms upto 30 per cent of the vegetation. The plants remain leafless for a long period during later winter. Other associates of Sal which keeps the forest green are 1. *Terminalia* species, 2. *Anogeissus latofolia*, 3. *Dalbergia latifolia*, 4. *Gardenia gunamifera*, 5. *Haldina cordfolia*, 6. *Lagerstroemia parviflora*, 7. *Semicarpus anacardium*. The grass herbs and shrubs area abundant in these areas. Sal forest was observed in hilltops and plateaus above 850m Amsl, characterized by pure stand of Sal with common association of 1. *Terminalia fomentosa*, 2. *Dillemia pentagyna*, 3. *Syzgium Ceresoideum*, 4. *Ptrpcarpus marsipium*. These beautiful compositions of plants and trees helps to find important niche to the animals in this area. Similipal-Kuldiha-Hadgarh Elephant Reserve is one of the important reserves for elephants. The Asian elephants prefer hot, humid and grassy areas within forest. These fields have grass, trees and shrubs on which the elephants prefer to feed. Asian elephants eat various vines, roots and leaves. During dry season, from January to April, the elephants move to river valleys close to water. In rainy season, they remain in the forest where availability of plenty of grasses. During the months of September through December, the second wet season, the elephants move on to the short grass of the open forests around the jungle.

Habitat Loss and Habitat Fragmentation: A Major Threat to Elephant Population

Based upon site visit and interviews with local inhabitants it has been observed that several factors play roles in the decline of elephants' population. People of 68 villages within and 1200 villages outside of Similipal Biosphere Reserve directly or indirectly depends on the forest. The activities include cattle grazing, collection of fire woods, sal leaves and other NTFPs. Massive cutting of Sal trees, illegal fire wood collection, and mass cutting of trees and digging of stones restricted animal movements in these areas. This is also a big question government never clarified the total area of mining which came under digging every year and loss of forest areas included which are creating danger to elephant movements. If the information of losing forests came into the facts, more constructive planning will be made for forest managements.

Besides cutting of trees and burning of forests by people has been documented. The major reasons for habitat loss are Illegal mining, including blasting, is continuing in several places of Similipal Biosphere Reserve, which has tremendous impacts on the wild life. Around fifteen places of illegal mining sites were documented. Besides, increasing human populations, urbanization of the forest related areas, deforestation due to illegal wood cutting activities, globalization and first going communication where the poachers are interacting with first communications and using battery saws for cutting of ivories and valuable trees now is a challenge for the State Forest Systems. Due to rapid migration of rural to urban's and gradually villager's natural life changed into a mechanical and preferably need more money in illegal way encouraged elephant poaching in various areas. The poachers from outer states are attracted by the local informers and insecure living of the poor animals easily targeted by the outstate poachers is the biggest problem for Odisha Elephants. t has been observed that at the time of paddy cultivation and crop collection elephants come to the villages adjacent to forests.

Un-Natural Death and Poaching Play Important Role an Declining Elephant Population

In Odisha data shows that 18 jumbos died due to natural causes, 19 of unknown reasons, 4 were killed by poachers, 9 were electrocuted, 11 by sagging lines and 6 elephants were died by trains. These deaths were reported between April 1, 2018 and December 18, 2018. This shows that 45 per cent of death of elephants due to unnatural causes. Male elephants are more vulnerable to be killed by the poachers. Odisha loses around 18-20 such adult male elephants every year threatening the sustainability of elephant population. The clear-cut intension of the poachers could not be revealed. But the local tribes are not directly linked to the poaching rather connecting themselves outside poachers for killing the male elephants for selling the ivories in international markets. Sometime forest officers remain silent because of fear of mafias activated in these areas. The forest guards also working in this forest areas are not all of them are local people.

Among the total deaths of elephants in Odisha, most of the poaching cases are coming from Athagarh, Dhenkanal reserve forests. Here poachers are killing elephants by electrocution, using poisonous arrow, by poisoning in food *etc.* The local people could not participate in killing as there is a chance arrest as local polices are known to the villages and the local hunters but the most unfortunate thing is that the locals are working as the informer and helped the outstate hunters and inform them and direct the local map's position and the elephants availability from time to time besides that they informed about the elephant's corridor and the way when the male elephants are hiding in night areas, which has targeted by the poachers to kill the elephants. In Athagarh, Dehenkanal Researve Forests poaching are repeatedly found where the role of "Villagers" of these communities are not properly functioning or unaware of the situations due to various reasons. Due to their non-involvement's poachers targeting easily and success about their plans.

Elephant Migration and its Obstruction

Due to habitat loss and mining activities, elephants are coming to Similipal from Ranchi and West Bengal in searching for foods with their herds. However, this time poaching is not possible as local volunteers are very much actives and by the by community leaders lead for an awareness to protect of these animals. Last two decades elephant's habitat is heavily affected by construction, irrigation projects and mining. Recently construction of Manjhor Dam has obstructed the movements of elephants between Satkosia and Khalasuni. Construction of the Talcher –Sambalpur railway lines, irrigation canal, mining, deforestation have fragmented this important habitat and increased conflict. Similipal–Kuladiha, Hadgarh and adjoining areas of Sukinda and Badampahad supports about five hundred elephants or more than that number. The elephant movement between Hadgarh and Kuladiha has been severely affected by chromites mining at Baula Reserve. Forest expansion of settlement and agricultural lands Movement from Similipal, Jharkhand and West Bengal states is also impacted badly by mining and agriculture.

Human Elephant Conflict

Elephant census conducted by Similipal Tiger Reserve during 2015-16, report there are all together 59 numbers of elephants of different ages were recorded. In the same period there were 21 death of elephants in different reasons were detected. Each year elephants damage houses, paddy fields, causes injury and even kill human in the nearby villages. During 2015-16, ten cases of human injury, seven cases of human death by elephant attack were recorded. The main objectives of killing elephants by human is like ivory business and in International markets. Locals of Similipal had identified several interstate poachers particularly from North East region of India in Similipal Biosphere Reserve. The "Lisu Tribes" from Manipur area are visiting regular interval with the help of the local hunters who are regularly give information about the number of male elephants and the areas where they stay and peculiarly the male elephant is not moving with the herds where little far from the team they are easily targeted by the hunters and the process of killing is using poisonous arrow. The poisons are prepared from particular types of herbs which are so poisonous that, could kill the majestic animal within minutes through single hit of arrow. The smuggler easily collects the ivory by cutting partially the head of the elephant.

Role of Local Community and NGOs in Elephant Conservation

Currently several Non-Government-Organization (NGO) actively working in the areas close to Similipal. These NGOs really working hard for the conservation of Similipal forest. Noted among them, "Gram Swaraj", an NGO that works with the tribals for regeneration of Similipal biosphere. "Sabuja Bahini" another NGO motivated youths for preserving the forests and not to kill elephants

CONCLUSION

A negative interaction between human beings and wild elephant over a period of time led to conflict in the wild landscapes. Population growth, urbanization, mining and shortening of forest land causing this majestic animal to get shortage of space. Keonjhar and Sundargarh district have higher mining activities, causing loss of habitats of elephants. Similarly, in the absence of strong local community in Athagarh Forest areas the poachers are very much active. With a comparisons Mayurbhanj Forest areas are more protected and properly connected to the community organisers. Similipal last twenty years. But, increasing human populations, unplanned developments and land use changes, forest fires conducted by the locals are major thereat to Elephants at Similipal Biosphere Reserve. Illegal hunting and poaching in various elephants corridors are major threat in the state of Odisha. Synchronization of various corridors where elephants could move freely and could share common habitat will help in the elephants population. For the same time, enforcement of radical forest laws, appointing more and more forest lovers as community watchers selecting from the local villages and provision of remuneartion to protect the jungles will be another solution or this problem. Involvement of more and more village youths are proper educated and awareness camp must be conducted which prohibited the loss of the forests.

REFERENCES

Barua, P. (1993). "Managing a problem population of elephants". *IN J.C Daniel and S.H Datye,editors: A weak with elephants,proceedings of the International Seminar in Asian Elephants* (*Oxford University Press, New Delhi,* 150-161

Barua, P. and Bist, S. S. (1993). "Changing Pattrern in the distribution and movement in the distribution and movement of wild elephants in North Bengal". J.*C Daniel and S.H Datye,Oxford University Press, New Delhi,* 66-84

Behura, N. K. (1982). "Forest and Tribal World- view". *Jr of Social Research,* 1, 1-8

Barnes, R. F. W. (1983). "The Elephant Problem in Ruaha National Park". *Tanzania Biological Conservation,* 26, 127-148.

Chadha, S. K. (2010). "Reduction of Elephant–Human Conflicts in Orissa Through Change of Land, Research works analysis".

Chauhan, N. P. S. and Choudhury, S. (2002). "Evaluation of Electric Fences for their efficacy in controlling elephant damage in Northern West Bengal". *The Indian Forest Journal,* 128(2), 179-188.

Mishra, S.R., Bisht, H.K., Sahoo, D.P., Behera, D.R. and Pradhan, R.N. Status Survey of Asiatic Elephant in Baripada Forest Division, Odisha (2014). *Journal of Wildlife Research,* 2(4): 27-30

Mohanty, B. (2019). "The Similipal Tiger Biosphere Reserve, comments in articles". *The Similipal Biosphere Reserve, net article.*

Ramakrishna, Siddiqui, S.Z., Sethy, P. and Dash, S. 2006. Faunal Resources of Similipal Biosphere Reserve, Mayurbhanj, Orissa, Conservation Area Series, 28: 1-87 + 8 plates (Published by the Director, Zool. Surv. India, Kolkata)

Chapter 14

Understanding Spatial and Temporal Patterns of Human-Sloth Bear Conflict in Northern Odisha, India

Himanshu Shekhar Palei and Hemanta Kumar Sahu*

Department of Zoology, North Orissa University, Takatpur, Baripada – 757 003, Mayurbhanj, Odisha
**e-mail: himanshu.palei@gmail.com*

Designing and optimising management strategies for human–carnivore coexistence is the greatest challenge and it requires detailed understanding of the ecological and sociological circumstances associated with conflict, particularly in multi-use, human-dominated landscapes. The sloth bear is a key example of a large-bodied, wide-ranging, conflict-prone species that is gradually threatened by various anthropogenic activities across its geographical range. To understand the temporal and spatial pattern of sloth bear attack on human, records of sloth bear attacks on humans were collected from seven divisional forest offices of northern Odisha. Sloth bear attack victims were interviewed to collect conflict details. Victim's age, sex, occupation, and activity at the time of attack was recorded, as well as date and time of the incident, followed by physical verification of the incident location. From 2001 through 2016, 529 sloth bear attack incidents were recorded in the study area. Forty five people were killed and 685 injured due to sloth bear attacks. Seventeen sloth bears were killed as a result of retaliation by the local villagers. There was a significant increase in incidents of conflict during 2009–2016 compared with 2001–2008. Although sloth bear attacks were reported throughout the year, they were more frequent during monsoon season and August month. Attacks were more frequent during morning between 0601 and 0800 hours. Most of the sloth bear attack incidents were occurred along the forest fringes than other habitat types. Sloth bear attacks were recorded in six forest divisions and most of the attacks were recorded in Keonjhar Forest Division (37.88 per cent) followed by Balasore Wildlife Division (36.89 per cent) and Keonjhar Wildlife Division (17.05 per cent). Sloth bear attack hotspot maps from spatially explicit models showed a patchy probability of high conflict risk with presence of definite clusters or hot spots of attack. High risk sites were noticeable in Balasore Wildlife Division and Keonjhar Forest Division. Although incidents of sloth bear attack were reported in all age groups, majority of victims were 31–45 years of age. Attacks occurred most frequently when victims were using forest fringes for toilet. Approaches to mitigate human–sloth bear conflict should emphasize avoiding sloth bears when they are sighted and conducting activities near forests during morning and evening and in groups when possible. Avoid sloth bear encounter by remove or not cultivated sloth bear attractants or food near home. Strategies such as improving sanitation facilities, developing effective compensation schemes, forming conflict management teams, and establishing sloth bear conservation awareness programs for villagers, as well as improving cooperation among various stakeholders, are necessary to foster human–sloth bear coexistence.

INTRODUCTION

Human-wildlife conflicts (HWC) are universal, and peoples near ubiquitous negative behavior to wildlife are a major challenge to biodiversity conservation (Woodroffe *et al.*, 2005). The conflicts are difficult to understand and manage, because they are influenced by many factors, including religious values, the cultural and economic value of animal and their body parts, and the economic loss is imposed by wildlife damage (Dickman, 2010, Michel and Ashey, 2004). A major reason for human-wildlife conflicts is human population growth, which is especially an issue in India, where the population has more than twice from 1950 until now. This has led to an increase wildlife habitat fragmentation and destruction (Rangrajan, 2001). Conflicts undermine human well-being and have socio-economic costs. In other hands animal are often killed or injured due to property and human life protection (Treves *et al.*, 2006). HWC exist throughout the world but are most severe in developing countries where livestock, agriculture and Non-Timber Forest Produces (NTFP) are an important part of rural villagers' livelihood and income (Karanth *et al.*, 2012). HWC hotspots are particularly found near forest areas, which often border densely populated farmlands (Treves *et al.*, 2006).

India is one of the mega-biodiversity countries in the world and is projected to be the most populous nation by 2028, with approximately 1.45 billion people (Karanth and DeFries 2010). Large-scale land-use change is occurring as a result of mining and industrial development, which ultimately affects wildlife and their habitats (Karanth and DeFries 2010). As a consequence of India's increasing population and development, human–wildlife conflicts are simultaneously increasing. Species contributing to human–wildlife conflict in India include elephants (*Elephas maximus*), tigers (*Panthera tigris*), lions (*Panthera leo*), sloth bears (*Melursus ursinus*), leopards (*Panthera pardus*), and wild pigs (*Sus scrofa*), with human–sloth bear conflict among the most important (Rajpurohit and Krausman 2000, DeFries *et al.*, 2010, Can *et al.*, 2014, Ratnayeke *et al.*, 2014).

The sloth bear is endemic to the Indian subcontinent and has been reported throughout India, occupying diverse habitats from wet or dry tropical forests to savannas, shrublands, and grasslands (Garshelis *et al.*, 1999, Yoganand *et al.*, 2013). However, its historical range has been fragmented and decreased overall because of largescale deforestation to facilitate human development. In addition, diminished food resources, illegal killing for gall bladders, and taking of young sloth bears from the wild for bear dancing resulted in severe population declines and some local extirpations (Yoganand *et al.*, 2006). In general, sloth bears are omnivorous with adaptation of myrmecophagy and their diet composition varies according to season and habitat (Garshelis *et al.*, 1999, Joshi *et al.*, 1997, Bargali *et al.*, 2004, Palei *et al.*, 2014a, Palei *et al.*, 2020). Mostly, their diet is characterized by feeding on social insects (predominantly ground living ants and termites) and sugar rich fruits (Garshelis *et al.*, 1999).

Most of the current sloth bear range occurs in forests outside protected areas (Jhala *et al.*, 2011, Sathyakumar *et al.*, 2012), where people largely depend upon these forests for domestic and subsistence resources such as fuel wood, fodder, thatch grass, building materials, medicinal plants, wild fruits, and grazing land for livestock (Bargali *et al.*, 2005). These situations may encourage competition for resource use and increase the likelihood of human–sloth bear encounters (Yoganand 2005), which can lead to conflict. As a result, sloth bears can elicit fear and anger in rural communities, often leading to killing of sloth bears (Chauhan, 2006). Studies on human–sloth bear conflicts have been carried out across the sloth bear range (Rajpurohit and Krausman 2000, Bargali *et al.*, 2005, Ratnayeke *et al.*, 2014, Garcia *et al.*, 2016). However, most of these studies have largely focused on understanding the nature, frequency of human injuries and casualties. Information about human attitudes and perceptions of people toward sloth bears are lacking and are necessary because effective conflict mitigation requires knowledge of the underlying human and environmental drivers (Thorn *et al.*, 2012).

Odisha is one of the eastern coastal states of India and represents prime habitat for sloth bear in the country (Yoganand *et al.*, 2006, Palei and Sahu, 2017). Journalistic reporting about human–sloth bear conflict in Odisha is common in print and electronic media. We conducted this study to gather reliable information on ecological and social aspects of conflict, which will be useful in improving human–sloth bear coexistence. This information is also expected to provide insights for other locations with similar conservation challenges pertaining to coupled human and natural systems. In this context, our objectives were to investigate (1) the extent of sloth bear attack on humans, (2) spatial and temporal patterns of attacks, and (3) identify human activities that make people vulnerable to attack.

Study Area

Our study area lies in the northern part of the Odisha state (extending over the districts of Keonjhar, Mayurbhanj and Balasore) in eastern India between 20° 55′ 14.94″ N to 22° 39′ 22.27″ N and 85° 13′ 55.63″ E to 87° 27′ 43.30″ E and covering an area of 22,519 km^2 (Figure 14.1). The northern and western part of the study area are hilly and undulating and it is an extension of Chhotanagpur Plateau of the Deccan Peninsula and rest of the region comes under the Gangetic Plains (Rodgers and Panwar 1988). Elevation of the area ranges from 145 m to 1100 m above sea level. The climate is described as tropical with three distinct seasons: summer (Mar–Jun), monsoon (Jul–Oct), and winter (Nov–Feb). Temperatures range between 10 °C during winter to 40 °C during summer and the annual rainfall varies between 1200 and 1400 mm. The region is a complex of various land-use types consisting of forests,

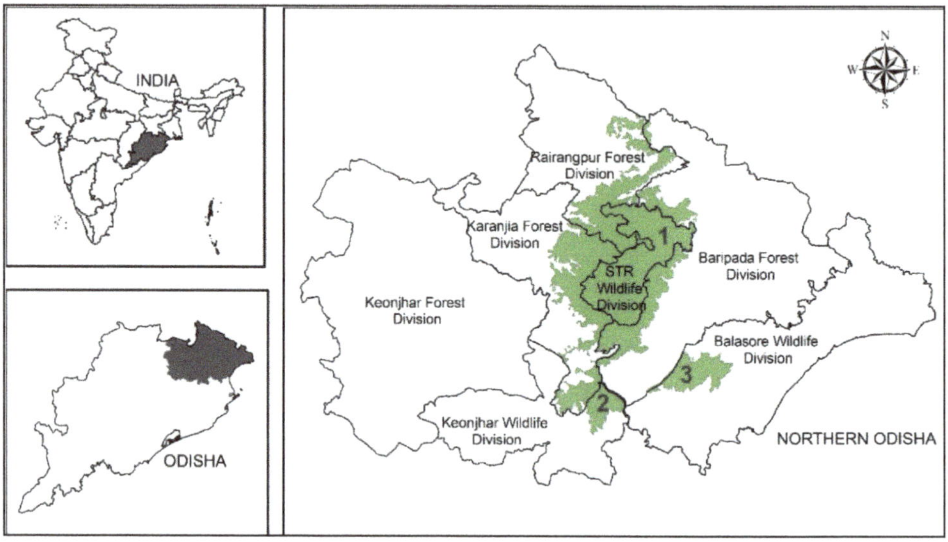

Figure 14.1. Study Area Showing the Forest Division of Northern Odisha, India.

Highlighted area showing the protected areas, 1. Similipal Tiger Reserve 2. Hadagarh Wildlife Sanctuary 3. Kuldiha Wildlife Sanctuary.

agricultural lands and human developments. The area represents a stronghold for sloth bears and other threatened species of large-bodied mammals (Sahu *et al.*, 2012, Sahu *et al.*, 2014, Mishra *et al.*, 2016). Conversion of forest into agricultural farmlands is the most significant threat of the study area (Reddy *et al.*, 2013).

Methodology

Collection of Records of Sloth Bear Attacks on Human

Records of sloth bear attacks (*i.e.* attacks resulting in physical injury or death) on humans were collected from seven divisional forest offices of northern Odisha. Villagers who suffer as a result of sloth bear attacks are entitled to claim compensation from the State Forest Department. Application of compensation claims made by villagers were collected from Forest Range Offices for the period 2001–2016. Since the Forest Range Offices are well established and has a good road network in the study area, assumed that distance or accessibility not being limitations to submit compensation applications. In these applications, victim's age, sex, occupation, and activity at the time of attack, as well as date and time of the incident were mentioned. Such information were collected from these applications.

Interviewing Victims

From 2012 to 2016, victims were interviewed to collect conflict details. If the attack resulted in human death, it was verified by interviewing a family member or an individual who had witnessed the incident. To avoid underestimating the actual number of victims (as compared with claimants), detail inquired during the interview as to whether the respondents knew of other victims that had not claimed, or had not received, compensation for a sloth bear attack (Ogra and Badola 2008). Victim's age, sex, occupation, and activity at the time of attack was recorded, as well as date and time of the incident, followed by physical verification of the incident location. The habitat of attack sites were classified as forest, forest fringe, agricultural, and human habitations. The age groups of the victims were categorized as <15 years old, 16–30 years old, 31–45 years old, 46–60 years old, and >60 years old. Sloth bear attacks were categorized by season and 2-hour intervals of the day.

During the interview, victims were explained their activities and the activities of the sloth bear before the attack. Based on their explanations, potential factors were identified responsible for the attack and categorized it as provoked (the sloth bear was intentionally approached or teased by people), accidental encounter (a person accidentally came into close proximity to a sloth bear, which provoked an aggressive response from the bear), or adult females defending dependent young. The intensity of human injuries were categorized as permanent injury (permanent damage to any body part, singly or in multiple locations, which has not recovered fully and negatively affects their physical abilities) or minor injury (small wounds or other injuries, which have fully healed and do not adversely affect the victims' physical abilities).

Distribution of Sloth Bear Attack Locations and Conflict Hotspot

In order to understand the characteristics of sloth bear attacks on people, sloth bear attack locations were used by visiting the actual site of attack or using the centroid of the village where the attack occurred as an approximate location of the attack. The mean distance between centre of the village and its boundary was nearly 1156 m indicating that the estimated location of attacks had a maximum error of 1.2 km. QGIS version 2.18 (http: //qgis.osgeo.org accessed on Jan 2018), Google hybrid layers from the Open Layers plugin, and the district map of all forest divisions were used to map the centroids of the villages.

To identify the spatial clusters of attacks on people by sloth bears, sloth bear attack locations were used to generate kernel density maps using the *HEATMAP* plugin (http: //docs.qgis.org/2.18/en/docs/user_manual/plugins/plugins_heatmap.html, accessed on Jan 2018) in QGIS version 2.18. In order to generate a kernel density map or 'hotspots' of sloth bear attack locations, 4 km radius was used to around each attack location, which corresponded to a mean home range size of a sloth bear of 15 km^2 (Laurie and Seidensticker 1977, Joshi *et al.*, 1995). Kernel density estimates through biweight method was used as it gives more weight to points which are nearby compared to points that are far away from each other (Jones 1993).

Data Analysis

Survey contained 'yes' and 'no' options for wherever specific quantitative data could not be obtained. Raw data were entered into Microsoft Office Excel and analysed as needed with excel and *PAST* software. Qualitative data were recorded as 'one' for yes and 'zero' for no, and a comparison between data sets was made in terms of percentage. Considering the total dataset on sloth bear attacks since 2001, first assessed whether the number of attacks varied over time, on a yearly basis, with the number of attacks against year. Mann–Whitney– Wilcoxon test (*W*) was performed to understand the trend of conflict incidents across years and Chi-square test (χ^2) to understand differences in nature and factors associated with conflict.

Results

Magnitude of Sloth Bear Attacks

From 2001 through 2016, 529 sloth bear attack incidents were recorded in the study area (mean ± SD; 33.06 ± 12.05, range = 7–52). Forty five people were killed and 685 injured due to sloth bear attacks. Seventeen sloth bears were killed as a result of retaliation by the local villagers. There was a significant increase in incidents of conflict during 2009–2016 compared with 2001–2008 ($W_{14} = 11.5$, $P = 0.015$; Figure 14.2). Sloth bear attacks were recorded in six forest divisions, Balasore Wildlife Divisions, Keonjhar Forest Division, Keonjhar Wildlife Division, Karanjia Forest Division, Rairangpur Forest Division and Bariapada Forest Division. Permanent injuries to victims included damage to eyes, mouth, ears, nose, and head; and skeletal fractures. Minor injuries included scratches on the body, which healed but often caused scarring. The study did not document any victim attacked more than once and up to 22 people injured in a single attack.

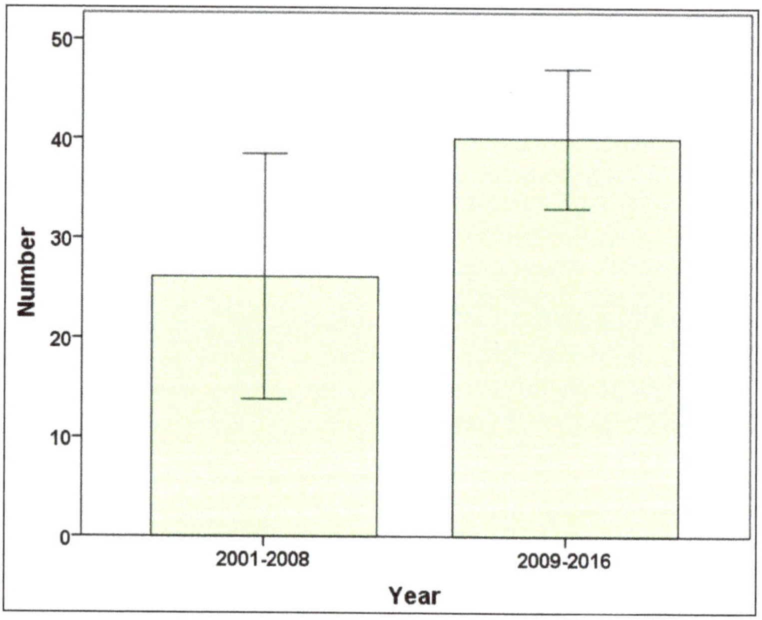

Figure 14.2. Graph Showing Comparison of Sloth Bear Attack on People in 2001-2008 and 2009-2016.

Temporal Patterns of Sloth Bear Attacks

Although sloth bear attacks were reported throughout the year, they were more frequent during monsoon season (49.84 per cent, n=161), followed by summer (28.17 per cent, n=91) and winter (21.98 per cent, n=71) ($n = 323, \chi^2_2 = 41.48, P < 0.01$; Figure 14.3). Most attacks occurred in August (16.1 per cent, n= 52), followed by September (14.86 per cent, n=48) and October (10.22 per cent, n=33) ($n = 323, \chi^2_{11} = 68.26, P < 0.01$). Attacks were more frequent between 0601 and 0800 hours (26.66 per cent, n=88) followed by 0401-6000 hours (17 per cent, n=56) and 1801-2000 hours (12.22 per cent, n=40) (n=330, χ^2_{11} =245.85, P<0.01).

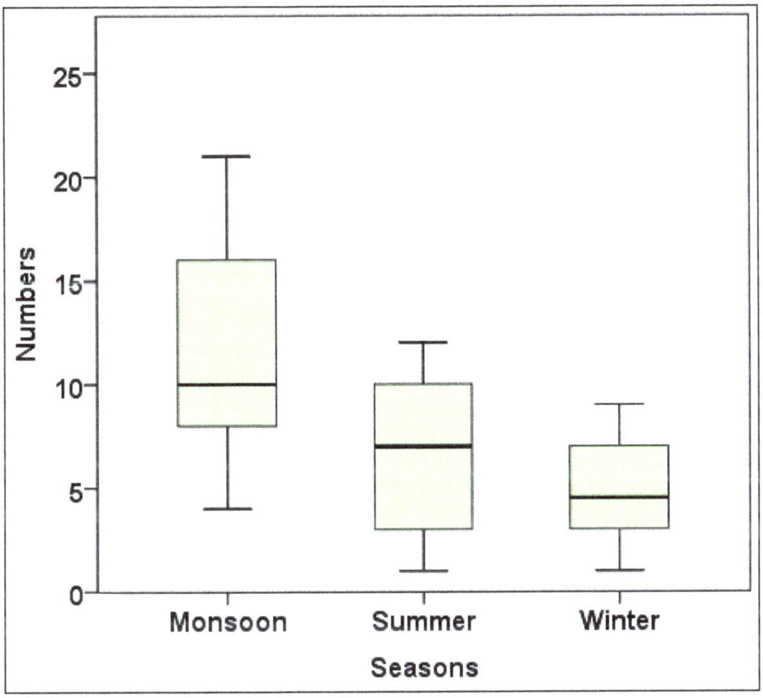

Figure 14.3. Box Plot Showing the Seasonal Distribution of Sloth Bear Attack on People.

Spatial Patterns of the Sloth Bear Attacks

Of all the habitat types, most incidents occurred along the forest fringes (n= 189, 41.90 per cent) followed by forest (n= 133, 29.49 per cent) and village (n=129, 28.60 per cent) (n=451, χ^2_2=14.97, P<0.01). Of the 529 sloth bear attack incidents, most of the attacks were recorded in Keonjhar Forest Division (n = 211, 37.88 per cent) followed by Balasore Wildlife Division (n = 206, 36.89 per cent) and Keonjhar Wildlife Division (n = 95, 17.05 per cent) (Figure 14.4). There seems to be a clustering of sloth bear attacks on human along the western part of Balasore Wildlife Division and south-eastern part of Keonjhar Forest Division (Figure 14.5). Sloth bear attack intensity was low in Baripada Forest Division, Rairangpur Forest Division, Karanjia Forest Division and Similipal Tiger Reserve Division of the study area.

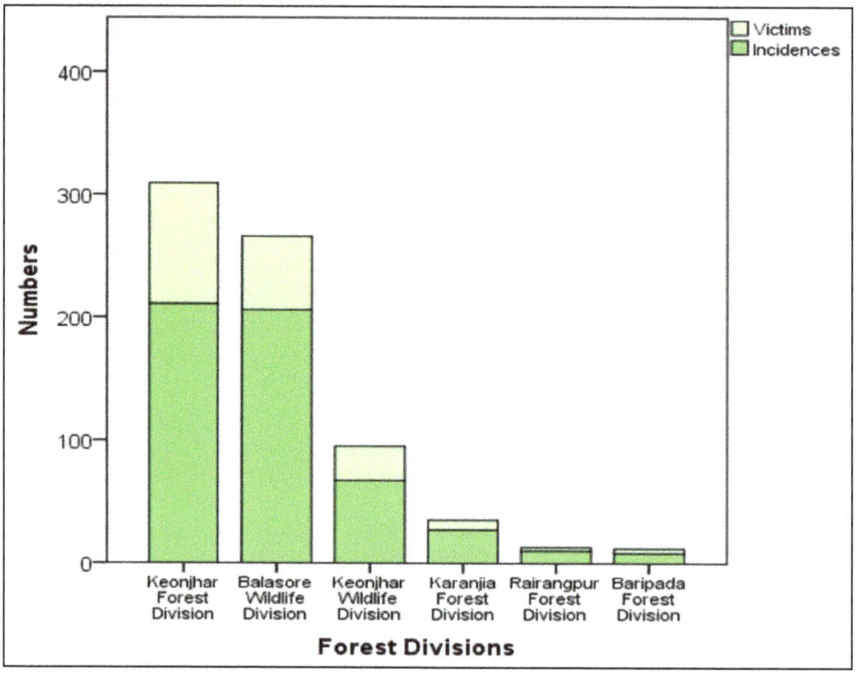

Figure 14.4. Graph Showing Sloth Bear Attack on People in different Forest Divisions.

Sloth Bear Attack Risk Map

Risk maps from spatially explicit models offered an additional locations specific perspective on human killed or injured by sloth bear (Figure 14.5). Maps visually showed a patchy probability of high conflict risk with presence of definite clusters or hot spots of attack. High risk sites were noticeable in Balasore Wildlife Division and Keonjhar Forest Division. Small swathes of high risk were observed in Keonjhar Wildlife Division, different part of Keonjhar Forest Division. Northern portion of the study area including Baripada Forest Division, Karanjia Forest Division, Rairangpur Forest Division and Similipal Tiger Reserve Forest Division had low conflict risk probability.

Demography and Activity of Victims During Attack

Majority of victims (n=213, 52.59 per cent) were 31–45 years of age. Others 84 (20.74 per cent), 78 (19.25 per cent), 23 (5.67 per cent) and 7 (1.72 per cent) of victims were 16-30, 46-60, >60 and <15 years of age, respectively (n=405, χ^2_4 =324.46, P<0.01). Although incidents of sloth bear attack were reported in all age groups, victims' ages ranged from 8 to 86 years (mean= 39) at the time of attack. Of the 405 victims, 78.76 per cent (n = 319) were men and 21.23 per cent (n = 86) were women. Attacks occurred most frequently when victims were using forest fringes for toilet (n=153, 39.43 per cent), followed by during collection of non-timber forest products (n=84, 21.64 per cent), when returning from a nearby markets or work places during late

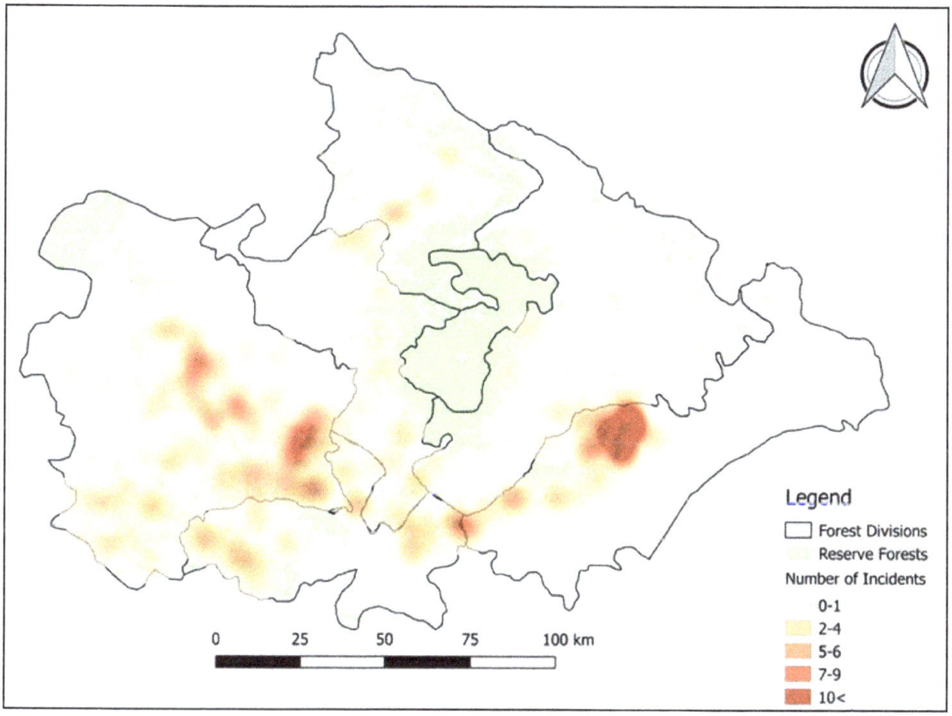

Figure 14.5. Map Showing Clusters of Location of Sloth Bear Attack on People in the Study Area.

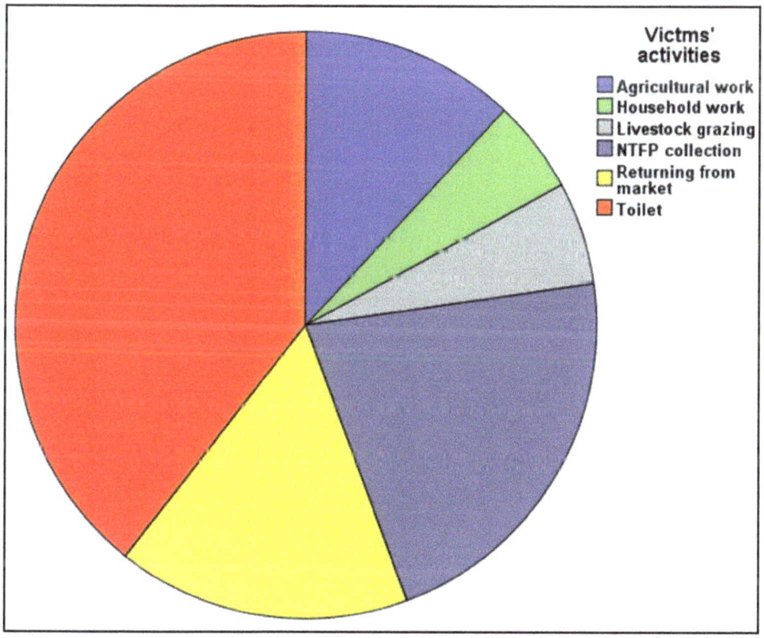

Figure 14.6. Figure Showing Activity of the Victim Prior to the Sloth Bear Attack.

evening (n=63, 16.23 per cent), during agricultural work (n=46, 11.85 per cent), when grazing livestock (n=22, 5.67 per cent), and while conducting household activities near their home (n=20, 5.15 per cent) (n=388, χ^2_5 = 190.87, $P < 0.01$, Figure 14.6).

DISCUSSION

The conservation of large carnivores like wolves, bears, tigers and lions is always associated with various challenges in human dominated landscape. Humans have modified and fragmented habitats and often experience a diversity of conflicts with large predatory neighbours, such as damage to human property and attacks on humans (Vijayan and Pati 2002, Chapron *et al.*, 2014, Palei and Sahu 2015, Kshettry *et al.*, 2017). Therefore, conservation of large carnivores in human dominated landscapes requires balancing of the safety of humans and their property with carnivore conservation goals (Bhattacharjee and Parthasarathy 2013, Chapron *et al.*, 2014). The study area is a mosaic of anthropogenic habitats dominated by agriculture field and villages, interspersed with degraded forests and protected areas with populations of large, potentially dangerous wild animals such as sloth bears, wolves, leopards and elephants. Sloth bears attacks on people were reported in the Keonjhar district including Keonjhar Forest Division and Keonjhar Wildlife Division district since 1990 (Raipurohit and Krausman 2000). Also, Keonjhar and Deogarh Forest Divisions had great sloth bear problem with 67 sloth bear related casualties between 1990 and 1995 in 20 Forest Divisions. However, within the sixteen-year period between January 2001 and December 2016, the year 2009 had the highest reports of sloth bear attacks on people in the study area.

Geographic Distribution of Sloth Bear Attack on Human

Analysing attacks by Forest Divisions indicated that some localities showed the increase in bear attacks on human. The study speculate that two factors may be responsible. First, last three decades of forest loss and habitat fragmentation in Keonjhar Forest Division and Keonjhar Wildlife Division (Reddy *et al.*, 2013). A consequences of forest loss, sloth bear attack rates increased in the forest fragmented areas. These Forest Divisions had a high forest loss and history of and high sloth bear attack rates but still have substantial forest cover. Secondly, attack rates may have increased in some localities because of increase of human population. In Nilgiri area of Balasore Wildlife Division, sloth bears are confined in a few isolated patches and most are surrounded by high human population densities approximately two times the national average and fourteen times the global average. Most of the sloth bear attacks occurred outside protected areas, suggesting national parks and sanctuaries may be effective, to some extent, at protecting sloth bears.

Causes of Conflict

HWC is becoming an emerging challenge for wildlife conservation worldwide and ample examples are there to prove the hypothesis (Rajpurohit and Krausman 2000, Packer *et al.*, 2006, Ambarlý and Bilgin 2008, Bhattarai and Fischer 2014, Kabir *et al.*, 2014, Ratnaeyke *et al.*, 2014, Singh *et al.*, 2015). Although India is one of the developing countries in the world, majority of the population of the study area is thriving under Below Poverty Line and about 89 per cent of the total population is predominantly rural based (District Annual Plan 2012). Their economy is mainly derived from crop farming or daily wage labour and dependent upon forest resources (Chattopadhyay 2012). The human population density of Balasore has risen from 291 individuals/km^2 during 1971 to 609 individuals/km^2 in 2011 (District Annual Plan 2012). This type of human growth has led to encroachment on forest and sloth bear habitats, leading to increased competition for space and resources. Therefore, rising of human-sloth bear conflicts in the study area are unsurprising.

Sloth bear has been portrayed as one of the most problematic and dangerous animals throughout its range because of its frequent interaction with human leading to casualties (Rajpurohit and Krausmann 2000, Ratnayeke *et al.*, 2014). Various studies have revealed that the extensive destruction to habitats and exploitation of the resources compelled sloth bears to come in interaction with human most frequently (Bargali *et al.*, 2005, Rajpurohit and Krausman, 2000). In northern Odisha most of the sloth bear habitats are fragmented, isolated and interspersed with human habitations, agricultural fields and stone quarries, thus creating a situation for local inhabitants and sloth bears to frequently come in interaction with each other.

It was observed that during the present study the average age of sloth bear victim was 39 years, which is similar to findings of Dhamorikar *et al.* (2017), where most of the average age of people attacked by sloth bear were 41 years age category in Madhya Pradesh, which is similar to the findings of the present study. From all the victims, majority are adult males between 31-45 years age and it is possibly they are the earning heads of the family and have the responsibility to get maximum active in the areas for earning livelihood and other activities. They were attacked by sloth bear during open toilet, farming activities during morning or on the road while returning from their work places during evening hours which corresponds to maximum sloth bear activity period.

Being monsoon one of the most productive season, people get mostly active in their agricultural fields and venture into nearby forests for livestock grazing to take the advantages of newly grown grasses and it is probably the main cause of encounter with sloth bear (Bargali *et al.*, 2005). The majority of sloth bears attacks in the neighbouring state of Chhattisgarh and Madhya Pradesh occurred during the monsoon season (Bargali *et al.*, 2005, Dhamorikar *et al.*, 2017). Attacks by Asiatic black bear (*Ursus thibetanus*) in Dachigam National Park (Kashmir) were mostly reported during May to November (Charoo *et al.*, 2011), while the encounters with Asiatic black bear and Himalayan brown bear (*Ursus arctos isabellinus*) in the Great Himalayan National Park Conservation Area (Himachal Pradesh) took place when villagers ventured into forests for fuelwood, fodder, medicinal plants or to graze livestock, irrespective of the season (Chauhan 2003). Attacks by bears during late August to September were recorded in the Sichuan Province in southwestern China where the Asiatic black bears confronted wild mushroom harvesters (Liu *et al.*, 2010).

Majority of the local inhabitants, being very poor (98 per cent of the total households) are unable to access toilet facilities (District Annual Plan 2012) for which they have to venture into nearby forested areas for defecation and get attacked by sloth bears. Summer is the prime season of various wild fruits, which constitute the seasonal diet of sloth bears (Baskaran *et al.*, 1997, Joshi *et al.*, 1997, Akhtar *et al.*, 2004, Yoganand 2005, Palei *et al.*, 2014b). Collection of these wild fruits is also equally important for the local villagers as a seasonal livelihood (Mahapatra and Panda 2009). So most of the people enter into the forest during the early morning hours for collection of NTFPs and fuel wood and get attacked by sloth bear due to sudden encounter with the animal.

Sloth bear is the most unpredictable animals and may attack without any provocation (Garshelis *et al.*, 1999, Rajpurohit and Krausman 2000, Yoganand 2005). The lack of awareness on the behavior of sloth bear among the local people might have encouraged them to fight with the sloth bear to escape, which resulted severe injury or death. As majority of the families live below poverty line, the loss of working potentiality of any earning head out of sloth bear attack leads to further economic crises. Although the victims or the families (in case of death) are eligible to get financial compensation from the state forest department, it is difficult to compensate the moral loss they experience. These situations are enough in developing negative attitude among local community towards sloth bears and loosing local support for wildlife conservation in human dominated landscapes.

Implications for Co-Existence

Understanding the factors involved in carnivore attacks on people allows the formation of recommendations for how people should behave during encounters with bears to reduce the risk of a negative outcome, and also inform people about what to expect from the bears. These are two important aspects of reducing fear (Johansson and Karlsson 2011, Johansson *et al.*, 2012). Mitigating risk factors is essential for ensuring the coexistence of people and the increasing large carnivore populations in human-dominated landscapes.

In reference to the present scenario, we here recommend to minimize the ill effects of socio-ecological conditions through appropriate interventions. Presently, Government of India is running different projects to provide sanitation facilities to the poor people (www.mdws.gov.in). The study proposes to implement such initiatives by concerned department or agencies in the villages of study area not only to provide better hygienic environment to the villagers but also to minimize the risk of sloth bear attack while going into fields for defecation.

The ecology, behavior and conservation of sloth bears need to be explained to the local people through continuous education and awareness programs throughout the affected villages. The study proposes for formation of a fully equipped, well trained and motivated conflict management team (Wildlife staffs, wildlife NGO representatives and local people) at the study area level. They should stay alert towards any human-wildlife conflict situations through networking and respond to the same at an earliest as well as take care of the first aid treatment or shifting the victims to the nearby hospital. They should also be well trained to deal with the emergencies born out of straying of sloth bear and other wild animals into human habitations including rescue, treatment (if required), translocation and monitoring the wild animals.

The local forest department in collaboration with different agencies or insurance companies should take initiative to make insurance of the conflict affected local villagers and their properties. All these initiatives are expected to build trust on forest department and gain local support for imparting various wildlife conservation interventions including sloth bear. The already constituted village level Forest Protection Committees otherwise known as *Vana Samraksana Samiti* (VSS) need to be involved in conflict management for long term period. With their support, the fragmented and degraded forest patches need to be protected and restored to sustain the existing sloth bear population. The village level committee should also be involved in education and awareness programs. Besides that, the committee should also take major decisions on regulating the collection of NTFPs from the forests, by which over exploitation of forest resources can be checked. People involved in implementing such kind of activities should be suitably rewarded. The stone quarries operating in and around sloth bear habitats need be immediately banned for minimizing the disturbance and protection of the habitat.

The study also propose to develop women Self Help Groups (SHGs) throughout the affected villages for development of micro-enterprises on dairy farming, goatry, poultry or fishery to make them economically stable and minimize their dependency on forest resources. Further research on behavior and ecology aspects of sloth bear along a densely human populated landscape will create much reliable information for better management of the situation and aid in long term conservation of sloth bear.

ACKNOELEDGEMENTS

We are grateful to Principal Chief Conservator of Forests (Wildlife) and Chief Wildlife Warden, Odisha Forest Department for supporting the study. We are thankful to Divisional Forest Officers and all forest officials for supporting the study. We also thank Subrat Debata for help to improve the manuscript.

REFERENCES

Akhtar, N., Bargali, H.S. and Chauhan, N.P.S. (2004). Feeding ecology of sloth bears in a disturbed area in central India. *Ursus,* 15(2): 212–217.

Ambarlý, H. and Bilgin, C.C. (2008). Human–brown bear conflicts in Artvin, north-eastern Turkey: Encounters, damage, and attitudes. *Ursus,* 19: 146–153.

Bargali, H.S., Akhtar, N. and Chauhan, N.P.S. (2005). Characteristics of sloth bear attacks in North Bilaspur forest division. *Ursus,* 16: 263–267.

Baskaran, N., Sivaganesan, N. and Krishnamoorthy J. (1997). Food habits of sloth bear in Mudumalai wildlife sanctuary, Tamil Nadu, southern India. *Journal of Bombay Natural History Society,* 94: 1–9.

Bhattacharjee, A. and Parthasarathy, N. (2013). Coexisting With Large Carnivores: A Case Study from Western Duars, India. *Human Dimension of Wildlife,* 18: 20–31.

Bhattarai, B.R. and Fischer, K. (2014). Human–tiger *Panthera tigris* conflict and its perception in Bardia National Park, Nepal. *Oryx,* 48: 522–528.

Can, Ö.E., D'Cruze, N., Garshelis, D.L., Beecham, J. and Macdonald, D.W. (2014). Resolving Human Bear Conflict: A Global Survey of Countries, Experts, and Key Factors. *Conservation Letters,* 7(6): 501–513.

Chapron, G., Kaczensky, P., Linnell, J.D.C. *et al.* (2014). Recovery of large carnivores in Europe's modern human-dominated landscapes. *Science*, 346: 1517–1519.

Charoo, S.A., Sharma, L.K. and Sathyakumar, S. (2011). Asiatic black bear-human interactions around Dachigam National Park, Kashmir, India. *Ursus*, 22(2): 106–113.

Chattopadhyay, A. (2012). The present scenario of socio economic condition of Balasore District of Orissa. *International Journal of Physical and Social Sciences*, 2: 420–427.

Chauhan, N.P.S. (2003). Human casualties and livestock depredation by black and brown bears in the Indian Himalaya, 1989-98. *Ursus*, 14(1): 84–87.

Chauhan, N.P.S. (2006). The status of sloth bears in India. In: T. Oi, T. Mano, K. Yamazaki, T. Aoi, M. Carr, M. Durnin, C.B. Imaki, A. Takayanagi, T. Tsubota, (Ed.), *Understanding bears to secure their future*. p. 26–34. Japan Bear Network, Gifu, Japan.

DeFries, R., Karanth, K.K. and Pareeth, S. (2010). Interactions between protected areas and their surroundings in human-dominated tropical landscapes. *Biological Conservation*, 143: 2870–2880.

Dhamorikar, A.H., Mehta, P., Bargali, H.S. and Gore, K. (2017). Characteristics of human-sloth bear (*Melursus ursinus*) encounters and the resulting human casualties in the Kanha-Pench corridor, Madhya Pradesh, India. *PLoS ONE*, 12(4): e0176612.

Dickman, A.J. (2010). Complexities of conflict: The importance of considering social factors for effectively resolving human-wildlife conflict. *Animal Conservation*, 13: 458–466.

Garcia, K.C., Joshi, H.M. and Dharaiya, N. (2016). Assessment of human–sloth bear conflicts in North Gujarat, India. *Ursus*, 27: 5–10.

Garshelis, D.L., Joshi, A.R., Smith, J.L.D. and Rice, C.G. (1999). Sloth bear conservation action plan. In: C. Servheen, and B. Peyton (Ed), *Bears: Status survey and conservation action plan* (p. 225-240). IUCN/SSC Bear and Polar Bear Specialist Groups. IUCN, Gland, Switzerland.

Jhala, Y.V., Qureshi, Q., Gopal, R. and Sinha, P.R. (2011). Status of the tigers, co-predators, and prey in India. *Technical report.* Wildlife Institute of India, Uttarakhand, Dehradun.

Johansson, M., Karlsson, J., Pedersen, E. and Flykt, A. (2012). Factors governing human fear of brown bear and wolf. *Human Dimension of Wildlife*, 17: 58–74.

Jones M.C. (1993). A Simple Nonnegative boundary correction for kernel density estimation. *Statistica Sinica*, 6: 1005–10136.

Joshi, A.R., Garshelis, D.L. and Smith, J.L.D. (1995). Home ranges of sloth bears to Nepal: Implications for conservation. *Journal of Wildlife Management*, 59: 204–214.

Kabir, M., Ghoddousi, A., Awanand, M.S. and Awan, M.N. (2014). Assessment of human leopard conflict in Machira National Park, Azad Jammu and Kashmir, Pakistan. *European Journal of Wildlife Research*, 60: 291–296.

Karanth, K.K. and DeFries, R. (2010). Conservation and management in human-dominated landscape: Case studies from India. *Biological Conservation,* 143: 2865–2869.

Karanth, K.K., Gopalaswamy, A.M., Defries, R. and Ballal, N. (2012). Assessing patterns of human-wildlife conflicts and compensation around a central Indian protected area. *PLoS ONE,* 7(12): e50433.

Kshettry, A., Vaidyanathan, S. and Athreya, V. (2017). Leopard in a tea-cup: A study of leopard habitat-use and human-leopard interactions in north-eastern India. *PLoS ONE,* 12(5): e0177013.

Laurie, A. and Seidensticker, J. (1977). Behavioural ecology of the sloth bears (*Melursus ursinus*). *Journal of Zoology,* 182(2): 187–204.

Liu, F., McShea, W.J., Garshelis, D.L., Zhu, X., Wang, D. and Shao, L. (2010). Human-wildlife conflicts influence attitudes but not necessarily behaviours: factors driving the poaching of bears in China. *Biological Conservation,* 144(1): 538–547.

Mahapatra, A.K. and Panda, P.C. (2009). Wild edible fruit plants of eastern India. Bhubaneswar: Regional Plant Resource Center.

Michel, J.M. and Asley, A.D. (2004). Concepts for exploring the social aspects of human-wildlife conflict in a global context. *Human Dimensions of Wildlife,* 9: 1–20.

Mishra, R.K., Nayak, A.K., Sahu, H.K. and Palei, H.S. (2016). *Ecology of Fauna and Flora of Similipal Tiger Reserve.* Daya Publishing House Pvt. Ltd.

Ogra, M. and Badola, R. (2008). Compensating human–wildlife conflict in protected area communities: Ground-level perspectives from Uttarakhand, India. *Human Ecology,* 36: 717–729.

Packer, C., Ikanda, D., Kissui, B. and Kushnir, H. (2005). Lion attacks on humans in Tanzania. *Nature,* 436: 927–928.

Palei, H.S., Debata, S. and Sahu, H.K. (2020). Diet of sloth bear in an agroforest landscape in eastern India. *Agroforest System,* 94: 269–279.

Palei, H.S. and Sahu, H.K. (2015). Human-Sloth bear Conflicts: A Conservation Challenge for Coexistence. In: H.K. Sahu, J. Sethy and R.K. Mishra (Ed), *Biodiversity Conservation and Management* (p. 175–181). Himalaya Publishing House, Mumbai, India.

Palei, H.S. and Sahu, H.K. (2017). Deadly Crossing: Roads are the Emerging Threats to Sloth Bears in India. *International Bear News,* 26(3): 26–27.

Palei, H.S., Mohapatra, P.P. and Sahu, H.K. (2014a). Dry season diet of the sloth bear (*Melursus ursinus*) in Hadagarh Wildlife Sanctuary, eastern India. *Proceedings of Zoological Society,* 67: 67–71.

Palei, N.C., Palei, H.S., Rath, B.P. and Kar, C.S. (2014b). Mortality of endangered Asian elephant *Elephas maximus* by electrocution in Odisha, India. *Oryx,* 48: 602–604.

Rajpurohit, K.S. and Krausman, P.R. (2000). Human-sloth bear conflicts in Madhya Pradesh, India. *Wildlife Society Bulletin,* 28: 393–399.

Rangarajan, M. (2001). *India's wildlife history*. Permanent Black, New Delhi.

Ratnayeke, S., Van Manen, F.T., Pieris, R. and Pragash, V.S.J. (2014). Challenges of Large Carnivores Conservation: Sloth Bear Attacks in Sri Lanka. Human Ecology, (42): 467–479.

Reddy, C.S., Jha, C.S. and Dadhwal, V.K. (2013). Assessment and monitoring of long-term forest cover changes in Odisha, India using remote sensing and GIS. *Environmental Monitoring and Assessment*, 185(5): 4399–4415.

Rodgers, W.A. and Panwar, S.H. (1988). *Biogeographical classification of India*. New Forest, Dehra Dun, India.

Sahu, H.K., Mohapatra, P.P., Parida, B.K., Palei, H.S. and Dutta, S.K. (2012). *Vertebrate fauna of Hadagarh Wildlife Sanctuary*. Himalaya Publishing House. Mumbai, India.

Sahu, H.K., Udgata, H.B., Debata, S. and Palei, H.S. (2014). *Vertebrate Fauna of Kuldiha Wildlife Sanctuary*. Himalaya Publishing House. Mumbai, India.

Sathyakumar, S., Kaul, R., Ashraf, N.V.K., Mookerjee, A. and Menon, V. (2012). *National Bear Conservation and Welfare Action Plan*. Ministry of Environment and Forests, Wildlife Institute of India and Wildlife Trust of India. p. 359.

Singh, R., Nigam, P., Qureshi, Q., Sankar, K., Krausman, P.R., Goyal, S.P. and Nicholoson, K.L. (2015). Characterizing human–tiger conflict in and around Ranthambhore Tiger Reserve, western India. *European Journal of Wildlife Research*, 61: 255–261.

Treves, A., Wallece, R.B. Naughton-Treves, L. and Morales, A. (2006). Co-managing human-wildlife conflicts: A review. *Human Dimensions of Wildlife*, 11: 383–396.

Vijayan, S. and Pati, B.P. (2002). Impact of changing cropping patterns on man-animal conflicts around Gir protected area with specific reference to Talala sub-district, Gujarat, India. *Population and Environment*, 23: 541–559.

Woodroffe, R., Thirgood, S. and Rabinowitz, A. (2005). The impact of human-wildlife conflict on natural systems. *In*: Woodroffe R, Thirgood S, Rabinowitz A, editors. People and Wildlife: Conflict or Coexistence? Cambridge University Press, New York, p. 1–12.

Yoganand, K. 2005. Behavioral Ecology of Sloth Bear (*Melursus ursinus*) In Panna National Park, Central India, thesis PhD, Saurashtra University, p. 322.

Yoganand, K. 2005. *Behavioral Ecology of Sloth Bear (Melursus ursinus) in Panna National Park, Central India*, thesis PhD, Saurashtra University, p. 322.

Yoganand, K., Rice, C.G. and Johnsingh, A.J.T. (2013). SLOTH BEAR *Melusus ursisnus*. In: A.J.T. Johnsingh and N. Manjrekar, Vol: 1 (Ed), *Mammals of South Asia* (p. 438-456). Universities Press (India) Private Limited, Hyderabad. ISBN: 9788173715907

Yoganand, K., Rice, C.G., Johnsingh A.J.T. and Seidensticker, J. (2006). Is the sloth bear in India secure? A preliminary report on distribution, threats and conservation requirements. *Journal of Bombay Natural History Society*, 103(2–3): 172–181.

Chapter 15

Assessment of Mugger Crocodiles (*Crocodylus palustris*) Population Status and Distribution in the Similipal Tiger Reserve, Mayurbhanj, Odisha

Srustidhar Rout[1] *and Sukanta Kumar Nayak*[2]*

[1]*Post-Graduation Department of Wildlife and Biodiversity Conservation,*
[2]*Post-Graduation Department of Biotechnology,*
North Orissa University, Takatpur, Baripada – 757 003, Odisha
e-mail: sukantanayak@rediffmail.com

Odisha is the unique state in the country where all three species of crocodiles are inhabited. The mugger crocodiles (*Crocodylus palustris*) conservation was started in Odisha along with the other two species of crocodiles. The captive breeding programme for *C. palustris* was started at Ramatirtha crocodile breeding centre of Jashipur, Odisha. Data revealed 812 individuals of mugger have been released in various phases in the river systems of Similipal since 1981. Although some success achieved through this project but adverse factors like Fishing activities, use of explosives, use of nets especially nylon nets to catch fishes, use of chemicals, DDT, insecticides, damage of the nesting sites due to fire, predation and or exhuming of mugger nests by wild boar and monitor lizard are believed to play an important role in non-propagation of muggers.

Keywords: *Crocodile, Crocodylus palustris, Conservation, Endanger, Rehabilitation, Ecology, Mugger.*

INTRODUCTION

In the Indian subcontinent, three species of crocodiles occur, namely gharial (*Gavialis gangeticus*), which belongs to the family Gavialidae, saltwater crocodile

(*Crocodylus porosus*) and mugger crocodile (*Crocodylus palustris*) belonging to the family Crocodylidae. The mugger is distributed in most parts of India except Jammu and Kashmir and some northern Indian States.

Odisha is the only state in the country where all the three Indian species of crocodilians - gharial (*G. gangeticus*), mugger (*C. palustris*) and saltwater crocodile (*Crocodylus porosus*) are found in their natural habitats. Among the three crocodile species, muggers are reported from over 10 states and the wild population is tentatively estimated as 2500 to 3500 non-hatchlings. Odisha is the distinction of having all three species of crocodiles. The mugger conservation programme at Ramatirtha in Odisha started in the during 1979 following the gharial and saltwater crocodile conservation programmes at Tikarpara and Dangmal in early 1975 had started in Odisha, respectively. It included the management of three crocodilian sanctuaries at Satkosia Gorge, Bhitarkanika and Hadgarh, and management of mugger crocodiles in the Similipal Tiger Reserve.

In 1979 as part of the Crocodile Conservation project, the field director of the Similipal Tiger Reserve and Hadgarh Sanctuary in the state of Odisha oversaw the project to restock and restore the mugger crocodile in this area of its natural habitat. This project started with a captive-breeding program at Ramatirtha with the procurement of young mugger crocodiles from Tamil Nadu. Subsequently, crocodiles have been released routinely to the rivers of Similipal from Ramatirtha crocodile breeding center of Jashipur. Till date, 812 individuals of mugger have been released in various phase in the river systems of Similipal from 1981. Several perennial rivers like river Budhabalanga in the North, Khadkei, Khairi and Bhandan in the north-west, West Deo in the south-west, Salandi in the south and East Deo and Sanjo in the east situated in Similipal were believed to be an ideal site for mugger rearing.

Crocodiles were endangered in India due to indiscriminate killing for commercial purposes and severe habitat loss until the enactment of the Wildlife (Protection) Act.1972. Earlier studies on mugger have generated or supplemented data on biology, growth variations, and territorial resource partitioning with the implication in sanctuary management, but data about their status, cause of death/ decline are meager. It was found that the mugger was extinct in Similipal by the year 1979. The state government has started the mugger conservation project in Similipal and monitoring of the rehabilitated population, after which, there has been a positive trend of growth of populations of the mugger in Similipal. But the findings of recent studies/surveys show a sharp decline non propagation in their population. Despite the state initiative and successes in first ever-captive breeding and other efforts initiated to conserve mugger, their status in the wild has declined and the status of mugger is still precarious and remains endangered. Unfortunately, due to the lack of systematic monitoring of released and declined numbers, the reasons for decline unnoticed, pushing these animals again to extinction.

Although the reasons for the decline of mugger from biosphere were not clear, adverse factors such as fishing activities, use of insecticides in agriculture, activities like fires on the river bank damage the nesting sites, natural activities such as predation and/or exhuming of mugger nests by wild boar and monitor lizard are very critical. Conducting census and estimating population status of mugger in the Similipal biosphere reserve is therefore important to know survivability of the species. Except for the census conducted periodically there have been no sustained long-term monitoring programme to look into ecological parameters like habitat use, movement patterns, territoriality, prey selection and breeding success of these crocodiles all of which are crucial in understanding the real impact of the rear and release programme in the sustained conservation of mugger in Similipal.

While very little work has been carried out on mugger especially in the biosphere in particular there is an urgent need for intense research activities to pinpoint the actual causes for their dwindling status in the biosphere. Therefore, this study has undertaken to study the entire biosphere reserve and its riverine systems to assess the exact status, habitat use and food of this species, causes of its depletion in the wild and suggest remedial measures.

With objectives to make a status survey of mugger in the Similipal biosphere reserve and different released sites and to identify the major limiting factors for which the mugger species is in danger and threats to known habitat in this specific locality.

Materials and Methods

Survey

The Similipal Tiger Reserve (STR) spreads over 2750 km^2 constituted during 1973, Similipal is declared as a wildlife sanctuary during 1979 covering an area of 2271.78 km^2. It is also part of Similipal biosphere reserve and an area measuring 845.70 km^2 in the core of Similipal has been proposed as Similipal National Park since 1980. Similipal is located between 21° 28′ to 22° 08′ North Latitude and 86° 04′ to 86° 37′ East longitude in the district of Mayurbhanj in Odisha state. At least twelve rivers cut across the plain area, all of which drain into the Bay of Bengal. The Perennial riverine systems like West Deo, Khadkei, Khairi, East Deo, Palpala, Budhabalanga*etc.*, were included in this study. The prominent among them are Budhabalanga, PalpalaBandan, Salandi, Kahairi and Deo. Map of the entire Similipal Biosphere Reserve was studied in detail along with its major river systems, tributaries, and canals situated in the biosphere reserve.

Direct field survey of muggers was carried out taking advantage of their sun-basking activities, signs of tunnels, body spoor, and excreta. The population size of the mugger was estimated by binocular observation from long distance. Binoculars were used to locate the species while they are swimming in the river. The GPS (Garmin), Map of study area 1: 50;000 with 30-second interval, DCLR Camera (Nikon D 90), Binocular (7×10), Survey sheet (Questionnaire), Scale; Pen; Pencil with Notebook, and First aid kit were used during the survey. Studies were also done to find out the bank conditions, suitable habitat for nesting, and the anthropogenic pressure which needs immediate attention.

Interviews and previous data can be helpful for the initial visit to a site as a method to obtain background information or in determining current research needs (Thorbjarnarson and Hernandez 1992). Collection of all information from local people on mugger, sighting, or death in the last two decades through an indirect collection of information was also done.

Results and Discussion

The study revealed that a total number of 812 individuals of mugger have been released in various phases in the river systems of Similipal since 1981 to date. However, as per the latest survey, the total population of mugger was 84 spreading across seven rivers such as West Deo (63), Khadkei (5), Khairi (10), East Deo (3), Palpala (2), Budhabalanga (2) and Thakthaki (1) running through seven ranges of Similipal Tiger Reserve. While the survey showed that the population was 85 in 2011 and currently (2018) it was 84.

Earlier, Mishra *et al.* (2013) recorded a total of 85 numbers of mugger crocodiles from the 6 rivers named East Deo, West Deo, Palpala, Khadkei, Khairi, and Budhabalanga running through seven ranges of Similipal Tiger Reserve in 2011. Out of 85 individuals, highest 64 numbers of muggers were recorded from West Deo followed by 9 individuals from Khadkei, 6 from Khairi, 3 from East Deo, 2 from Palpala and 1 from Budhabalanga (Table 15.1). As per literature, their population was maximum *i.e.*, 67 in West Deo river during 2016. The population size of mugger was varied from a minimum 42 in 2004 to the highest 67 in 2016 in West Deo. However, since then their number is fluctuating in and around 60 (Figure 15.1).

Table 15.1. Assessment of Mugger in different Rivers of Similipal during the Period (2004, 2012-2018)

Year	*Total*	*West Deo*	*East Deo*	*Khairi*	*B. Balanga*	*Khadkei*	*Palpala*	*Thakthaki*
2004	83	42	5	20	9	7	0	0
2012	82	60	7	6	0	9	0 (2 in 2011)	0
2013	80	56	12	2	6	4	0	0
2014	79	60	6	6	3	4	0	0
2015	81	63	4	7	3	4	0	0
2016	79	67	3	1	3	5	0	0
2017	77	63	3	4	3	4	0	0
2018	84	63	3	10	2	5	0	1

On the contrary in rest of the river systems *i.e.*, Khadkei, Bhandan, Khairi, Salandi, Budhabalanga, Palapa, East Deo, Sanjo, and Thakthaki, the mugger number was reduced to single-digit Although, these river systems are believed to be an ideal site for the mugger, over the years the number of muggers is not increasing especially since last 7 year. In the Palapala river system, the last reported mugger (2 in number) was in 2011. Since then no mugger has been recorded. During 2018 in the river Thakthaki one single mugger was sighted.

The reasons for the non-propagation/decline of muggers in the major riverine system of STR were not clear. However, agreeing with the view of Mishra *et al.*

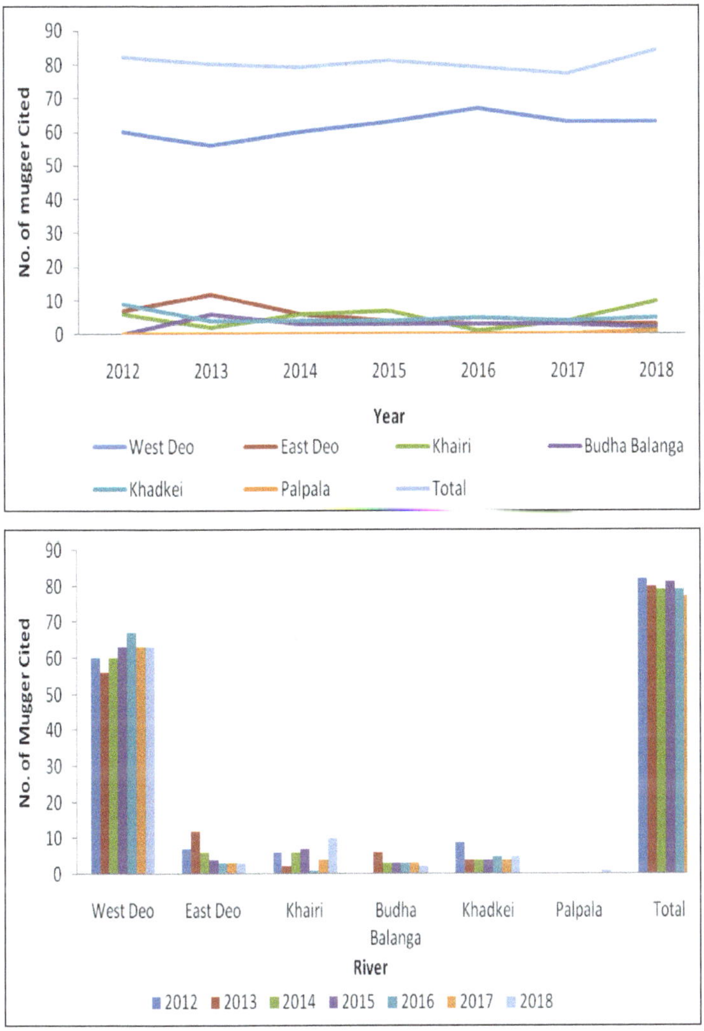

Figure 15.1. Distribution of Mugger Crocodiles in different Rivers of Similipal Tiger Reserve (2012-2018).

(2013), several adverse factors like fishing activities, use of explosives to catch fish, use of nylon nets to catch fishes, use of chemicals, DDT, insecticides, fires on the river bank damage the nesting sites, natural affects of activities such as predation and/or exhuming of mugger nests by wild boar and monitor lizard are believed to play an important role in non-propagation of muggers.

Apart from some individuals released in the rivers in core areas like East Deo and West Deo, most of the others have dispersed widely some migrating far downstream into the plain area in search of suitable habitats. Except for the census conducted periodically during the summer season there has been no sustained long-term monitoring programme to look into ecological parameters like habitat use, movement patterns, territoriality, prey selection and breeding success of these crocodiles all of which are crucial in understanding the real impact of the rear and release programme in the sustained conservation of *Crocodylus palustris* in Similipal.

CONCLUSION

Today crocodiles are mostly threatened by human activities. Foremost and the most significant among these is the destruction or alteration of wild habitat. In the past, commercial over-exploitation and indiscriminate killing have resulted in many species suffering a drastic decline in numbers and reduction in distribution, but no species has become extinct because of direct human exploitation. However, overexploitation combined with severe habitat loss has brought several species to the brink of extinction.

Crocodilians of all species depend on wetland habitats. Different species have varying preferences and requirements and crocodilians have adapted to most available tropical and subtropical wetland types (marshes, mangroves, rivers, lakes, lagoons, *etc.*). Because crocodiles are quite large animals and as they grow from hatchling to adult, they require a habitat that is both large and diverse. Rural people are often intolerant of large and potentially dangerous crocodiles and deliberate destruction of both nests and adults has been widely reported from the Andamans, China, Bangladesh, and Madagascar. Conservation of the crocodilian population is therefore highly dependent upon providing incentives to maintain crocodiles and their habitats in a relatively undisturbed state, and a willingness to accept management practices that allow crocodiles and humans to co-exist.

REFERENCES

Mishra S.R., Nayak A.K. and Nandi D. (2013). Population status of mugger crocodile (*Crocodylus palustris*) in Similipal Tiger Reserve, Odisha, India. *Int. Res. J. Env. Sci.*, 2(5): 92-94.

Sahu H.K. and Swain D. (2007). Status of mugger *Crocodylus palustris* in Similipal Tiger Reserve, Orissa, India, J. *Bombay Nat. Soc.*, 104: 2.

Vijayakumar V., Vyas R. and Choudhury B.C. (1999). Status of mugger and its conservation problems in Gujarat. *ENVIS* (*Wildlife and Protected Areas*) 2 (1): 69-76

Dani, C.S., Sagar, S.R. and Singh, LA.K. (1991). Mugger crocodile research at Ramatirtha: A review. *Indian Forester*, 117(10): 881-891.

Prusty, B.C. and Singh, L. A.K. (1994). Status of mugger crocodile in Similipal Tiger Reserve, Orissa, India. *Paper presented at Zool. Soc. of Orissa*, Utkal Univ., Vani Vihar, April 1994.

Rath, R.L., Singh, L.A.K. and Kar, S.K. (1990). Work plan for crocodilian management in Odisha, India during 1990-95. In: Crocodiles. *Proceeding of the 10th Working Meeting of the Crocodile Specialist Group. IUCN*-The World Conservation Union, Gland, Switzerland. Vol.2, 345p.

Rout, S.D. (2008). Anthropogenic threats and biodiversity conservation in Similipal biosphere reserve, Odisha, India. *Tiger Paper*, 35(3): 24-26.

Sagar, S.R. and Singh, L.A.K. (1990). Rehabilitation of mugger crocodile (*Crocodylus palustris*) in Similipal Tiger Reserve, Odisha, India. 188- 200. *In: Crocodiles. Proceedings of the 10th Working Meeting of the Crocodile Specialist Group.IUCN*-The World Conservation Union Gland, Switzerland.Vol.2,. 345p.

Singh, L.A.K. and Rout, S.D. (1998). Environment enrichment conditions and their preference by hatchling mugger (*Crocodylus palustris*). *Zoos' Print*, 13(1): 6-8

Singh, L. A.K. and Rout, SD. (1995). Twinning abnormality in *Crocodylus palustris* (Reptilia, Crocodilia). *Zoos' Print*,: 37-38.

Sahu, H.K., Dutta S.K., and Rout S.D. (2007). Survey of mugger crocodile (*Crocodylus palustris*) in Similipal Tiger Reserve, Odisha, India. *Tiger Paper*, 34(1): 27-32.

Thorbjarnarson, J.B. and Hernandez G. (1992). Recent investigations into the status and distribution of the Orinoco Crocodile *Crocodylus intermedius* in Venezuela. DBiol. Conserv. 62(3): 179-188.

Whitaker, R. and Whitaker, Z. (1984). Reproductive biology of the mugger (*Crocodylus palustris*). *J. Bomb. Nat. Hist. Soc.*, 81: 317.

Whitaker, R. and Andrews, H.V. (2003). Crocodile conservation: Western Asia Region: An update. *J. Bomb. Nat. Hist. Soc*, 100(2 and 3): 433-445.

Chapter 16

Effect of Zinc Chloride and Nickel Chloride on Morphology of Tadpole of *Duttaphrynus melanostictus*, *Euphlyctis cyanophlyctis* and *Microhyla ornata*

Puspanjali Parida and Maheswar Sahoo*

Post-Graduate Department of Zoology,
North Orissa University, Baripada – 757 003, Odisha
**e-mail: p_parida2000@yahoo.co.in*

Amphibians are a class of vertebrates that generally undergo dramatic metamorphosis involving that transformation from an aquatic larva to a terrestrial and semi-aquatic adult. Amphibians are more threatened and declining more rapidly than birds and mammals. Amphibian tadpoles are sensitive to aquatic pollutants and are used as good bioindicator. In the present study tadpole of three different species such as *Duttaphrynus*, *Euphlyctis* and *Microhyla* were exposed to different concentrations of heavy metals (Zinc chloride and Nickel chloride) at different time intervals and changes such as pigmentation pattern, intestine coiling, tail deformities, and notochord bending were observed.

Keywords: *Tadpole, Zinc chloride and Nickel chloride, Duttaphrynus, Euphlyctis, Microhyla.*

INTRODUCTION

Habitats of frog and toad are usually small, temporary ponds and the surrounding forested area. They are usually suffered by many stressors in their surroundings, such as UV-radiation (Cummins 2003, Hatch and Blaustein 2003), pesticides (Gendron *et al.*, 2006, Fellers *et al.*, 2004) and industrial chemicals (Bishop and Gendron 1998, Sower *et al.*, 2000), urbanization (Barrett *et al.*, 2010), climate change (Corn 2005). Amphibians in comparison to other aquatic species are more vulnerable to contaminants because of their aquatic larval stage, greater permeability of the skin, and the presence of gills during the larval stage (Blaustein *et al.*, 2003). Since frogs and toads are sensitive to the alterations of their environment, they could be used as bioindicator organisms to follow changes in their habitats and in ecotoxicological studies (Henry, 2000). In the present study, tadpole of three different species such as *Duttaphrynus, Euphlyctis,* and *Microhyla,* were exposed to different concentrations of heavy metals (Zinc chloride and Nickel chloride) at different time intervals and are examined under a stereomicroscope to observe the morphological changes.

Materials and Methods

Animal

Free swimming and independently feeding tadpole larva of *Duttaphrynus melanostictus, Euphlyctis cyanophlyctis,* and *Microhyla ornate* were collected from ponds and shallow water around the University campus. All tadpoles used in this experiment were collected from the same spawn of the respective species. After collection, the tadpoles were kept in the laboratory for one day before the experiment. In the laboratory, the tadpoles were maintained at 25±3°C. The tadpoles were separated into six different trays and each tray containing 10 numbers of tadpole filled with 500 ml tap water.

Experimental Set up

Healthy tadpoles were segregated into two groups: - (i) Control C_{Zn}/C_{Ni} and (ii) Experimental ($E1_{Zn}/E1_{Ni}$, and $E2_{Zn}/E2_{Ni}$). The experimental tray E1 and E2 exposed with 0.01 ppm and 0.1 ppm concentration of Zinc chloride and Nickel chloride. Tadpoles were feeded with aquatic vegetation. On 7th and 14thdays, two number of tadpoles from each control (C) and Experimental (E1, E2) tray were collected and fixed in 4 per cent formalin solution.

Microscopy and Photomicrography

Malformation like tail deformity, intestine coiling, notochord bending, visceral edema was photographed using the Stereomicroscope and U-EYE software. The experimental groups were compared with that of the control group.

Results and Discussion

There was no mortality occurred in the control group and no morphological abnormalities were recorded. The tadpoles of *Duttaphrynus, Euphlyctis, Microhyla* were exposed to heavy metals ($ZnCl_2$and $NiCl_2$)., were observed that after 7 days five morphological abnormalities were seen like tail deformity, notochord bending, visceral edema, change in intestine coiling and change in pigmentation pattern of the tail (Tables 16.1–16.8 and Figures 16.1–16.8).

Table 16.1. Morphological Changes of *Duttapharynus exposed* to Nickel Chloride

Conc. (ppm)	*Days*	*Tail Deformity*	*Intestine Coiling*	*Visceral Odema*	*Notochord Bending*
0	7	Normal	Normal, S- Shaped	Not Found	Straight
	14	Normal	Normal, S- Shaped	Not Found	Straight
0.01	7	Breakage	Changed	not found	Straight
0.01	14	Breakage	not visible	not found	Straight
0.1	7	Breakage	Scattered	Present	Bending
0.1	14	Breakage	Slightly Changed	Not Found	Straight

Table 16.2. Morphological Changes of *Duttapharynus* Exposed to Zinc Chloride

Conc. (ppm)	*Days*	*Tail Deformity*	*Intestine Coiling*	*Visceral Odema*	*Notochord Bending*
0	7	Normal	Normal, S-Shaped	Absent	Straight
	14	Normal	Normal, S-Shaped	Absent	Straight
0.01	7	Absent	Not visible	Present	Straight
	14	Absent	Slightly changed	Absent	Bending
0.1	7	Breakage	Not visible	Absent	Bending
	14	Dead			

Table 16.3. Morphological Changes of *Euphlyctis* Exposed to Nickel Chloride

Conc. (ppm)	*Days*	*Tail Deformity*	*Intestine Coiling*	*Visceral Odema*	*Notochord Bending*
0	7	Not found	Normal, S-Shaped	Normal	Straight
	14	Not found	Normal, S-Shaped	Normal	Straight
0.01	7	Breakage	Slightly changed	Increases	Bending
	14	Breakage	Not visible	Increases	Bending
0.1	7	Breakage	Normal, S-Shaped	Decreases	Bending
	14	Breakage	Change in pattern	Decreases	Bending

Table 16.4. Morphological Changes in *Euphlyctis* Exposed to Zinc Chloride

Conc. (ppm)	*Days*	*Tail Deformity*	*Intestine Coiling*	*Visceral Odema*	*Notochord Bending*
0	7	absent	Normal	Normal	Straight
	14	absent	Normal	Normal	Straight
0.01	7	absent	Slightly changed	Decreases	Straight
	14	Breakage	Normal	Increases	Straight
0.1	7	Absent	Slightly changed	Increases	bending
	14	Breakage	Normal	Increases	bending

Table 16.5. Morphological Changes in *Microhyla* Exposed to Nickel Chloride

Conc. (ppm)	*Days*	*Tail Deformity*	*Intestine Coiling*	*Visceral Odema*	*Notochord Bending*
0	7	Normal	Normal, S-shaped	Normal	Straight
	14	Normal	Normal, S-shaped	Normal	Straight
0.01	7	Breakage in lower region of tail fin	Normal, S-shaped	Slightly dispersed	Straight
	14	Breakage in of lower tail fin	Normal, S-shaped	Dispersely placed	Straight
0.1	7	Breakage in lower region of tail fin	Pattern changed	Decreases	Straight
	14	Breakage in lower region of tail fin	Not visible	Decreases	Present

Table 16.6. Morphological Changes in *Microhyla* Exposed to Zinc Chloride

Conc. (ppm)	*Days*	*Tail Deformity*	*Intestine Coiling*	*Visceral Odema*	*Notochord Bending*
0	7	Normal Tail	Normal S Shaped	Normal	Straight
	14	Normal Tail	Normal S Shaped	Normal	Straight
0.01	7	Breakage In Tail Fin	Changed	Decreases	Straight
	14	Dead			
0.1	7	Breakage in lower region of tail fin	Normal	Decreases	Bending
	14	Breakage in tail Fin	Normal	Normal	Straight

Table 16.7. Comparison of Morphological Changes Observed in Three different Species of Tadpole when Exposed to 0.1 ppm of Nickel Chloride

Species	*Days*	*Tail Deformity*	*Visceral Odema*	*Intenstine Coiling*	*Pigmentation in Tail Region*	*Notochord Bending*
Duttaphrynus	7	Present	Present	Not changed	Not changed	Present
	14	Present	Absent	Changed	Not changed	Present
Euphlyctis	7	Present	Absent	Not changed	Changed	Present
	14	present	absent	Changed	Slightly changed	Absent
Microhyla	7	Present	Absent	Not changed	Slightly changed	Absent
	14	Present	Absent	Not changed	Slightly changed	Present

Table 16.8. Comparison of Morphological Changes Observed in Three different Species of Tadpole when Exposed to 0.1 ppm of Zinc Chloride

Species	*Days*	*Tail Deformity*	*Visceral Odema*	*Intenstine Coiling*	*Pigmentation in Tail Region*	*Notochord Bending*
Duttaphrynus	7	Present	Absent	Not Vissible	Not changed	Present
	14	Present	Absent	Not Vissible	Not changed	Present
Euphlyctis	7	Absent	Absent	Normal	Changed	Present
	14	Present	Absent	Not Vissible	Changed	Present
Microhyla	7	Present	Absent	Normal	Not changed	Present
	14	Present	Absent	Normal	Not changed	Absent

From the Tables 16.1–16.8, it is found that tadpoles were sensitive towards the exposure of heavy metals like Zinc chloride and Nickel chloride. Both the heavy metals cause tail breakage (Figures 16.6 and 16.8) and notochord bending (Figures 16.2, 16.3, 16.4 and 16.7) in all three species (*Duttaphrynus, Euphlyctis* and *Microhyla*). Pigmentation pattern (Figure 16.5) not changed on 7th or 14 th day in case of *Duttaphrynus* exposed to heavy metals but visceral edema (Figure 16.1) observed in response to Nickel chloride. Intestine coiling has no significant effect.

CONCLUSION

Both Zinc chloride and Nickel chloride caused tail breakage and notochord bending in tadpole larvae of *Duttaphrynus, Euphlyctis,* and *Microhyla;* However, intestine coiling and pigmentation pattern were not changed significantly. Nickel

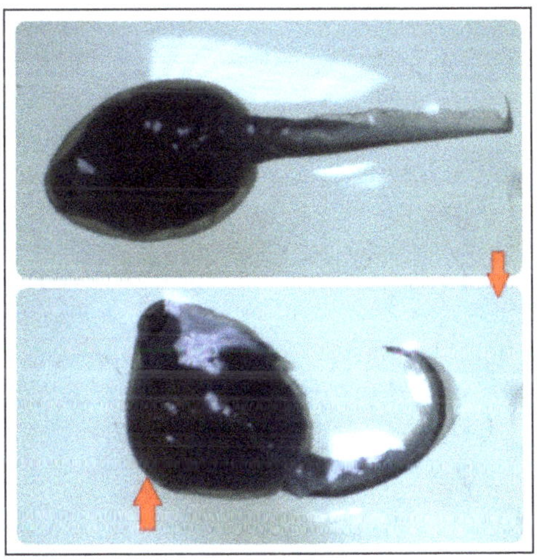

Figure 16.1. (A) Representative image of Normal Tadpole of *Duttaphrynus* observed after 7 days; (B) Shows presence of visceral odema in tadpole of *Duttaphrynus* exposed to 0.1ppm $NiCl_2$ for 7 days.

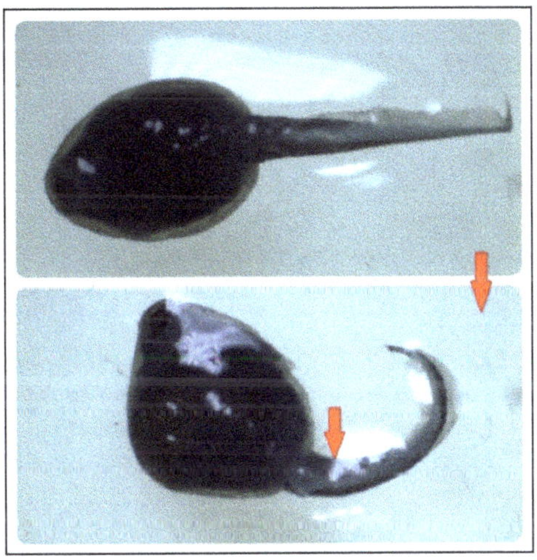

Figure 16.2. (A) Representative image of Normal Tadpole of *Duttaphrynus;* (B) Shows bending in notochord of *Duttaphrynus* when exposed to 0.1ppm $NiCl_2$for 14 days in comparison to normal.

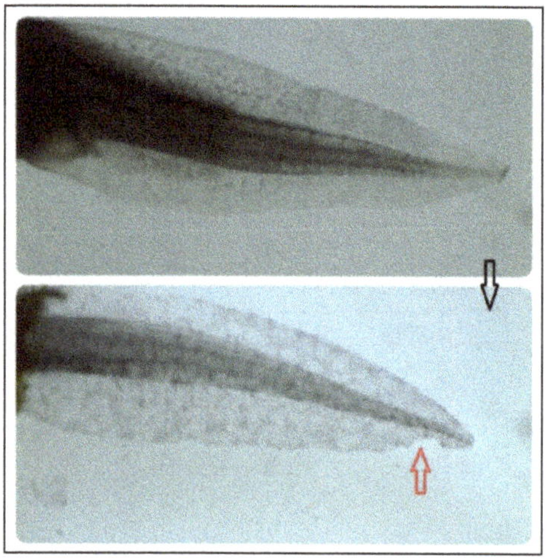

Figure 16.3. (A) Shows normal tadpole observed for 7 days; (B) Shows breakage in tail fin of *Euphlyctis* when exposed to 0.1 ppm $ZnCl_2$ for 7 days.

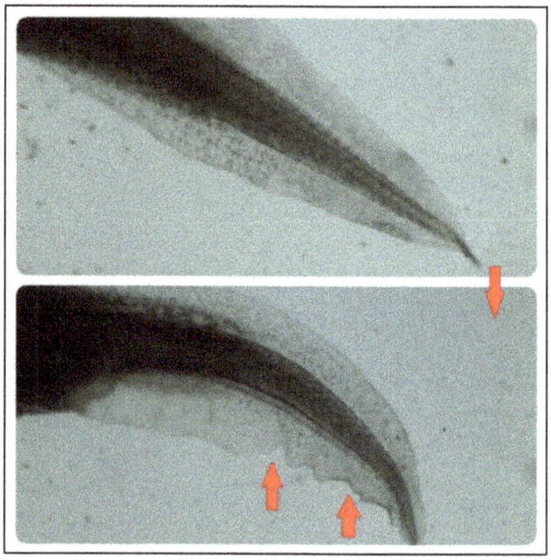

Figure 16.4: (A) Shows normal tadpole observed for 14 days; (B) Shows breakage in tail fin of *Euphlyctis* when exposed to 0.1 ppm $ZnCl_2$ for 14 days.

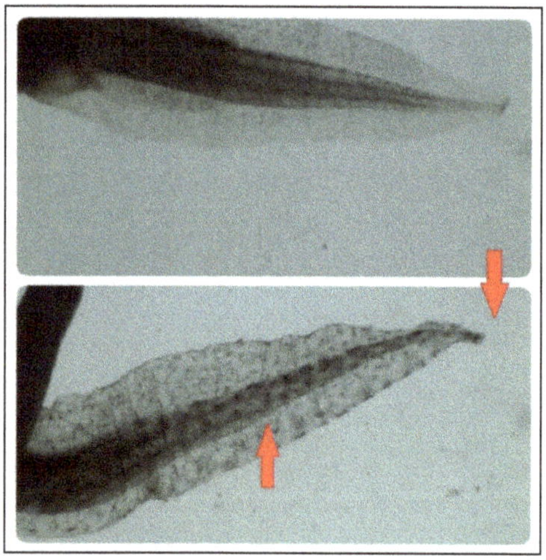

Figure 16.5: (A) Shows normal tadpole observed for 7 days; (B) Shows change in pigmentation in tail region of *Euphlyctis* when exposed to 0.1 ppm $NiCl_2$ for 7 days.

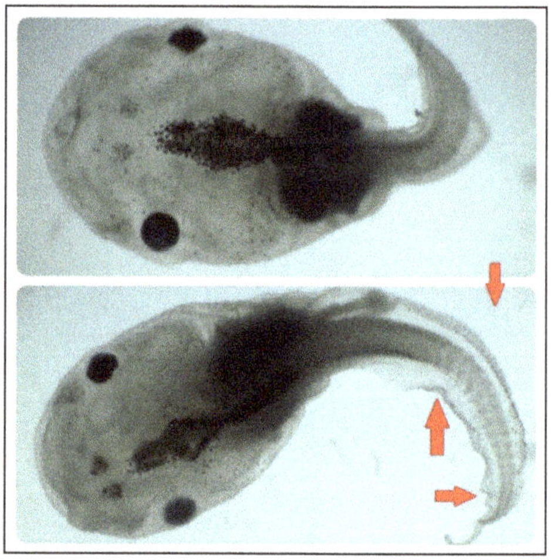

Figure 16.6. (A) Shows normal tadpole observed for 7 days; (B) Shows breakage in tail fin of *Microhyla* when exposed to 0.1 ppm $NiCl_2$ for 7 days in comparison to normal.

chloride was more toxic than Zinc chloride for the tadpole larvae. Visceral edema was found in the case of *Duttaphrynus* which indicates that the species was more prone to the toxic effect of heavy metal in comparison to the other two species; it was also observed that the mortality rate of *Duttaphrynus* was higher in comparison to *Euphlyctis* and *Microhyla*.

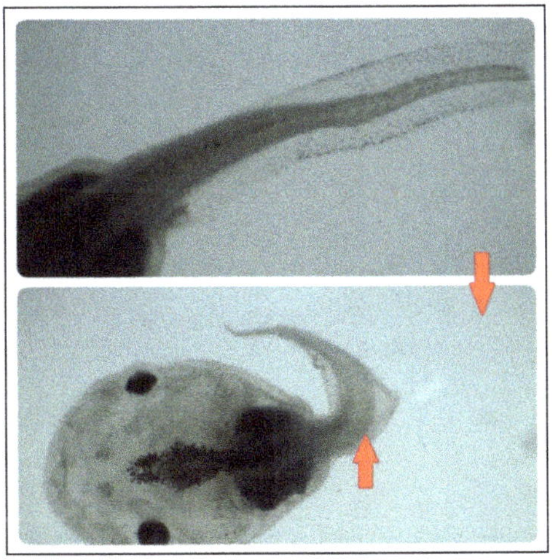

Figure 16.7. (A) Shows normal tadpole observed for 7 days; (B) Shows Bending of notochord in *Microhyla* when exposed to 0.1 ppm $NiCl_2$ for 7 days.

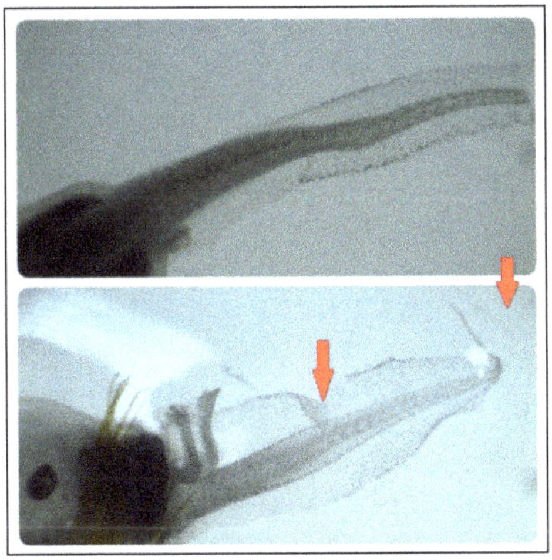

Figure 16.8. (A) Shows normal tadpole observed for 7 days; (B) Shows breakage in tail fin of *Microhyla* when exposed to 0.1 ppm $NiCl_2$ for 14 days.

Tail breakage might reduce the swimming efficiency of the tadpole larva which hampers to get sufficient food for its survival; maybe a reason for declining of the amphibian species. The above morphological character of tadpole may be used as a bioindicator of heavy metal toxicity for aquatic animals.

REFERENCES

Barrett, K., Guyer, C. and Watson, D. (2010). Water from Urban Streams Slows Growth and Speeds Metamorphosis in Fowler's Toad (*Bufo fowleri*) Larvae. *Journal of Herpetology*,44: 297-300.

Bishop, C.A. and Gendron, A.D. (1998). Reptiles and amphibians: shy and sensitive vertebrates of the Great Lakes basin and St. Lawrence river. *Environmental Monitoring and Assessment*, 53: 225-244.

Blaustein, A.R., Romansic, J.M., Kiesecker, J.M. and Hatch, A.C. (2003). Ultraviolet radiation, toxic chemicals, and amphibian population declines. *Diversity and Distribution*, 9: 123-140.

Corn, P.S. (2005) Climate change and amphibians. *Animal Biodiversity and Conservation*, 28(1): 59-67.

Cummins, C.P. (2003) UV-B radiation, climate change and frogs - the importance of phenology. *Annales Zoologici Fennici*, 40: 61-67.

Fellers, G.M., Mcconnell, L.L., Pratt, D. and Datta, S. (2004). Pesticides in mountain yellow legged frogs (*Rana muscosa*) from the Sierra Nevada Mountains of California, USA. *Environmental Toxicology and Chemistry*, 23: 2170–2177.

Gendron, A.D., Marcogliese, D.J., Barbeau, S., Christin, M.S., Brousseau, P., Ruby, S., Cyr, D. and Fournier, M. (2006). Exposure of leopard frogs to a pesticide mixture affects life history characteristics of the lungworm Rhabdiasranae. *Oecologia,* 135: 469-476.

Hatch, A.C. and Blaustein, A.R. (2003). Combined effects of UV-B radiation and nitrate fertilizer on larval amphibians. *Ecological Application,* 13: 1083-1093.

Henry, P.F.P. (2000): Aspects of amphibian anatomy and physiology. In: Sparling, D. W., Linder, G. and Bishop, C.A. (Eds.). 2000. *Ecotoxicology of amphibians and reptiles Pensacola,* F.L: Society of Environmental Toxicology and Chemistry (SETAC). pp. 71-111.

Chapter 17

Traditional Value and Antibacterial Activity of Microtermite (*Odontotermes obesus*) of Mayurbhanj District, Odisha, India

Subhrakanta Jena, Sushree Sudha Das and Hemanta Kumar Sahu*

Department of Zoology, North Orissa University, Baripada – 757 003, Mayurbhaj, Odisha
**e-mail: jena.subhrakanta@gmail.com*

Insects are tiny creatures and more fascinating diversified groups of organism in nature. Many researches have been conducted worldwide regarding the beneficiary roles of insects with respect to biodiversity, ecology, economic zoology, parasitology and ethno-entomology. The techniques of practicing insects for food, medicine and other economic use were integral parts of the ancient culture of indigenous people of the world including India. However, the studies reveals with ethno-entomological use come into existing approximately two decades back to find out alternative food and medicinal source for humankind. Therefore, a study has been carried out on an edible micro-termite species (*Odontotermes obesus*) in Mayurbhanj district of Odisha regarding its traditional use and antibacterial property against four selected human pathogenic bacterial strains. It is found that the fried form of termite is medicinally used to treat lactation problems in women and the healing of an ulcer. The raw extract of the species also has shown significant antibacterial activity in *B. licheniformis* (19.916 ± 0.352) followed by *V. cholerae* (17.823 ± 0.438), *P. aeruginosa* (16.766 ± 0.185) and *S. typhimurium* (15.256 ± 0.339). More research in documenting the traditional practice in a wide area and the potency against large no of pathogenic bacteria and fungus species should be conducted to conserve more medicinal information for the future benefits.

Keywords: *Traditional value, Antibacterial activity, O. obesus, Mayurbhanj district.*

INTRODUCTION

Nature has bestowed humankind with many resources for its existence. The technique of utilizing both the living and non-living matters have adopted by humans since it comes into existence. Early reports of human civilization concerned with Vedic culture revels the use of plants and animals for treating various ailments. A lot of traditional plants and animal species are reported from different parts of the world, which are being used by the different ethnic tribes. Entomophagy is now a practice in different cultures as a preferred nutritious food and effective source for medication to overcome malnutrition and health hazards of daily life respectively. Documentations of different traditional insect species have been reported with respect to nutritional, medicinal, and socioeconomic use. Mayurbhanj district is the largest district of Odisha state and has dense forest areas, rich floral and faunal diversity. The district is inhabited by various tribes like Kolha, Munda, Santhal, Majhi and Lodha, which are the key characteristic features of the district. These tribes are also called the ethnic tribes as they are experts in the traditional medicinal knowledge in treating a variety of diseases by using the plant and animal products in their raw forms. Many researchers have documented the traditional use of plants (Rout and Panda, 2010; Kar *et al.*, 2013; Routray and Nayak, 2017) and their scientific evaluation with respect to different bioactive properties *e.g.* antimicrobial, antihelmintic, antimalarial and antiviral activity (Nahak and Sahu, 2015; Panda *et al.*, 2017). As like the plants, the insects are also consumed in 80 per cent countries of the world and serving as an important food (human nutrition), medicinal and commercial resource in many regions such as Africa, Asia and Latin America (Alletor, 1995; Siriamornpun and Thammapat, 2008; Ekop, 2010; Carrington, 2010). As the record suggests there are about 2000 edible insects are exist worldwide and being used in different cultures until date (CAC, 2010).In India, a long history is lying behind in North and East India represents ethnic rural communities possessing techniques of utilizing a large number of edible insects in their culture. About 255 species of the edible insect so far recorded from different parts of India, out of which 158 species from Arunachal Pradesh followed by in Manipur, Assam, Nagaland, and Meghalaya (16 to 40 insect species) and to a lesser extent in Kerala, Madhya Pradesh, Odisha Tamil Nadu, and Karnataka, this number limits only six insect species (Chakravorty, 2014). Termites belong to the family: Termitidae represents about 2 per cent of the total insects consumed in India (Chakravorty, 2014). However, very few records have been reported on the ethno-entomophagy practice along with their scientific values from Mayurbhanj district. Macrotermite (Jharipoka) is the only edible termite species is used as food, consumed together with rice in the fried form (Srivastava *et al.*, 2009; Jena and Sahu, 2017). Besides the macro-termite, the other two species like Red weaver ant (*Oecophylla smaragdina*) and pupae of *Anthera myllita* are also reported from Mayurbhanj district of Odisha, which are used for different ailments and food by different tribes (Jena and Sahu, 2017). The significant antibacterial activity of the subterranean termites of South Indian folk medicines has studied against 9 species of bacteria (Solavan *et al.*, 2007).

In this investigation, the traditional use and antibacterial activity of the subterranean microtermite (*Odontotermes obesus*) have carried out against human pathogenic bacteria in in-vitro condition by using the raw extract as used by the tribes.

Methodology

The traditional information was collected by questioner survey method (Figure 17.1) in different local areas of the Mayurbhanj district and in the same time termite sample (Figure 17.2) was collected for laboratory use.

Figure 17.1. Local Interaction in Field.

Figure 17.2. Termite Sample Collected from Field.

After collection of termite samples, the identification was done by using a proper identification manual (Shanbhag and Sundararaj, 2011) and then sterilized with 70 per cent alcohol. The aqueous extract was prepared at a proportion of 1: 10

ratio (one gram of termite sample in 10 ml of sterile distilled water *i.e.* at of 0.1gm/ml concentration as suggested by Vidhu and Evans, 2015) and homogenized to make it into a crude extract. Then the extract was centrifuged by using Cooling Centrifuge (REMI Company) at 4000rpm and the supernatant was collected and kept in the freezing condition for further use. The bacterial cultures were maintained on nutrient agar NA slants and stored at 4°C. Activation of the bacterial cultures was carried out by a streak plate culture method. A single colony was picked from each plate and transferred to 2ml of NB broth medium to get pure broth culture of each strain. The agar cup well diffusion method (Barry, 1980) was used to study the antibacterial activity against four selected bacterial species (Table 17.1). The termite aqueous extract at different amount (*e.g.* 10µl, 20µl, 40µl, 60µl, 80µl, and 100µl) was administrated in the cultured agar plate of each strain along with the control antibiotic Chloramphenicol (5µl) and the plate was incubated at 37°C for 24 hours. Then zone of inhibition was measured by using digital slide caliper.

Table 17.1. Showing the Bacterial Species Details Used for Antibacterial Activity

Sl. No.	*Bacterial Code*	*Strain Name*	*Gram Positive/ Negative*	*Growth Temperature*	*Patho-genicity*
1	NCTC 74	*Salmonella typhimurium*	Gram-Negative	37°C	Pathogenic
2	MTCC 854	*Vibrio cholerae*	Gram-Negative	37°C	Pathogenic
3	MTCC S1	*Pseudomonas aeruginosa*	Gram-Negative	37°C	Pathogenic
4	MTCC 10341	*Bacillus licheniformis*	Gram-Positive	37°C	Pathogenic

MTCC: Microbial Type Culture Collection, Chandigarh; NCTC: National Collection of Type Cultures, United Kingdom.

Results

The *Odontotermes obesus* (Figure 17.3) species belong to the family: Termitidae and mostly called the Microtermite category due to its small size but in local Oriya language it is named as "Uwe".

Systematic Position

Kingdom: Animalia

Phylum: Arthropoda

Class: Insecta

Order: Isoptera

Family: Termitidae

Genus: *Odontotermes*

Species: *obesus*

Figure 17.3: *Odontotermes obesus.*

The traditional information reveals that this species of the termite is used as food and medicine among the tribes. According to the tribal healers, it is an effective source for treating the Lactation problem in women and used in Ulcer. The significance of this species is that it is available in all seasons and the method of administration during treating is quite easier *i.e.* it is consumed orally either in the fried condition or in the crude aqueous extract form.

Table 17.2. Detail of Traditional Use of Termite by Tribes of Mayurbhanj District

Insect Name	*Parts Use*	*Aliments (Used for disease)*	*Used as*	*Models of Preparation*	*Result*
Termite	Whole organism and queen	Increases the lactation in women and ulcer	Both as food and medicine	Fried and directly taken	Long processes

The antibacterial activity of *O. obesus* species is presented (Table 17.3). It showed effective antibacterial activity against all the bacterial strains and increased with an increase in the amount of the extract (Figure 17.4). The antibacterial activity was most apparent in all the four bacterial strains. The extract was highly active against *Vibrio cholerae* and *Bacillus licheniformis* as the inhibition was found in the entire administrated amount.

Table 17.3. Antibacterial Activity of the Termite Extract and Chloramphenicol Antibiotic

Sl. No.	*Bacterial Code*	*Zone of Inhibition of Termite Extract (Avg. ±SD mm)*						*Chloram-phenicol (Avg. ±SD mm)*
		10μl	*20μl*	*40μl*	*60μl*	*80μl*	*100μl*	
1	NCTC 74	0	0	10.660 ± 0.38	12.983 ± 0.304	13.683 ± 0.486	15.256 ± 0.339	24.576 ± 2.042
2	MTCC 854	11.096 ± 0.484	12.113 ± 0.311	13.800 ± 0.505	15.940 ± 0.431	17.513 ± 0.474	17.823 ± 0.438	24.83 ± 1.028
3	MTCC S1	0	9.920 ± 0.664	12.016 ± 0.302	13.923 ± 0.195	15.290 ± 0.397	16.766 ± 0.185	33.293 ± 1.349
4	MTCC 10341	9.763 ± 0.394	10.820 ± 0.449	8.540 ± 6.050	15.993 ± 0.295	18.173 ± 0.319	19.916 ± 0.352	35.906 ± 0.691

Avg.: Average; SD: Standard deviation.

In *Salmonella typhimurium* and *Pseudomonas aeruginosa*, the inhibition of termite extract was lesser than *Vibrio cholerae* and *Bacillus licheniformis*. The inhibition of termite extract in comparison to the commercial antibiotic Chloramphenicol is quite less in *P. aeruginosa* and *Bacillus licheniformis* but the effectiveness in the case of *Vibrio cholerae* and *Salmonella typhimurium* can be comparable.

DISCUSSION

Considerable antibacterial activity is found in the termite extract it may be due to the production of different metabolites produced inside this species, which signifies the scientific value of this termite. The early report suggests the presence of antimicrobial peptides like termicin, cysteine and spingerin in isolated fugal species

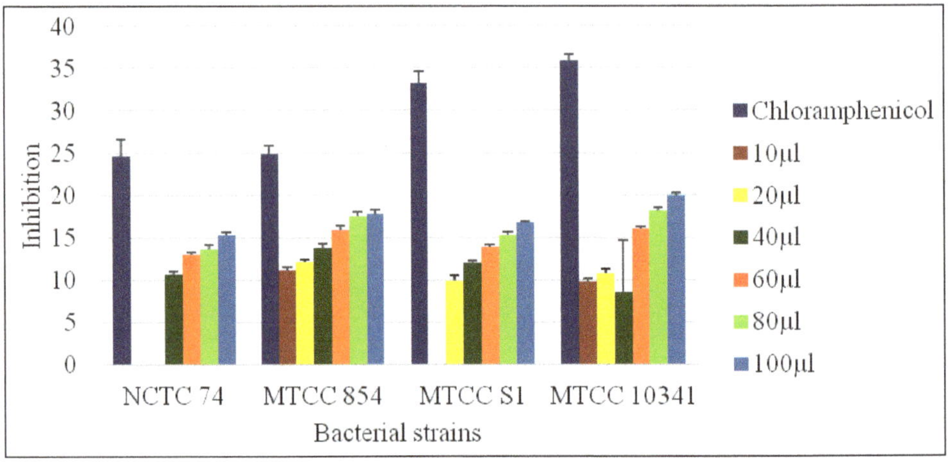

Figure 17.4. Antibacterial Activity of Termite Extract and Control Antibiotic (Chloramphenicol).

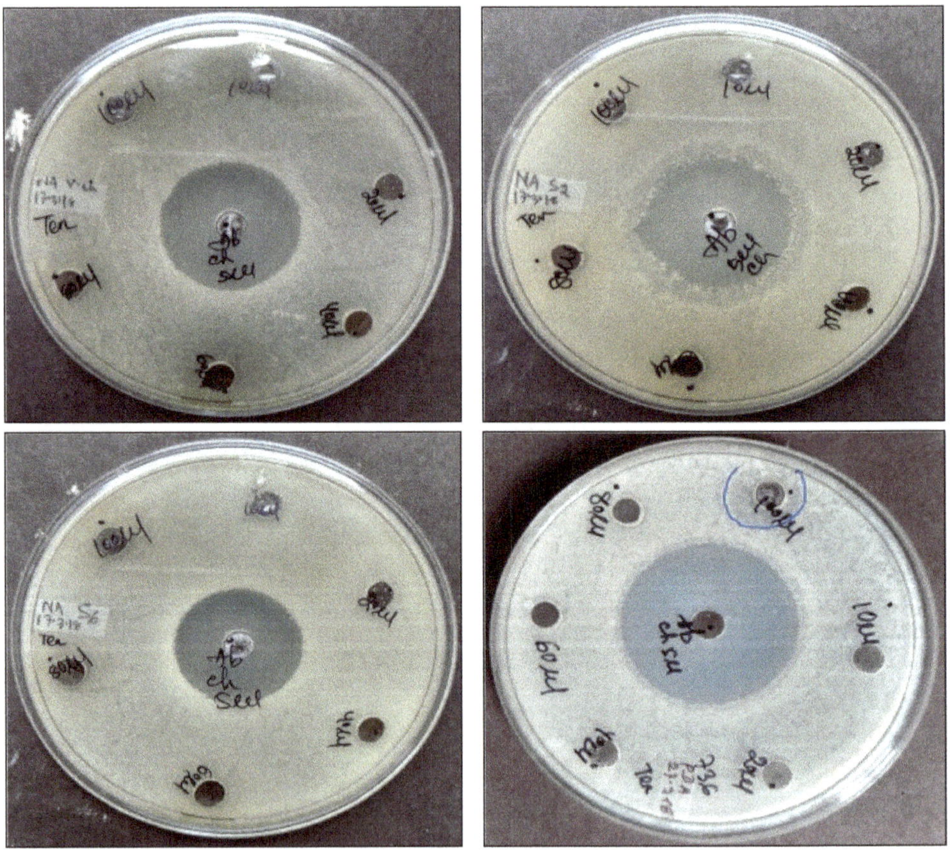

Figure 17.4: Agar Cup Well Diffusion of Termite Extract against Bacterial Strains.

(a) *Bacillus licheniformis*; (b) *Pseudomonas aeruginosa;* (c) *Vibrio cholerae*; (d) *Salmonella typhimurium.*

found in *Pseudacanthotermes spiniger* (Lamberty *et al.*, 2001; Da Saliva *et al.*, 2003). Interestingly, the tribes use this termite (*O. obesus*) for the treatment of lactation and ulcer problems. The potency of termite extract indicates that proximity arthropods have always a chance to be infected by the microbes from their surrounding environment. Therefore, whenever the microbes encounter these arthropods, they may incorporate some antimicrobial compound synthesized only inside their own body to minimize the level of microbial infection (Beattie *et al.*, 1986). This defense mechanism found in arthropods not only helpful to revive them from microbial infection but also protect their nest or colony as found in soil-dwelling ant's defensive behavior in protecting their nest from bacteria and fungus (Holldobler and Wilson, 1990). However, the mystery behind its antibacterial activity is yet to be discovered. Maybe the isolation and enumeration of bioactive antimicrobial compounds or metabolites present in this termite extract will be an effective one and could have more bio-pharmacological properties. Once these desired beneficiary bioactive metabolites are isolated and examined based on, the bio-pharmacological properties will positively contribute a way to design new effective medications for future benefits. Similar studies should be conducted on a wide range of edible arthropods, which could be beneficial in discovering the new source for antiviral, antifungal, anticancer drugs for biomedical research.

CONCLUSION

The study provides preliminary information on the antibacterial property of the subterranean termite *O. obesus* used as folk medicine in the Mayurbhanj district. Since the traditional knowledge of endogenous tribes inhabiting different parts of the world playing an important role in identifying natural resources worthy of commercial exploitation and pharmacological values. The documentation of this ethnobiological information will helpful to conserve the traditional knowledge and techniques of utilizing edible insects. In forthcoming days, the study should be carried out on a greater number of edible insects for ethno- entomological values. As scientific studies require more samples of these tiny creatures, which may affect the nature's integrity so that the researchers should sustainably conduct the experiments for a better future of humankind.

REFERENCES

Alletor, V.A. (1995). Compositional studies on edible tropical species of mushrooms. *Food Chem*, 54: 265-268.

Barry, A.L. (1980). *Procedure for testing antimicrobial agents in agar media*. In: Lorin, V. (ed.). Antibiotics in Laboratory Medicine, Williams Wilkins Co. Baltimore, pp. 1-23.

Beattie, A.J., Turnbull, C.L. and Hough, T. (1986). Antibiotic production: a possible function for the metapleural glands of ants (Hymenoptera: Formicidae). *Annals of the Entemological Society of America*, 79: 448-450.

CAC, (2010). Development of regional standard for edibles crickets and their products. 17th CCASIA-CRD 8, Seventeenth Session held at Bali, Indonesia, 22-26 November 2010. pp. 1-9.www.codexalimentarius.net

Carrington, D. (2010). Insects could be the key to meeting food needs of growing global population. The Guardian, Retrieved 27 February 2011.https: //www. theguardian.com

Chakravorty, J. (2014). Diversity of Edible Insects and Practices of Entomophagy in India: An Overview. *Journal of Biodiversity Bioprospecting and Development*, 1: 124.

Da Saliva, P., Jouvensal, L., Lamberty, M., Bullet, P., Caille, A. and Vovelle, F. (2003). Solution structures of tremicin, an antimicrobial peptide from termite *Pseudacanthotermes spiniger. Protein Science,* 12: 438-446.

Ekop, E.A., Udoh, A.I. and Akpan, P.E. (2010).Proximate and anti-nutrient composition of four edible insects in Akwalbom state, Nigeria. *World Journal of Applied Science and Technology*, 2: 224-231.

Holldobler, B. and Wilson, E.O. (1990). The ants. Belknap Press of Harvard University, Cambridge, pp. 732.

Jena, S. and Sahu, H.K. (2017). Traditional medicinal use of insects among the tribes of Baripada, Mayurbhanj, Odisha. In: Tribal Health: A regional prospective, Edited by L. Sahoo, pp. 194-210. New Delhi Serial Publisher, New Delhi.

Kar, T., Mandal, K.K., Reddy, C.S. and Biswal, A.K. (2013). Ethnomedicinal plants used to cure Diarrhoea, Dysentery and Cholera by some tribes of Mayurbhanj district, Odisha, India. *Life Sciences Leaflets*, 2: 18-28.

Lamberty, M., Zachary, D., Lanot, R., Bordereau, C., Robert, A., Hoffmann, J. and Bulet, P. (2001). Insect immunity, constitutive expression of a cysteine-rich antifungal and a linear antibacterial peptide in a termite insect. *Journal of Biological Chemistry*, 276, pp. 4085-4092.

Nahak, G. and Sahu, R.K. (2015). Review on ethnomedicinal plants of Odisha for the treatment of Malaria. *Indian Journal of Pharmacognosy and Phytochemical Research,* 7(1): 156-165.

Panda, S.K., Padhi, L., Leyssen, P., Liu, M., Neyts, J. and Luyten, W. (2017). Antimicrobial, Anthelmintic and Antiviral activity of plants traditionally used for treating infectious disease in the Similipal Biosphere Reserve, Odisha, India. *Frontiers in Pharmacology*, 8: 658.

Rout, S.D. and Panda, S.K. (2010). Ethnomedicinal plant resources of Mayurbhanj district, Odisha. *Indian Journal of Traditional Knowledge,* 9(1): 68-72.

Routray, A. and Nayak, S. (2017). Ethnomedicinal plants resources of Stakosia RF of Karanjia Division, Mayurbhanj, Odisha. *Imperial Journal of Interdisciplinary Research,* 3(4): 2159-2161.

Shanbhag, R.R. and Sundararaj, R. (2011). An identification guide to the wood destroying termites of south India. *Journal of the Indian Academy of Wood Science,* 8(2): 148-151.

Siriamornpun, S. and Thammapat, P. (2008). Insects as a Delicacy and a Nutritious Food in Thailand. In: Using Food Sciencce and Technology to improve Nutrition and Promote National Development: Selected case studies. Edited by G. Robertson and J. Lupien (Eds), *International Union of Food Science and Technology.*

Solavan, A., Paulmurugan, R. and Wilsanand, V. (2007). Antibacterial activity of subterranean termites used in South India folk medicine. *Indian Journal of Traditional Knowledge*, 6(4): 559-562.

Srivastava, S.K., Babu, N. and Pandey, H. (2009). Traditional insect bioprospecting as human food and medicine. *Indian Journal of Traditional Knowledge*, 8: 485–494.

Vidhu, V.V. and Evans, D.A. (2015). Ethnoentomological values of *Oecophylla smaragdina* (Fabricius). *Current Science*, 109 (3): 572-579.

Chapter 18

Freshwater Cyanobacterial Diversity of Similipal Biosphere Reserve, Odisha, India

Harisankar Dey[1]* *and Akshaya Kumar Bastia*[2]

[1]*Post-Graduate Department of Botany,*
Ramakrishna Mission Vivekananda Centenary College (Autonomous),
Rahara, Kolkata – 700 118, West Bengal
[2]*Post-Graduate Department of Botany,*
North Orissa University, Takatpur, Baripada – 757 003, Odisha
**e-mail: harisankardey@gmail.com*

Cyanobacteria are an extraordinarily diverse group of Gram-negative, oxygenic photosynthetic prokaryotes that originated 3.5 billion years ago and distributed in all possible biotopes of the world. The Similipal Biosphere Reserve, a unique ecosystem in the eastern part of India, occupy the northern boundary of Odisha state, lies between 21°28′ to 22°08′ North latitude and 86°04′ to 86°37′ East longitude, covering a vast area of 5578 sq. km. It is the richest watershed in the state of Odisha giving rise to many perennial freshwater bodies rich in diverse microalgae including cyanobacteria. In the present investigation, cyanobacterial samples were collected from different freshwater habitats of Similipal Biosphere Reserve including river, stream, canal and rice fields. In this study altogether 58 cyanobacterial forms belonging to 23 genera were characterized. Out of these 27 forms were heterocystous and 31 were non-heterocystous. They belonged to 2 orders namely Chroococcales and Nostocales of Cyanophyta. The most common cyanophycean members were *Aphanothece, Gloeocapsa, Microcystis, Gomphosphaeria, Merismopedia, Oscillatoria, Phormidium, Lyngbya, Microcoleus, Microchaete, Anabaena, Nostoc, Aulosira, Raphidiopsis, Cylindrospermum, Calothrix, Gloeotrichia, Scytonema, Hapalosiphon, Nostochopsis, Mastigocoleus, Mastigocladopsis* and *Westiellopsis. Oscillatoria* was the most abundant and dominant cyanobacteria followed by *Phormidium, Lyngbya, Anabaena, Calothrix* and *Nostoc etc.*

Keywords: *Cyanobacteria, Diversity, Freshwater and Similipal Biosphere Reserve.*

INTRODUCTION

Cyanobacteria or blue-green algae are a fascinating group of Gram-negative photosynthetic prokaryotes that are distributed in all possible biotopes of the world and have played a significant role in Earth history as primary producers and the ultimate source of atmospheric oxygen. Due to their occurrence in diverse habitats, these organisms are excellent materials for investigation by ecologists, physiologists, biochemists, microbiologists and biotechnologists. Cyanobacteria have tremendous potentials in production of biofertilizers, biomonitors of soil fertility and water quality, environmental management as soil conditioners, amelioratory agents aiding in the reclamation of saline and usar lands, biosorption of heavy metals, feed for animals and protein supplements, production of biofuels, production of NH_3 for scrubbing the excess atmospheric CO_2 and other gases causing the "Greenhouse" effects, production of hydrogen, production of hydrocarbon, production of various secondary metabolites, vitamin, toxin and other therapeutic substances Singh, (1950 and 1961); Borowitzka, (1995); Rai *et al.* (1998); Whitton and Potts, (2000). Survey of literature reveals that very little work has been done on freshwater cyanobacterial diversity of Similipal Biosphere Reserve as well as Odisha state Parija*et al.*,(1969); Mohanty and Padhi (1982 and 1984); Sahu*et al.*,(1996); Adhikary (1998); Sahu (2000); Das (2002); Jena *et al.* (2006); Bhakta *et al.* (2006); Bhuyan *et al.* (2007); Dey and Bastia, (2008 and 2012); Dey *et al.* (2008 and 2010); Dash *et al.* (2011); Deep *et al.* (2013) *etc.* and most of the investigations are confined to rice fields

The Similipal Biosphere Reserve, a unique ecosystem in the eastern part of India, occupy the northern boundary of Odisha state, lies between 21°28´ to 22°08´ North latitude and 86°04´ to 86°37´ East longitude, covering a vast area of 5578 sq. km. The climate of the biosphere is tropical with humid, warm summer and severe winter. The maximum temperature during summer is 40°C and in winter it falls to 4°C. It is the richest watershed in the state of Odisha giving rise to many perennial freshwater streams and drainage systems radiating all around like Budhabalanga, Baitarani systems *etc.* Though scattered human settlements are present on the buffer zone, the freshwater cyanobacterial diversity found in rivers and streams remains undisturbed. As no detailed information is available on freshwater cyanobacterial diversity of this unexplored habitat, an attempt has been made by us to explore the freshwater cyanobacterial diversity of different water bodies of this important biosphere reserve.

Materials and Methods

Study Area

Similipal Biosphere Reserve, a unique ecosystem in the eastern part of India, occupy the northern boundary of Odisha state, lies between 21°28´ to 22°08´ North latitude and 86°04´ to 86°37´ East longitude, covering a vast area of 5578 sq. km (Plate 18.1). The studies were conducted from different freshwater bodies including river, stream, canal and rice fields during 2007 to 2011.

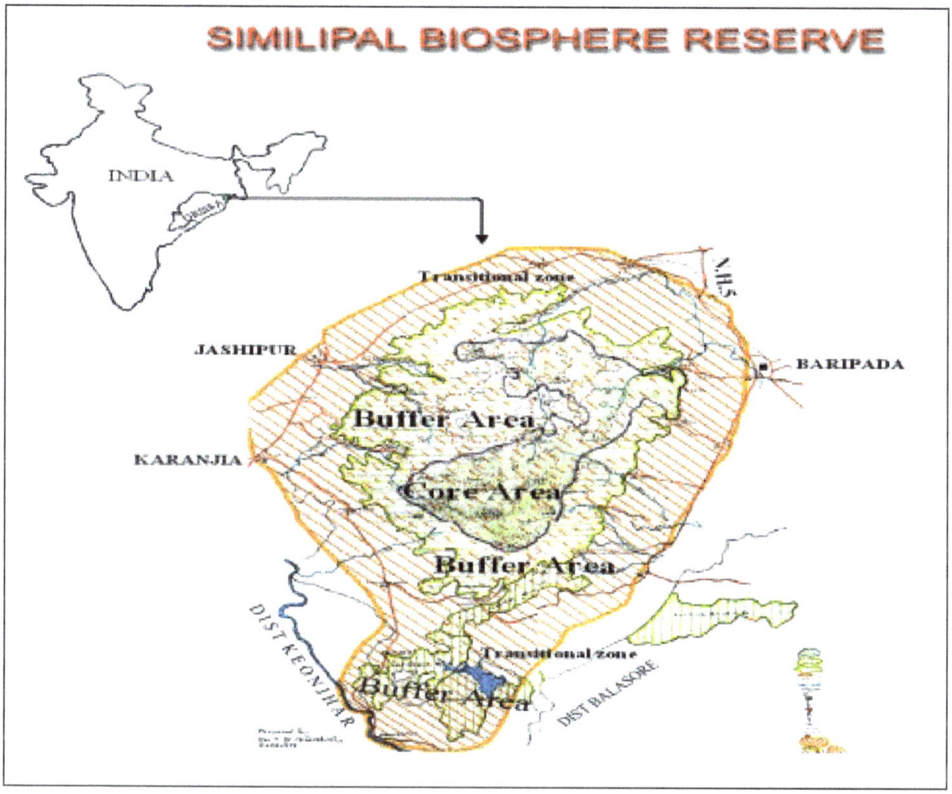

Plate 18.1: Photograph of Similipal Biosphere Reserve, Odisha, India.

Collection, Identification, Maintenance and Preservation of Samples

The samples were collected randomly from both soil and water bodies of different locations. Samples were collected with the help of fine forceps and planktonic net. The collected samples were put into sampling bottles or clean polythene bags, then immediately brought to the laboratory. The samples were assigned with collection numbers and date for field record.

Temporary slides were prepared for each sample for identification. The strains were identified, based on their morphological features and cell structure, following the standard literature and monographs of Desikachary (1959), Anand (1989) and Komarek and Anagnostidis (1989).

The collected samples were preserved in 4 per cent Formalin solution for future reference and few dominant species were maintained by culturing in freshly prepared BG_{11}±N medium Rippka*et al.* (1979) and incubated at 28±2°C with illumination at 25-30 µmol photon $m^{-2}s^{-1}$ white continuous light and aeration. The old cultures were periodically transferred to freshly prepared BG_{11}±N medium.

Results and Discussion

In this study altogether 58 cyanobacterial forms belonging to 23 genera were characterized. Out of these 27 forms were heterocystous and 31 were non-heterocystous (Plates 18.2–18.4). They belonged to 2 orders namely Chroococcales and Nostocales of Cyanophyta. The most common cyanophycean members were *Aphanothece, Gloeocapsa, Microcystis, Gomphosphaeria, Merismopedia, Oscillatoria, Phormidium, Lyngbya, Microcoleus, Microchaete, Anabaena, Nostoc, Aulosira, Raphidiopsis, Cylindrospermum, Calothrix, Gloeotrichia, Scytonema, Hapalosiphon, Nostochopsis, Mastigocoleus, Mastigocladopsis* and *Westiellopsis*. *Oscillatoria* was the most abundant and dominant cyanobacteria followed by *Phormidium, Lyngbya, Anabaena, Calothrix* and *Nostoc etc.*

Systematic Enumeration

1. *Microcystis aeruginosa* Kutz. (Desikachary, 1959, P.90, Pl.19, Figure 3)

Colonies longer than broad, solid, with distinct hyaline colonial mucilage; cells 4-6 μm in diameter; spherical with gas vacuoles.

2. *Gloeocapsa magma* (Breb.) Kutz. (Desikachary, 1959, P.122, Pl.24, Figure 13)

Colonies spherical or irregularly arranged, mostly 45 μm diameter with 2.5 cm. thick sheath; cells blue green, spherical, 4-6 μm diameter.

3. *Gloeocapsa* sp. Kutz. (Desikachary, 1959)

Cells spherical, 5-7 in colonies, colonies single or some time united together forming an expanded mass, cells in large colonies often with secondary colonies, arranged irregularly.

4. *Aphanothece microscopica* Nag. (Desikachary, 1959, P.104, Pl.22, Figure 4)

Thallus small, gelatinous, 1.5-2.0 mm diameter; cells oblong cylindrical, more or less 4-5 μm broad, 1.5-2.0 times as long broad, sheath present, blue-green, colorlessnannocytes present.

5. *Gomphosphaeria* sp. Kutz. (Desikachary, 1959)

Cells subglobose or spherical, often in groups of four but also in 2-8, with or without individual envelopes, arranged in a hollow spherical mucilaginous free swimming colony, blue green to olive green in colour.

6. *Merismopedia glauca* (Ehrenb.) Nag. (Desikachary, 1959, P.151, Pl.29, Figure 5)

Colonies mostly small with 16 cells, 45-50 μm diameter, cells oval or spherical, closely arranged, 5-6 μm broad, pale blue-green.

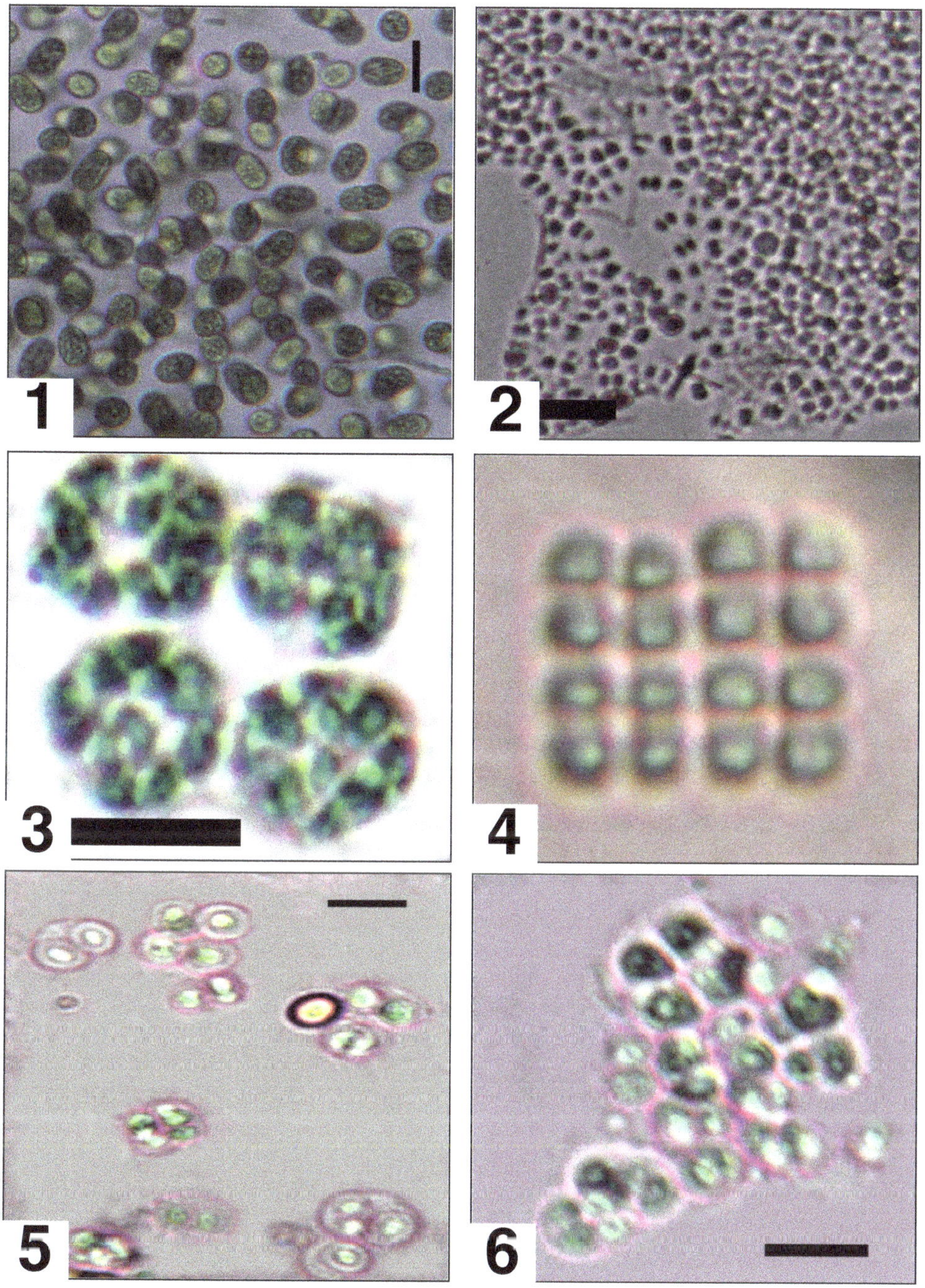

Scale Bar: -1, 3, 4, 5, 6 = 10 µm, 2= 40 µm.

Plate 18.2. Photomicrographs of Cyanobacteria Species in Similipal Biosphere Reserve.

1: ***Aphanothece microscopica***; 2: ***Microcystis*** sp.; 3: ***Gomphosphaeria*** sp.; 4: ***Merismopedia glauca;*** 5: ***Gloeocapsa*** sp.; 6: ***Gloeocapsa magma***

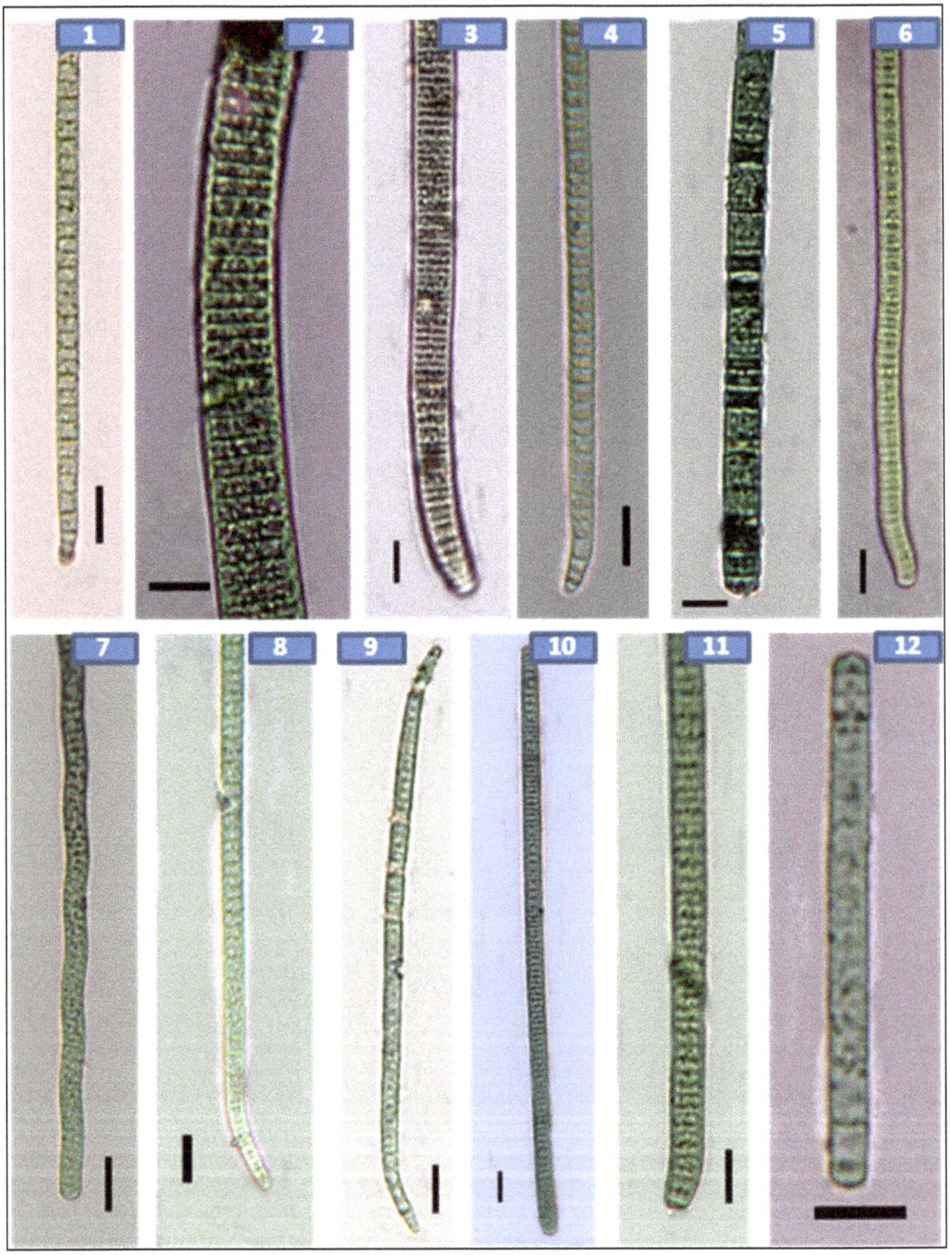

Scale Bar: - 1, 2, 3, 4, 5, 6, 7, 8, 9, 10, 11 and 12 = 10 µm

Plate 18.3: Photomicrographs of Cyanobacteria Species in Similipal Biosphere Reserve.

1: ***Oscillatoria amoena***; 2: ***Oscillatoria perornata***; 3: ***Oscillatoria formosa;*** 4, 8 and 9: ***Oscillatoria acuta***; 5: ***Oscillatoria vizagapatensis***; 6: ***Oscillatoria chalybea;*** 7: ***Oscillatoria curviceps***; 10: ***Oscillatoria subbrevis***; 11: ***Oscillatoria proteus;*** 12: ***Oscillatoria limosa.***

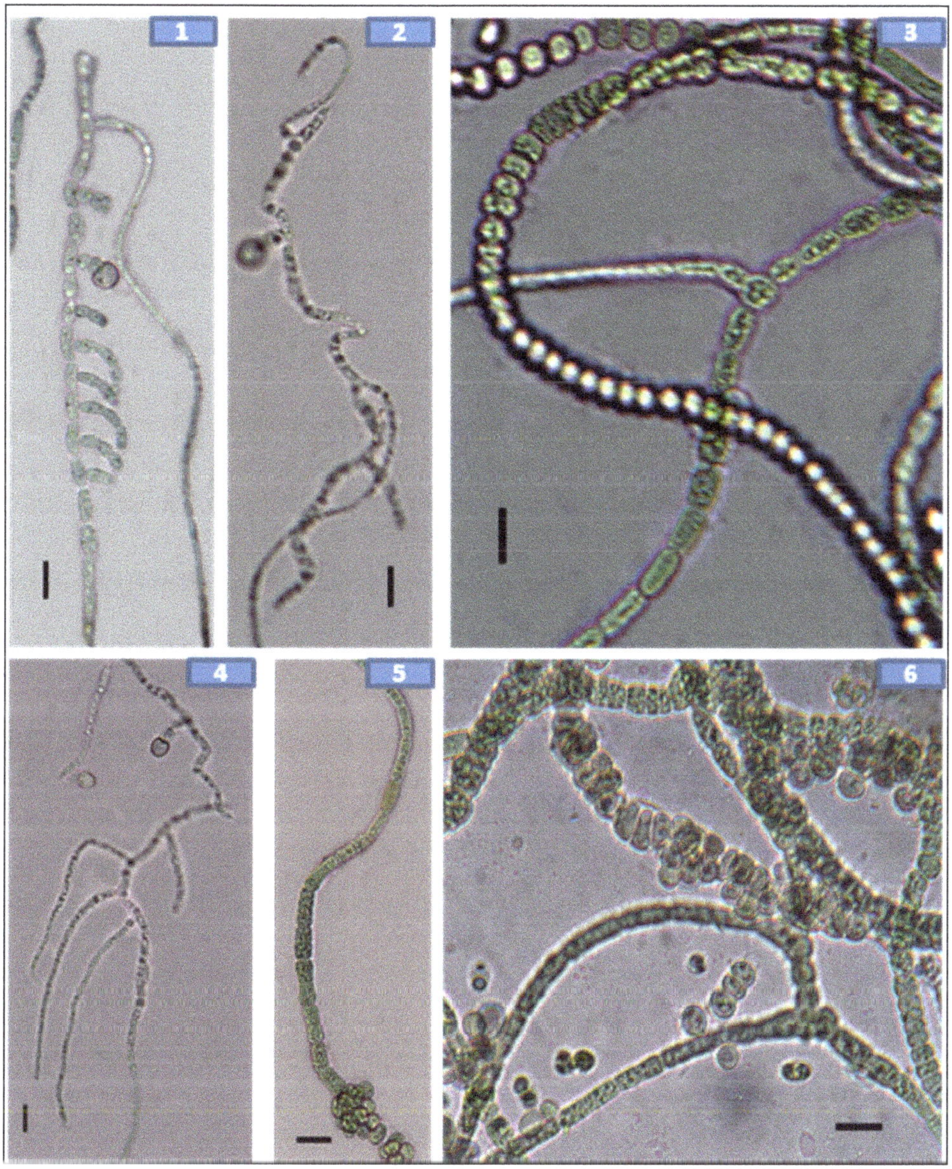

Scale Bar: - 1, 2, 3, 4, 5, 6 = 10 μm

Plate 18.4: Photomicrographs of Cyanobacteria Species in Similipal Biosphere Reserve.

1: *Nostochopsis lobatus*; 2 and 4: *Nostochopsis hansgirgi*; 3: *Hapalosiphon welwitschii*; 5 and 6: *Westiellopsis prolifica*.

7. *Oscillatoria acuta* Bruhl et Biswas (Desikachary, 1959, P.212, Pl.39, Figure 5)

Trichomes solitary, some time number of them aggregated parallel to each other into bundles of moderate size, 5-6 µm thick.

8. *Oscillatoria amoena* (Kutz.) Gomont (Desikachary, 1959, P.212, Pl.39, Figure 6)

Thallus more or less blue green, trichomes straight, slightly constrictd at the crosswalls 3.5 – 4.5µm broad, 2.5 – 3.5µm long.

9. *Oscillatoria curviceps* Ag. ex Gomont (Desikachary, 1959, P.208, Pl.38, Figure 2)

Thallus light or dark blue green, trichome more or less straight, bent at the end or spirally coiled, 10-11 µm broad, 3-4 µm long.

10. *Oscillatoria formosa* Bory ex Gomont (Desikachary, 1959, P.212, Pl.39, Figure 4)

Thallus blue green, trichome straight and at the end slightly bent, slightly constricted at the crosswalls, 7.5-10.8 µm broad, 3.5-4.0 µm long.

11. *Oscillatoria laete-virens* (Crouan) Gomont (Desikachary, 1959,P.212,Pl.39, Figure 2)

Thallus thin, membranous, green, trichome yellowish green, straight, 3.0-3.5 µm broad, 2.5-3.0 µm long.

12. *Oscillatoria chalybea* (Mertens) Gomont (Desikachary, 1959, P.208, Pl.38, Figure 3)

Thallus dark blue green, trichome straight, some timeslightly bent at the end, 7-8 µm broad, 3-4 µm long (Pl.10, Figure 6 and Pl.12, Figure3).

13. *Oscillatoria limosa* Ag. ex Gomont (Desikachary, 1959, P.208, Pl.38, Figure 1)

Thallus dark blue green, trichome straight, not constricted at the cross walls, 13-14 µm, broad 3-4 µm long.

14. *Oscillatoria perornata* Skuja (Desikachary, 1959, P.220, Pl.41, Figure 7, 8, 9)

Trichome straight or slightly bent, attenuated, constricted at the cross walls, 12-13 µm broad, single or aggregated in masses; cells commonly one-third to one-fourth as long as broad, 2.5-4.0 µm long, septa more or less granulated.

15. *Oscillatoria proteus* Skuja (Desikachary, 1959, P.220, Pl.41, Figure 15, 16)

Trichome straight, apices briefly attenuated and slightly bent, 6.5-7.3 µm broad, well constricted at the cross walls; cells one-third to one-fourth times as long as broad, 2.0-2.2 µm long, blue green; apical cell hemispherical to rounded conical.

16. *Oscillatoria ornate* Kutz. ex Gomont (Desikachary, 1959, P.204, Pl.37, Figure 12)

Thallus dark blue green, trichomes coiled at the end, 8.0-8.5 µm broad, 3.0-3.5 µm long.

17. *Oscillatoria princeps* Vaucher ex Gomont (Desikachary, 1959, P.24, Pl.37, Figure 1)

Thallus blue-green, brownish, straight, 25-55 µm broad, 4-8 µm long; slightly bent at the apices; end cell flatly rounded, with or without slightly thickened membrane.

18. *Oscillatoria subbrevis* Schmidle (Desikachary, 1959, P.204, Pl.37, Figure 2)

Trichome single, 6-8 µm broad, nearly straight, 1-2 µm long, end cell rounded.

19. *Oscillatoria vizagapatensis* Rao, C.B. (Desikachary, 1959, P.212, Pl.39, Figure 16)

Thallus blue green, trichomes straight, some time coiled, uniformly broad except at the extreme apex, 8-10 µm broad, 1.5-2.5 µm long.

20. *Phormidium ambiguum* Gom. (Desikachary, 1959, P.270, Pl.45, Figure 5-8)

Thallus more or less expanded, bright blue-green; trichomes slightly constricted at the cross-walls, not attenuated at the ends, 4.5-6.0 µm broad, blue green; sheath thin and more or less lamellated; cells shorter than broad, 1.0-2.5 µm long; end cell rounded.

21. *Phormidium lucidum* (Kutz.) Gom. (Desikachary, 1959, P.264, Pl.44, Figure 17, 18)

Thallus thick, dull green, inner colourless; filaments flexuous; sheath diffluent; trichome slightly constricted at the cross-walls, 7-8 µm broad; ends straight; cells half to one-fourth as long as broad, 2.0-2.5 µm long, end cell with a rounded calyptra.

22. *Phormidium purpurascens* (Kutz.) Gom. (Desikachary, 1959, P.264, Pl.44, Figure 4)

Thallus compact, purple to brownish violet; trichome nearly straight, not constricted at the cross-walls, 5-8 µm broad, 4.2-6.8 µm long; sheath diffluent; cells nearly quadrate; end cell rounded.

23. *Phormidium rotheanum* Itzigs. (Desikachary, 1959, P.270, Pl.45, Figure 14, 15)

Thallus thin, dark blue-green; trichome straight, slightly constricted at the cross-walls, 9-10 µm broad, blue green; sheath thin, cell much shorter than broad, 3-4 µm long, end cells obtuse and conical.

24. *Phormidium* sp. Kutz. (Desikachary, 1959)

Filaments straight, sheath present, very thin and mucilaginous, colour deep blue green to yellowish green, cells 1.8-2.7 µm long and 4.3-4.7 µm broad.

25. *Phormidium uncinatum* (Ag.) Gom. (Desikachary, 1959, P.270, Pl.45, Figure 9-10)

Filaments straight or slightly bent, sheath mucilaginous; trichome blue-green, not constricted at the cross walls, 7-8 µm broad, ends briefly attenuated; cells half to one-third as long as broad, 3-5 µm long; end cell with a rounded conical calyptras.

26. *Lyngbya ceylanica* Wille (Desikachary, 1959, P.312, Pl.54, Figure 4)

Thallus olive green; filaments 9-11 µm broad, straight; sheath thin, colourless; trichome blue green, unconstricted at the cross walls, not attenuated at the end, 9-10 broad; cells quadrate, end cell rotund, without calyptras.

27. *Lyngbya contorta* Lemmermann (Desikachary, 1959, P.295, Pl.50, Figure 5, 9)

Filament single, free floating, 3.5-5.0 µm broad; sheath narrow, colourless; cells 3.8-4.5 µm broad, 2.5-3.0 µm long, not constricted at the cross walls; end cell rounded.

28. *Lyngbya majuscula* Harvey ex Gomont (Desikachary, 1959, P.304, Pl.52, Figure 10)

Thallus expanded, up to 3 cm long, dull blue green to brown; sheath colourless; trichome blue green, not constricted at the cross walls, not attenuated at the ends; cells 3.7-4.5 µm long, cross walls not granulated; end cells rounded, calyptras absent.

29. *Lyngbya porphyrosiphonis* Fremy (Desikachary, 1959, P.306, Pl.53, Figure 11)

Filaments long and straight, about 2.7-3.1 µm broad; sheath hyaline, homogeneous, thin; trichome about 2.5-3.0 µm broad, not constricted at the cross walls; cells 2-3 time as long as broad, contents bluish grey; apical cell rounded or obtusely conical, not capitate.

30. *Microcoleus chthonoplastes* Thur. (Desikachary, 1959, P.342, Pl.60, Figure 7)

Filaments single, dirty to dark green, coiled, branched, sheath gelatinizing, 2-6 µm broad trichome, blue-green, 3-10 µm long, end cell not capitate, pointed.

31. *Microcoleus lacustris* (Rabenh.) Farlow (Desikachary, 1959, P.342, Pl.60, Figure 4)

Thallus blackish, blue-green, filament branched; sheath colourless, slimy, trichome distinctly constricted, 4-5 µm broad, cells cylindrical, 6-12 µm long, end cell more or less rounded, not capitate.

32. *Cylindrospermum indicum* Rao, C. B. (Desikachary, 1959, P.361, Pl.64, Figure 4)

Trichome single, dark blue green, with deep constrictions at the joints, 4 µm broad; cells almost quadrate, more or less barrel shaped, 3-4 µm long; heterocysts spherical or ellipsoidal.

33. *Cylindrospermum muscicola* Kutz. ex Born. et Flah. (Desikachary, 1959, P.361, Pl.64, Figure 3, 5, 12)

Thallus mucilagaginous, blakish-green; trichome 3-4 µm broad, constricted at the crosswalls; cells cylindrical or nearly quadrate, 4-5 µm long; heterocysts sub-spherical, 4-5 µm broad, 5.0-5.6 µm long.

34. *Cylindrospermum* sp. Kutz. (Desikachary, 1959)

Thallus mucilaginous, mostly blue green; trichome nearly straight; heterocysts terminal, oval or barrel shaped, hair like out growth are present in the periphery of heterocysts, heterocysts about 5.3 µm broad and 10.6 µm long; cells 3.6-3.8 µm broad and 4.6-6.2 µm long.

35. *Nastoc linckia* (Roth) Bornet ex Born. et Flah. (Desikachary, 1959, P.376, Pl.68, Figure 1)

Thallus varying in size, gelatinous, blue-green or violet, filaments densely entangled, trichomes 3-4 µm broad; cells barrel shaped, 3.0-3.6 µm long and 2.8-3.4 µm broad; heterocysts sub-spherical 5-6 µm broad.

36. *Nostoc piscinale* Kutz. (Desikachary, 1959, P.382, Pl.69, Figure 3)

Thallus mucilaginous, light blue green, filaments loosely entangled, sheath distinct at the periphery of the thallus; trichome 4-6 µm broad; heterocysts sub-spherical or oblong, 4-5 µm broad.

37. *Nostoc punctiforme* (Kutz.) Hariot (Desikachary, 1959, P.382, Pl.69, Figure 1)

Thallus scattered or confluent; filaments densely entangled, sheath hyaline; trichome 2.3-4.3 µm broad, cells short ellipsoidal, blue green; heterocysts 5-7 µm broad; spores sub-spherical 4.8-6.3 µm broad and 5-7 µm long.

38. *Anabaena ambigua* Rao, C.B. (Desikachary, 1959, P.407, Pl.76, Figure 2)

Trichomes enclosed in mucilaginous sheath; sheath usually 250-400 µm long, 15-40 µm broad, trichome bent, tapering at the end, cells barrel shaped with deep constrictions at the joints; cell contents deep blue green, cells 3.0-4.5 µm length and 4.8-7.1 µm breadth; heterocysts almost spherical or with slightly flattened ends.

39. *Anabaena circinalis* Rabneh. ex Born. et Flah. (Desikachary, 1959, P.412, Pl.77, Figure 2, 5)

Thallus floating, trichome circinate, without sheath; cells barrel shaped or spherical, 3.5-4.2 µm broad and 4.5-5.2 µm long; somewhat shorter than broad; heterocysts subspherical 3.9-4.3 µm broad and 5.8-7.2 µm long.

40. *Anabaena oryzae* Fritsch (Desikachary, 1959, P.397, Pl.72, Figure 3)

Thallus blue-green, gelatinous, trichomes nearly straight, densely aggregated; cells barrel shaped, 6.5-8.4 µm long and 4.0-4.5 µm broad; heterocysts terminal and intercalary, 5.6-8.4 µm long and 4.6-5.0 µm broad.

41. *Anabaena* sp. Bory (Desikachary, 1959)

Trichomes broad, blue green to olive green, sheath absent, thallus mucilaginous; heterocysts oval, intercalary, about 5.5-5.8 µm broad and 7.6-7.9 µm long; cells 2.7-3.0 µm broad and 2.2-4.1 µm long.

42. *Raphidiopsis* sp. Fritsch et Rich (Desikachary, 1959)

Trichomes short, more or less bent, without sheath, attenuated at ends, ends pointed with mucilaginous bristles; heterocysts absent.

43. *Aulosira bombayensis* Gonzalves (Desikachary, 1959, P.429, Pl.82, Figure 1-12)

Filaments yellowish to deep blue green in colour; 4-5 µm broad; sheath firm and hyaline; trichomes constricted at the joints, 2-3 µm broad; heterocysts intercalary with rounded ends.

44. *Aulosira fertilissima* Ghose (Desikachary, 1959, P.425, Pl.80, Figure 1-6)

Thallus dark blue-green; trichomes straight or densely intricate; cells 7-10 µm broad, 6-9 µm long; cylindrical or barrel shaped; sheath thick; heterocysts intercalary, oblong or elliptical, 7-8 µm broad and 8-11 µm long.

45. *Scytonema ocellatum* Lyngb. ex Born. et Flah. (Desikachary, 1959, P.470, Pl.92, Figure 3)

Thallus cushin like, greyish blue; filaments 12-14 µm broad, up to 3 mm long; false branched, branches short; sheath firm and brownish; trichomes 8-11 µm broad, olive green; cells shorter than broad; heterocysts sub-quadrate and yellowish.

46. *Microchaete aequalis* (Fremy) comb. Nov. (Desikachary, 1959,P.427,Pl.81, Figure 16)

Thallus expanded, blackish blue; filaments elongated, 4-5 µm broad; sheath thin and colourless, not constricted at the cross-wall; cells 2-4 times long as broad.

47. *Microchaete tenera* Thuret ex Born. et Flah. (Desikachary, 1959, P.512, Pl.105, Figure 1-2)

Filaments single, 6-7 µm broad, nearly straight; sheath thin and hyaline; trichome 5 µm broad, blue green to brownish green; cells at base twice as long as broad and at apices as long as broad; heterocysts intercalary, cylindrical.

48. *Calothrix braunii* (A. Br.) Born. et Flah. (Desikachary, 1959, P.541, Pl.114, Figure 3)

Thallus blue green; filaments curved, 6-9 µm broad; sheath thin and colourless;

trichome 6-7 µm broad, ending in a long hair, constricted at the cross-walls; cells 4.8-6.6 µm long and 3.0-3.8 µm broad; heterocysts basal.

49. *Calothrix clavatoides* Ghose (Desikachary, 1959, P.539, Pl.113, Figure 9)

Trichome single or in groups, straight and 12-13 µm broad; sheath narrow, thin and colourless; cells at the base discoid, 10-11 µm thick, in the middle 2-3 times as long as broad, 3-4 µm in diameter; heterocysts basal, single and globose.

50. *Calothrix gardneri* J. De Toni (Desikachary, 1959, P.547, Pl.115, Figure 5)

Filaments erect, swollen at the base and attenuated at the apex; sheath 2 µm thick and colourless; trichome cylindrical 5-6 µm broad, ending in a short hair, cells quadrate to halfas long as broad; heterocysts basal.

51. *Calothrix marchica* Lemm. (Desikachary, 1959, P.539, Pl.113, Figure 1-4)

Filaments brownish, slight bent, single, 8-9 µm broad at base with thin colourless sheath; trichome brownish green, constricted at the cross-walls; cells 6.0-6.8 µm long and 8.6-10.2 µm broad; heterocysts single, basal and spherical, 10-12 µm diameter.

52. *Gloeotrichia raciborskii* Woloszynska (Desikachary, 1959, P.559, Pl.118, Figure 4, 5)

Thallus dark green, spherical; trichome 7-8µm broad, ending in a long hair, dull brown; cells 7.8-11.5 µm long and 6.8-8.0 µm broad; heterocysts spherical, 11-13 µm diameter.

53. *Nostochopsis hansgirgi* Schmidle (Desikachary, 1959, P.571, Pl.121, Figure 6-13)

Thallus firm, yellowish brown; filaments sparsely branched, 2.0-3.2 µm broad, tapering to a rounded apex; cells cylindrical, occasionally constricted at the joints, light blue green; intercalary heterocysts long ellipsoidal, 4-5 µm broad and 7-9 µm long, lateral heterocysts spherical or hemispherical, 4-6 µm broad.

54. *Nostochopsis lobatus* Woodem. Geitler (Desikachary,1959, P.569, Pl.120, Figure 1-8)

Thallus spherical, irregularly lobed, blue green to olive green; trichomes radially arranged seemingly straight, branches lateral, broadly stretching upwards by a short bent and with its end being parallel to the main filaments; cells in the branching zone barrel shaped, 3.6-7.1 µm long and 2.7-4.2 µm long, trichome pointed at end; heterocysts spherical, terminal on 2-3 or 4 celled lateral branch, mostly 6.3-6.7 µm broad, brownish.

55. *Mastigocoleus testarum* Lagerh. (Desikachary, 1959, P.574, Pl.122, Figure 1-5)

Thallus thin, blue green; filaments variously curved, 7-9 µm broad; sheath thin and hyaline; trichome 4-5 μm broad; cells cylindrical or subcylindrical; heterocystsbroder than the trichome, lateral and terminal, 6-7 μm broad and 7-9 μm long.

56. *Mastigocladopsis jogensis* Iyengar et Desikachary (Desikachary, 1959, P.576, Pl.123, Figure 1-13)

Filaments flexuous, branches profuse; sheaths hyaline; trichome 4.4-5.1 μm broad and 4.7-5.6 μm long; cells cylindrical, 3.1-3.6 μm broad and 8.3-11.4 μm long; heterocysts intercalary and cylindrical.

57. *Hapalosiphon welwitschii* W. et G.S.West (Desikachary, 1959,P.589, Pl.130, Figure 4)

Filaments single or sometime groups, 5.4-5.6 μm broad; sheath colourless; cells subsperical or elongate, 5.2-6.3 μm broad and 5.4-7.2 μm long; heterocysts intercalary and cylindrical, about 6-7 μm broad and 6.5-7.5 μm long.

58. *Westiellopsis prolifica* Janet (Desikachary, 1959, P.595, Pl.131, Figure 1-12)

Filaments long, branched, deep blue green; cells elongate, 3.5-3.8 μm broad and 3.6-4.4 μm long; heterocysts oblong-cylindrical, 4.8-5.6 μm broad and 7.3-8.1 μm long.

ACKNOWLEDGEMENTS

The authors are thankful to Head, Department of Botany, North Orissa University, Baripada, Odisha and Principal, Ramakrishna Mission Vivekananda Centenary College (Autonomous), Rahara, Kolkata, West Bengal for providing necessary laboratory and other facilities.

REFERENCES

Adhikary, S.P. (1998). Cyanobacteria germplasm of Odisha state maintained at the department of Botany, Utkal University. *Plant Sci. Res.* 20: 57-63.

Anand, N. (1989). Hand book of blue-green algae. Pub. Bishen Singh Mahendra Pal Singh, Dehra Dun. Pp.77.

Bhakta, S., Dey, H., and Bastia, A.K. (2006). Study of algal diversity from rice-fields of Baripada, Mayurbhanj, Odisha. In: *Environmental Biotechnology and Biodiversity conservation.* (Ed. Das, M.K.) Daya publishing House, New Delhi, India. pp. 154-163.

Bhuyan, S., Dey, H.S., and Bastia, A.K. (2007). Cyanobacterial diversity in selected patches of Similipal Sanctuary. Similipal Biosphere Reserve, Vol.2 (1 and 2): 30-31.

Borowitzka, M.A. (1995). Microalgae as sources of pharmaceuticals and other biologically active compounds. *J. Appl. Phycol.* 7: 3-15.

Das, M. K. (2002). Cyanobacterial diversity in the rice fields of Sambalpur District of Odisha. *Phycos,* 41(1 and 2): 71-76.

Dash, P. K., Mohapatra, P. K., and Kar, M. (2011).Diversity of Cyanobacteria from freshwater bodies of Similipal Biosphere Reserve, Odisha, India. *E-Planet*, 9(1): 1-14.

Deep, P. R., Bhattacharyya, S., and Nayak, B. (2013). Cyanobacteria in Wetlands of the Industrialized Sambalpur District of India. *Aquatic Biosystems*, 9: 14

Desikachary, T.V. (1959). Cyanophyta. Indian Council of Agric. Res., New Delhi, pp. 686.

Dey, H.S. and Bastia, A.K. (2008). Cyanobacterial Flora from Rice Growing Areas of Mayurbhanj. *Plant Sci. Res.*30 (1 and 2): 22-26.

Dey, H.S. and Bastia, A.K. (2012). Taxonomic Enumeration of Nostocacean Cyanobacteria from Rice Field Soils of Northern Odisha, India. *Journal of Seaweed Research and Utilisation*, 34: 191-204.

Dey, H.S., Bhuyan, S., Pradhan, B., and Bastia, A.K. (2008). Freshwater algal diversity of Similipal Biosphere Reserve, Odisha. *Indian Hydrobiology*, 11: 167 – 172.

Dey, H.S., Tayung, K., and Bastia, A.K. (2010). Occurrence of Nitrogen-Fixing Cyanobacteria in Local Rice Fields of Odisha, India *Ecoprint*, 17: 77- 85.

Jena, M., Ratha, S.K., and Adhikary, S.P. (2006). Algal diversity of Similipal Biosphere Reserve, Odisha, *Indian Hydrobiology*, 9: 103-113.

Komarek, J. and Anagnostidis, K. (1989). Modern approach to the classification system of Cyanophytes, 4 - Nostocales.*Arch. Hydrobiol.*Suppl.82: 247-345.

Mohanty, R.C. and Padhi, B. (1982). Blue-green algae of Bhubaneswar and its adjoining region II. *Phykos*, 21: 96-98.

Mohanty, R.C. and Padhi, B. (1984). Blue-green algae of Bhubaneswar and its adjoining region-III. *Phykos*, 23: 90-95.

Parija, P., Pattnaik, H., and Rao, D.S. (1969). Blue-green algae in soils of Cuttack. *Prakruti*, Utkal University Journal Science, 6: 183-186.

Sahu, J.K., Nayak, H.,andAdhikary, S.P. (1996). Blue green algae of rice fields of Odisha state I. Distributional pattern in different agroclimatic zones, *Phykos*35: 93-110.

Sahu. J.K. (2000).Distribution and monthly succession of blue-green algae in the rice fields of Puri district of Odisha. *J. Eco. Taxon.* Bot. 24: 191-195.

Singh, R.N. (1950).Reclamation of usar lands in India through blue- green algae. *Nature*, 165: 325-326.

Singh, R.N. (1961). Role of blue- green algae in nitrogen economy of Indian agriculture. Indian Council of Agricultural Research, New Delhi.

Rai, L.C., Singh S. and Pradhan, S. (1998). Biotechnological potential of naturally occurring and laboratory growth Microcystis in biosorption of Ni^{2+} and cd^{2+}, *Curr. Sci.*74: 461 – 464.

Rippka, R., Deruelles, J., Waterbury, J.B., Herdman, M. and Stainer, R.Y. (1979). Generic assignments, strains histories and properties of pure culture of cyanobacteria. *J. Gen. Microbiol.*111: 1-61.

Whitton, B.A. and Potts, M. (2000). The Ecology of Cyanobacteria. Kluwer Academic Publishers, Dordrecht, The.

Chapter 19

Microbial Diversity of Similipal Biosphere Reserve, Odisha: An Update

Sukanta Kumar Nayak and Hrudayanath Thatoi*

Department of Biotechnology,
North Orissa University, Takatpur, Baripada – 757003, Mayurbhanj, Odisha
**e-mail: sukantanayak@rediffmail.com*

Similipal Biosphere Reserve of Odisha spreading over 555 sq km area is located on the northernmost part of Odisha. The biosphere is known for its rich in biodiversity of flora and fauna. However, few studies have been carried in the recent past to explore the diversity of microorganisms in this biosphere reserve. Most of the studies on this aspect were started with the establishment of North Orissa University. The research works were mostly carried out in the areas of algae, bacteria, fungi, (both micro and macro-fungi) from the soil, plants, and aquatic environment. Apart from biodiversity, studies were undertaken to assess the biotechnological potentials of these microbes. The present paper provides an account of the microbial diversity studied so far in the Similipal Biosphere Region.

Keywords: *Algae, Bacteria, Similipal Biosphere Reserve, Microbiota, Fungi, Endophytic fungi, Mushrooms.*

INTRODUCTION

Similipal Biosphere Reserve (SBR), the representative ecosystem under the Eastern Ghats has a unique assemblage of several ecosystems, like mountains, forests, grasslands, and wetlands, often referred as "Himalayas of Odisha" and is responsible for controlling the climatic regime of parts of Odisha, Jharkhand, West Bengal, and Eastern India (Saxena and Brahmam, 1989). It is located in the central part of Mayurbhanj district in Odisha, close to the interstate boundary with West Bengal state in the North-East direction and Jharkhand state in the North-West (Misra, 1997; Girach *et al.*, 1999; Behera, 2006; Reddy *et al.*, 2007; Thatoi *et al.*, 2008;

Rout *et al.,* 2009). It is located between 21°28′ to 22°08′N latitude and 86° 04′ to 86° 37′ E longitude spreading over 5,578 km^2 of forest land in three zones, *i.e.* Core zone (845 km^2), Buffer zone (2,174 km^2) and transitional zone (2,559 km^2). The average elevation is 559.31 meters (Mohanta *et al.,* 2006; Misra *et al.,* 2011; Panda *et al.,* 2011).

The biosphere reserve has varied topography and geologic formations, tropical climatic conditions with warm and humid, summer temperatures around 40 °C. Likewise, in the rainy season, it receives a rainfall of about 125 mm and during winter temperature drops up to 5 °C in many parts of the hill. SBR has idiosyncratic biodiversity harboring the number of endemic, threatened, rare, medicinal, and economically important plants. Recent reports indicate SBR has around 1254 species of vascular plants. Likewise, it has a unique assemblage of wildlife resources.

SBR provides an ecological niche which is a massive scale suite of microbial provinces but to date, very few studies have been made to document the SBR microorganisms and to explore their role in the biosphere as well as other types of bio-potential applications for improving the livelihood of people. The research works were mostly focused on the areas of algae, bacteria, fungi, (both micro- and macro-fungi) from the soil, plants, and aquatic environment. Here in this review, detailed information on the microbial diversity of SBR has been illustrated to explore their possible biotechnological utilization.

Microbial Diversity of SBR

The microbiome especially soil microbiota plays a dominant role in the maintenance of any ecosystem. Soil microorganisms play an important role in regulating nutrient cycling and also control the number of available nutrients to plants (Hernot and Robertson 1994; Singh and Rai 2004; Jain *et al.,* 2005). They are responsible for decomposing and mineralization of plant and animal residues into organic matter, which influences the physico-chemical properties of soil (Olson *et al.,* 2000). Several factors like physical and chemical properties of soil, organic carbon, P and K contents, *etc.,* are critical that the microorganisms of forest ecosystems (Wardle *et al.,* 2004; Ogram *et al.,* 2006). The population size and distribution of these organisms are often influenced by the abundance and nature of the organic content of the soil, climatic conditions, surface vegetation, and soil texture (Marschner *et al.,* 2003).

Algal Diversity

Algae are aquatic habitat, but some genera of this group able to colonize on a variety of terrestrial habitats (Hoffman, 1989). Cyanobacteria are common in terrestrial habitats and have important biological activity in the ecosystem acting to adding the nitrogenous compounds to the soil (Singh, 1961; Whitton, 2000). Corticolous algae are inhabitants of the bark of trees from the areas of different altitudes. They occur as dark, gelatinous, red dark or brown patches, streaks, or velvet masses. They are exposed to air and absorb water, minerals, and other nutrients directly from the atmosphere. Most studies on taxonomy and ecology of aerophytic algae were conducted in temperate regions (Harload and Schlichting, 1975; Metting, 1981; Starks *et al.,* 1981; Ortega-Calvo *et al.,* 1991; Broady and Ingerfeld,

1993; Flechtner *et al.*, 1998; Salleh and Millow, 1999; Gaylarde and Gaylarde, 1999; Rindi and Guiry, 2003; Rindi *et al.*, 2005; López-Bautista *et al.*, 2006; Uher, 2007; Lemes *et al.*, 2010; Zammit *et al.*, 2011).

Limited attempts have been made on aerophytic algae of SBR (Table 19.1). Earlier, Bhakta *et al.* (2014) isolated 19 corticolous algae species from 10 sampling sites of SBR. Out of these, 18 were cyanobacteria with the highest number of isolates belonged to *Nostoc* (4) followed by 2 each belonging to *Gloeocapsa, Chroococcidiopsis; Aphanocapsa; Phormidium; Scytonema* and 1 each of *Myxosarcina, Stigonema; Tolypothrix* and *Fischerella*. They have recorded only one green algae namely *Desmococcus olevaceoous*. Furthermore, they have succeeded in isolating several sub-aerial algae like *Nostoc, Phormidium, Scytonema, Stigonema; Chlorella, Cylindrocystis, Euglena, Mesotaenium, Physolmum, Pleurococcus, Trentepohila, Characiopsis chlorothesium, Ophiocytium* and diatoms like *Cymbella, Hantzschia, Navicula, Nitzschia* and *Pinnularia*.

Table 19.1. Various Algal Species Isolated from different Sites in the Similipal Biosphere Reserve

Algal Species Isolated	*Source of Isolation*	*Reference*
Gloeocapsa lignicola, Gloeocapsa alpina, Myxosarcina spectabilis sensu, *Chroococcidiopsis indica, C. kashaypi Aphanocapsa testacea, A. grevillei, Nostoc punctiforme, N. commune, N. piscinale, N. parmelioides, Phormidium rubritericola, P. holdenii, Stigonema turfaccum, Tolypothrix byssoidea, Scytonema mirabile, S. pseudopunctatum, Fischerella muscicola, Desmococcus olivaceus*	Tree bark	Bhakta *et al.*, 2014
Microcystis aeruginosa, Gloeocapsa magma, Gloeocapsa sp., *Aphanothece microscopic, Gomphosphaeria, Merismopedia glauca, Oscillatoria acuta, Oscillatoria amoena, Oscillatoria curviceps, Oscillatoria Formosa, Oscillatoria laete-virens, Oscillatoria chalybea, Oscillatoria limosa, Oscillatoria perornata, Oscillatoria proteus, Oscillatoria ornate, Oscillatoria princeps, Oscillatoria subbrevis, Oscillatoria viz.agapatensis, Phormidium ambiguum, Phormidium lucidum, Phormidium purpurascens, Phormidium rotheanum, Phormidium* sp., *Phormidium uncinatum, Lyngbya ceylanica, Lyngbya contorta, Lyngbya majuscule, Lyngbya porphyrosiphonis, Microcoleus chthonoplastes, Microcoleus lacustris, Cylindrospermum indicum, Cylindrospermum muscicola, Cylindrospermum* sp., *Nostoc linckia, Nostoc piscinale, Nostoc punctiforme, Anabaena ambigua, Anabaena circinalis, Anabaena oryzae, Anabaena* sp, *Raphidiopsis* sp., *Aulosira bombayensis, Aulosira fertilissima, Scytonema ocellatum, Microchaete aequalis, Microchaete tenera, Calothrix braunii, Calothrix clavatoides, Calothrix gardneri, Calothrix marchica, Gloeotrichia raciborskii, Nostochopsis hansgirgi, Nostochopsis lobatus, Mastigocoleus testarum, Mastigocladopsis jogensis, Hapalosiphon welwitschii, Westiellopsis prolific*	River, stream, canal and rice fields	Dey *et al.*, 2008

Earlier, Dey *et al.* (2008) isolated 27 forms were heterocystous and 31 were non-heterocystous cyanobacterial forms belonging to 2 orders namely Chroococcales and Nostocales of Cyanophyta and 23 genera. The most common cyanophycean members were *Aphanothece, Gloeocapsa, Microcystis, Gomphosphaeria, Merismopedia, Oscillatoria, Phormidium, Lyngbya, Microcoleus, Microchaete, Anabaena, Nostoc, Aulosira, Raphidiopsis, Cylindrospermum, Calothrix, Gloeotrichia, Scytonema, Hapalosiphon, Nostochopsis, Mastigocoleus, Mastigocladopsis* and *Westiellopsis. Oscillatoria* was the most abundant and dominant cyanobacteria followed by *Phormidium, Lyngbya, Anabaena, Calothrix, Nostoc etc.*

Bacterial Diversity

Bacteria by their physiological and metabolic ability can adapt to extreme ecological habitats including SBR. Information's regarding the bacterial diversity of SBR are very scanty (Table 19.2). However, the limited studies that have been done on soil bacteria of SBR indicated that the total bacterial load of soil samples of SBR varies from 10^6 - 10^9 CFU per gm soil (Behera *et al.*, 2008; Jena and Rath 2013). Jena and Rath (2013) demonstrated that the total bacterial load of SBR ranged between 1.15×10^5 - 5.8×10^7 CFU per gm soil.

Table 19.2. Various Bacterial Species Isolated from different Sites in the Similipal Biosphere Reserve

Name of Bacteria Isolated	*Source of isolation*	*Reference*
Bacillus sp. (*Bacillus* Sp., *B.subtilis, B. megaterium*), *Pseudomonas* sp.,	Soil	Jena and Rath, 2013
Bacillus sp., *Micrococcus* sp., *Geomicroccus* sp., *Pseudomonas* sp.	Soil	Behera *et al.*, 2008

Among bacteria, Behera *et al.* (2008) found *Bacillus* species to be most predominant (76.0 per cent) for bacteria in SBR followed *Pseudomonas* (16.0 per cent), *Micrococcus* (4.0 per cent) and *Geomicrococcus* species (4.0 per cent). A similar type of result was also recorded by Jena and Rath (2013). They found that among Bacillus species, 28.0 per cent belong to *B. subtilis* followed *B. magaterium* (16.0 per cent), *B. coagulans* (6.0 per cent) and *B. thuringensis* (6 per cent) while the percentage of other bacteria like *Pseudomonas* spp., *Staphylococcus* spp. and *Geomicroccus* spp., were 25.0 per cent, 13.0 per cent, and 6.0 per cent, respectively. The findings of both the studies corroborated each other although the number of isolates was varied from 50 isolates collected from 15 different sampling sites in SBR by former researchers to 160 isolates collected from 50 soil samples in SBR by later investigators.

Jena and Rath (2013) determined the total bacterial load of SBR soil samples to vary from 1.15×10^5 -5.8×10^7 CFU per gram of soil. Further, they have found that 20-60 per cent of total bacterial populations *i.e.*, around 3.0×10^4 -9.3×10^5 CFU per gm soil to be phosphate solubilizing bacteria. In another study of Behera *et al.* (2008) a total of 50 bacteria were isolated from 15 soil samples collected from SBR. The total plate count ranged from 10^6 to 10^9 CFU per gram of soil and the isolates belonged to the genera *Bacillus* (38, 76 per cent), *Pseudomonas* (8, 16 per cent), *Micrococcus* (2, 4 per cent) and *Geomicrococcus* (2, 4 per cent). While these isolates grew well at acidic pH and a wide pH range, they have recorded a high degree of multiple antibiotic resistance.

Fungal Diversity

Fungi like bacteria form an important part of the ecosystem along with other microbes in turnover of the biomass (James and Hyde, 1998). Earlier several investigations have been conducted to document the fungal diversity of SBR (Table 19.3).

Table 19.3. Various Fungal Species Isolated from different Sites in the Similipal Biosphere Reserve

Name of Fungi Isolated	*Source of Isolation*	*Reference*
Absidia sp., Alternaria sp., Aspergillus alliceus, Aspergillus flavus, Aspergillus fonscecous, Aspergillus humicola, Aspergillus luchuensis, Aspergillus niger, Aspergillus panamensis, Aspergillus awamori, Aspergillus sp., *Aspergillus tamarine, Aspergillus terrus, Aspergillus untii, Aspergillus ustus, Curvularia* sp., *Emencellopsis* sp., *Epicoccum nigrosu, Epicoccum* sp., *Fusarium oxysporium, Fusarium* sp., *Gliocladium catenulatum, Gliocladium* spp., *Homicola brevis, Mucor circinelloides, Mucor* sp., *Penicillium coryphilum, Penicillium cumminie, Penicillium granulatum, Penicillium implicatum, Penicillium regulosum, Penicillium* sp., *Penicillium stectii, Penicillium tardum, Penicillium rubrum, Rhizoctonia* sp., *Rhizopus* sp., *Scopulariopsis brevicaulis, Talaromyces* sp., *Trichoderm aliguorum, Trichoderma* sp., *Zygorhynchus expenes*	Soil	Jena *et al.*, 2015
Aspergillus niger, Aspergillus flavus, Aureobasidium pullulans, Tubercularia vulgaris	Soil	Panda *et al.*, 2012
Ganoderma applanatum, Lentinus sajor-caju, Ganoderma lipsience, Pycnoporus cinnabarinus, *Microporus xanthopus, Lenzytes* sp., *Russula emtica, Lentinus torulosus, Ganoderma chalceum, Amanita vaginata, Boletus* sp., *Phellinus* sp., *Boletus* sp., *Ganoderma tsugae, Tyromyces* sp., *Russula* sp., *Polyporus* sp., *Gymnopilus* sp., *Termitomyces clypeatus, Russula pseudodelica, Coprinus comatus, Ganoderma sessiliforme, 0Boletus edulis, Russula* sp., *Ganoderma* sp., *Russula* sp., *Lactarius* sp., *Astreatus hygrometricus, Termitomyces microcarpus, Russula delica, Russula albonigra, Ruccula vesca, Entoloma* sp., *Lactarius* sp., *Sclerodorma* sp., *Xylaria polymorpha, Daldinia concentric, Trametes* sp., *Tricholoma* sp., *Agaricus* sp., *Amanita* sp., *Cantharellus lateritius, Oxyporus populinus, Pleurotus ostreatus, Volvorella volvacea, Lentinus tuber-regium, Pleurotus pulmonarius, Coprinus plicatilis, Coprinus disseminates, Sparasis crispa, Psathyrella candolleana, Coprinus fimetarius, Leucoagaricus* sp., *Panaeolus acuminatus, Marasmius clavatus, Filoboletus manipularis, Agaricus trisulphuratus, Psathyrella* sp., *Auricularia* sp., *Termitomyces eurhizus, Amanita bisporigera, Crepidotus applanatus*	Soil/Plant/Termite nest/Decomposed Paddy straw	Sachan *et al.*, 2013; Sachan *et al.*, 2014; Sachan *et al.*, 2017;
Ganoderma applanatum, Ganoderma tsugae, Volvorella volvacea, Astraeus hygrometricus, Geastrum sp., *Lentinus sajor-caju, Russula delica, Russula vesca, Termitomyces eurrhizus*	Soil	Panda *et al.*, 2019

Name of Fungi Isolated	*Source of Isolation*	*Reference*
Fusarium sp.	Plant tissue	Mohanta *et al.*, 2008
Curvularia sp., *Fusarium* sp., *Acremonium* sp., *Aspergillus* sp., *Aspergillus flavus, Aspergillus fumigatus, Cladosporium* sp., *Colletotrichum* sp., *Curvularia* sp., *Rhizopus* sp., *Trichoderma* sp., *Trichoderma viridae, Trichoderma harzianum*	Plant tissue	Tayung *et al.*, 2012
Aspergillus niger, Aspergillus fumigates, Aspergillus versicular, Aspergillus sydowi, Aspergillus fonsecaeus, Aspergillus terricola, Curvularia lunata, Curvularia geniculata, Penicillium purpurogenum, Penicillium lanosum, Penicillium oxalicum, Trichoderma viridae, Trichoderma lignorum, Colletotrichum sp., *Morphotype* sp., *Aspergillus clavatus, Aspergillus flavus, Curvularia interseminata, Curvularia subulata, Penicillium albidum, Trichoderma koningi, Trichoderma* sp., *Cladosporium herbarum, Nigrospora zimmermann, Nigrospora sphaerica, Torula* sp., *Chaetomium dolichotrichum, Chaetomium globosum*	Leaf, stem and fruit tissues	Jena and Tayung, 2013
Aspergillus sp., *Penicillium* sp., *Fusarium* sp., *Colletotrichum* sp., *Nigrospora* sp., *Neurospora* sp., *Cladosporium* sp., *Acremonium* sp., *Xylaria* sp., *Geotrichum* sp., *Phomopsis* sp., *Pestalotiopsis* sp., *Cochilobolus* sp., *Talaromyces* sp., *Mucor* sp., *Alternaria* sp., *Curvularia* sp., *Trichoderma* sp., *Cytospora* sp., *Quambalaria* sp.	Lichen thalli	Padhi *et al.*, 2013; Padhi *et al.*, 2015; Padhi *et al.*, 2016; Padhi *et al.*, 2017

A. Microfungi

i) Rhizospheric Fungi

Rhizosphere, which is a complex environment of forest ecosystems, is an intense site of microbial activity that harbours a great diversity of microorganisms affecting plant growth and health (Khan *et al.*, 2007). The role of microbes especially those occurring in rhizospere in the decomposition of various macromolecules such as celluloses, hemicelluloses, *etc.* is well demonstrated (Berg and McClaugherty, 2003; Hanson *et al.*, 2008; McGuire *et al.*, 2010). Earlier few researchers have studies rhizophore microbes in general and rhizophoric fungi in particular of SBR.

Jena *et al.* (2014) collected soil samples from ten selected sites inside SBR and studied for their physicochemical parameters and culturable soil fungal diversity. Their investigation revealed that the fungal population was dependent on moisture and organic carbon (per cent) of the soil but its diversity was found to be regulated by sporulating species like *Aspergillus* and *Penicillium* (Jena *et al.*, 2014). They have demonstrated that soil samples to be acidic with a pH ranged from 5.1-6.0. The highest percentage of organic carbon and moisture content was observed in the samples collected from the sites, Chahala-1, and Chahala-2.

They have demonstrated that fungal population ranged from 3.6×10^4-2.1×10^5 and 5.1×10^4-4.7×10^5 cfu per gram of soil in summer and winter seasons, respectively. Among the soil fungus, *Aspergillus niger* was found to be the most dominant species with a Species Important Values Index (SIVI) of 43.4 and 28.6 in summer and winter season, respectively. Similarly, they have found highest fungal diversity indices during summer in Natto-2 and Natto-1 sites of SBR with the Shannon-Wiener and Simpson indices of 3.12 and 3.022 and 0.9425 and 0.9373 respectively. However, the highest Fisher's alpha was observed during winter in the sites Joranda, Natto-2, Chahala-1 and Natto-1 and the values were 3.780, 3.683, 3.575 and 3.418, respectively.

ii) Endophytic Fungi

Endophytes are microorganisms often fungi and bacteria, that inhabit hosts plant cells for all or part of their life cycle. As per Bacon and White (2000) definition which is most inclusive and widely recognized endophytes are microbes that colonize living, internal tissues of plants without causing any immediate, overt negative effects. Thus endophytes play an important role in physiological and ecological role in their host life and can be transmitted from one generation to the next through the tissue of the host seed or vegetative propagules (Carroll, 1988). Endophytic fungi are ubiquitous and believed that every plant species is colonized by endophytes (Arnold *et al.*, 2001). The distribution of endophytic mycobiota differs among the species and it has been found that a single plant species may harbor hundreds of endophytes in plant leaves, petioles, stems, bark, xylem, roots, flowers, fruit, seeds, *etc.* (Saikkonen *et al.*, 1998; Chapela and Boddy, 1988; Fisher *et al.*, 1993).

Earlier, Behera *et al.* (2013) isolated 50 endophytic fungi belonging to 9 different species from the root and leaf tissues of three orchid species such as *Acampe praemorsa, Cymbidium aloifolium* and *Vanda testacea.* They have recorded a sterile endophytic fungus from all the three Orchid species with highest colonization frequency in *Acampe praemorsa* (7.5 per cent) followed by *Vanda testacea* (3 per cent) and *Cymbidium aloifolium* (2 per cent). Similarly, they have demonstrated highest number of endophytes (28) colonized in leaf of *A. praemorsa* while 22 isolates in root tissue. Species of *Colletotrichum* was found to be colonized only in root tissues of *A. praemorsa* and *C. aloifolium*. Further, 2 endophytic species belonging to *Rhizoctonia* with variable colonizing frequency were isolated from the root and leaf tissue of all the three orchid species. Further, *Xylaria* sp. and *Fusarium* sp. were reported from the ethno-medicinal plants *A. praemorsa* and *V. testacea* with antimicrobial potentials (Mohanta *et al.*, 2008). Moreover, *Ceratobasidium* sp. was also identified from both leaf and root tissue of all the three orchid species except leaf of *Vanda testacea*. The variation in the number of endophytic fungus species inside *A. primorsa* might be the cause of the one dominant and rigid orchid species in this locality.

Jena and Tayung (2013) endophytic fungi associated with two plant species namely *Solanum rubrum* and *Morinda pubescence* of SBR. Out of 458 endophytes, isolated from leaf, stem and fruit tissues of both the plants, they have demonstrated the dominance of *Aspergillus, Colletotrichum, Curvularia* and *Mycelia sterilia* with maximum endophytic isolates from leaves segments followed by stem and fruit tissues.

iii) Endolichenic Fungi

A lichen is an ecologically obligate, stable mutualism between an exhabitant fungus (the mycobiont) and inhabitant population of extracellularly located unicellular or filamentous alga or cyanobacteria (the photobiont) (Hawksworth and Honegger 1994). It refers to a symbiotic association between 'mycobiont', and 'photobiont'. Lichen thalli provide a suitable ecological niche for a wide variety of microorganisms and distribution of fungi in endolichenic form seems to be ubiquitous among lichens (Li *et al.*, 2007; Petrini *et al.*, 1990; U'Ren *et al.*, 2010). Endolichenic fungi living in close association with the alga in the thalli of lichen are analogous to that of plant endophytes inhabiting the intercellular spaces of the hosts. Few studies were also been made on this aspect in SBR. Padhi *et al.* (2017) succeeded in isolating an endolichenic fungus *Aspergillus tubingensis* was successfully isolated from lichen thallus of Parmelia caperata. Similarly in a recent study, Padhi *et al.* (2019) found the association of an endolichenic *Aspergillus niger* from lichen thallus of *Parmotrema ravum*

Earlier, Padhi and Tayung (2015) succeeded in isolating 19 endolichenic fungi from the surface tissue fragments of three different thalli of *Parmelia* species. They have recorded that highest percentage of thalli found to colonize with fungal isolates belong to mycelia sterilia (47.4 per cent), the isolates that did not sporulate in culture followed by *Phomopsis*, *Aspergillus* and *Penicillium* species. Besides, *Phomopsis* (15.7 per cent), *Aspergillus* (10.5 per cent), *Penicillium* (10.5 per cent), *Trichoderma* (5.2 per cent), *Fusarium* (5.2 per cent) and *Cytospora* (5.2 per cent) species were other dominant colonizers.

B. Macrofungi

i) Ectomycorrhizal Fungi

Ectomycorrhizal macrofungi, such as mushrooms, are the important constituent of forest ecosystem. The climatic condition of SBR is conducive for the growth and development of a wide range of these macrofungi. The relative humidity is normally high during the monsoon and post monsoon months, sometimes being more than 85 per cent often favours the growth of mushrooms. Unlike other microbiota, several researchers were found have done extensive studies on both edible and non-edible mushrooms diversity of SBR (Sachan *et al.*, 2013; Panda *et al.*, 2015; Singdevsachan *et al.*, 2017; Panda *et al.*, 2019). The occurrence of several wild medicinal mushrooms like *Ganoderma applanatum*, *Ganoderma tsugae*, *Volvorella volvacea*, *Astraeus hygrometricus*, *Geastrum* sp., *Lentinus sajor-caju*, *Russula delica*, *Russula vesca*, *Termitomyces eurrhizus* *etc.* (Panda *et al.*, 2019). Panda *et al.* (2015), identified 24 different fleshy mushroom species out of which 14 mushroom species have been documented to have food and ethnomedicinal uses from the northern region of Odisha especially in SBR region.

Biotechnologial Potency of SBR Microbiota

Natural resources are one of the major assets of developing countries. There is a correlation between biodiversity and biotechnology. Countries were divided into groups like biodiversity-rich but biotechnology-poor, and biodiversity-rich and biotechnology rich. India belongs to the former case. In nature, microorganisms have been endowed with vast biopotentials.

Industrially Important Enzymes

In nature, microorganisms have been endowed with vast potentials. They produce an array of enzymes, which have been exploited commercially over the years. Various enzymes like are extensively important due to their industrial application for example in the hydrolysis of cellulose, starch processing, feed production, fermentation, malting and brewing, extraction of fruit and vegetable juices, pulp and paper industry, textile industry. Behera *et al.* (2008) studied the ability of bacterial isolates derived from SBR for production of industrially important enzymes and found t that 51.02 per cent, 48.97 per cent, 36.73 per cent, 36.73 per cent and 57.14 per cent of the isolates were positive for amylase, protease, lipase, phosphatase and DNase, respectively.

i) Phosphatase

Certain microorganisms possess the ability to mineralize and solubilize organic and inorganic phosphorus by extracellular phosphatase (alkaline and acid phosphatase). Phosphorus solubilizing activity is determined by the ability of microbes to release organic acids, or/and proton extrusion which through their hydroxyl and carboxyl groups chelate the cation bound to phosphate, the latter being converted to soluble forms (Surange, 1995; Dutton and Evans, 1996, Nahas, 1996). In a study, Jena and Rath (2013) earlier demonstrated that the phosphate solubilizing bacterial population ranged between 3.0×10^4 - 9.3×10^5 CFU per gm of soil constituting 20-60 per cent of the total bacterial populations. Furthermore, they have demonstrated that isolates belonging to *Bacillus* sp. (*B. subtilis, B. megaterium*), *Pseudomonas* sp., were the highest phosphate solubilizing activity.

Similarly, certain researchers have studied the phosphatase activity of fungi isolated from SBR. Jena and Rath (2014) found the production of more acid phosphatase in comparison to alkaline phosphatase by *Aspergillus awamori*, a cryptic species of *A. niger* with optimal pH, incubation period, temperature were reported to be pH 6, 6 days and 29 °C respectively. Maximum enzyme activity was observed at 3 per cent of sucrose and 0.1 per cent of ammonium nitrate, in the medium.

ii) Cellulase

It is a complex enzyme composed of cellobio-hydrolases, endoglucanases and β-glucosidases which all act synergistically to convert complex carbohydrates present in lignocellulosic biomass into glucose efficiently. These are inducible enzymes which are synthesized by microorganisms during their growth on cellulosic materials. Panda *et al.* (2012) demonstrated that SBR fungus *Aspergillus* strains like *A. niger, A. terrus, A. stellatus, A. flavus, A. fumigates* to possess high enzymatic activity. They have found the highest potency index value of *A. niger* to produce cellulase.

iii) Pectinases

These are important for plants as they help in cell wall extension and softening of some plant tissues during maturation and storage. They also aid in maintaining ecological balance by causing decomposition and recycling of waste plant materials.

It catalyses the degradation of the pectins *via* depolymerization and de-esterification reactions. In a study, Panda *et al.* (2012) recorded the highest pectinase producing potential of *A. flavus* (4.1 for IFcel) among the 25 fungal isolates from SBR.

Bioactive Compounds

Bioactive molecules/substances from microbes are extremely important as a potential source of antioxidant, antimicrobial, antitumor, antiviral agents. Studies involving SBR indicate that various microorganisms have high bioactivity potential. For example medicinal mushrooms such as *Ganoderma* sp. (G *applanatum, G. lipsiense, G. chalceum* and *G. tsugae*) were found to have very good antioxidant and antibacterial potential (Sachan *et al.*, 2017). Sachan *et al.* (2017) have demonstrated that per hundred gram of the above-stated mushrooms, protein ranged from 17.10–20.50g, carbohydrate range 41.80g – 59.16g and least fats content of 1.2g – 3.0g along with micronutrients like vitamins, carotenoids, *etc.* The high phenol (27.40mg – 220.41mg catechol per gram extract) and flavonoid (3.42mg – 171.44 mg quercetin per gram extract) content was mostly attributed to the antioxidant activity of these mushrooms.

Therapeutic Substances/Compounds

The antimicrobial activity of mushrooms of SBR is well documented (Sachan *et al.*, 2013; 2014 and 2017). Several researchers have found the antibacterial and antifungal activity of mushrooms like *Lentinus sajor-caju* and *Lentinus torulosus* against *Streptococcus aureus* and *Vibrio cholera* (Sachan *et al.*, 2013); *Russula vesca, R. delica* and *T. eurrhizus* against human pathogens (Sachan *et al.*, 2014). Furthermore, recently Panda *et al.* (2017) found that the metabolites of wild edible mushroom species *Geastrum* sp., against human pathogens. They have demonstrated that mushroom possesses considerable antimicrobial activity against the fungal pathogen like *Trychophyton mentagrophytes* and bacterial pathogens like *Shigella flexneri, Klebsiella pneumonia* and *Staphylococcus aureus* at 7th, 14th and 21st days incubation.

Sachan *et al.* (2013) recorded the antioxidant and antibacterial activities of 2 wild edible mushroom species *Lentinus sajor-caju* and *Lentinus torulosus*. Sachan *et al.* (2014) analyzed for their nutritional antioxidant and antibacterial potential of *Russula vesca, R. delica* and *T. eurrhizus* of SBR. Their findings showed that these three mushrooms are rich sources of antioxidants with scavenging potential up to 89 per cent at concentration 100 mg/ml. Total phenol content was found between 21.92–41.99 mg catechol/g extract and flavonoid 2.53–7.52 mg quercetin/g extract. The studied mushrooms possess moderate antibacterial properties with zones of inhibition ranging from 13 to 30 mm. Being a source of nutrients and molecules with medicinal potential, the studied mushrooms can be used in the human diet as nutraceuticals/functional foods for maintaining and promoting health, longevity, and life quality.

Sachan *et al.* (2017) also evaluated the antioxidant and antibacterial potential of various medicinal mushrooms *viz. Ganoderma applanatum, Ganoderma lipsiense, Ganoderma chalceum* and *Ganoderma tsugae.* They have recorded that the macronutrient content of these mushrooms revealed high amounts of protein (17.10–20.50/100 g), carbohydrate (41.80–59.16/100 g) and low fats (1.2–3.0/100 g) along with substantial amount of micronutrients like vitamins, carotenoids *etc.* Mushrooms showed strong antioxidant properties due to the presence of high phenol (27.40–220.41 mg catechol/g extract) and flavonoid (3.42–171.44 mg quercetin/g extract) content. All these mushrooms possess moderate antibacterial properties with *a* zone of inhibition ranging from 11.76 to 18.83 mm against five human pathogenic bacteria.

Nonetheless, Padhi and Tayung (2015) recorded that 10.52 per cent isolates from 19 isolated endolichenic fungi, *Parmelia* sp. thalli, possessed varying degrees of antimicrobial activity against bacterial pathogens, *Bacillus subtilis*, *Staphylococcus aureus*, *Pseudomonas aeruginosa*, *Proteus vulgaris*, *Shigella flexneri*, and *Klebsiella pneumonia* and pathogenic fungi, *Candida albicans*, *Candida krusei*, and *Trichophyton mentagrophytes* while 31.57 per cent and 10.52 per cent of the isolates inhibited all the above bacterial and fungal pathogens, respectively.

Recently, Padhi *et al.* (2018) succeeded in isolating antimicrobial secondary metabolites of an endolichenic *Aspergillus niger* from lichen *Parmotrema ravum* thallus. They have isolated a new 6-benzyl-γ-pyrone, named aspergyllone together with six other known metabolites identified as aurasperones A and D, asperpyrone A, fonsecinone A, carbonarone A and pyrophen from the culture filtrates of an endolichenic fungus *A. niger* with antimicrobial activity against the various human, plant, foodborne and fish pathogens. Further they have found that while, Aspergyllone showed strong selective antifungal activity against *Candida parapsilosis*, with an IC_{50} of 52 μg/ml, Aurasperone A and pyrophen exhibited moderate to strong antimicrobial activity.

Bioremediating, Biofertilizing, Probiotic Agents

The biosphere is often acting as a major reservoir of many potential candidate bioremediating, biofertilizers, probiotics agents. Although, microorganisms isolated from SBR with various enzymatic, antimicrobial, activity by various researchers could be exploited as biofertilizers, to date no systematic attempt has been made to develop such bioagents from SBR for the welfare of mankind.

CONCLUSION

SBR is an interesting and unique place for ecological studies. To date, this virgin ecosystem/habitat has not been explored from the microbial wealth perspective. Therefore, there is a need for considerable competence and skills in a whole range of subjects starting with microbial ecology, taxonomy, genetics, molecular biology, biochemistry, fermentation technology, biochemical and process engineering, economics, law, and training of local communities in modem conservation skills. Exploring less studied ecosystems and habitats for microbes could be a profitable endeavor both in terms of understanding their biology, biotechnological exploration, novel genes for technology, *etc.*

REFERENCES

Bhakta, S., Pattnaik, L., Dutta, P., Sahu, E. and Bastia, A.K. (2014). Diversity of corticolous algae from Similipal biosphere reserve, Mayurbhanj, Odisha. *Phykos*, 44 (1): 9-16.

Behera, D., Tayung, K. and Mohapatra, U.B. (2013). PCR-based identification of endophytes from three orchid species collected from Similipal Biosphere Reserve, India. *Amer. Int. J. Res. Form. Apl. Nat. Sci.*, 3(1): 10-17.

Behera, M., Dandapat, J. and Rath, C.C. (2008). Isolation, characterization and screening of bacteria isolates from Similipal Biosphere Reserve forest soil for their metal tolerance capacity and extracellular enzymatic activities. *Biorem. Biodiv. Bioavail.*, 3: 72-78.

Broady, P.A. and Ingerfeld, M. (1993). Three new species and a new record of Chaetophoracean (Chlorophyta) algae from terrestrial habitats in Antartica. *Eur. J. Phycol.*, 28: 25-31.

Dey, H.S., Bhuyan S., Pradhan B. and Bastia, A.K. (2008). Freshwater algal diversity of Similipal Biosphere Reserve, Odisha, India. *Indian Hydrobiol.*, 11(1): 167-172.

Flechtner, V.R., Jeffrey, J.R. and Clark, W.H. (1998). Algal composition of the microbiotic crusts from the central desert of Baja California, Mexico. *The Great Basin Naturalist*, 54: 295-311.

Gaylarde, P.M. and Gaylarde, C.C. (1999). Algae and cyanobacteria on painted buildings in Latin America. *Int. Biodeterior. Biodeg.*, 46: 93-97.

Harlod, E. and Schlichting, Jr. (1975). Some subaerial algae from Ireland. *Brit. Phycol. J.*, 10: 257-261.

Islam, N. (1960). Some subaerial green algae from East Pakistan. *Trans. Am. Microsoc. Soc.*, 79: 471-479.

Jena, S.K. and Tayung, K. (2013). Endophytic fungal communities associated with two ethno-medicinal plants of Similipal Biosphere Reserve, India and their antimicrobial prospective. *J. Appl. Pharmac. Sci.*, 3: S7-S12.

Jena, S.K. and Rath, C.C. (2013). Optimization of culture conditions of phosphate solubilizing activity of bacterial sp. isolated from Similipal Biosphere Reserve in solid-state cultivation by response surface methodology. *Int. J. Curr. Microbiol. App. Sci.*, 2(5): 47-59.

Jena, S.K. and Rath, C.C. (2014). Effect of environmental and nutritional conditions on phosphatase activity of *Aspergillus awamori* and viability test of the strain. *Curr. Res. Envtl. Appl. Mycol.*, 4(1): 45–56.

Jena, S.K., Tayung, K., Rath, C.C. and Parida, D. (2015). Occurrence of culturable soil fungi in a tropical moist deciduous forest Similipal Biosphere Reserve, Odisha, India. *Braz. J. Microb.*, 46(1): 85-96.

Lemes-Da-Silva. N. M., Zanini Branco, L.H., and Necchi-Junior, O. (2010). Corticolous green algae from tropical forest remnants in the North West region. *Revista Brasil Bot.*, 33: 215-226.

López-Bautista, J.M., Rindi, F. and Guiry, M.D. (2006). Molecular systematics of the subaerial greenalgal order Trentepohliales: An assessment based on morphological and molecular data. *Int. J. Syst. Evol. Microbiol.*, 56: 1709-1715.

Metting, B. (1981). The systematics and ecology of soil algae. *The Bot. Rev.*, 47: 195-312.

Mikter, S.S. and Shukla, S.P. (2006). Study on algal flora of algae. Moss Association on Barks. *Bull.Arunachal Forest Res.*, 22: 1-8.

Mohanta, J., Tayung, K. and Mohapatra, U.B. (2008). Antimicrobial potentials of endophytic fungi inhabiting three ethno-medicinal plants of Similipal Biosphere Reserve, India. *The Internet J. Microbiol.* 5: 2-6.

Neustupa, J. and Skaloud, P. (2008). Diversity of sub-aerial algae and cyanobacteria on tree bark in tropical mountain habitats. *Biologia*, 63: 806-812.

Neustupa, J. and Skaloud, P. (2010). Diversity of sub-aerial algae and cyanobacteria on tree bark and wood in the low land tropical forests of Singapore. *Plant Ecol. Evol.*, 143: 51-62.

Ortega-Calvo, J.J., Hernandez-Marine, M. and Saiz-Jimenez, C. (1991). Bio-deterioration of building material by cyanobacteria and algae. *Int. Biodet. Biodeg.*, 28: 165-185.

Padhi, L.P., Panda, S.K., Satapathy, S.N. and Dutta, S.K. (2011). *In vitro* evaluation of antibacterial potential of *Annona squamosa* L. and *Annona reticulata* L. from Similipal Biosphere Reserve, Odisha, India. *J. Agril. Teach.*, 7(1): 133-142.

Panda, S., Dubey, D. and Dutta, S. (2010). Anticandidal activity of *Diospyros melanoxylon* Roxb. bark from Similipal Biosphere Reserve, Odisha, India. *Int. J. Green Pharm.*, 4(2): 102-104.

Panda, S.S., Sahoo, K., Das, R. and Dhal, N.K. (2012). Pectinolytic and cellulolytic activity of soil fungal isolates from Similipal Bioreserve Forest. *World Env.*, 2(2): 1-3.

Panda, M.K. and Tayung, K. (2015). Documentation and ethnomedicinal knowledge on wild edible mushrooms among ethnic tribes of northern Odisha, India. *Asian J. Pharmac. Clin. Res.*, 8: 139-143.

Panda, M.K., Thatoi, H.N. and Tayung, K. (2017). Antimicrobial potentials of spore culture of geastrum sp., a rare wild edible mushroom of Simlipal biosphere reserve, Odisha, India against some significant human pathogens. *Int. J. Advance Res.*, 5: 9.

Padhi, S., Das, D., Panja, S. and Tayung, K. (2017). Molecular characterization and antimicrobial activity of an endolichenic fungus, *Aspergillus* sp. isolated from *Parmelia caperata* of Similipal Biosphere Reserve, India. *Interdisc. Sci.*, 9(2): 237-246.

Padhi, S., Masi, M., Panda, S.K., Luyten, W., Cimmino, A. and Tayung, K. (2019). Antimicrobial secondary metabolites of an endolichenic *Aspergillus niger* isolated from lichen thallus of *Parmotrema ravum*. *Nat. Prod. Res.*, 34(18): 2573-2580.

Rindi, F., Sherwood, A.R. and Guiry, M.D. (2005). Taxonomy and distribution of Trentepohlia and Printzina (Trentepohliales, Chlorophyta) in the Hawaian Islands. *Phycologia*, 44: 270-284.

Rindi, F., Guiry, M.D. and López-Bautista, J.M. (2006). New records of Trentepohliales (Ulvophyceae, Chlorophyta) from Africa. *Nova Hedwigia*, 83: 431-449.

Sachan, S.K.S., Patra J.K., Tayung, K., Sarangi, K. and Thatoi, H.N. (2014). Evaluation of nutritional and nutraceutical potentials of three wild edible mushrooms from Simlipal Biosphere Reserve, Odisha, India. *J. Cons. Prot. Food Saft.*, 9: 111-20.

Sachan, S.K.S., Patra, J.K. and Thatoi, H.N. (2013). Indigenous knowledge of ethnic tribes for utilization of wild mushrooms as food and medicine in Simlipal biosphere reserve, Odisha, India. *J. Agril. Tech.*, 9: 335-348.

Sachan, S.K.S., Patra, J.K. and Thatoi, H.N. (2013). Nutritional and bioactive potential of two wild edible mushrooms (*Lentinus sajor-caju* and *Lentinus torulosus*) from Simlipal Biosphere Reserve, India. *Food Sci. Biot.*, 22: 137-145.

Sachan, S.K. and Patra, J.K. (2017). Chemical constituents, antioxidative and antibacterial properties of medicinal mushrooms collected from Similipal Biosphere Reserve, Odisha, India. *Proc. National Acad. Sci., India, Sect. B Biol. Sci.*, 87: 559.

Salleh, A., and Millow, P. (1999). Notes on *Trentepohlia dialepta* (Nylander) Hariot (Trenteopohliaceae, Chlorophyta) and sporangia of some other species of Malaysia. *Micronesica*, 31: 373-378.

Shuvashish, Behera and Swain, Manas Ranjan (2013). A survey of post-harvest spoilage of mahula (*Madhuca latifolia* L.) flowers. *Int. J. Agril. Tech.*, 9(1): 227-235.

Starks, L.T., Shubert, L.E. and Trainor, F.R. (1981). Ecology of soil algae: A review. *Phycology*, 20: 65-80.

Uher, B. (2007). Morphological characterization of three subaerial *Calothrix* species (Nostocales, Cyanobacteria). *Fottea*, 11: 187-201.

Zammit, G., Billi, D., Shubert, E., Kastovsky, J., and Albertano, P. (2011). The biodiversity of subaerophytic phototrophic bioflims from Maltase hypogea. *Fottea*, 11: 187-201.

Chapter 20

Evaluation of Antimicrobial Potential of *Nyctanthes arbor-tristis* Leaf Extracts

*Himadri Gourav Behuria, Gayatree Panda and Santosh Kumar Sahu**

Department of Biotechnology, North Orissa University, Baripada – 757 003, Odisha
**e-mail: drsantoshnou@gmail.com*

Nyctanthes arbor-tristis, an important medicinal plant, widely distributed across the Indian subcontinent is commonly used for its antimicrobial and anti-plasmodial properties. Its phytochemical extracts are effective against bacteria, fungi, and plasmodium. However, many of its active principles and their mechanism of action remain elusive. Further, the composition of its secondary metabolites varies with its geographical location and climate. In the present study, we extracted phytochemicals from the leaves of *Nyctanthes arbor-tristis* collected from the Similpal Biosphere Reserve (SBR), Odisha, India, by sequential extraction with petroleum benzene, chloroform, and methanol. Phytochemical analysis of the extracts revealed the presence of alkaloids, anthraquinones, flavonoids, saponins, cardiac glycosides, tannins, terpenoids, and steroids. All three extracts exhibited antimicrobial activity against *Pseudomonas aeruginosa* (MTCC-1035), *Shigella sonnei* (MTCC-586), *Shigella flexinery* (MTCC-1185), *Bacillus licheniformis* (MTCC-2617) and *Staphylococcus aureus* (MTCC-212). However, the chloroform extract exhibited highest antimicrobial activity against all above pathogens. The petroleum benzene extracts induced change in membrane phospholipid composition of *Bacillus licheniformis* (MTCC 2617) by decreasing the amino-phospholipid content. The extract also led to destabilization of giant unilamellar vesicles prepared from the polar lipids of *Bacillus licheniformis* (MTCC2617), indicating that its components contain membrane disrupting active ingredients. Our results show that the *Nyctanthes arbor-tristis* extract exhibits antibacterial activity on *Bacillus licheniformis* by inhibiting amino-phospholipid synthesis and destabilizing its plasma membrane.

INTRODUCTION

Since time immemorable, man has exploited nature for food, clothes, shelter and medicine. Rigveda, the holy book of Hindus, supplies curious information on this subject (Uniyal *et al.*, 2006). Medicinal plants represent a rich source of novel molecules with pharmacological properties, which are lead compounds for the development of new drugs. During recent years, increasing research on herbal products has led to the discovery of many antimicrobial, anticancer and antidiabetic therapeutics. Herbal drugs are gaining momentum in the field of pharmacology because of their nontoxic nature and absence of deleterious side effects. Despite the revolution in western medicines towards the end of the 20th century, the world has seen a revival of interest in traditional medicines (Sittiwet and Puangpronpitag, 2009). The resurgence of research on herbal drugs is mainly due to the increasing evidence of the health hazards associated with the indiscriminate use of modern medicine such as antibiotics, steroids and other synthetic drugs (Aggarwal *et al.*, 2013). Increasing resistance of pre-existing microbes to commercially available synthetic antimicrobials, leading the appearance of drug-resistant varieties is one of the most prevalent global health problems (Li and Zhang, 2014). Besides, the high cost of chemotherapeutics and their adverse side effects such as hypersensitivity, allergic reactions and immunosuppression are some of the major disadvantages associated with synthetic drugs (Schinor *et al.*, 2007).

In order to find novel natural antimicrobials with new modes of action, medicinal plants have been explored as reservoir of novel bioactive compounds (Okigbo and Omodamiro, 2006; Kumar and Yadav, 2013). The mode of antimicrobial activity exhibited by phytochemicals involves (i) inhibition of peptidoglycan synthesis, (ii) damage to plasma membrane, (iii) inhibition of protein synthesis and damage to DNA. *Nyctanthes arbor-tristis* (Oriya: Gangasiuli or Sephali) commonly known as Harsinger or Night jasmine is a well-documented medicinal plant of India, belongs to family *Oleaceae.* It is widely distributed across Indian subcontinents including sub-Himalayan, eastern and western Ghat ranges region and also cultivated in gardens as an ornamental plant (Hukkeri *et al.*, 2006). Juice of its leaves is used as a digestive stimulant, antidote to reptile venoms, mild anti-diabetic tonic, laxative, anti-splenomegaly therapeutics, anti-diaphoretic, anti-diuretic and antimicrobials (Kartikar and Basu, 2000; Saxena *et al.*, 1987). The plant extract is also used as an anti-cancer, anti-pyretic and anti-anorexic agent. Its extract is also used for the treatment of arthritis, malaria and intestinal infections (Khandelwal *et al.*, 1999). However, their mechanism of action and cellular targets remain elusive.

Bacillus licheniformis is one of the most common human pathogens that inhabit the human intestine and other visceral epithelia. It is associated with diseases such as septicemia, peritonitis, ophthalmitis and food poisioning (Adimpong *et al.*, 2020). The bacterium is also known to exhibit resistance against cationic peptide drugs and chloramphenicol (Agerso *et al.*, 2019). Recent studies reveal that resistance against cationic peptide drugs mostly arises from change in membrane PL composition. In *B. licheniformis*, similar alteration in plasma membrane PL composition is known to contribute to its drug resistance. However, the effect of herbal phytochemicals on the bacterial drug resistance remains unknown. The aim of our present study is to evaluate the efficacy of its extract against common human bacterial pathogens those exhibit resistance to common drugs. In addition, the presents study also attempts to identify it's cellular targets and mechanism of action.

Materials and Methods

Collection of Plant Materials and Extraction Method

Healthy plant leaves were collected from Similipal Biosphere Reserve, Mayurbhanj, Odisha in a clean sterilized polyethylene bag. The plant was identified at Regional Plant Research Center, Bhubaneswar, Odisha. Leaves were cleaned thoroughly with distilled water, dried at room temperature and homogenized to make fine powder for the extraction process. 30 g of leaf powder was weighed, packed in a soxhlet apparatus and extracted sequentially with 300 ml each of petroleum benzene, chloroform and methanol. Solvent extraction was carried out for 30-35 cycle at 45 °C till a clear solvent observed in the thimble. All three extracts were dried in a rotary evaporator below 40 °C and dissolved in their respective solvent at 10 mg/ml and stored at 4 °C for phytochemical screening (Tiwari *et al.*, 2011).

Antibacterial Assay

Antimicrobial activity of phytochemical extracts was determined on five human pathogenic strains following agar well diffusion method (Magaldi *et al.*, 2004). Briefly, 100 µl of fresh microbial culture (OD ~ 1.0) was spread homogenously on nutrient broth agar (NBA) plates. 50 to 100 µl of phytochemical extract at 1 mg/ml in DMSO was applied in wells of 8 mm diameter and 5 mm height. The plates were incubated at 37 °C for 8 h. Diameter of the zone of inhibition around each well was measured to determine the antimicrobial activity of the phytochemical extract.

Determination of the Effect of Phytochemicals on *B. licheniformis* Membrane Lipid Composition

B. licheniformis (MTCC2617) were grown in 5 ml NB containing 0, 7.5 or 75 µg/ml of PB extract for 12 h at 25 °C and 200 rpm. Cells were collected at 4500 rpm for 10 mins at 25 °C and total lipid was extracted using aqueous two-phase method (Bligh and Dyer, 1959). Phosphlipid (PL) composition in the lipid extracts was determined by separating the total lipids on Silica Gel GF-254 TLC plates (Schneiter and Daum, 2006). PL was identified using PL standard and phosphate estimation (Sahu and Behuria, 2018). Amino-PL were identified using ninhydrin reagent (Slavetinsky *et al.*, 2012).

Determination of Membrane-destabilizing Effect of Phytochemicals

Effect of phytochemicals on bacterial plasma membrane was determined by treating the giant unilamellar vesicles (GUV) prepared from the polar lipids extracted from *Bacillus licheniformis* (MTCC-2617) (Pereno *et al.*, 2017). Briefly, 1 mg polar lipid (corresponding to 1300 nmol PO_4) in chloroform was coated on a thin copper electrode (5cm x 0.5cm x 1mm) and dried at 45 °C for 1 hour in a hot air oven. The dried lipid film was hydrated by placing inside a small (1.2 cm x 5 cm) round bottomed glass bottle containing 3 ml low salt solution (LSS) (10 mM KCl and

100 mM Sucrose). The GUVs were formed by placing the electrodes 0.5 cm apart and passing an alternating current of 10 Hz and 30 V for 3 h at 25 °C. Formation of GUVs was analyzed under a Nikkon compound light microscope using 100X objective. Briefly, phase-contrast images of GUVs were analyzed for their shape, morphology and numbers after a dose-dependent and time-dependent treatment with different phytochemical extracts. All statistical analyses were performed using Graph pad prism (version 6).

Results and Discussion

Identification of Major Phytochemicals

The result of phytochemical screening of *Nyctanthes arbor-tristis* leaf extracts is given in Table 20.1. Results show that the petroleum benzene extract is enriched in alkaloids, glycosides, terpenoids, resin and carbohydrate. Similarly chloroform extract is enriched in alkaloids, glycosides, terpenoids, phenol, resin, and tannins and methanol extract is enriched in glycosides, carbohydrate, phenol, flavonoid and quinine. The overall polarity of phytochemicals in the extracts increased in the order of PB (polarity index = 0.1) < Chloroform (polarity index = 4.1) < methanol (polarity index = 5.1). Similar composition of phytochemicals was also obtained from *Nyctanthes arbor-tristis* leaf extract using solvents of similar polarity (Kumari and Charya, 2017). However, we did not observe any protein, amino-acids, saponin and lipids in the phytochemicals (Table 20.1). Presence of these secondary metabolites has also been reported in flower and root extracts of *Nyctanthes arbor-tristis* using hot water, ethanol and benzene (Aggarwal and Goyal, 2013). However, differences in phytochemical composition of the extracts may vary with solvent polarity, geographic location of plant, climatic variation, presence of endophytes and temperature of extraction (Liu *et al.*, 2015).

Table 20.1. Analysis of Phytochemicals in different Extracts of *Nyctanthes arbor-tristis* Extact[a]

Phytochemicals (Tests)	*Petroleum Benzene*	*Chloroform*	*Methanol*
Alkaloid (Mayer's test, Wagner's test, Tannic acid test)	+	+	–
Glycoside (Keller killiani test, Molisch's test, Conc. H_2SO_4 test)	+	+	+
Quinone (Conc.H_2SO_4, Conc. HCl)	–	–	+
Carbohydrates (Fehling's test, Molisch's test)	+	+	+
Tannin ($FeCl_3$ test, Alkaline regent test)	–	+	–
Protein (Biuret test, Ninhydrin test)	–	–	–
Flavonoid	–	–	+
Saponin	–	–	–
Phenol	–	+	+
Terpenoid	+	–	–
Resin	+	+	–

+: Indicates the presence –: Indicates absence of the phytchemicals.

Antimicrobial Activity of the Phytochemical Extracts

All three extracts (PB, chloroform and methanol) exhibited antimicrobial activity against all five pathogenic strains tested as indicated by their zone of inhibition (Table 20.2). However, the overall antimicrobial activities of the extracts were in the order chlorofrom > PB > methanol. Hence, active ingredients responsible for antimicrobial activity are moderately hydrophobic. The chloroform extract shows highest inhibition zone against *S. flexineri.* However, incubation for > 12 h led to resistance of all pathogens tested against all three crude extracts. The chloroform and methanol extract showed highest activity against *S. aureus. S. aureus* shows resistance, whereas, *B. lichenifromis* exhibited highest sensitivity against PB extract. Both PB and $CHCl_3$ extract exhibited comparable antimicrobial activity against *P. aeruginosa*. However, none of the extracts exhibited antimicrobial activity on *Shigella sonnei* (Table 20.2).

Table 20.2. Zone of Inhibition for Various Extracts of *Nyctanthes arbor-tristis* against the Pathogenic Bacteria

Bacterial Strains	*Zone of Inhibition (mm) on Nyctanthes arbor-tristis (0.1 mg/ml)*				
	Methanol	*Chloroform*	*Petroleum benzene*	*+ve Control (Azithromycin) (0.1 mg/ml)*	*–ve Control (DMSO) (50 µl)*
Shigella flexineri	10.33±1.53	20.67 ±1.53	10.67± 2.08	17.33± 1.15	0.8
Bacillus licheniformis	10.67± 1.15	16.67± 1.53	16.67± 2.52	14.00± 3.00	0.8
Staphylococcus aureus	16.67±1.53	17.33± 1.15	11.00± 2.00	10.50± 2.12	0.8
Shigella sonnei	11.00± 2.00	10.33± 1.15	11.00±2.00	18.67± 2.52	0.8
Pseudomonas aeruginosa	10.67± 0.58	15.00 ±1.00	15.33± 3.51	18.33± 1.53	0.8

Our results reveal that all 5 bacterial strains responded differently to various extracts. This differential response of tested microbes might have resulted from (i) different composition of various active ingredients in PB, chloroform and methanol extracts, (ii) differential permeability of phytochemicals into the bacteria due to their unique membrane composition and (iii) differential drug resistance of various bacteria. Our investigation reveals that all three crude extracts are potent antimicrobial that shows comparable activity at 0.05 to 0.1 mg as Azithromycin (0.1 mg/ml). In addition, extract of other parts of the plant such as bark, root and flower is also reported to exhibit antimicrobial, antifungal and antiplasmodial activity (Saha *et al.*, 2012, Vyas and Sarin, 2013). This observation reveals that the purified active principle from all three extracts will have more activity compared to the activity determined here.

Phytochemicals Induce Inhibition of Amino-phospholipid Synthesis in *Bacillus licheniformis* and Destabilize the Membrane

Drug resistant bacteria are known to induce changes in their membrane lipid composition, leading to increase in zwitter-ionic PLs in their membrane that makes their plasma membrane more positive. In addition, many drug resistant strains can add positively charged-amino acids (*e.g.* Arg, Lys and Ala) to their negatively charged PLs (*e.g.* CL and PG) to change them into cationic (Hebecker *et al.*, 2015). These cationic PLs make their plasma membrane more positive as indicated by an increase in membrane zeta (ζ) potential. This Aminoacylation of anionic PLs in drug resistant bacteria is facilitated by a membrane protein termed as multiple peptide-resistance factor (MPRF) that acts as a aminoacyl-PL synthase as well as a aminoacyl-PL translocase involved in both synthesis and flipping of these modified lipids.

We observed that in *B. licheniformis*, many amino-acylated lipids already exist that might contribute to its resistance to peptide drugs. Addition of PB extract of *N. arbor-tristis* to its growth medium decreased the amino-acylated lipids in a dose-dependent manner (Figure 20.1).

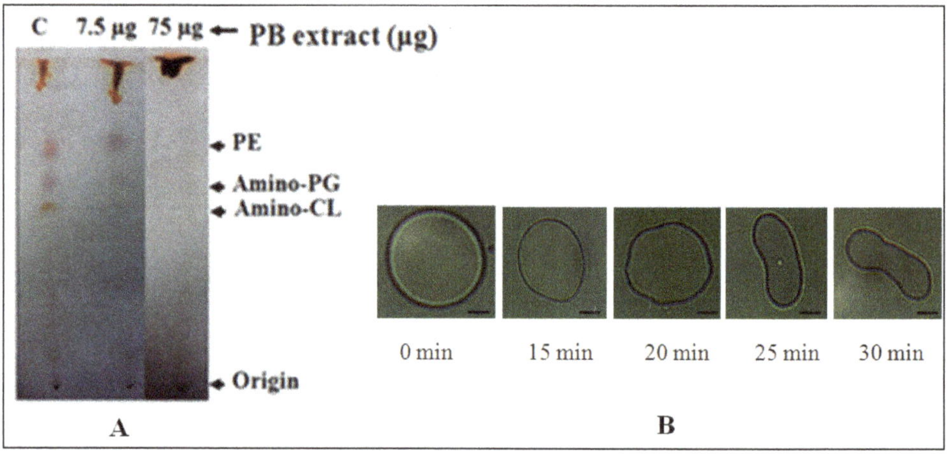

Figure 20.1. Change in Phospholipid Composition of *B. licheniformis* after Treatment with different Concentrations of Petroleum Benzene (PB) Extract.

(A) 20 µl of total lipid extract (1 mg/ml) in chloroform was loaded on TLC GF-254 plate at origin and separated using mobile phase $CHCl_3$: CH_3OH: 28 per cent ammonia = 65: 35: 5 by volume. PLs were identified using PL were identified by the presence of phosphate and amino-acylated lipids were visualized by treating the plate with ninhydrin. Amount of PB extract in the growth medium of *B. licheniformis* is shown on the top of the plate (B) Effect of PB extract on giant unilamellar vesicles (GUVs) made from polar lipid extract of *B. licheniformis*. Time of incubation of GUVs with the phytochemical is shown below. Scale = 10 µm.

Increased concentration of PB extract in growth medium of *B. licheniformis* up to 7.5 µg/ml completely inhibited aminoacylation of CL and reduced aminoacylation of PG. Further increase in PB extract up to 75 µg/ml in growth medium completely inhibited the synthesis of amino-PLs. These results show that the PB extract contains active principles that interfere with synthesis of amino-PLs leading to their inhibition. As depletion of amino-PLs in plasma membrane makes the membrane more negative, it enhances the attachment of cationic peptides to the plasma membrane, leading to membrane destabilization and lysis.

Our results show that the PB extract of *Nyctanthes arbor-tristis* contains inhibitors of amino-PL synthesis those could be potential pro-drugs to inhibit growth of *B. licheniformis* as well as to be used as adjuvant therapeutics to increase the efficacy of cationic peptides. In order to investigate the effect of phytochemical constituents of *Nyctanthes arbor-tristis* leaf extract on plasma membrane, we used GUVs prepared from *B. lichenifrmis* polar lipids as an *in vitro* model system. GUVs prepared were rounded, unilamellar and 50-100 µm in diameter (Figure 20.2B). However, treatment with 75 µg PB extract led to time-dependent change in morphology of the vesicles, leading to formation of irregular membrane structures. These results indicate that components of the *Nyctanthes arbor-tristis* PB extract interact with the membrane, leading to its destabilization and lysis (Rowlett *et al.*, 2017). Membrane-destabiling effect of phytochemicals are now being exploited as novel drug leads (Ingolfsson *et al.*, 2014, Tsuchiya, 2015). Hence, our study shows that the *Nyctanthes arbor-tristis* could be further exploited as a source of membrane-active antimicrobial drugs.

CONCLUSION

In conclusion, our study revealed that the *Nyctanthes arbor-tristis* is a rich source of antimicrobials that exhibits antimicrobial activity against wide varieties of human pathogens. The drug resistant pathogen *B. licheniformis* shows susceptibility to the extracts of petroleum benzene, chloroform and methanol. The components of PB extract inhibited synthesis of amino-PLs in the plasma membrane of *B. licheniformis*, that plays an important role in it's susceptibility to the extracts. Further, the extract contains active principles that interact with the bacterial membrane and leading to its destabilization. Further studies need to be conducted for purification and identification of the active principles in the *Nyctanthes arbor-tristis* extract that is an effective pro-drug against *B. licheniformis*.

REFERENCES

Adimpong D.B., Sorensen K.I., Thorsen L., Stuer-Lauridsen B., Abdelgadir W.S., Nielsen D.S., Derkx P.M. and Jespersen L. (2012). Antimicrobial susceptibility of Bacillus strains isolated from primary starters for African traditional bread production and characterization of the bacitracin operon and bacitracin biosynthesis. *Applied Environmental Microbiology*, 78(22): 7903-7914.

Agerso Y., Bjerre K., Brockmann E., Johansen E., Nielsen B., Siezen R., Stuer-Lauridsen B., Wels M. and Zeidan A.A. (2019). Putative antibiotic resistance genes present in extant *Bacillus licheniformis* and *Bacillus paralicheniformis* strains are probably intrinsic and part of the ancient resistome. *PloS one*, 14(1): 567-77.

Aggarwal S.G. and Goyal S. (2013). *Nyctanthes arbor-tristis* against pathogenic bacteria. *Journal of Pharmacognosy and Phytochemistry*, 2(3): 124-127.

Bligh E.G. and Dyer W.J. (1959). A rapid method for total lipid extraction and purification. *Canadian Journal Biochemical Physiology*, 37(1): 911-917.

Hebecker S., Krausze J., Hasenkampf T., Schneider J., Groenewold M., Reichelt J., Jahn D., Heinz D.W. and Moser J. (2015). Structures of two bacterial resistance factors mediating tRNA-dependent aminoacylation of phosphatidylglycerol with lysine or alanine. *Proceedings of the National Academy of Sciences,* 112(34): 10691-10696.

Hukkeri V., Akki K., Sureban R.R., Gopalkrishnan B., Byahatti V.V. and Rajendra S.V. (2006). Hepatoprotective activity of the leaves Nyctanthes arbor-tristis Linn. I*ndian Journal of Pharmaceutical Sciences,* 4(1): 542-543.

Ingoìlfsson H.I., Thakur P., Herold K.F., Hobart E.A., Ramsey N.B., Periole X., De Jong D.H., Zwama M., Yilmaz D., Hall K. and Maretzky T. (2014). Phytochemicals perturb membranes and promiscuously alter protein function. *ACS Chemical Biology,* 9(8): 1788-1798.

Khandelwal K.R., Kadam S.S. and Singhama. R. (1999). Antibacterial acivity of the leaves of *Nyctanthes arbor-tristis* Linn. *International Journal of Natural Prodocut,* 15(5): 18-20.

Kirtikar K.R. and Basu B.D. (2000). Indian Medicinal Plants. Sri Satguru Publications, New Delhi, 2(1): 2110-2113.

Kumar A. and Yadav P.D. (2013). Antibacterial activity of ethanolic extracts of *Nyctanthes arbortristis* and *Nerium oleander. Indian Journal of Research in Pharmacy and Biotechnology,* 1(3): 311-313.

Kumari T.S. and Charya M.S. (2017). Phytochemistry, Anti-cancer and anti-inflammatory activities of solvent leaf extracts of *Nyctanthes arbor-tristis. International Journal of Pharmaceutical Sciences and Research,* 8(4): 1654.

Li S. and Zhang W. (2014). Ethnobotany of camptotheca decaisne: new discoveries of old medicinal uses. *Pharmaceutical Crops,* 5(9): 140-145.

Liu W., Liu J., Yin D. and Zhao X. (2015). Influence of ecological factors on the production of active substances in the anti-cancer plant *Sinopodophyllum hexandrum* (Royle) TS Ying. *PLoS One,* 10(4): 17-35.

Magaldi S., Mata-Essayag S. and Capriles C.H. (2004). Well diffusion for antifungal susceptibility testing. *International Journal of Infectious Disease,* 8(4): 39-45.

Okigbo R.N. and Omodamiro O.D. (2006). Antimicrobial effect of leaf extracts of Pigeon Pea (CajanuscajanL Mill sp) on some human pathogens. *Journal of Herbs, Spices and Medicinal Plants,* 12(3): 117-127.

Pereno V, Carugo D, Bau L, Sezgin E, Bernardino de la Serna J, Eggeling C, Stride E. (2017). Electroformation of giant unilamellar vesicles on stainless steel electrodes. *ACS omega* 29(1): 994-1002.

Rowlett, V.W., Mallampalli, V.K., Karlstaedt, A., Dowhan, W., Taegtmeyer, H., Margolin, W. and Vitrac, H. (2017). Impact of membrane phospholipid alterations in Escherichia coli on cellular function and bacterial stress adaptation. *Journal of Bacteriology,* 199(13): 00849-16.

Saha K.M., Acharya S., Shoven S.H., Apu A.S. and Rroy P. (2012). Biochemical investigation and biological evalution of the methanolic extract of the Nyctanthes arbortristis in vitro. *Asian Pacific Journal of Tropical Biomedicine,* 1: 1534-1541.

Sahu S.K. and Behuria H.G. (2018). Biochemical characterization of H_2O_2-Induced oxidative stress in *E. coli. Journal of Applied Microbiology and Biochemistry,* 2(3): 1-10.

Saxena R.S., Gupta B., Saxena K.K., Srivastava V.K. and Prasad D.N. (1987). Analgesic, antipyretic and ulcerogenic activities of Nyctanthes arbortristis leaf extract. *Journal of Ethnopharmacology,* 19: 193-200.

Schinor E.C., Salvador M.J., Ito I.Y. and Dias D.A. (2007). Evaluation of the antimicrobial activity of crude extracts and isolated constituents from Chrestascapigera. *Brazilian Journal of Microbiology,* 38(2): 145-149.

Schneiter R. and Daum G. (2006). Analysis of yeast lipids. *Methods of Molecual Biology,* 313: 75-84.

Sittiwet C. and Puangpronpitag D. (2009). Antimicrobial properties of Derris scandens aqueous extract. *Journal of Biological Science,* 9(11): 607–611.

Slavetinsky C.J., Peschel A. and Ernst C.M. (2012). Alanyl-phosphatidylglycerol and lysyl-phosphatidylglycerol are translocated by the same MprF flippases and have similar capacities to protect against the antibiotic daptomycin in Staphylococcus aureus. *Antimicrobial Agents and Chemotherapy,* 56(7): 3492-3497.

Tiwari P., Kumar B., Kaur M., Kaur G. and Kaur H. (2011). Phytochemical screening and Extraction: A Review. *Internationale Pharmaceutica Sciencia,* 1(1): 98-106.

Tsuchiya H. (2015). Membrane interactions of phytochemicals as their molecular mechanism applicable to the discovery of drug leads from plants. *Molecules,* 20(10): 18923-18966.

Uniyal S.K., Singh K.N., Jomwal P. and Lal B. (2006). Traditional use of medicinal plants among the tribal communities of Chota Bengal, Western Himalaya. *Journal of Ethnobiological and Ethnomedicine,* 2(6): 1-14.

Vyas A. and Sarin R. (2013). Analysis of the phytochemical contant and antimicrobial activity of Nyctanthes arbor tristis. *International Journal of Pharmacology and Biological Science,* 4 (1): 201-206.

Chapter 21

Antimicrobial Secondary Metabolites of Endolichenic Fungi Associated with Lichen Thalli of Similipal Biosphere Reserve, Odisha

Srichandan Padhi[1]*, Marco Masi*[2]*, Alessio Cimmino*[2]*, Antonio Evidente*[2] *and Kumanand Tayung*[3]*

[1]*Institute of Bioresources and Sustainable Development, Sikkim Centre, Gangtok – 737 102, Sikkim*

[2]*Deprtment of Botany, Gauhati University, Guwahati – 781 014, Assam*

[3]*Department of Chemistry, University of Naples Federico II, Via Cintia 4, 80126, Naples, Italy*

**e-mail: kumanand@gauhati.ac.in*

Endolichenic fungi have been recognized as repository of bioactive metabolites with wide range of pharmacological properties, including antioxidant, anticancer, antifungal, antibacterial, antiquorum sensing, antiviral, and anti-Alzheimer's disease properties. Nevertheless, a few studies have been carried out on endolichenic fungi and the metabolites that they produced remained vastly unexplored for therapeutic applications. In this study, 117 endolichenic fungal isolates were obtained from surface sterilized fragments of 16 different lichen species collected from Similipal Biosphere Reserve. The isolated fungi belong to class Ascomycetes, Hyphomycetes and Zygomycetes with the former being the dominant. The fungal genera comprises of *Aspergillus* (10 per cent) followed by *Penicillium* (9 per cent) and *Fusarium* (8 per cent) while 24 per cent were of sterile mycelia. The EtOAc organic extracts obtained from culture filtrates of the isolates were determined for antimicrobial activity against several clinically significant human pathogens. The results indicated 5.12 per cent of the isolates displayed antimicrobial activity inhibiting most of the test pathogens. Only 3.14 per cent of the isolates were found to show antibacterial activity and 0.85 per cent of them were able to inhibit the fungal pathogens. Among five dominant strains that showed promising activity, organic extracts of two strains were further purified to get 10 pure homogenous compounds. Aurasperones A and D, asperpyrone A, fonsecinone A, carbonarone A, pyrophen and a new 6-benzyl-γ-pyrone named aspergyllone were isolated from *Aspergillus niger* (GYT3). Similarly mangrovamide J, ravenilin including another new substituted dihydroxanthene-1,9-dione named funiculosone were isolated from *Talaromyces funiculosus* (BBWT6). The pure metabolites isolated from *A. niger* were evaluated for their antimicrobial activity against different pathogens comprising plant, human and food borne and fish pathogens. Aurasperone A and pyrophen showed moderate to strong antimicrobial activity inhibiting seven different test pathogens, being pyrophen active with IC_{50} ranging from 35 to 97 µg/mL. Aspergyllone showed strong selective antifungal activity against *Candida parapsilosis*. The other compounds displayed antibacterial activity with an IC_{50} range 23–104 µg/mL when assayed against *Escherichia coli* and *Staphylococcus aureus*. Funiculosone also showed anticandidal activity against *Candida albicans* with an IC_{50} 35 µg/mL. The findings of the study suggest the endolichenic fungi colonizing lichen thalli could be good sources of new and effective antimicrobial metabolites.

Keywords: *Endolichenic fungi, Aspergillus niger, Talaromyces funiculosus, Antimicrobial metabolites, Aspergyllone, Funiculosone.*

INTRODUCTION

Fungi, conveniently called moulds are heterotrophic; depend on energy rich carbon compounds synthesized by other organisms. Most of the fungi break dead organisms down into detritus and return inorganic nutrients to the ecosystem, some cause diseases to plants and animals; and others exist as mutualists. Mutualists constitute one-third of total fungi either as mycorrhizae or lichens (Alexopolus *et al.*, 1996). Mycorrhizal fungi live on the roots of plants and provide inorganic nutrients to the plants in exchange for organic sugars. The lichens, meanwhile, are fungi living in symbiotic relationships with algae or cyanobacteria. The fungus typically provides water and minerals for the algae, in exchange for organic food from photosynthesis. Besides, there exist a group of fungi in the intercellular spaces of plants and do not possess any threat to the host. These are termed as endophytic fungi and occur within the apparently healthy tissues of all major lineages of land plants (Arnold and Lutzoni, 2007). Fungal endophytes are noteworthy for their potential to produce antibiotic and industrially important chemicals. Recently, fungi resembling endophytes called endolichenic fungi have also been discovered within living, apparently healthy lichen thalli. These endolichenic fungi coexist with the fungal partner of the symbiosis and lives in close association with the alga or cyanobacteria (Suryanarayan *et al.*, 2005; Arnold *et al.*, 2009). Preliminary studies have suggested that endolichenic fungi bear similar prospective for industrial, pharmaceutical and agricultural applications (Ding *et al.*, 2009). However, the pervasiveness of such fungi has not been deliberated expansively with regards to their metabolites and bioactivity.

In the present investigation, endolichenic fungi associated with some corticolous lichens collected from Similipal Biosphere Reserve were studied for their antimicrobial secondary metabolites, Odisha is studied. The organic EtOAc extracts of some of the fungi displayed promising antimicrobial activity against a panel of clinically important bacterial and fungal pathogens. The potent extracts were further characterized through bio-guided purification and spectroscopy methods to isolate the active pure compounds.

Materials and Methods

Sample Collection, Isolation and Identification of Endolichenic Fungi

Frequent field tours were conducted to the study sites for collection of lichens and the lichen thalli were cut from the plant barks using a disinfected knife. The samples were brought to the laboratory in sterile plastic bags, kept at 4 °C and processed within 24 - 72 h after collection. The lichen species were identified referring the standard identification manuals (Moncada *et al.*, 2013). For isolation of endolichenic fungi, thalli were first washed with tap water to remove the debris followed by surface sterilization. The sterilization was carried out as per the Li *et al.* (2007) and Bates *et al.* (2011) with slight modifications wherever necessary. The lichen thalli were sequentially immersed in 3 per cent H_2O_2 solution for 4–5 min followed by rinsing with 70 per cent alcohol for 5 s and 1 per cent NaOCl for 6 min and finally rinsed with sterile distilled water. The surface sterilized thalli were soaked with sterile filter paper to remove the excess moisture and air dried under laminar air flow chamber. The efficacy of the sterilization was evaluated by rubbing sterilized thalli on the surface of solid culture medium (Potato Dextrose Agar) for growth of contaminant if any. The dried thalli were dissected aseptically to small fragments of 1 mm^2 size and inoculated onto petri plates containing potato dextrose agar (PDA) medium. The plates were sealed with parafilm and incubated in biological oxygen demand (BOD) incubator at 29±1 °C. The plates were observed once a day for fungal growth and incubated for several days till appearance of mycelium. Fungal hypha growing out of surface-sterilized fragments were immediately sub-cultured and maintained at 4°C in PDA slants. The pure isolates were stained with lactophenol cotton blue and identified based on the colonial morphology and microscopic observation referring standard identification manual (Gilman 1971; Barnett and Hunter, 1998).

Fungal Culture and Screening for Antimicrobial Activity

Pure endolichenic fungal isolates were cultivated on Potato Dextrose Broth (PDB) by placing agar blocks of actively growing culture (3 mm in diameter) in 250 mL Erlenmeyer flasks containing 100 mL of the medium. The flasks were incubated in BOD shaking incubator for 14 days at 29 ± 1 °C with periodic shaking at 150 rpm. The cultures were filtered through sterile cheesecloth to remove the mycelial mats. The liquid broth were collected and extracted with equal volume of ethyl acetate in a separating funnel by vigorous shaking for 15 min. The organic fractions were left for evaporation and the resultant compounds were dried with magnesium sulphate (MgSO4) to yield the concentrated organic extracts. The extracts were then dissolved in 15 per cent Dimethyl sulfoxide (DMSO) for antimicrobial activity evaluation. Antimicrobial assay were carried out by agar cup diffusion method against seven bacterial pathogens namely *Bacillus subtilis* (MTCC 736), *Staphylococcus aureus* (MTCC 737), *Pseudomonas aeruginosa* (MTCC 424), *Proteus vulgaris* (MTCC 426),

Shigella flexneri (MTCC 1457), and *Klebsiella pneumonia* (MTCC 3384), *Escherichia coli* (MTCC 443) and three pathogenic fungi, *Candida albicans* (MTCC 227), *Candida krusei* (MTCC 9215), and *Trichophyton mentagrophytes* (MTCC 8476). The test pathogens were obtained from the Institute of Microbial Technology (IMTECH), Chandigarh, India. For determination of antimicrobial activity, Muller Hinton Agar (MHA) plates were inoculated with overnight culture of each bacterial suspension. Similarly for the fungal pathogens, Sabourad Dextrose Agar (SDA) plates were inoculated with each fungal suspension. The plates with the inoculated organisms were evenly spread out with sterile cotton swabs. Agar cups were prepared by scooping out the media with a sterile cork borer (9 mm). The cups were filled with 100 µL of the organic extracts (1 mg/mL). The plates were then incubated at 36 ± 1 °C for 24 h and 48 h for bacterial and fungal pathogens respectively. Antimicrobial activity was determined as growth inhibition of the target organism around agar cup as appearance of clear zones.

Purification and Structure Elucidation of Metabolites

Purification of bioactive organic extracts obtained from some potent isolates was done by analytical and preparative chromatographic techniques. Optical rotation was measured on a Jasco P-1010 digital polarimeter; IR spectra were recorded as glassy film on a Thermo Nicolet 5700 FT-IR spectrometer; UV spectra were recorded in MeCN solution on a Jasco V-530 spectrophotometer 1H and 13C NMR spectra were recorded in CDCl3 at 500/125 MHz on a Varian or at 400/100 MHz on a Bruker spectrometer. The same solvent was used as internal standard. Carbon multiplicities were determined by DEPT spectra (Berger and Braun, 2004); DEPT, COSY-45, HSQC, HMBC and NOESY experiments (Berger and Braun, 2004) were performed using Bruker or Varian microprograms. HRESI and ESI mass spectra and liquid chromatography (LC)/MS analyses were performed using the LC/MS TOF system AGILENT 6230B, HPLC 1260 Infinity. The HPLC separations were performed with a Phenomenex LUNA (C18 (2) 5u 150×4.6 mm). Analytical and preparative TLC were performed on silica gel (Merck, Kieselgel 60 F254, 0.25 mm) and on reversed phase (Merck, Kieselgel 60 RP-18, F254, 0.20 mm) plates; the spots were visualized by exposure to UV light and/or by spraying first with 10 per cent H2SO4 in MeOH and then with 5 per cent phosphomolybdic acid in EtOH, followed by heating at 110 °C for 10 min. Column chromatography was performed on silica gel (Merck, Kieselgel 60, 0.063–0.200 mm).

Minimum Inhibitory Concentration (MIC_{50})

MIC_{50} of the metabolites was determined against a range of test organisms by micro-broth dilution assay in sterile 96-well plate. A two-fold dilution of metabolites (initial conc. for crude 10 mg/mL, for pure 5 mg/mL, upto 16 fold) in different concentrations was made and the wells were filled up with 190 µL of each test bacterial and fungal suspension (approx. 10^6 CFU/mL). Metabolite extracts (10 µL) of different concentrations were added into each well to give a final volume of 200 µL and medium containing 5 per cent DMSO was used as negative control. Control wells were filled with 190 µL of culture broths (NB for bacteria, PDB for fungi) and 10 µL test of the metabolite solution. The microplate was then placed in a shaking incubator at 37 °C for 24 hrs and the absorbance was determined at 595 nm in a microplate reader (550-rad laboratories, S/N 11583). The relative inhibition (per cent) of the test extracts/metabolites was calculated by the formula;

Relative inhibition (per cent) =

$$\frac{\text{(OD value of the test sample – OD value of non-inoculated extract control)}}{\text{Average OD of the solvent control}} \times 100$$

Data from the dose - response experiments were analyzed with Prism (GraphPad Prism 5.0). The IC_{50} for each growth condition was calculated by fitting the data to a non-linear least squares sigmoid regression curve.

Results

Isolation and Identification of Endolichenic Fungi

In the present study, 16 different lichen species were selected for isolation of endolichenic fungi (Figures 21.1, A-D). The lichen species were collected from different forest reserves of SBR, Odisha, India. Altogether, 117 endolichenic fungal isolates were isolated from the surface sterilized lichen thalli. The isolates were identified based on their colonial morphology and reproductive structures through lactophenol cotton blue staining and referring standard fungal identification-manuals. The fungi which did not produce spores were categories as sterile mycelia and each distinct sterile isolate were designated as morphotypes. The overall fungi groups belong to class Ascomycetes, Hyphomycetes and Zygomycetes. The dominant fungal class was Ascomycetes which were represented by 10 species followed by Hyphomycetes (8 species) and Zygomycetes (1 species). The dominant fungal genera were *Aspergillus* (10 per cent) followed by *Penicillium* (9 per cent) and *Fusarium* (8 per cent). Among the isolated fungi, 24 per cent of the isolates were of sterile mycelia in which no spore were observed throughout the investigation. The other fungi belong to species of *Colletotrichum, Nigrospora, Neurospora, Cladosporium, Acremonium, Xylaria, Geotrichum, Phomopsis, Pestalotiopsis, Cochilobolus, Talaromyces, Mucor, Alternaria, Curvularia, Trichoderma* and *Cytospora* (Figures 21.2, A and B).

Determination of Antimicrobial Activity

All the isolates were grown in PDB for about 2 weeks as described earlier and the culture filtrates were extracted with ethyl acetate to yield crude organic extracts. The organic extracts were dissolved in DMSO (1 mg/mL) and antimicrobial activities of the same were evaluated against a panel of clinically significant pathogens. Antimicrobial activity was assayed following both disc diffusion and agar well method at 100 μL extract per well. The antimicrobial activity was confirmed by appearance of clear zones around the agar cup. The results showed 5.12 per cent of the isolates were able to inhibit all the test pathogens. Organic extracts of about 3.14 per cent of the isolates were found to show antibacterial activity whereas only 0.85 per cent of them were able to inhibit the fungal pathogens. However, 46.15 per cent isolates were found to have no effect on the growth of the test organisms (**supplementary material).** The preliminary antimicrobial assay showed

Figure 21.1. Photographs of different Lichen Thalli Used for Isolation of Endolichenic Fungi.

Parmelia caperata

Trapelia sp.

Diorygma hieroglyphicum

Parmotrema ravum

Parmelia sp.

Flavoparmelia sp.

***Cryptothecia* sp.**

***Lecanora* sp.**

Parmotrema ravum

Parmotrema tinctorum

***Cryptothecia* sp.**

***Parmelia* sp.**

Graphis sp.

Cryptothecia sp.

Xanthoparmelia sp.

Cryptothecia sp.

5 endolichenic fungal isolates (Figure 21.3) with good antimicrobial activity inhibiting all the test pathogens (Table 21.1) and among them two isolates namely *Aspergillus niger* (GYT3) and *Talaromyces funiculosus* (BBWT6) isolated respectively from *Parmotrema ravum* and *Diorygma hieroglyphicum* were selected for further investigation.

Bio-guided Purification and Characterization of Metabolites

The isolate *Talaromyces funiculosus* (BBWT6) was mass cultured and the organic extract obtained (471 mg) was characterized using different analytical, chromatography and spectroscopy methods. The extract was subjected to column chromatography on silica gel eluted with $CHCl_3$-*i*-PrOH (9: 1, v/v) obtaining eleven homogeneous fractions. The residue of 2nd fraction (54.7 mg) was further purified on preparative TLC using $CHCl_3$-*i*-PrOH (98: 2, v/v) as eluent to yield

Table 21.1. Antimicrobial Activity of the Organic Extracts (Conc. 1 mg/mL, 100 µL in each well) Obtained from Selected Endolichenic Fungi against Test Pathogens

Endolichenic Fungi	*Antimicrobial Activity of the Selected Isolates against Test Pathogens Zone of inhibition (mm)*									
	Bs	*Sa*	*Pa*	*Pv*	*Sf*	*Kp*	*Ec*	*Ca*	*Ck*	*Tm*
Aspergillus niger (2GT4)	18.3±1.52	19.3±0.57	16±0	15.3±1.15	16±1	15±0	NA	21.3±1.15	16±1	20±0
Aspergillus niger (SGL32)	16.0±0.00	13.0±0.00	15.0±0.00	13.3±2.51	13.0±0.00	15.0±0.00	NA	18.0±1.73	16.0±0.57	22.0±0.57
Talaromyces funiculosus (BBWT6)	21.8±0.76	20.6±0.57	20±0	19.3±0.57	24.3±1.54	23±0	NA	31.3±0.28	19.3±0.57	29.06±0.11
Aspergillus tubingensis (FGRK7)	20.30±0.57	16.50±0.50	18.03±0.45	15.66±2.51	13.00±0.00	15.56±0.32	NA	25.30±1.73	25.00±0.00	26.60±0.57
Aspergillus niger (GYT3)	18.0±0.00	NA	18.0±0.00	NA	16.0±0.00	16.6±0.57	21.6±0.57	31.0±0.00	31.0±0.00	28.3±1.10

* Values are are mean ± SD, n = 3; NA: Indicates No activity.

Bs: Bacillus subtilis; Sa: Staphylococcus aureus; Pa: Pseudomonas aeruginosa; Ec: Escherichia coli; Pv: Proteus vulgaris; Sf: Shigella flexneri; Kp: Klebsiella pneumoniae; Ca: Candida albicans; Ck: Candida krusei and *Tm: Trychophyton mentagrophytes.*

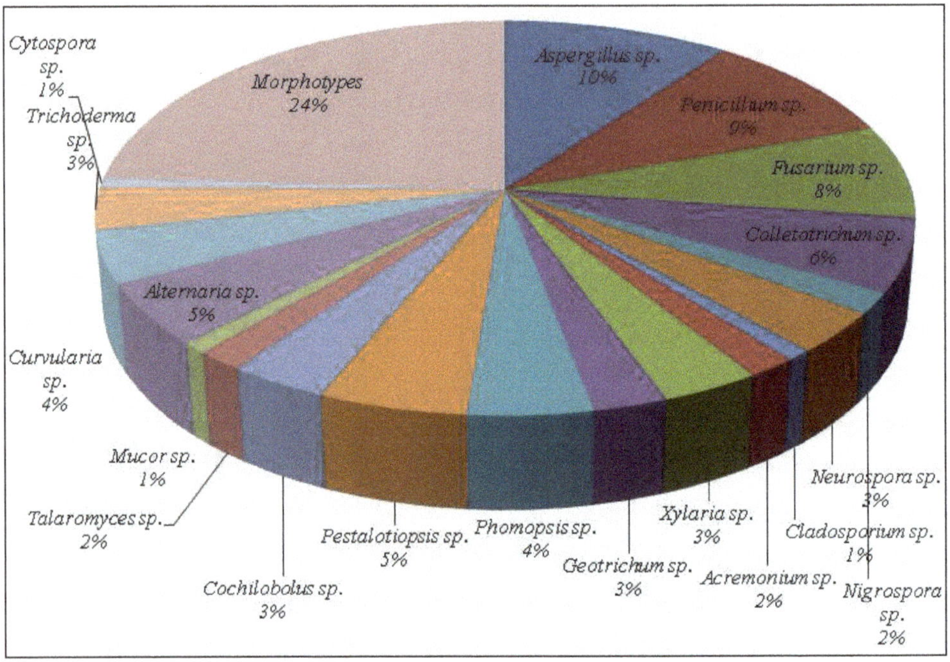

Figure 21.2. Percentage of Endolichenic Fungi Belonging to different Genera Isolated from different Lichen Thalli.

three homogeneous pure compounds identified as mangrovamide J (1.6 mg), ravenilin (1.4 mg) and a new substituted dihydroxanthene-1,9-dione (2.2 mg) named as funiculosone (Padhi *et al.,* 2019). Similarly, the organic extract of *Aspergillus niger* (GYT3) (332 mg) was purified on a silica gel column chromatography using $CHCl_3$-*i*-PrOH (9: 1, v/v) as eluent and thirteen homogeneous fraction groups were collected. The residue of 2nd fraction (26.7 mg) was further purified on preparative TLC on silica gel using *n*-hexane-EtOAc (5.5: 4.5, v/v) as eluent to yield four pure compounds identified to be aurasperone A (4.1 mg), fonsecinone A (2.8 mg) (Fang *et al.,* 2016), aurasperone D (1.5 mg) (Ghosal *et al.,* 1979), asperpyrone A (1.6 mg) (Fang *et al.,* 2016). The residues of fourth (17.6 mg) and fifth fraction (20.6 mg) were purified on analytical TLC petroleum ether-Me_2CO (7: 3, v/v) as eluent to obtain three homogenous solids identified to be a new 6-benzyl-γ-pyrone (2.6 mg) named aspergyllone (Padhi *et al.,* 2018), carbonarone A (2.2 mg) (Zhang *et al.,* 2007), and pyrophen (3.1 mg) (Zhang *et al.,* 2010). The chemical structures of the identified compounds are presented in Figure 21.4.

Funiculosone. [α] 25 D: +60.5 (c=0.01 MeOH); IR λ max 3463, 1763, 1724, 1637, 1538, 1410, 1268 cm^{-1}; UV λ max nm (log ε) 317 (4.12), 225 (4.08); HR ESIMS (+) m/z: 313.0690 $[M + Na]^+$ (calculated for $C_{15}H_{14}NaO_6$ 313.0688), 291.0871 $[M + H]^+$ (calculated for $C_{15}H_{15}O_6$ 291.0869).

Aspergyllone. UV (MeOH) λ max (log ε) 249 (3.0) nm; IR í max 1730, 1566 cm^{-1}; ^1H and ^{13}C NMR data; HR ESIMS m/z 373 $[2M+H]^+$,395 $[2M+Na]^+$, 209.0582

Figure 21.3. Photographs Showing Colonial Morphology and Microscopic View of Selected Isolates Described in this Study.

A. *Aspergillus niger* (2GT4), B. *Talaromyces funiculosus* (BBWT6), C. *Aspergillus niger* (SGL32), D. *Aspergillus tubingensis* (FGRK7) and E. *Aspergillus niger* (GYT3).

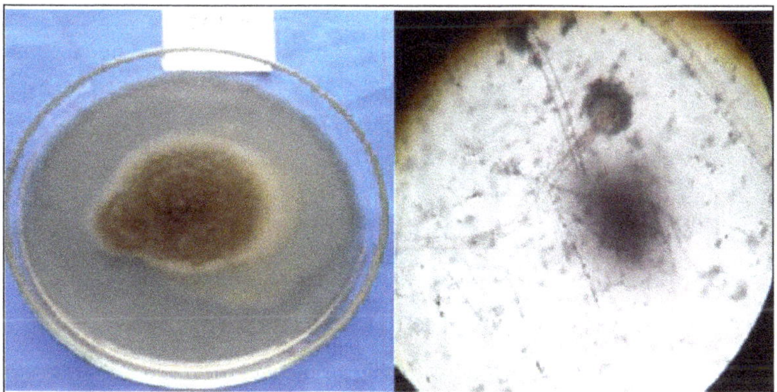

A

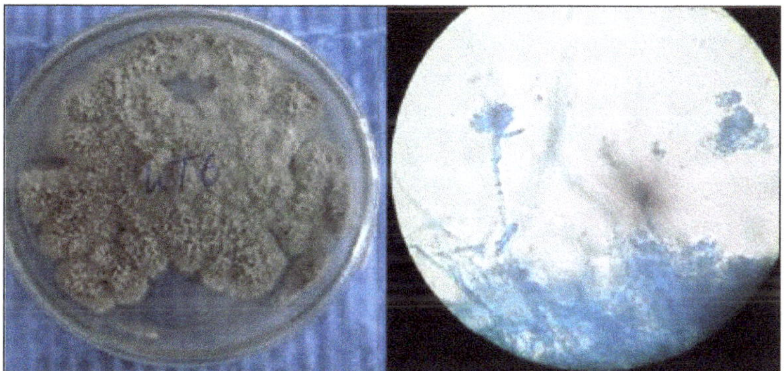

B

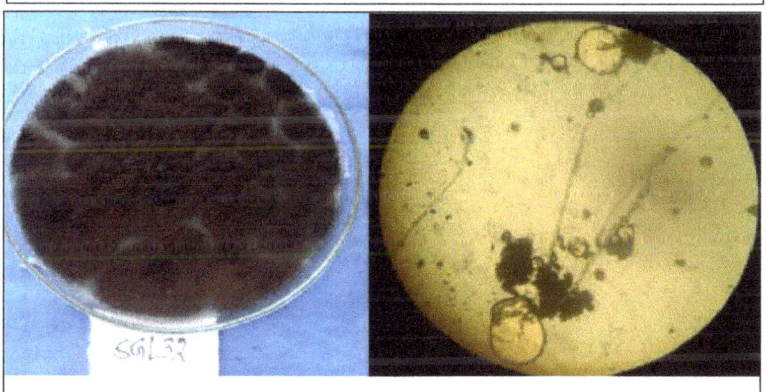

C

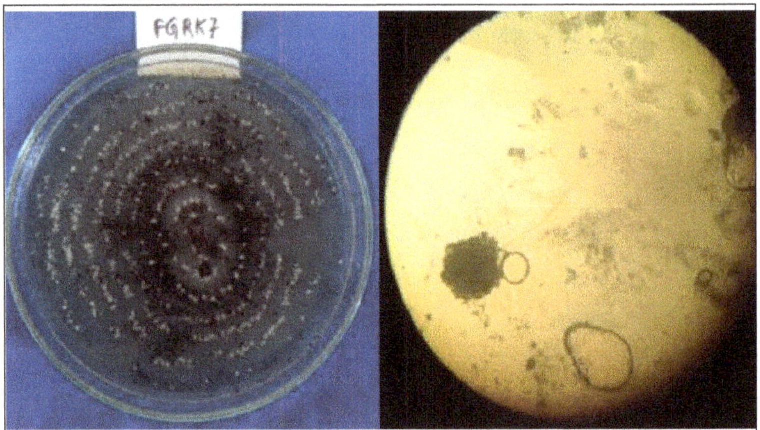

D

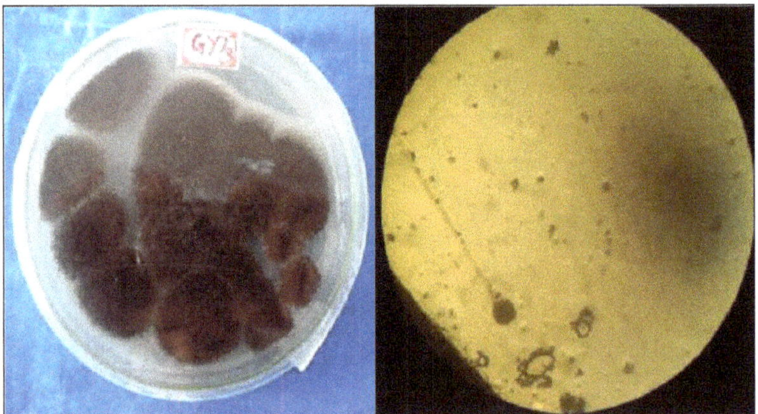

E

[M+Na]$^+$ (calculated for $C_{12}H_{10}NaO_2$, 209.0578),187.0763, [M+H]$^+$ (calculated for $C_{12}H_{11}O_2$, 187.0759).

Minimum Inhibitory Concentration (IC_{50})

The metabolites obtained from selected endolichenic fungi were determined for antimicrobial activity by calculating the efficacy based on their IC_{50} values. A two-fold dilution of the metabolites (starting conc. 5 mg/mL) was prepared to achieve different concentrations and antimicrobial activity was determined against the pathogens in a 96-wells microplate. The purified metabolites of *T. funiculosus* (BBWT6) were tested for antimicrobial activity against selected pathogens namely *S. aureus, E. coli, P. aeruginosa* and *C. albicans*. Among the metabolites, funiculosone displayed antimicrobial activity inhibiting both the bacterial and fungal pathogens. Funiculosone was able to inhibit the growth of *E. coli, S. aureus* and *C. albicans* with an IC_{50} 25 µg/mL, 58 µg/mL and 35 µg/mL respectively. However, the compound did

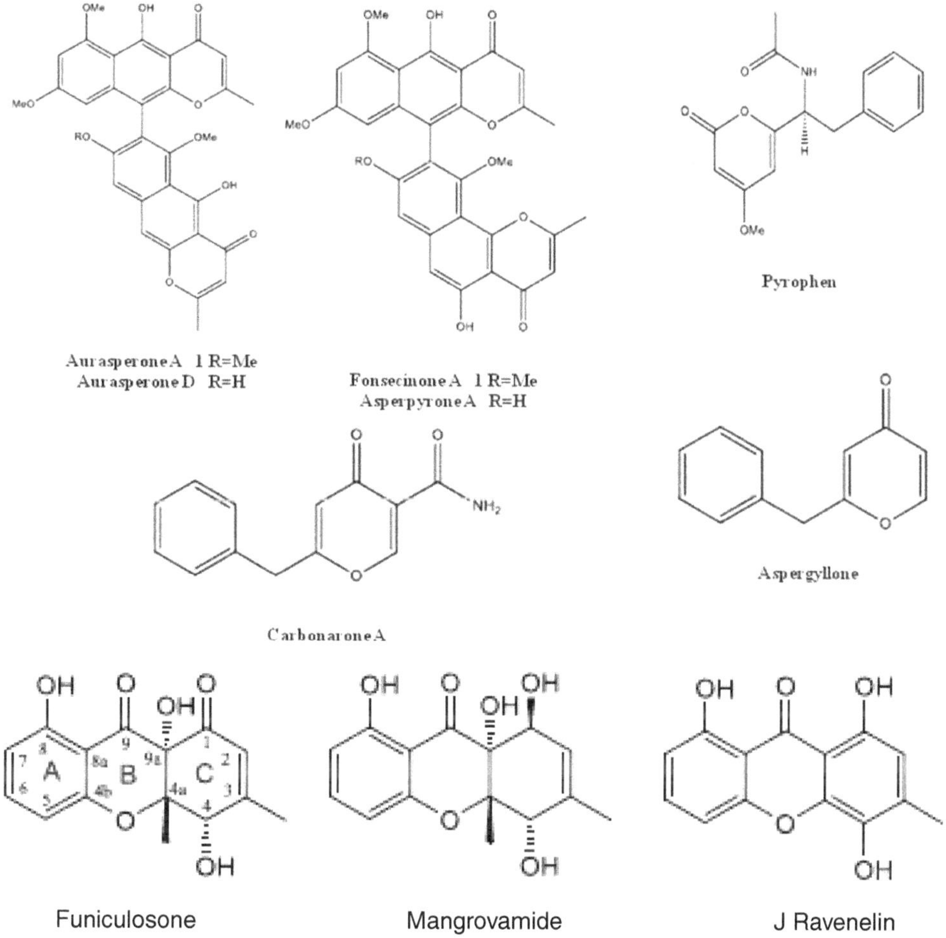

Figure 21.4. Structures of Pure Metabolites Isolated from the Organic Extracts of Endolichenic Fungi *A. niger* and *T. funiculosus.*

not show any activity against *P. aeruginosa*. Mangrovamide J displayed antibacterial activity only against E. coli and S. aureus with an IC_{50} 65 µg/mL and 104 µg/mL. Similarly, the other compound ravenilin was able to inhibit all the bacteria with an IC_{50} of 23 µg/mL against *E. coli*, 25µg/mL against *S. aureus* and 96 µg/mL against *P. aeruginosa*. Similarly the metabolites of *A. niger* (GYT3) were tested against a panel of test pathogens comprising human, plant, fish and food borne pathogens, aurasperone A displayed antimicrobial activity against a range of the test pathogens. The compound showed antibacterial activity against *P. aeruginosa* and *S. aureus* respectively with IC_{50} 160 µg/mL and 135 µg/mL. It also showed anti- candidal activity against *C. krusei* with IC_{50} 373 µg/mL. However, the compound did not show activity against the other *Candida* species. Similarly, fonsecinone A showed antibacterial activity against some few test pathogens. The metabolite was shown to inhibit *S. aureus* and *E. coli* with IC_{50} 120 μg/mL and 47μg/mL respectively. It also showed antibacterial activity against one of the test plant pathogen, *Pseudomonas syringae* with IC_{50} 154 μg/mL. Asperpyrone A showed antibacterial activity against *E. coli* with IC_{50} 112 μg/mL while no antimicrobial activity was observed by aurasperone D. Aspergyllone and carbonarone A showed antimicrobial activity against some few test pathogens. Aspergyllone showed antifungal activity against *Candida parapsilosis* with promising IC_{50} 52 μg/mL. However, no activity was observed against the other test pathogens. Similarly, carbonarone A showed antifungal activity against *C. albicans* and *C. krusei*, exhibiting IC_{50} 103 μg/mL and 31 μg/mL respectively. The compound also showed activity against a plant pathogenic bacterium, *Dickeya solani* with IC_{50} 88 μg/mL. Pyrophen showed promising antimicrobial activity inhibiting 50 per cent of the test organisms of human, fish and food borne pathogens. The pure metabolite showed antifungal activity against the entire test *Candida* species except *C. krusei*. Significant activity was observed against *C. parapsilosis* with IC_{50} 35 μg/mL followed by *Candida utilis* (IC_{50} 62 μg/mL) and *C. albicans* (IC_{50} 74 μg/mL). The metabolite also showed antibacterial activity against some food borne and fish pathogens. Considerable activity was observed against *Micrococcus luteus* (IC_{50} 63 μg/mL) followed by *Aeromonas hydrophila* (IC_{50} 78 μg/mL) and *Listeria innocua* (IC_{50} 86 μg/mL). The results are presented in Table 21.2.

DISCUSSION

Lichen thalli provide an unexplored ecological niche for a wide variety of microorganisms including fungi. Other than the obligate symbiotic mycobionts, fungi resembling plant endophytes are frequently encountered in lichen thalli. These fungi are highly diverse, asymptomatic, transmit horizontally and colonize the intercellular tissue space of the lichen. This distinct ecology group of fungi is termed as "endolichenic fungi" and has been isolated from every lichen species investigated till date (Paranagama *et al.*, 2007). In the present study, endolichenic fungi inhabiting different lichen thalli of Similipal Biosphere Reserve, Odisha, India was investigated. Altogether, 117 isolates of endolichenic fungi were isolated from different surface sterilized lichen tissues. Many researchers have also reported endolichenic fungi colonizing several lichen species from the Arctic to the tropics (Suryanarayanan *et al.*, 2005; Arnold *et al.*, 2009; Tripathi and Joshi 2015). This suggests that they are ubiquitously distributed among the lichen thalli. Altogether, 117 isolates of endolichenic fungi belonging to class Ascomycetes, Hypomycetes and Zygomycetes were isolated from different surface sterilized lichen tissues. The dominant endophytic fungi were found to be the class Ascomycetes. This collaborates with the finding of Arnold *et al.* (2009) who were of the opinion that endolichenic fungi belong to the major lineages of Ascomycota. However, our result did not agree with the findings of Tripathi and Joshi (2015) who reported the class Hypomycetes as dominant endolichenic fungi from some healthy macro lichens of Kumaun Himalaya. The occurrence of class Zygomycetes was found to be lowest in our present study which was similar with the result obtained by Tripathi and Joshi (2015). The dominant endolichenic fungi comprised of genera belonging to species of *Aspergillus* followed by *Penicillium*, *Fusarium* and several isolates of sterile fungi. Some of the isolates with distinct colonial morphology were categorized as

Table 21.2 Antimicrobial Activity of the Purified Compounds Obtained from the Crude Organic Extract of Endolichenic Fungi Showing IC_{50} Values against the Test Pathogens

Name of the Pathogens Metabolites	*IC50 (μg/mL) Values against Test*														
	Human Pathogens								*Plant Pathogens*			*Foodborne and Fish Pathogens*			
	Pa	*Sa*	*Ec*	*Ca*	*Ck*	*Cg*	*Cp*	*Cu*	*Ps*	*Pb*	*Ds*	*Ah*	*Ss*	*Ml*	*Li*
Aurasperone A	160	135	–	–	373	–	–	–	80	76	63	–	–	–	141
Asperpyrone A	–	112	–	–	–	–	–	–	–	–	–	–	–	–	–
Fonsecinone A	–	120	47	–	–	–	–	–	154	–	–	–	–	–	–
Aurasperone D	–	–	–	–	–	–	–	–	–	–	–	–	–	–	–
2-benzyl-*4H*-pyran-4-one	–	–	–	–	–	–	52	–	–	–	–	–	–	–	–
Carbonarone A	–	–	–	103	31	–	–	–	–	–	88	–	–	–	–
Pyrophen	–	–	–	74	–	97	35	62	–	–	–	78	–	63	86
Ravenilin	96	28	23												
Mangrovamide J		104	65												
Funiculososne		58	25	35											

Pa: Pseudomonas aeruginosa MTCC 424; *Sa: Staphylococcus aureus* MTCC 737; *Ec: Escherichia coli* MTCC 443; *Ca: Candida albicans* MTCC 227; *Ck: Candida krusei* MTCC 9215; *Cg: Candida galbrata* ATCC 2001; *Cp: Candida parapsilosis* ATCC 22019; *Cu: Candida utilis* IHEM 4005; *Ps: Pseudomonas syringae pv. maculicola* I11004; *Pb: Pectobacterium sp.* GBBC 1529; *Ds: Dickeya solani* GBBC 1502; *Ah: Aeromonas hydrophila* ATCC 7966; *Ss: Shigella sonnei* LMG 10473; *Ml: Micrococcus luteus* DPMB3; *Li: Listeria innocua* LMG11387.

*— indicates no activity.

morphotypes. Since these non-sporulating mycelia sterilia cannot be provided with taxonomic names without reproductive structures in conventional classification, they are now generally categorized as "morphotypes" based on similar cultural characteristics (Promputtha *et al.*, 2005). In many instances these fungal groups were isolated as endolichenic fungi from various lichen species (Zhao *et al.*, 2014; Dou *et al.*, 2014; Samanthi *et al.*, 2015). In recent years, there is an urgent need for new antimicrobial agents to combat drug resistance in bacteria and for effective treatment of systemic infection by fungi. It is believed that the less explored microbial diversity might provide new and interesting metabolites with wide therapeutic applications. Endolichenic fungi are relatively unexplored and understudied group of microbial symbiont which has been recognized as a new source of bioactive secondary metabolites (Kellog and Raja, 2017). In the present study the isolates were determined for their antimicrobial potentials against a panel of clinically significant human pathogens. Some of the isolates showed promising antimicrobial activity against the test pathogens. Many workers have also reported that endolichenic produced new and interesting bioactive molecules with antimicrobial, antitumor, antiviral and antioxidant activity (Ding *et al.*, 2009; He *et al.*, 2012; Wang *et al.*, 2010). It is believed that production of bioactive metabolites by these fungi may be similar to the mechanism of fungal endophytes that colonize higher plant species. Further, the organic extracts of some selected fungi were purified using bioassay guided fractionation. Altogether 10 homogenous compounds were purified from the active fractions of two potent isolates, 3 of them were isolated from an endolichenic fungus which was identified as *T. funiculosus* (BBWT6) and identified as ravenilin, mangrovamide J and a new substituted ubstituted dihydroxanthene-1,9-dione named as funiculosone. The purified metabolites were tested for antimicrobial activity against selected pathogens namely *S. aureus, E. coli, P. aeruginosa* and *C. albicans.* Funiculosone was able to inhibit the growth of E. coli, S. aureus and C. albicans with an IC50 25 µg/mL, 58 µg/mL and 35 µg/mL respectively. Mangrovamide J displayed antibacterial activity only against E. coli and S. aureus with an IC50 65 µg/mL and 104 µg/mL. Similarly, the other compound ravenilin was able to inhibit all the bacteria with an IC50 of 23 µg/mL against *E. coli*, 25 µg/mL against *S. aureus* and 96 µg/mL against *P. aeruginosa*. The other compounds were isolated from *Aspergillus* niger (GYT3) and identified as aurasperone A, asperpyrone A, fonsecinone A, aurasperone D, carbonarone A, pyrophen and a new 6-benzyl-γ-pyrone named aspergyllone. These compounds were evaluated for antimicrobial activity against a broad range of test pathogens including some selected humans, plants, food borne and fish pathogens. Most of the compound were known and reported earlier from *Aspergillus niger* except aspergyllone (Nielsen *et al.*, 2009). The compound showed antifungal activity against *Candida parapsilosis* (ATCC 22019) with a promising IC_{50} 52 µg/mL. Among the known compounds, aurasperone A showed good antimicrobial activity against some of the test organisms. He *et al.* (2016) reported this compound to possess antibacterial activity. However, our compound showed wider spectrum of antimicrobial activity inhibiting both bacterial and fungal test pathogens indicating its broad spectrum nature. Another compound, pyrophen also showed promising anticandidal and antibacterial activity against some food borne and fish pathogens. Earlier workers reported this compound to possess antifungal activity against *Candida albicans* and cytotoxic properties against some cancer cell lines (Shaaban *et al.*, 2012; Astuti *et al.*, 2016). Other known compounds also showed some biological activity inhibiting at least one of the test pathogens but no activity was detected in assaying Aurasperone D. The present work elucidated the characterization of bioactive metabolites from endolichenic fungi for the first time in India.

CONCLUSION

The present study suggests that endolochenic fungi colonizing lichen thalli can be good sources of antimicrobial metabolites. Considering the myriad of lichen species and only a few of them have been studied for endolichenic fungi, studies on this fungal group might results into discovery of potent strains with multiple therapeutic applications. Such studies become more significant as over the past decade, endolichenic fungi have received increasing research attention because of their potential to produce unique secondary metabolites with promising pharmacological properties (Kellogg and Raja, 2016). Therefore research priorities should be directed to study endolichenic fungi from various lichen thalli targeting new and effective antimicrobial metabolites to combat emerging drugs resistance pathogenic microorganisms.

REFERENCES

Alexopoulos, C.J., Mims, C.W. and Blackwell, M. (1996). Introductory Mycology, fourth ed. John Wiley, New York.

Arnold, A.E. and Lutzoni, F. (2007) Diversity and host range of foliar fungal endophytes: Are tropical leaves biodiversity hotspots. *Ecology*, 88(3): 541–549.

Arnold, A.E., Miadlikowska, J., Higgins, K. L, Sarvate, S. D., Gugger P, Way A, Hofstetter V, Kauff, F. and Lutzoni, F. (2009) A phylogenetic estimation of trophic transition networks for ascomycetous fungi: are lichens cradles of symbiotrophic fungal diversification. *Syst Biol* 58: 283–329.

Astuti, P, Erden W, Wahyuono S, Hertiani T (2016) Pyrophen produced by endophytic fungi *Aspergillus* sp., isolated from *Piper crocatum* Ruiz and Pav exhibits cytotoxic activity and induces s phase arrest in T47D breast cancer cells. *Asian Pac J Cancer Prev* 17(2): 615-618.

Barnett HL and Hunter B.B. (1998) Illustrated genera of imperfect fungi. APS Press, St. Paul.

Bates ST, Cropsey GWG, Caporaso JG, Knight R, Fierer N (2011) Bacterial communities associated with the lichen symbiosis. *Appl Environ Microbiol* 77(4): 1309–1314.

Berger S, Braun S (2004) 200 and More Basic NMR Experiments: a Practical Course. Wiley-VCH, Weinheim.

Ding G, Li Y, Fu S, Liu S, Wei J and Che Y (2009) Ambuic acid and torreyanic acid derivatives from the endolichenic fungus *Pestalotiopsis* sp. *J Nat Prod* 72(1): 182– 186.

Dou Y, Wang X, Jiang D, Wang H, Jiao Y, Lou H, Wang X (2014) Metabolites from Aspergillus versicolor, an endolichenic fungus from the lichen *Lobaria retigera. Drug Discov Ther* 8(2): 84–88,

Fang W, Lin X, Wang J, Liu Y, Tao H and Zhou X (2016) Asperpyrone-type bis-naphtho-y-pyrones with COX-2–inhibitory activities from marine-derived fungus *Aspergillus niger. Molecules* 21(7): 941- 948,

Ghosal S, Biswas K and Chakrabarti DK (1979) Toxic naphtho-gamma-pyrones from *Aspergillus niger. J Agri Food Chem* 27(6): 1347-1351,

Gilman JC (1971) A manual of soil fungi. Iowa State College Press, Iowa,

He JW, Chen GD, Gao H, Yang F, Li XX, Peng T, Guo LD, Yao XS (2012) Heptaketides with antiviral activity from three endolichenic fungal strains *Nigrospora* sp., *Alternaria* sp. and *Phialophora* sp. *Fitoterapia* 83: 1087–1091.

He Y, Tian J, Chen X, Sun W, Zhu H, Li Q, Lei L, Yao G, Xue Y, Wang J, Li H and Zhang Y (2016) Fungal naphtho-γ-pyrones: Potent antibiotics for drug-resistant microbial pathogens. *Sci Rep* 6: 24291.

Kellogg JJ and Raja HF (2017) Endolichenic fungi: a new source of rich bioactive secondary metabolites on the horizon. *Phytochem Rev* 16: 271–293.

Li WC, Zhou J, Guo SY, Guo LD (2007) Endophytic fungi associated with lichens in Baihua mountain of Beijing, China. *Fungal Divers* 25: 69–80.

Moncada B, Lawrey J, Chaves JL, Dal-Forno M, Lucking R (2013) Tropical lichen guides. *The field museum*, Chicago, US.

Nielsen KF, Mogensen JM, Johansen M, Larsen TO, Frisvad JC (2009) Review of secondary metabolites and mycotoxins from the *Aspergillus niger* group. *Analyt Bioanalyt Chem* 395: 1225–1246.

Padhi S, Masi M, Cimmino A, Tuzi A, Jena S, Tayung K and Evidente A (2019) Funiculosone, a new substituted dihydro-xanthene-1,9-dione and its two analogues produced by an endolichenic fungus *Talaromyces funiculosus* and their antimicrobial activity. *Phytochem* 157: 175-183.

Padhi S, Masi M, Panda SK, Luyten W, Cimmino A, Tayung K, Evidente A (2018) Antimicrobial secondary metabolites and molecular investigation of an endolichenic *Aspergillus niger*, isolated from the lichen thallus *Parmotrema ravum. Nat Prod Res* 1 – 8.

Paranagama PA, Wijeratne EMK, Burns AM, Marron MT, Gunatilaka MK, Arnold AE, Gunatilaka AAL (2007) Heptaketides from *Corynespora* sp. inhabiting the cavern beard lichen, *Usnea cavernosa*: first report of metabolites of an endolichenic fungus. *J Nat Prod* 70: 1700–1705.

Promputtha I, Jeewon R, Lumyong S, McKenzie EHC, Hyde KD (2005) Ribosomal DNA fingerprinting in the identification of non sporulating endophytes from *Magnolia liliifera* (Magnoliaceae). *Fungal Divers* 20: 167–86.

Samanthi KA, Wickramarachchi S, Wijeratne EM, Paranagama PA (2015) Two new bioactive polyketides from Curvularia trifolii, an endolichenic fungus isolated from *Usnea* sp., in Sri Lanka. *J Natl Sci Found Sri Lanka* 43: 201–224.

Shaaban M, Shaaban KA, Abdel-Aziz MS (2012) Sevennaphtho-y-pyrones from the marine-derived fungus *Alternaria alternata*: structure elucidation and biological properties. *Organic Medicinal Chem Lett* 2: 61 – 68.

Suryanarayan TS, Thirunavukkarasu N, Hariharan GN, Balaji P (2005) Occurrence of non-obligate microfungi inside lichen thalli. *Sydowia* 57: 120–130.

Tripathy M and Joshi Y (2015) Endolichenic fungi in kumaun himalaya: A case study In: *Recent Advances in Lichenology*. Upreti DK, Divakar PK, Shukla V, Bajpai R (Eds.) Springer.

Wang Y, Zheng Z, Liu SC, Zhang H, Li EW, Guo LD and Che YS (2010) Oxepinochromenones, furochromenone, and their putative precursors from the endolichenic fungus *Coniochaeta* sp. *J Nat Prod* 73(5): 920–924.

Zhang Y, Li XM, Feng Y and Wang BG (2010) Phenethyl-α-pyrone derivatives and cyclodipeptides from a marine algous endophytic fungus *Aspergillus niger* EN–13. *Nat Prod Research* 24(11): 1036-1043.

Zhang Y, Zhu T, Fang Y, Liu H, Gu Q and Zhu W (2007) Carbonarones A and B, new bioactive y-pyrone and α-pyridone derivatives from the marine-derived fungus *Aspergillus carbonarius. The J antibiotics* 60(2): 153 – 157.

Zhao H, Wang GQ, Tong XP, Chen GD, Huang YF, Cui JY, Kong MZ, Guo LD, Zheng Y Z, Yao XS and Gao H (2014) Diphenyl ethers from *Aspergillus* sp. and their anti- Aþ42 aggregation activities. *Fitoterapia* 98: 77–83.

Chapter 22

Traditional Knowledge for Biodiversity Conservation

Nabin Kumar Dhal, Nirad Chandra Rout and Sandeep Kumar Kabi

Environment and Sustainability Department, CSIR-IMMT, Bhubaneswar – 751 013, Odisha
e-mail: nkdhal@immt.res.in

Loss of biodiversity in almost all the ecosystems is an emerging global threat. Several species of the plants and other organisms are disappearing and biodiversity is more threatened now than at any time in the past. Though deforestation is generally believed as the maincause behind the current crisis, global climatic change, industrialization and pollution, habitat loss due to shifting cultivation, mining, soil erosion, unchecked expansion of urban areas *etc.* are the other causes of this problem. Biodiversity loss not only endangers the sustainability of ecosystem and flow of ecosystem services but also undermines the food and livelihood security of the growing population. Community participation is the key for sustainable management of biodiversity and traditional knowledge in terms of indigenous resources plays a vital role to protect our environment and conserve the bio resources. Traditional knowledge system available with the rural community and forest dwellers not only ensures the judicious use of the locally available resources but also preserve them through sustainable harvesting and community based conservation practices. Traditional knowledge compels the rural people as well as the forest dwellers to protect them by sustainable management. The current rate of extinction demands immediate concerted efforts for conservation of biodiversity for future generations. Conservation of biodiversity could be accomplished using both in situ and ex situ methods and educating the community for sustainable use of the valuable bio-resources

Keywords: *Biodiversity, Traditional knowledge, Climatic condition, In situ and ex situ conservation.*

Introduction

Traditional knowledge innovations, practices of indigenous and local communities have received increasing attention at international level since the adoption of the Convention on Biological Diversity (CBD). This knowledge came through a long series of observations, tried, tested and transmitted from generation to generation. The indigenous people depend on environment over a long period of time for their provision of a variety of resources; they have developed a stake in conserving the biodiversity. The potential of traditional and local ecological knowledge to contribute towards biodiversity conservation has been widely recognized, but the actual application of this knowledge is not implemented yet. The traditional knowledge provides a valuable input to the modern industries such as pharmaceuticals, herbal medicines, cosmetics, sustainable agriculture, eco-friendlypest management and organic farming. Most of the industries used traditional knowledge for developing novel products having commercial acceptability. They make huge profit by greater commercial use of this knowledge. With the advent of civilisation, the construction of new roads in jungle areas, indiscriminate cuttings of forest blocks, and acquisition of forest lands for agriculture and population intrusion from neighbouring rural and urban areas destructs the natural habitats. There is every possibility that the valuable knowledge acquired by the tribal community on diseases and medicine, food and nutrition, forestry and agriculture will be vulnerable in near future (Kate, 2002).

The tribal communities are living in the areas that are immensely rich in biodiversity and have learnt how to live in the most hostile environmental condition. It is estimated that about 300 million indigenous people are living in world, out of which nearly half *i.e.* 150 million are living in Asia, about 30 million of which are living in Central and South America and a significant number of them are living in Australia, Europe, New Zealand and Africa and Soviet Union. These communities played a vital role in conservation of biodiversity, environmental management and development process by their traditional knowledge.

In India, 68 million people belong to 227 ethnic groups and comprising 573 tribal communities (Pushpgandhan, P. 1994). These indigenous tribal people live in the close vicinity of forests and have been conserving and sustainably utilising the local natural resources and biodiversity for food, shelter and medicine. India is a country with a large ethnic society and rich in biodiversity. There are 45000 species of wild plant out of which 9500 species are ethno botanically important species and out of this among them 7500 species have medicinal value and have been playing a significant role in the system of traditional medicines importance and are used for indigenous health practices. About 3900 plant species are used by tribal people as food (among which 145 species comprise of root and tubers, 521 species of leafy vegetables, 101 species of bulbs and flowers, 647 species of fruits), 525 species are used for fibre, 400 species are used as fodder, 300 species are used in preparation and extraction of chemicals which are used as naturally occurring insecticides and pesticides, 300 species are used for extraction of gum, resins, dyes and perfume (Arora, RK. 1997). In addition to these a number of plants are used as timber, building material and about 700 species are culturally important from moral, cultural, spiritual, religious, and aestheticvalue. Indian subcontinent is one of the twelve mega-centres of biodiversity representing two of the eighteen hotspots of biological diversity, one occurring in Eastern Ghat and another in North- Eastern Himalaya (Zeven and Zhikovsky, 1973). There are 141 floristically endemic genera belonging to over 47 families of higher plant occur in India. Botanical Survey of India (BSI) has reported that 46214 plant species are found in India and among these 17500 represents flowering plants (Arora, RK. 1997). This composition makes the conservation of this diversity is most important and community participation could be proved as an important tool for in situ conservation.

Importance of Traditional Knowledge

Traditional knowledge is invaluable and has wide range of application in several sectors includingthe conservation of biodiversity. Indian systems of medicine (Ayurveda, Siddha, Unani) are also the part of the official healthcare system in India and it mostly depends on the biological resources and traditional knowledge system. Industries have often used traditional knowledge for developing commercial products without the consent and acknowledgement of the original holders of the knowledge. These conflicts arise when such knowledge is commercially used without consent or when IPRs and exclusive rights are claimed over such resources/ knowledge (Tripathy, SK. 2003). The Convention on Biological Diversity (CBD) is the principal international instrument which explicitly acknowledges the role of traditional knowledge, innovations and practices of indigenous and local communities in biodiversity conservation and its sustainable development. According to Article 8(j) of the convention, each contracting party subject to its national legislation is required to respect, preserve and maintain knowledge, innovations and practices of indigenous and local communities tangible or visible lifestyles relevant for conservation and sustainable use of biological resources and promote the wider application of such knowledge, innovations and practices with the approval and involvement of their holders and also encourage the equitable sharing of benefits arising from the utilization of such knowledge, innovations and practices (Kannaiyan, S., 2007). Plants and plant based medications have been used for a long time for prolonging the human life by combating various ailments. Ancient ethnic communities around the world had learnt to utilise the herbal wealth for curative purposes. Due to lack of literacy and scientific knowledge about plantsoften at the cost of their dear life through their century's old experience could not be documented and therefore is at the risk of complete disappearance. it had rather descended from one generation to another as a domestic cultural heritage (Rama Shankar, G.S., *et al.*, 2012).

Modern Drug Discovery from Traditional Knowledge

Drug discovery strategies for pharmaceutical and agrochemical application are in a rapid development stage. Modern approach to drug discovery includes random high-volume screening of thousands of chemical entities. These are generated from chemical libraries, combinatorial approaches and natural products, all of which feed into various test systems. The technological advances of robotics for high-throughput, random screening in the 1980's gives the industry an ability to handle a very large number of samples. This has stimulated a renaissance of activities in the areas of plant natural products chemistry, pharmacognosy and ethno-medical research. This approach includes a random collection of plants that are incorporated into high-throughput screening programmes using a variety of mechanism based assay with specific applications to numerous therapeutics areas. All positive hints are subjected to an initial *in vitro* evaluation. This methodology requiring screening of plants for natural products and chemical entities is well suited to the drug discovery (Gupta, R. 2005)

Modern Therapeutics Originated Through Ethnic Leads

Some of the claims clinically pursued for the drug development programme in the recent past by obtaining clues from tribals. Some of the clues are – *Atropa belladonna, Cinchona ledgeriana, Rauvolfia serpentine, Digitalis lanata, Dioscoreadeltoidea*etc; *Acanthus illicifolius* seeds for analgesic and anti-inflammatory activities; *Vernoniaanthelmintica* for rheumatism, conjunctivitis, dysuria *etc.*; *Garcinia morella* and *G. xanthocymus* for antiprotozoal and anti-bacterial activities (morellin and neomorellin); *Rhussemialata* for cardio tonic activity; *Butea monosperma* seeds for antifertility and anthelmintic activities *etc.* Diabetes mellitus is a chronic hereditary disease from which millions of peoples are suffering all over the globe. A number of plants and plant-products are mentioned as antidiabetic agent in literature. Plants like *Momordicacharantia* L. (Karela), *Gymnemasylvestre* (Gudumari), *Syzygiumcumini, Tinosporacordifolia* (Gulu), *Cleodendronphlomidis,Pterocarpus marsupium etc.* are used for treatment of diabetic. The primitive tribal practice is a fascinating area of research with more promise for modern drug development programme. As a result of sophisticated isolation and pharmacological testing procedures, many new drugs of plant origin have found their way into the modern medical world (Gupta, R. 2005 and Si Yuan-Pan, 2013)

Traditional Knowledge for Sustainability

The concept has developed on the basis of the recognition that the well-being of human society is closely related to the well-being of natural ecosystems. Sustainability seeks to comprehend the fundamental interactions between nature and society, specifically the interaction of global processes with the ecological and social characteristics of particular places and sectors. The intellectual resources on which the sustainability science is building on needs to take into account the knowledge of local people as well (Pandey, 2001).Application of scientific research and local knowledge contributes both to the equity, opportunity, security and empowerment of local communities as well as to the sustainability of the natural resources. Local knowledge helps in scenario analysis, data collection, management planning, designing of the adaptive strategies to learn and get feedback and institutional support to put policies in to practice. On the other hand, science provides new technologies that help in improvement to the existing ones. It also provides tools for networking, storing, visualizing, and analysing information, as well as projecting long-term trends so that efficient solutions to complex problems can be obtained (Pandey, 2002a). All the developmental approaches described above are essential for mankind. Traditional knowledge provides legitimate clues for them and deciphers the importance of biodiversity for the sustenance of civilization. However in the process the over exploitation of such resources leads to catastrophic effects on the resources.

Biodiversity Conservation

Conservation is the protection, preservation, management, or restoration of wildlife and natural resources such as forests and water. Through the conservation of biodiversity and the survival of many species and habitats which are threatened due to human activities can be ensured. The main functions of biodiversity reserves are:

1. **Conservation:** conservation of species, ecosystem and genetic resources.
2. **Development:** promotes economic development with maintaining the cultural, social and ecological identity.
3. **Scientific Research:** provides support related to research by monitoring education, local, national and global issues.

Types of Conservation

Biodiversity Conservation is broadly divided into two types *i.e. in situ* conservation and *ex situ* conservation.

b) *In situ* Conservation

In situ conservation is the onsite conservation of the genetic resources in natural populations of plant or animal species, such as forest genetic resources in natural populations of tree species. It is the process of protecting an endangered plant or animal species in its natural habitat either by protecting or cleaning up the habitat itself and defending the species from predators. *In situ* conservation is being done by declaring area as protected area

b) *Ex situ* Conservation

This is the preservation of components of biological diversity outside their natural habitats. It involves the conservation of genetic resources as well as wild and cultivated species. Such strategies include establishment of botanical gardens, zoos, conservation of pollen seed, seedling, tissue culture gene and DNA banks

In situ Conservation	*Ex situ Conservation*
1. Protected Areas like	1. Seed banks and Gene banks
National Parks	2. Long term captive breeding
Sanctuaries	3. Animal Translocations
Biosphere Reserves	4. Tissue culture Banks
2. Scared Forests and Lakes	5. Cryopreservation
	6. Botanical Gardens

In situ Conservation

The following types of natural habitats are being maintained in India:

1. National Parks

This is an area strictly reserved for the betterment of the wildlife, where activities like forestry, grazing on cultivation are not permitted. They are usually small reserves spreading in an area of 100 Sq. km. to 500 sq. km. Their boundaries are well marked and circumscribed. They are mainly emphasises on the preservation of plants or animal species.

2. Sanctuaries

It is a protected area reserved for the conservation of only animals and human activities like harvesting of timber, collecting minor forest products and private ownership rights are allowed as long as they do not interfere with well-being of animals. Boundaries of sanctuaries are not well defined and controlled biotic interference is permitted, *e.g.*, tourist activity.

3. Biosphere Reserves

Biosphere reserves serve as 'living laboratories' for testing out and demonstrating integrated management of land, water and biodiversity. It is a protected area where human population also forms a part of the system. They are large protected area of usually more than 5000 sq.km. A biosphere reserves has 3 parts- core, buffer and transition zone.

1. Core zone is the inner zone which is undisturbed and legally protected area.
2. Buffer zone lies between the core and transition zone and in this area research and educational activities are permitted.
3. Transition zone is the outermost part of biosphere reserves where cropping, forestry, recreation, fishery and other activities are allowed.

Advantages

1. It is less expensive and easy to manage.
2. The life cycles of the organisms and their evolution progresses in a natural way.
3. It provides the required green coverage and associated benefits to our environment.
4. The flora and fauna live in natural habitats without human interference.

Ex situ Conservations

1. Seed Gene Bank

Here seeds are kept under controlled temperature and humidity. This process stores the germ plasma of plants at low temperature (minus temperature) remain viable for long durations of time.

2. Gene Bank

Genetic variability is also preserved by gene bank/ genetic resources under normal growing conditions. These are cold storages where germplasm were kept under controlled temperature and humidity for storage.

3. Cryopreservation

This is the newest technology for preservation of biotic parts. This type of conservation is done at very low temperature (196°C) in liquid nitrogen. The metabolic activities of the organisms are suspended under low temperature, which are later used for research purposes.

4. Tissue Culture Bank

Long term culture of excised roots and shoots are maintained. Meristem culture is very popular in plant propagation as it's a virus and disease free method of multiplication.

5. Botanical Gardens

A botanical garden is a place where flowers, fruits and vegetables are grown. The botanical gardens provide proper environment for growth and development plants. It contains exotic plants for educational and research purposes.

Advantages

1. It is useful for declining population of species.
2. Endangered animals on the verge of extinction are successfully breeded. Threatened species are breeded in captivity and then released in the natural habitats.
3. *Ex situ* centres offer the possibilities of observing wild species, which is otherwise not possible.
4. It is extremely useful for conducting research and scientific work on different species.

Traditional Knowledge on Biodiversity Conservation

Traditional societies have accumulated a wealth of local knowledge, transmitted from generation to generation. Experience has taught them how the water, trees, and other natural resources should be used and managed to last a long time. By capitalizing on the collective wisdom of formal and traditional sciences, we shall be able to help people address the problem of global warming as well as to manage the risks they face because of the destruction of the local resources. Effective biodiversity conservation can be learned from the context-specific local knowledge and institutional mechanisms such as cooperation and collective action; intergenerational transmission of knowledge, skills and strategies; concern for well-being of future generations; reliance on local resources; restraint in resource exploitation; an attitude of gratitude and respect for nature; management, conservation and sustainable use of biodiversity outside formal protected areas; and, transfer of useful species among the households, villages and larger landscape. These are some of the useful attribute of local knowledge systems (Pandey, 2002a).

Conservational Prospects

The 1992 Convention on Biological Diversity requires that every Contracting Party should respect, preserve and maintain knowledge, innovations and practices of traditional and local communities and promote the wider application with the approval and involvement of the holder of such knowledge, innovations and practices and encourage the equitable sharing of the benefits. As nations implement the Convention on Biological Diversity (CBD) work programs, apply its guidelines, and execute national strategies, its influence on science is likely to grow. CBD-compliant national laws and policies already set priorities for research and affect the way in which scientists can access and use genetic resources (Kate, 2002).

This may be useful from a different perspective as well: that the useful knowledge is not lost. Identification of science behind traditions (Arunachalam, 2001) is a more constructive endeavour than entering into the 'indigenous vs. scientific' or 'traditional vs. western' arguments (Agrawal, 1997). Scientists need not encounter traditional knowledge systems uncritically, just as local people need not approach formal science uncritically.

Nonetheless, it needs to be reiterated that formally trained scientists as well as researchers on traditional knowledge systems have often misinterpreted the process of what is often referred to as validation. The term 'validation' need not be understood from a narrow reductionist perspective of disciplinary confines. It can, and should, draw on complimentarily and the "*consilience*" across local and formal systems. Thus, both formal and local methods, as well as local people and formally trained scientists, shall contribute to comprehend the data, information and knowledge. In collaborative efforts of such kind perhaps everyone involved may stand to benefit. Both local people as well as external experts need access to the latest scientific developments and see if it can help improve existing conservation knowledge and practices. The policy makers need ready access to the science as well as understanding the difficulties of its application (Kohm *et al.* 2000).

Indeed, there are numerous examples where local knowledge derived from long-term nature-society interaction has been extremely useful in validating scientific hypotheses and suggesting new research directions (Kimmerer, 2002 and Robertson and Hull, 2001). Likewise, formal scientific methods have been extremely valuable in validating the traditional ethno-pharmacological knowledge by identifying the active ingredients (chemicals) in plants used in ethno medicine. One such example of significant contribution that established the ancient-modern concordance came with the isolation of the hypertensive alkaloid from the sarpagandha plant (*Rouwolfia serpentina*), valued in Ayurveda for the treatment of hypertension, insomnia, and insanity. Several such isolations of active ingredients have been made since then (Dev 1999; Mishra *et al.*, 2001). Another example pertains to the conservation of ethno medicinal species that are also globally traded, and, therefore, have become endangered in India. "A reasonable degree of scientific rigour" is required to assess the threat status of species to be banned in trade (Ved *et al.*, 1998) as well as to monitor, learn and craft strategies for context specific adaptive management by using formal and local sciences. The important issue to be guarded here is that the benefits must go to the community.

Intellectual Property Rights are now being extended to beyond the conventional domain of mechanical and chemical innovations to include biological resources. National Biological Diversity Act of India in response to our commitment to the Convention on Biological Diversity and intellectual property rights must, therefore, devise operational mechanisms to share benefits of commercial applications of traditional knowledge on biodiversity with local communities. Also useful shall be to ensure a harmonized basket of rules made under the Patent Act, Protected Plant Varieties Act, and the Biological Diversity Act (Utkarsh *et al.*, 1999).

Scientists shall have to collaborate with people to put forth new hypotheses that incorporate aspirations of formal and local systems of knowing and modify their methodologies accordingly. Humanity needs to go beyond disciplinary divide and find a common ground across cultures, faiths and disciplines (Pandey, 2002a). Collective wisdom of humanity for conservation of biodiversity, embodied both in formal science as well as local systems of knowledge, therefore, is the key to pursue our progress towards sustainability.

Conclusion and Future Outlook

This chapter summarizes how traditional knowledge is useful for the development of human civilization and at the sametime being instrumental for inflicting irreversible changes to the environment. Biodiversity provides clues for collective innovation and economic sustainability. It depends on ecological sustainability. The local people do not conserve all Earth's species for pragmatic reasons. Conservation of diversity, sustainable management, propagation of such valued flora and their *in situ* as well as *ex situ* conservation are the need of this century. According to the indigenous belief of some cultural groups, the natural environment also has a value given to it by God, and no one has the power to destroy it. They emphasise humankind's guardianship role and respect for and appreciation of God's creations. This study also suggests that the multidimensional perspectives of the life, culture, tradition, and knowledge systems of tribal peoples should be carefully studied, documented and inventorised before they disappear from nature. Local people should be involved in the conservation of biodiversity. Children and the people need to be sensitised to their biodiversity heritage and its values. Traditional knowledge systems should be integrated into modern knowledge systems. As equity is a critical component for sustainable development of biodiversity. The people with indigenous knowledge should benefit from their own knowledge. Therefore various disciplines like Genetics, Pollen biology, Tree Breeding, Ecology, Botany, Physiology, Eco - restoration, Taxonomy, Ethno -botany, Taxonomy, Phyto-chemistry, Biometrics, Bio-stat should work at one platform and linkages have to be established.Fair compensation should guide the conservation and sustainable development of bio cultural diversity. In order to encourage the tribal peoples towards biodiversity conservation we have tomanage, conserve and develop social and natural diversity, the existing international and national intellectual property law should be revised and include local knowledge entitlements.

References

Agarwal, A., and Narain, S. eds. (1997). *Dying Wisdom: Rise, Fall and Potential of India's Traditional Water Harvesting Systems.* Centre For Science and Environment, New Delhi.

Agrawal, D.P. (1997). Traditional knowledge systems and western Science. *Current Science* 73: 731-733.

Anon. (1991) Atlas of Enviroment. WWP. Oxford Publication. Oxford, London

Anon. (1998) Status report on biodiversity. Ministry of Enviroment and Forest. GOI, New Delhi

Arora R.K. (1991). Conservation and Management concept and Approach in Plant Genetic resources.(eds.) Paroda R.S.&R.K.Arora, IBPGR, Regional office South and Southeast Asia, New Delhi, p25.

Arora R.K. (1997). Ethnobotany and its role in Conservation and Use of Plant genetic resources in India. Journal of ethnobotany

Arunachalam, V. (2001). The science behind tradition. *Current Science* 80: 1272-1275.

Dev, S. (1999). Ancient-modern concordance in Ayurvedic plants: some examples. Environ. *Health Perspect.* 107: 783-9.

Jain S.K. (1996) Ethnobiology in Human Welfare, Deep Publications New Delhi.P-519 -520.

Kate, K.T. (2002). Science and the Convention on Biological Diversity. *Science* 295: 2371-2372.

Kimmerer, R.W. (2002). Weaving traditional ecological knowledge into biological education: A call to action. *BioScience* 52: 432-438.

Kohm, K., Boersma, P.D, Meffe, G.K., Noss, R. 2000. Putting the science into practice and the practice into science. *Conservation Biology* 14: 593-594.

Mishra, L, Singh, B.B., Dagenais, S. (2001). Ayurveda: a historical perspective and principles of the traditional healthcare system in India. *Altern. Ther. Health Med.* 7: 36-42.

Nature. (1999). Caution: traditional knowledge. *Nature* 401: 623-623.

Pandey, D.N. (2001). A bountiful harvest of rainwater. *Science* 293: 1763-1763.

Pandey, D.N. (2002a). Cultural resources for conservation science. *Conservation Biology* (in press).

Pushpagandhan P. (1994). Ethnobiology in India. A status report, Ministry of Enviroment and Forest, GOI, New Delhi.

Rama Shankar, G. S. Lavekar,S. Deb and B. K. Sharma Traditional healing practice and folk medicines used by Mishing community of North East India, *J Ayurveda Integr Med.* 2012 Jul-Sep; 3(3): 124–129. doi: 10.4103/0975-9476.100171.

Ranjan Gupta, Bjarne Gabrielsen and Steven M. Ferguson, Nature's Medicines: Traditional Knowledge and Intellectual Property Management. *Case Studies from the National Institutes of Health (NIH), USA.*Curr Drug Discov Technol. 2005 Dec; 2(4): 203–219.

Robertson, D. P., and R. B. Hull. (2001). Beyond biology: toward a more public ecology for conservation. *Conservation Biology* 15: 970-979.

S. Kannaiyan, Traditional knowledge and biodiversity conservation. 2007.

Si-Yuan Pan,Shu-Feng Zhou, Si-Hua Gao, Zhi-Ling Yu, Shuo-Feng Zhang, Min-Ke Tang, Jian-Ning Sun, Dik-Lung Ma, Yi-Fan Han, Wang-Fun Fong, and Kam-Ming Ko. New Perspectives on How to Discover Drugs from Herbal Medicines: CAM's Outstanding Contribution to Modern Therapeutics.Evidence-Based Complementary and Alternative Medicine.Volume 2013, https://doi.org/10.1155/2013/627375.

Tripathy SK., Traditional Knowledge: its Significance and implications. *Indian journal of traditional knowledge.* Vol.2, No. 2. 2003.

Utkarsh, G., Gadgil, M., and Rao, PRS. 1999. Intellectual property rights on biological resources: Benefiting from biodiversity and people's knowledge. *Current Science* 77: 1418-1425.

Vartak KVD (1996) Sacred groved of tribals for *in situ* conservation of biodiversity. In S.K.Jain (eds.) Ethnobiology in Human Welfare, p -300-302.

Ved, D. K., Mudappa., A. and Shankar, D. 1998. Regulating export of endangered medicinal plant species – Need for scientific rigour. *Current Science* 75: 341-344.

Zeven, A.C.and Zhukovsky, P.M. (1973) Dictionary of cultivated plants and their centers of diversity. PUDOC, Wageningen, p-29.

— *Section 2* —

LIVELIHOOD MANAGEMENT

Chapter 23

Similipal Biosphere Reserve: An Anthropology of a Lost Paradise

Jagannath Dash

Department of Anthropology,
Utkal University, Bhubaneswar – 751 004, Odisha
e-mail: jdash1955@gmail.com

Forests has been an important resource for mankind since the time immemorial, for meeting all minimum requirements to survive. In course of time when forests lost their area and status as the feeding as well as breeding grounds for all living beings, a global responsibility emerged for its protection and conservation. Such issues gave rise to identify the major forest patches and protect them from the hazards of modernization and Similipal as a national park, as a sanctuary, as a project tiger reserve and as a biosphere reserve born out of such endeavors and attempts. As we all know, forests have maintained all the potentials for the growth and rise of human civilizations in different parts of the globe in the past but if one looks at the scenario now, it is quite surprising to mark that the so-called civilized man to-day is mapping for its destruction. For its restoration and protection which are highly essential for the existence of man and other living beings, right from Stockholm to Rio D Janeiro, environmentalists have emphasized on the conservation programme as a result of which UNESCO launch MAB programme (Man and Biosphere) in 1970 to minimize man and nature conflict and to protect as well as conserve the biospheres for the existence of the earth and mankind. In 1957, Similipal was declared a National Park which also functioned like a wildlife sanctuary. In 1973, it was declared a Tiger Reserve and in 1994, a Biosphere Reserve covering 5,578 sq. km. (or 4374 sq. km. in some other source) with a core area of 845 sq. km. In the year 2001, Similipal was also declared an Elephant Reserve.

Similipal, the Green Paradise

Similipal is not simply a wide green forest accommodating numerous biotic and abiotic resources, it is the green paradise of the country as well as the state of Odisha for so many reasons and specialties. It is well known that the Biosphere Committee, Government of India had chosen Similipal as the 8th Biosphere Reserve of the country on its following 5 reasons and specialties such as (a) Presence of genetic diversity, (b) Effectiveness as conservation unit, (c) Representativeness, (d) Naturalness, (e) Importance- unique richness in the genetic resource.

The above mentioned major specialties of Similipal are mainly based on certain ground realities for which it has been widely considered the green crown of Odisha. Thus, at the outset, everybody will feel that Similipal is like a green paradise due to its uniqueness as well as specialties which may be conceptualized from the following features:

1. Located between 21° 10′ to 22° 12′N latitude and 85° 58′ to 86° 42′ E longitude, Similipal enjoys the credit of a single patch compact forest in the whole country.
2. Similipal stands centrally in the District of Mayurbhanj on its own being away from Eastern Ghat as well as the Central Vindhya range.
3. Though popularly it is called the crown of Odisha, because of its geographical shape just like a 'human heart', it also controls the climatic balance of the state and other adjacent areas just like a functional heart.
4. The hills in Similipal comprise Precambrian rock belonging to the Similipal group. Geographically, it consists of the Archaean system of rock and the Magnetite is 1,200 million years old.
5. Similipal has the richest watersheds of the country to give rise to many perennial rivers like:
   (i) The Budhabalanga System: [Balanga (North -Northeast), Sona, East Deo, Sanjo, Gangahar, Palpala in Southeast]
   (ii) The Baitarani System: [Khairi, Bhandan in Northwest, West Deo in the West, Tel in the West, Salandi in South]
   (iii) The Subarnarekha System: [Khad Kei (originated from Eastern Similipal to flow northward and join Subarnarekha near Jamshedpur].
6. Another specialty of Similipal is that it is an integration of semi-evergreen to deciduous forests. Champion and Seth (1968) have outlined the forest types like,
   (i) Northern Tropical Semi-evergreen Forest,
   (ii) Northern Tropical Moist Deciduous Forest,
   (iii) Dry Deciduous Hill Forests,
   (iv) High-level Sal Forests, and
   (v) Grassland and Savannah.
7. The climate of Similipal is mainly tropical with warm and humid summer with a temperature hovers around 40°C. However, few areas like

Bhanjabasa and other adjacent points are there where one will experience a bit of climatic variation. Bhanjabasa whereas the day temperature will be around 40°C in Summer, at midnight it may come down to 20°-25°C.

8. The Similipal Plateau is underlined by an assemblage of Volcano-sedimentary rocks consisting of an alternating sequence of quartzite and splitic lavas. The age of the rocks is believed to be about 2200 million years confirming the early Protozoic era (Mahalik, 1997).
9. Similipal belongs to the Biogeographic zone of the Deccan Peninsula, Biogeographic province of Chhotnagpur, and Mahanadian Biogeographic region (Rodgers and Panwar, 1988).
10. Similipal contributes to 38 per cent area of the total Protected Area (PA) network in Odisha (Odisha is covered with 57183 sq. km. of forest area).
11. As RPRC (2006: 1) has reported, Similipal "forms the biological link and shows the similarity of floral and faunal composition between Northern and Southern India and considered as a transitional bridge for migration of species".
12. Because of certain induced programmes in the past, Kols from Singhbhum started inhabiting the core area of Similipal. Before them, the tribes like Santal, Bhuinya, Khadia, Birhor, Ujia, Bathudi, and few other non-tribal communities were living in the fringe or transitional zones that encircle buffer area. All of them were primarily and secondarily depend on the resources of Similipal (will be discussed in detail later in this paper). However, about 5,00,000 people of various castes and tribes live in the 1265 revenue villages in the core, buffer, and periphery areas of the reserve.
13. Similipal is also known for the 'largest tropical peninsular sal zone in the country' (RPRC, ibid.). In its northern tropical moist deciduous forests which extend over 1540 sq. km, sal (Shorea robusta) is a dominating species and it is also noticed in the dry deciduous hill forests which extend over 250 sq. km. area.

Human Intervention in Similipal Hills

Though Similipal is rich in great potentials with botanical and zoological species as well as mineral and water resources, as an anthropologist, I am confining this study only to the human groups inhabiting in and around of it. The tribe *Khadia* has been the most dominating name among other tribes to be mentioned in most of the earliest descriptions. In 1866 and 1872 Colonel Dalton, in 1868 V. Ball, in 1877 Risley and W. W. Hunter, and 1888 and 1891, A. P. Driver have provided few accounts of the tribal population especially *Khadia* in and around Similipal Hills. W. W. Hunter (afterward Sir William) has described that the hills of the Similipal are densely wooded to the summit and except at regular passes are quite inaccessible to the outsiders. The central hill tract was previously covered under the name of Similipal perganna includes 10 pirs namely Brahmangaon, Chakiri, Barehipani, Kasira, Nij Similipal, Rajnagar, Barghati, Jamuna Bardanda, Ram Raja and Olakudar.

At present these pirs are included under a number of gram panchayats belonging to four sub-divisions of the district.

The forest types of Mayurbhanj were previously categorized as Reserved, Protected, and Lakharaj (private). Since 1973-74, the Project Tiger had started with 303 sq. km of the core area (later extended to 845 sq. km.) and the rest area was managed under the Similipahar Forest Development Corporation Limited (SFDC) 1979 to 1991.

The land types especially for agriculture are classified as Pal, Jal, and Dahi. With reference to Odisha District Gazetteers of Mayurbhanj, the soil structure of the district is of different types. The area is mainly undulating and the topsoil is shallow. The bottom soil is either lateritic or non-calcareous gritty silt. Throughout Similipal through loam soil is dominating, the local people classify the soil as - (i) *Balimati* (sandy soil), (ii) *Dorasa* (mix of sand and clay), (iii) *Matial* or *Chikita mati* (strong sticky clay), (iv) *Tilak mati* (sticky and grey in colour), (v) *Lalmati* (sandy red loam of lateritic origin). In the revenue records, the soil structure of Mayurbhanj had a three-fold classification since the past. They are- (i) *Jal Soyem* (third class wet land), (ii) *Jal Doyem* (second class wetland) and (iii) *Jal Awl* (first-class wet land).

Even though the hill range is popularly said to maintain its virginity, truly speaking, no part of it is left unintervened by human beings. Its natural resources have been exploited by the surrounding communities since centuries. As a forest ecosystem, Similipal is said to be the cradle of many a wild animal and plant species. As per the available historical data and local saying of the people, the forest territory was not habitable only one and half centuries ago. The nomadic and semi-nomadic bands of foragers came to establish their settlements within the hill range after the outside forest contractors were invited to install their camps inside the forest and exploit the timber wealth of Similipal for increasing the state revenue. Khadias, (the so-called Hill Khadias) came to inhabit the hill range only after the Kols and Bathudis established their temporary camps in the forest to assist the contractors as wage labourers.

As a very dense forest though Similipal seems to be a forest full of wild animals and plants without any human habitation as soon as one will enter into it, many revenue villages along with the vast stretch of agricultural land will come to sight. The major tribes in the forest are the Kol, Bathudi, and the Khadia. On the forest-plains fringe, food gathering tribes like Birhor and Ujia are also seen.

As the history of Similipal witnessed, Late Mr. Saroj Raj Chaudhury, the founder field Director of the Project Tiger had duly considered the local man as a part of the project. In the management Plan of Similipal Tiger Reserve (1974) had emphasized that, "The tiger cannot be conserved in isolation. It has to be integrated into the local complex of the structural and functional attributes of the land, its clime, its biota and its people. The local man in and around Similipal is a part of such a total complex. His influence has significant bearing both on the range, structure, and its function, and hence the tiger too. The adverse effects of that influence have to be controlled. The question is: How will they be controlled? For policing alone, besides being expensive, is not enough, particularly when the tiger can be silently dispatched by pesticides if alone ever to eliminate its actual or presumptive conflict with domestic interests. One of the main works in the Tiger Project, therefore, will be to explore the socio-ecological facts to find the answer. And in doing so it will need the cooperation of all other departments, particularly the Revenue, the Public Relations, the Tribal and Rural Welfare, the Education and the Police"(1974: 18: 31).

Coming to the human settlements inside Similipal, three tribal communities mainly settled here. They are the Kol, the Bathudi, and the Khadia. Due to certain historical reasons, they have migrated to the present forest villages at different periods and few of them may be explored both from primary and secondary sources. As already mentioned, human settlements inside Similipal is less than 150 years old. Because of a dense invincible forest, depredation of tigers and elephants, attack of fever, people were not preferring Similipal as a habitable site, and they confined only to the fringe villages. The report of the annual administration of 1906-07 shows that during the reign of Maharaja Sriram Chandra Bhanja Deo (1882-91) for revenue collection from the untapped forest resources, the first systematic exploration of Similipal reserve was made through contractors following the work plan of C. C. Hatt. 8 contractors were working in the western circle and 4 contractors in the eastern circle (District Gazetteers of Mayurbhanj, 1967). During 1906-07, M/s. Borooah and Co. a respectable firm of Assam was invited to join forest exploitation. For the problems of procuring labourers from the locality, M/s Borooah and Co. started importing Kol labourers from Ranchi and Singhbhum for undertaking forest operations. It is said that after the work was over, some of the labourers went back to their native place and some settled back in Similipal. In addition to the grant of lease to the forest contractors, temporary leases were also granted to a few settlers. Peter Dubraj, a Kol postmaster at Chainbasa (Singhbhum) was granted 9,206.42 acres of land at an annual rent of Rs. 254 and nine annas only. He was declared a Sardar in Similipal and was permitted to establish villages. He invited his fellow tribesmen from Chainbasa and its adjacent places to Budhabalanga area of Similipal. In 1948, one village head Mana Ho invited some Munda people to stay in Similipal permanently. The history of Kol settlements continues to increase and has some historical consequences later on.

As regards the Bathudi, another tribal inhabitant of Similipal, the history of the settlement goes in a different direction. As there is no recorded account of migration of different communities to Similipal, based on certain historical incidences, folk sayings, and empirical information, the settlement of Bathudi can be established to some extent. The local history reveals that once upon a time, Bathudis were inhabiting the places like Jashipur, Gudgudia and Podadiha areas of Similipal. Culturally they were very much associated with the neighboring tribal communities like Khadia and Ujia. It is very interesting to mention that all the three tribal communities designate themselves as 'Sabara' category of tribals connecting to the cult of Lord Jagannath. As the legend says, Bathudis were the earliest tribal settlers in Similipal. Oudh (present Ayodhya) is said to be their ancestral home. Later they came to Chhotnagpur plateau. They were living in Barthali ('Bar' means twelve and 'thali' means place). As the people say, the term 'Bathudi' is derived from 'Barthali'. From Barthali Bathudis first of all came to Jamuna Kula in the central part of Similipal. According to Nagendra Nath Vasu (1911) owing to the ravages of tigers, they migrated in three groups to Bamuria, Adipurgarh, and Daspurgarh respectively. Though Bathudi leaders of the district date back their first arrival to the 10th century, it has no supporting evidence. However, it is a popular saying among the Bathudis that Madan Nayak, the so-called the first Bathudi to Similipal was belonging to Emperor Ashoka's army. He had 3 sons who were staying in Jamuna Kula, namely Nanda, Benu, and Beni. The present-day Bathudis is said to be their descendants. If this saying is taken into consideration, the Bathudis may be the earliest of all tribes in Similipal.

Regarding the coming of the Khadia tribe into Similipal, it is said that Khadias were, once upon a time, living in the fringe upland around Similipal. They were having a mud-fort at Jashipur, the entrance door to Similipal. They were ruled by a Khadia chief Jashu Khadia (with the title Digar). The outsider Bathidis tried to occupy Jashipurgarh of Khadia several times. Khadias defeated the Bathudis seven times consecutively. However, at the 8th fight Khadias were defeated and had to quit the Jashipurgarh. Both the tribal communities maintain this local saying till the date. As the Bathudis say, the battle occurred from 1100 to 1400 A. D. However, there is no written documentary evidence available in the district of Mayurbhanj. After the defeat, all the Khadias in Jashipur took refuge in the Similipal forest mainly in the peripheral parts. Bathudis after winning the battle started ruling over four forts at Adipur, Jamuna Bardanda, Karanjia and Jashipur.

Earlier, Khadias were mostly living in Panch-pir sub-division (Karanjia) mainly as agriculturists. Soon after the defeat, when they came to the peripheral forest region of Similipal, first of all, adhere to *Dahi* Cultivation (upland cultivation) along with the collection of minor forest resources. As stated earlier, only a century and a half before (about 150 years) they came to reside in the core area of Similipal in different phases. First of all, they went there temporarily for 2/3 months especially during the winter staying mostly at the side of hill streams both for the collection of arrowroots as well as their processing. After moving temporarily for a few years to such arrowroot processing sites (at the hill stream sides), some of them started staying there permanently. In this way, the Khadias became the inhabitants of the Similipal hill range though the majority of them were staying in the peripheral or buffer zones. The defeat by the Bathudis turned their way of life from an agricultural economy to a forest-based economy. They were gradually identified as food-gatherers or foragers in Similipal. Such an example is mostly denoted by anthropologists as the brightest example of the process of *Secondary Primitivization.*

Thus, the historical incidence of defeat by the Bathudis, made two significant changes in the way of life of the Khadias. First of all, they became forest-based people depending mainly on the food gathering activities like the collection of sal resin, honey and arrowroot (these are only major collections). Secondly, as the defeated community, Khadias had to also imbibe the culture of the Bathudis (mainly their Gods, Goddesses). Because of this both the communities were taken as the Badam-worshippers in and around Similipal. And Similipal is said to be the homeland of the Bathudi and the Khadia both. A little before the arrival of the Khadias, the Similipal hill range was inhabited by the Bathudis and few Gonds here and there in buffer zones. Later the Kols came from Ranchi and Singhbhum regions and permanent human habitations started in the Similipal hill range. With their association, the Khadias started entering into the core areas of Similipal and identified as the Hill Khadia. All of them accepted the oral and religious traditions of the Bathudis. From Jashipur, Khadias came to Gudgudia (in the buffer zone) and thereafter to Kabatghai, Khejuri towards the central part of Similipal. Finally, some of the Khadias from Gudgudia also came to Jenabil, the Core Tiger Project area of Similipal.

Another tribal community, namely Ujias also similarly maintains food gathering economy. They are mostly confined to Podadiha and Thakurmunda areas at the fringe of Similipal. Like the Bathudis and Hill Khadia, they also claim themselves to be a Sabara category of a tribe. They were enumerated as one of the tribes of Mayurbhanj up to 1931 census after which, the name was deleted from the list of Mayurbhanj and found in the list of Scheduled Castes in Balasore district as fishermen. However, on empirical inquiry, I establish them as a tribe who claim to be a fraternal community of the Hill Khadia and maintain cultural similarities in several aspects. At present, as forest-dependent people, they are found confined only to three villages — Mituani (near Thakurmunda) Badjunpal and Gandidar(near Podadiha). As the Ujias say, in these three villages which are located in the foothill region, people maintain a food-gathering economy whereas their kin members in Udala, Kaptipada and Nilagiri regions are already acculturated and known as Saaras. The Ujias are not forest-dwellers. Every day morning they enter into the Similipal forest, collect Sal resin, honey and several other edible as well as inedible resources for the livelihood and comeback in the evening. However, the Ujias are best known for the catching of Hill myna and hornbill which have a great demand in the local markets of Mayurbhanj and the nearby metropolitan city of Kolkata. But as regards the collection of arrowroots, the Hill Khadias are highly specialized. Both Hill Khadias and Ujias have maintained ritual friendship with the neighbouring Bathudis and by friendship; Bathudis learned the techniques of natural resource collections. At present Bathudis are competing with Khadia and Ujia in resource collections as a result of which Khadia and Ujia economy suffers greatly.

Another forest-dependent tribe of Similipal is Birhor. Bighorns are locally called as Mankidia (because they hunt monkeys and relish their meat). Like the Hill Khadia and Ujia, they are also landless people. Whereas for minor forest produce collection, Hill Khadias and Ujias maintain a semi-nomadic living, the Birhors go for a completely nomadic way of life. Unlike the Hill Khadia and Ujia, Birhors mainly live on the collections of Siali (*Bauhinia vahlii*) and other plant fibers from the Similipal forest for the rope making. They sell the ropes in the local peasant villages as well as weekly markets for cash. Besides, they are also highly specialized for monkey hunting. While the males collect plant fibres for rope making, the females collect dry firewood and few edible resources for the kitchen.

Birhors usually live in between the peasant village and the forest. They mostly prefer the foothill villages because of forest collections (fiber) and their marketing. As a nomadic community, they stay in a particular site for not more than 3 months. Soon after the plant fibers are exhausted in the local forest (of Similipal), they move to another site as per the availability of the required resources. Hill Khadia and Ujia, though in the past, are said to have their native dialects, at present they have already forgotten them and converse in local Odia language. But the Birhors are still conversing in their native Mundari language similar to that of the Santals. Birhors are specialized both in hunting and food gathering activities.

Sometimes the Hill Khadias and Ujias, during the offseason work as daily wage labourers in the agricultural fields of neighbouring communities but the Birhors are full-time hunter-gatherers in the forest. Around Similipal hill range in most of the fringe or foothill areas they always move throughout the year. Like the Ujia and other foothill agricultural tribes they were never forest-dwellers. Regarding their coming to the Similipal region, no definite local saying or myth are heard. But according to anthropological investigations by S. C. Roy and Surjeet Sinha, it is taken for granted that their place of origin was Chhotnagpur and Singhbhum. As the term 'Birhor' explains, 'Bir' means forest and 'hor' or 'hol' means man. They were primarily forest dependent people. But in course of time, the community was split into 2 groups, namely *Jaghi* and *Uthlu.* After developing an acquaintance with the cultivation, the group acquired land and remained settled in a particular settlement being denoted as *'Jaghi'* and the other group being landless moved around the forest for rope making and monkey hunting throughout the year and named as *'Uthlu'*. An intermediary group also developed which continued partly cultivation work during the agricultural season and after the harvest moved around the forest for forest collections. Such a third group always oscillates from one type (*Jaghi*) to another type (*Uthlu*) in a year signifying the oscillation process of change like the pendulum of a wall clock. In course of gradual movement in search of forest resources, the *Uthlu* group of Birhors moved to Similipal hill range and because of its dense forest and plentiful of resources they remained confined to the surrounding foothill fringe sites.

The history of human habitation in and around Similipal is so vast and complex that it is simply impossible to cover it completely. Therefore, discussing only a few important tribes whose names have been long associated with the Similipal Biosphere Reserve, the factor of human intervention in the hill range is complied with.

Exploitation and Destruction

It is a fact that tribal communities living inside and outside of Similipal are both, directly and indirectly, dependent on their natural resources. Such dependents may be categorized into three types (Table 23.1).

Table 23.1. Classification of Communities Based on Utilization of Natural Resources in Similipal

Categories	*People/Tribes*	*Purpose*
1. Primary Dependents	Hill Khadia, Ujia, and Birhor	Subsistence (Basic Need)
2. Secondary Dependents	Bathudi, Santal, Munda, Ho, Kol, Bhumij and many other caste people in the fringe area	Subsistence and Surplus (Beyond Basic Need)
3. Tertiary and Other Dependents	Nearby urban, industrial or modern elite groups	Commercial gain and accumulation.

As the above categorization shows, primary dependents consider the resources of Similipal as their bread and butter. At no cost, they desire for its destruction or over-exploitation. Some edible resources like greens, mushrooms, flowers, roots and tubers, fish, *etc.* are common to all the Hill Khadia, Ujia, and Birhor hunter-gatherers. But for the Hill Kharia and Ujia, Sal resin and honey collections are common. But Birhors specifically depend on the plant fibers for rope making and go for monkey hunting once or twice in a month only when there is rice at home. Hill Khadias beside the common resource collections specifically go for the catching of hill myna and hornbill as per the market demand. Thus, all three tribes maintain a subsistence economy and are very careful about the propagation of the required resources.

They never go for over-exploitation of the resources. But on the other hand, the secondary dependents of the forest mainly depend on agriculture or any other remunerative job. During the offseason or lean period of agriculture, they go for the exploitation of minor forest produces. As such produces are not their primary requirement, they often overexploit or destroy them without being careful. Such secondary dependents have permanent settlements in revenue villages. They exploit timber, bamboo and other resources for the making of luxurious wood furniture, and agricultural implements, which the non-agriculturist primary dependents do not do. The plants and trees are therefore denuded or destroyed basically by such tertiary and secondary dependents. The tertiary dependents never come to the forest directly, but their demands allure the secondary dependents to over-exploit plant resources. Thus, mostly the secondary and tertiary dependants are responsible for the severe destruction or depletion of forest wealth.

It is needless to maintain here that hunting of animals and birds in Similipal are mostly organized by the secondary dependents. They go for hunting tigers for its skin, nail, teeth, elephants for its tusk and several other animals/birds for various commercial purposes.

The greatness of the primary dependents is that they consider Similipal as sacred land. Mythically they believe that *'Badam'* is the presiding deity of the Similipal Biosphere Reserve. He has with him wife *Badam Budhi*, 15 sons (the eldest one *Ganapati* or *Ganesara*), and one daughter (name -*Sana Dei*). All his sons and daughter are in charge of different parts of Similipal. All are believed to be the creator and protector of plant, animal and other resources of Similipal. For every forest venture, the first of all pray and propitiate the in-charge gods/goddesses for their sacred approval after which only they exploit resources. It is very interesting to note that all the first plant resources or produce of the forest are offered to the Hill gods and goddesses first and then only exploited for personal use. In the month of Magha (Jan-Feb.) they organize their annual festival in which all the gods and goddesses of the forest are offered all the first production of the forest like honey, arrowroot, sal resin and other resources and take the supernatural permission as well as blessings for the exploitation of forest resources throughout the year. Besides, all such tribals consider each hill as a goddess and worship them before starting the collection work.

Renewal and Replenishment

Already the over-exploitation and destruction of the forests especially of Similipal have been discussed regarding secondary and tertiary dependents. The contexts of primary dependents need a special mention here as they contribute to the renewal and replenishment of various plants and other resources they collect for a living (Table 22.2). They manage the collection of their required resources very well based on their indigenous knowledge and technology. Very rarely they destroy the resources unless compelled. Before the collection or exploitation of any resource, they are extremely careful about their renewability and propagation so that in the next season they can avail them for the livelihood maintenance. Thus, several examples may be cited here to show the beneficial role of the primary dependents like Hill Khadia, Ujia and Birhor.

Table 23.2. Renewability and Non-renewability of Forest Resources

Sl. No.	*Items Collected*	*Renewable or Replaceable*	*Non-Renewable or Irreplaceable*	*Destroyed or Lost*
1	Honey	Yes	-	Trees (occasionally)
2	Arrow-root	Yes	-	No
3	Sal Resin	Yes	-	No
4	Gums	**Yes**	-	Trees (rarely)
5	Mango	**Yes**	-	No
6	Other Fruits, Nuts, Berries	**Yes**	-	No
7	Greens	**Yes**	-	No
8	Grasses	**Yes**	-	No
9	Flowers (Edible and Medicinal)	**Yes**	-	No
10	Mushrooms	**Yes**	-	No
11	Roots and Tubers	**Yes**	-	No
12	Sal and Siali leaves	**Yes**	-	No
13	Broom stick	**Yes**	-	No
14	Bamboo	**Yes**	-	No
15	Timber	-	-	Trees (often)
16	Fire-wood Usually Dead and Dried logs used			Trees (occasionally)

Sl. No.	Items Collected	Renewable or Replaceable	Non-Renewable or Irreplaceable	Destroyed or Lost
17	Catching of Birds			
	Hill Myna and	**Yes**	-	No (pet birds)
	Horn Bill	-	Yes	The female birds caught and consumed
	Jungle fowl	-	Yes	Caught and consumed
	Parrot	-	Yes	No (pet birds)
	Other birds	-	Yes	Caught and consumed
18	Fishing	-	Yes	Poisoned and caught
19	Hunting	-	Yes	Mostly small games killed
20	Chopped Fibres	Yes		The bark of the creeper on the tree is used.

First of all, we have to see that Khadia and Ujia share several natural resources and the technology of their collection in common (Table 23.3). As a result, the technology of replenishment and renewal of natural resources is also found to be the same among them. Some examples may be cited below as evidence.

Table 23.3. Year-round Major Forest Collections/Activities

Group	Items and Activities	Responsible Animal/Plant Species	Month/Season of Collection
Khadia and Ujia	Honey (*Mahu*)	*A. dorsata, A. cerana, A. florea, Trigona species* (Dammer bees)	Apr. to Jun. and Oct. to Nov.
	Sal Resin (*Jhuna*)	*Shorea robusta*	Oct. to Mar.
	Arrow root	*Curcuma angustifolia, Curcuma aromatic*	Dec. to Mar.
	Catching of		
	(a) Hill myna (*Sara*)	Sturnidae family	Mar. to Jun.
	(b) Horn bill (*Nakuala*)	Bucerotidae family	May to Jul.
	Gum (*Atha*)		
	(a) *'Bharal'*	Caesalpiniaceae family	Dec. to Mar.
	(b) '*Mankad - Pichhula*' or '*Gardei*'	*Storculia urens*	Available throughout the year but collected during Dec. to Mar. only
Birhor	Fibre (*Bayer* of *Siali*)	*Bauhinia vahlii*	Throughout the year
	Hunting of Monkeys (*langur*)		
	(a) '*Hanu*'	*Presbytis entellus*	Throughout the year (More during Summer)
	(b) *'Bander'* (*Gali*)	*Macaca mulatta*	Throughout the year (More during Summer)

(i) Hunting of Game Animals/Birds

Hunting is very common practice with most of the tribes in general. The secondary dependent tribes ruthlessly but Hill Khadia and Ujia very rarely go for it. Most of them because of totemic taboo didn't hunt certain animals and birds like tiger, monkey, *etc.* Few members also have few plants as their totems. It is also taboo on their part to exploit them.

(ii) Collection of Honey

Honey bees usually have their combs built on the big trees. The secondary dependents collect them simply by felling down the trees. For the sake of honey collection, a full-grown tree is fell which is a great loss to the Similipal Hills. But the primary dependents usually collect the honey of all the four species of honey bees namely - *Apis dorsata, Apis cerena, Apis florea,* and Trigona species (dammer bees). Locally *Apis dorsata* is known as *'Baghua'* or *'Bad mahu'*. The amount of honey from this swarm usually vary from 10 to 30 kgs, whereas the honey from combs of the swarm of *Apis cerena* vary from 5-6 kg. The honey production from the other two species of honey bee is very meagre amounting to 500 grams maximum. Therefore, they mostly collect the honey of *Apis dorsata* and *Apis cerena* for the market. The honey *of Apis dorsata,* as they say, is not easy to digest. Therefore, the honey of *Apis cerena* is in great demand because of its multiple-use as well as medicinal property. Such *Apis cerena* honey bees always prefer hollow trunks to build their combs. They usually build seven combs inside the hollow trunk. Only through a small slit like opening the bees enter into the hollow trunk. The slit-opening needs to be very small for the maintenance of a standard temperature inside the trunk.

For the collection of honeycombs, the tribals have to open the slit a little wider. But soon after the opening when the slit becomes bigger, honey bees do not prefer to have their combs once again there. The main reason goes that with a bigger slit, the temperature inside the tree trunk is difficult to maintain. Therefore, soon after the collection of honeycombs, Khadias and Ujias keep a small piece of stone (*tangana*) at the opening to maintain only a very small opening for the bees to enter and come out. As a result, the honey bees are found to build their combs inside the hollow trunk once again. And quite comfortably, in the next honey season, tribals come directly to the tree side and go for the honey collection.

(iii) Collection of Arrow-Root

Both the tribal communities(Khadia and Ujia) collect a variety of arrowroot tubers under two major categories. Under the first category, three local types are -*Bhandua* (Khadia) or *Mainsa* (Ujia), *Chauli* (Khadia) or *Gua* (Ujia), and *Haladia*. The second category of arrowroot shoots is a bit inferior in quality. The local types under the second category are *Balunga, Kanka* (or *Sadangia*). Mainly during the winter, arrowroot tubers are collected both by males and females. It grows naturally in the forest. The arrowroot tubers of Similipal are of the wild variety. Unlike the domestic varieties, it is bitter. It is very popular in the market mainly because people feel that it has more medicinal properties than the domestic varieties.

Once the arrowroot tuber is dug out completely, it will not grow further. Like honey, as arrowroot is another cash earning product of the Khadia and Ujia, very decisively they harvest them from the forest keeping in view the harvest next year. Therefore, while digging out the tubers, they consciously leave some shoot there in the soil, so that the crop grows once again next year. Thus, both tribes take extreme care for the renewal of the crop. But other tribal people never consider for the renewal of the crop.

(iv) Sal Resin

Like honey and arrowroot, sal resin is another major collection of Hill Khadia and Ujia. But the Birhor never collect either of the three. Though sal resin, unlike the other two, is an inedible seasonal resource, it greatly adds to the Khadia and Ujia economy. It is locally called *'Jhuna'* and produces only from the tree of Sal (*Shorea robusta*). Similipal especially Mayurbhanj district possesses the largest Sal forest of the country, therefore, the sal resin is collected in good quantity from Similipal every year.

Sal resin is produced only as a gum which comes out through the cracks of the bark and collected after dried. It is produced mostly on the mature sal trees which are more than 25 years old. It is produced on the upper part of the sal tree at a good height, mostly during the rainy season. As the tribals say, during the rainy season, sufficient rain and wind twist the Sal tree, and cracks come up in the upper tender part of the tree. Therefore, in the case of little rain and wind, Sal resin doesn't come in sufficient quantity. Khadia and Ujia also believe that due to rain a particular type of insect is born which looks like black bees. It makes holes or pores on the upper tender parts of the Sal trees. Through such outlets on the barks, usually, resin comes out and dried after a few days.

Indigenously, Khadia and Ujia divide sal resin into two types, namely black or dark brown resin (*Kala jhuna*) and white resin (*Dhala Jhuna*). When the resin oozes out of the bark, it remains attached to the bark of the tree. When the tribals collect them by the help of axe, as per the convenience or condition, few resins collected in strick form and others in broken or powdered form. The stick form which are known as *Khadi,* fetch a good price in the market. The white sal resin costs more than the brown sal resin in the market. The technology of sal resin collection primarily needs great expertise in climbing up the tall sal trees. While climbing up, a basket hangs from the waist by the help of a bifurcated wooden hook. The climber makes notches on the tree trunk by the axe and climbs the tree by gripping into them by feet. By one hand the climber has to hold tightly to the trunk of the tree and with the other hand to scrape out the hard resins from the bark by the help of an axe in such a manner that the resins will fall into the hanging basket. Therefore, the whole process needs great expertise because of which other neighboring tribes very rarely go for such collections.

After understanding the high skill required for sal resin collection, the next phase comes for its continuity and livelihood maintenance of Khadia and Ujia. At present few Bathudis and Kols are noticed to learn this skill through ritual friendship with Ujia and Khadia. However, due to the fast denudation of the forest especially sal trees, because of their high economic value and multipurpose utility, the sal resin collection is severely threatened. Earlier, the situation was quite favorable for the food gatherers like Hill Khadia and Ujia. Mainly because Khadias were living in the core area, they have no difficulty in reaching the matured sal trees. At present they have been displaced to fringe zones and severe restrictions are there to enter into the core area. Anyway, one thing that emerged from the narration that mainly because Hill Khadias and Ujias primarily depend on the resin collection, they never harm matured and other Sal trees. On the other hand, they not also allow other outside people as well as timber mafias to fell matured Sal trees. Thus, the hunter-gatherers mainly because of their subsistence economy, very rarely do any harm to minor forest produces of Similipal. During the end part of the rainy and whole winter seasons, they mostly depend on sal resin and arrowroot collection. During winter and summer, they depend on honey collection. Most of the rainy season, the forest becomes inaccessible and food gathering tribes simply collect firewood and few edibles like mushrooms, greens, tubers, *etc.* from the foothill region along with certain nuts and berries for personal consumption only.

In addition to the above-mentioned collection of honey, arrowroot and sal resin, both the tribes also collect various produces like a variety of edible gums, greens, fruits, flowers, mushrooms, sal leaves, siali leaves, fish from hill streams, a good variety of roots and tubers, red, white and yellow ochres, nuts and berries, firewood, sal, kusum and chara seeds, medicinal herbs, broomsticks, *etc.* for their daily living. It is to be taken for granted that the above-mentioned plant resources are mostly renewable. If not tapped or collected they are simply destroyed in the forest. By the process of collection, therefore, people maintained livelihood without causing any harm to Similipal.

As regards the Birhor (Mankidia) way of living, as already stated earlier, in addition to two major forest-based activities like rope making and monkey hunting, they also collect sal leaves, siali leaves, firewood, edible greens, mushrooms, fruits, nuts, berries *etc.* from the forest for their daily consumption.

(v) Collection of Plant Fibres

Birhors possess the most archaic economy in the country, so simple (technologically) so environment friendly which no one can ever imagine. The Birhor economy as stated earlier is dependent mainly on 'rope-making'. Though another favorite of all Birhor is monkey hunting, it has no economic contribution. With a very small residential hut (known as *Kumba*), and least resource exploitation, all nomadic Birhor make a historical contribution to the conservation of natural resources in Similipal.

The rope-making economy usually depends on (i) raw material collection (collection of plant fibers) and (ii) techniques of rope-making (Figure 22.1). Birhors collect a variety of plant fibres or barks mostly from the wild creepers. The most important one among them is the creeper of Bauhinia vahlii (in Odia-Siali, in Birhor language, *Lama nani*). Besides *Lama nani,* Birhors also collect a number of other plant fibres like - *Sisi Chhali* (from *Pandara* tree), *Udal bayer* (from *Udal* tree), *Talhet* bayer (from *Talhet* tree), *Damda bayer* (from *Damda* tree), *Tanja bayer* (from *Tanja* tree),

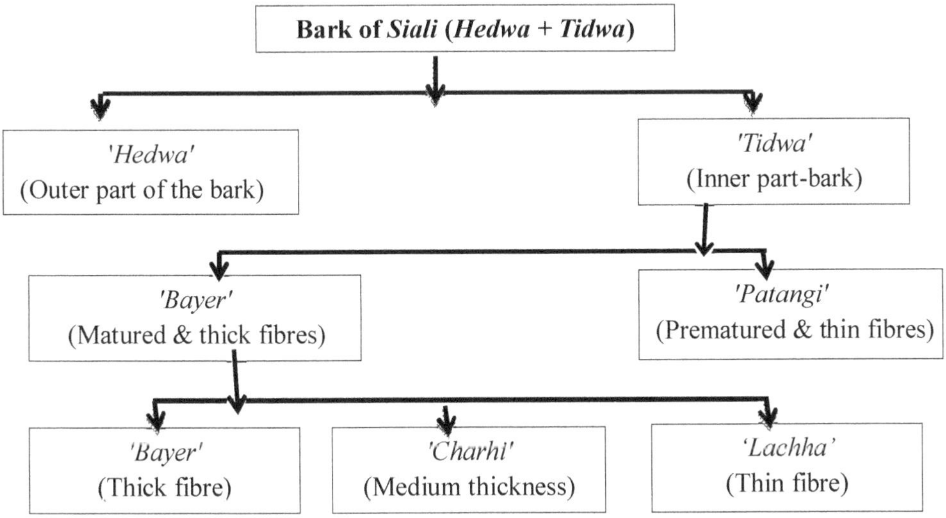

Figure 23.1. Birhor's Classification of *Siali* Fibres.

Jaja Bayer and *Kaskam Bayer*. Mainly because all such fibres are available in small quantities in interior and distant forests (Table 23.4). Birhors give utmost priority to Siali fibres for daily collection.

Table 23.4. Birhor Unit of Measurement of Chopped Fibre

Si. No.	*Birhor Unit of Measurement of Fibre or 'Bayer'*	*English Equivalent to*	*Relative Estimation or Measurement (approx.)*
1	Mih Barham	One shaftful of fibre	Can produce 4 nos. of '*Barhi*' (each one 20 'hat' or 30 ft. long)
2	Mih Habar	Half shaftful of fibre	Can produce 2 nos. of '*Barhi*' (each one 20 'hat' or 30 ft. long)
3	Mih Khatla	One fourth shaftful of fibre	Can produce one '*Barhi*' (20 'hat' or 30 ft. long)
4	Mih Sap	One eighth shaftful of fibre (usually means a handful)	Can produce one '*Barhi*' (each one 10 'hat' or 15 ft. long)

Structurally, the fibre or bark of Siali (Bauhinia vahlii) is composed of several parts. They call it *Lama nani*. *'Nani'* is a common term for all creepers. The term *'Bayer'* is used for the matured or thick fibres (bark) which is mainly used for rope making. Birhors use a common term *'Bakla'* for all kinds of barks. The *Bayer* (or *Bakla*) or bark, has several parts. The outer skin of the *'Bayer'* is called as *'Hedwa'* or *'Hedua'*. After the *'Hedwa'* is removed from the fibre, the remaining part which is used for rope-making is called as *'Tidwa'*. The bark or fibre of the creeper which is thin or premature is known as *'Patangi'* (instead of *'Bayer'*). Usually from the creeper of one-year-old, *'Patangi'* is available which after one year more, becomes matured and known as *'Bayer'*).

It is very important to note that Birhors usually collect *'Bayer'* or matured barks and leave *'Patangi'* or premature ones very decisively. After the exhaustion of all *'Bayer'* from the local forest, they move to another site around Similipal for *'Bayer'* collection. As a nomadic community they go on moving around Similipal and after more than a year come back to all previous settled sites and by that time, the *'Patangis'* left earlier, have already become matured enough to become *'Bayer'*. In a very planned way, nomadic movements or organized around the availability of the *'Bayer'* or *'Patangi'* in Similipal. It is needless to mention here that with the collection of the *'Bayer'*, the main stem of the creeper is to be completely cut down. However, the roots are not removed and several *'Patangis'* grown around are left for the exploitation next year.

Such a collection activity carries with it two essential conservation parameters. Firstly, when the matured barks or fibres are collected, the main creeper is cut down and it dies. But because the *'Patangis'* left at the root, the plant grows further. Next year *'Patangi'* is growing to become *'Bayer'* and the process goes on to allow the creeper to grow further and the subsistence economy of rope making to continue with no loss to the people and no loss to the forest. Secondly, as the forest officials say, once in every year, they have to clear the creepers in the forest in order to allow the plants and trees to grow comfortably. It is well known that among all the creepers, the creeper of *Bauhinia vahlii* is very long and strong. The moment the mature creepers are cut down by the Birhor, it helps immensely to the forest department as well as their conservation activities.

The Birhors, though very archaic/backward economically, they organise most of their activities in a very systematic/scientific manner. After the collection of plant fibres is over they carry them on a wooden shaft called *'Behenga'*. In this context, they use the native terms for the quantity of the fibres on the shaft or *'behenga'*. When the shaft becomes full, it is called *'Mih Barham'*. If it is half, the term is *'Mih Habar'*. The terms for one fourth shaftful is *'Mih Khatla'* and for one eighth, *'Mih Sap'*.

So far as the food gathering activities are concerned, all are connected to renewable resources of the forest. As such resources are their bread and butter, Hill Khadia, Ujia and Birhor take extreme care for their propagation and replenishment. But tribals other than the food gatherers without caring for the conservation, overexploit and often destroy them. But when we discuss about the hunting of game animals, catching of birds *etc.* matters go in the opposite direction of conservation. Even though Hill Kharias and Ujias are mostly food-gatherers, sometimes they join the hunting expeditions of the neighboring agriculturist tribes. There is a principle among all the tribes in India that people go for hunting only when they have something at least rice at home. Therefore, hunting for all tribes is a secondary activity. But it kills the animals ruthlessly incurring a great loss to the forest. However, for the Hill Kharia and Ujia it is only a leisure time activity only when they save something to eat at home. For the Birhor, on the other hand, the hunting activity is confined to monkeys only. After a week or two, when they save some rice to eat, they go for hunting monkeys.

(i) Catching of Birds by Khadia and Ujia

Both Hill Khadia and Ujia are well known for the catching birds (Table 23..5) mostly Hill myna and Hornbill from Similipal hill range. Some times they also catch parrots. Such activities though do not kill the birds, it helps in reducing their population in Similipal. Whereas Hill myna and parrots are kept as pet birds the horn bill is brutally killed for its meat as well as fat with a belief to cure rheumatic pain. People prefer to keep Hill myna and parrot as pets mainly because they imitate human voice very well and entertain. Therefore, popularly they are known as talking birds.

Table 23.5. People's Categorization of Hill Myna and Horn-bill

Birds	*Types (Native term)*	*Description*	*Use*	*Size and Weight*
(A) Sara (Sari) or Hill Myna	1. *Sunakania Sara*	Provided with two golden lobes on the both sides of head *Gracula religiosa*	As pet	Small and 250 to 300 gms.
	2. *Sada Sara*	No lobe is present *Gracula religiosa*	As pet	Small and 250 gms.
(B) *Nakuala* (*Kochilakhai*) or Horn-bill	1. *Kani Nakuala*	Malabar pied Horn bill	No use	Small and 500 to 700 gms.
	2. *Bada Nakuala*	*Anthracoceros coronatus*	Medicine	
	3. *Kajali Nakuala*	Common Grey Hornbill or *Tockus birostris*	Medicine	Big and 1 kg to 1500 gms.

From Hill Myna to Hornbill: The Untold Story

(a) Hill Myna

Though in all parts of Similipal, Hill mynas are found, they are concentrated more in Bhanjabasa region near Podadiha. Therefore, Ujias catch more Hill Mynas and Hornbills than the Hill Khadia.

Hill Mynas usually do not build their nest. They mostly use the holes made by the woodpeckers as their nest. Such rest-holes are located at a good height especially on the sal trees. The nature and behaviour of Hill Myna is very interesting. Out of all the wild birds people consider it as the most intelligent bird. It never changes it nest-hole unlike most other birds. During the month of March, Ujia and Khadia catch the nestlings from the nest-holes. Usually wild animals and birds go for breeding once in a year. But when due to some reason they loss their babies, they once again go for breeding second time. After a few months, the tribals also catch the second batch of nestlings for the market. Soon after losing the nestlings, Hill mynas also go for third time breeding. By the time nestlings come out of the eggs, it is already rainy season. During this season Similipal forest becomes very much inaccessible and tribals cannot catch the nestlings. So the third batch of nestlings survive by the natural process of raining and it helps in the propagation of Hill myna bird in Similipal Hills. Earlier the forest department has been raising the issue of Hill myna catching as a serious threat to the Hill myna population. But with the above empirical knowledge on wild birds and animals, the views of the forest department may be found baseless.

The technique of catching of horn bill is unique and different. The nature and behaviour of the bird is completely different from that of Hill myna. Like hill myna, it has also no permanent nest. It prefers to lay eggs inside a hollow tree trunk. Both the male and female horn bills search for a suitable hollow tree trunk with a suitable opening passage. After the tree trunk was selected, the female horn bill enters inside the hollow trunk through the opening. The male bird just like a human husband starts taking care of the female bird inside the trunk. Such a female and male bird relationship is unique and amazing too. The male bird very industriously brings mud from the stream side, and mixing it with its own saliva, seals the opening of hollow trunk leaving only a small slit for excretion. This is a very strange practice. Perhaps for safety purpose, the female bird is to remain inside a sealed chamber. Through the slit opening, the male bird supplies food to the female bird inside. Through the same slit also the female bird excretes outside. This practice of the female bird provides a clue to the food-gatherers in the forest. Looking at the hollow trunk, its sealed mouth and the excreta mark of the female bird around the tree, they confirm the presence of the bird in the hollow trunk. They climb up the tree, break the seal, and catch the female bird along with the nestlings. They sold them in the market only for cash. The most stunning fact is that the male bird because it is moving outside, it couldn't be caught but the mother bird and the innocent nestlings are caught and brutally slaughtered for the medicinal use of their meat. Therefore, the population of the horn bill in Similipal is in danger and depleting very fast.

Monkey Hunting: The Episode of an Anti-ethics

Whereas the Birhors prefer monkey hunting most, the Hill Khadia and Ujia consider it a taboo, due to the connection of monkeys to the sacred epic Ramayan. The Birhors also mythologically relate themselves to the dynasty of Ravan, the demon king who fought against Lord Rama and his monkey soldiers and finally killed. Therefore, they always hunt monkeys and relish their meat,in revenge.

Though hunting of monkey is a great threat to the monkey population of Similipal, the technique of monkey hunting is very interesting. It is operated through a series of systematic activities. As already stated, Birhors hunt once in a week only when they save adequate quantity of rice at home for consumption. The hunting expedition is started early in the morning under the leadership of the priest. Only the male members participate in the hunting expedition. The Birhors have a good knowledge about the habitats and behavior of the monkeys in Similipal and accordingly they proceed for hunting.

They usually categorize the monkey troops into 4 types:

1. *Saria* or *Mih-ala*: A troop of monkeys more than one male and many females. (Often males and females are also found in equal numbers)
2. *Enga Hapan Mih-ala*: A troop of monkeys with one male and many females.

3. *Min-a-Giae*: A lone male monkey.
4. *Sapani:* A troop of males only.

Birhors identify the males and females from a distance only by hearing their sounds. Male langurs are identified by the sound *itt, itt etc.* whereas females make the sound - *'khe', 'khe' etc.* As the Birhors explain, during the rainy season (*Jargi*) the monkey troups prefer to move in the plains or open space (*'Dalam'* in Birhor term) to get sunlight. During the winter (*'Rabang'*) monkey troups are found on the cliffs and plains (*'Ragha'* and *'Dalam'*) to get sunlight and during summer (*'Aghan' or 'Sitang'*), they are found on the hill stream sides (*'Chapal'*).

In the forest of Similipal, two types of monkeys are noticed - (i) Langurs (*'Hanu mankad'* in Odia and Birhor) -*Presbytis entellus* and (ii) Bander or Rhesus monkey (*'Pati mankad'* in Odia, *'Gali'* in Birhor dialect) - *Macaca mulatta*. Though Birhors hunt both, in practice Bander (*'Pati'*) or Rhesus monkeys are hunted more.

In the process of monkey hunting (*'Sendra'*), the monkey troops are usually chased from their habitat (*'Laga-rakap'*) to downwards (*Laga-algu*). In the beginning, after the priest performs the ritual in the names of Hill gods and goddesses, the hunting party is divided into following four groups.

1. *'Bebera'*- Those who chase the monkeys.
2. *'Bajhur'*-Those who guard both the sides to refrain the monkeys from escaping.
3. *'Atam-dap'* - Those who guard both the ends of the net.
4. *'Ath -wahan'* - Those who hide at the net and kill the monkeys by an axe or stick.

Accordingly the group is divided into four subgroups as mentioned above. After chasing the monkey troops by relatively young persons, the side-guards and net-guards refrain the monkeys from escaping and at last when the monkeys go forward to escape, they are caught in the net camouflaged by tightly fixed branches on the ground. At the net, the most experienced Birhors hide with axe. The moment the monkeys are caught by the net, immediately they are hit on the head by the axe and killed. Thus, Birhors on one hand through rope making help the forest to grow but on the other, through monkey hunting destroy the monkey population of the forest.

In a similar manner, Hill Kharia and Ujia through their collections of sal resin, arrowroot and honey, maintain their livelihood and conserve the plant resources. Similarly, in Hill myna catching, they help the process of conservation but through the catching of the Horn bill, they depopulate the Horn bill population severely.

Besides the above mentioned food gathering activities, Hill Khadias specially collect two types of gums from Similipal as per the demand of the outside merchants from Mumbai who use them in various food products. Such gums are - (i) *'Bharal' or 'Amthu'*(family - Caesalpiniacea) and (ii) *'Gardei' or Mankadpichhula* (*Sterculia urens*). Khadias make numerous deep cut marks on the barks of such trees in a dispersed manner and after about two weeks the gums ooze out and collected by the help of flat bamboo or wooden sticks. However, Khadias never make cut marks by ring method by which a tree dies. There are also several other plant resources which Hill Khadia and Ujia collect from the forest without doing any harm to the concerned plant species.

Fishing in Similipal

Earlier in the paper I have mentioned mostly about monkey hunting which is definitely a very harmful activity in the forest like most other hunting activities. But fishing is not so harmful activity in the forest. The fishing is usually done in the hill streams which are filled with pebbles, big rocks and often the dry branches of the trees. Therefore, fishing by net is never possible in hill streams. Hill Khadias, Ujias and sometimes other neighbouring agriculturist tribes fish using the fish-poisons collected from the plant resources (Table 23.6). In such a process of fishing, the tribes have to fish from the flowing water. In still water (like ponds, tanks), people fish using fishing traps, nets and few other devices. But in hill streams, about 13 types of plant poisons are used.

Table 23.6. Types of Fish Poison

Sl.No.	*Local Name*	*Scientific Name*	*Family*	*Part Used*
1	Patua	-	-	Fruit
2	Rulia, Rudhia or Khakada	-	-	Fruit
3	Asana	*Terminalia tomentosa*	Combretase	Bark
4	Jamu	*Syzygium cumini*	Myrtaceae	Bark
5	Gaurani, Gaharani or Gaudani	-	-	Bark
6	Jalei or Jaleita	-	-	Leaf
7	Kanti, Kantil or Kirikinchi	-	-	Bark
8	Sirisha	*Albizia lebeck*	Mimosaceae	Bark
9	Bana-Sijhu (wild)	*Euphorbia nivulla*	Euphoribiaceae	Latex
10	Kataka	*Strychnos potatorum*	Loganiaceae	Fruit
11	Gila (Beta)	*Entada prusaetha*	Mimosaceae	The stem of the creeper
12	Gudmana	-	-	Fruit
13	Jodi	*Ficus gibbosa*	-	Latex

More than 7 common varieties of fish from deep water and 7 varieties from shallow water are usually caught in the hill streams. However, Similipal is known for a very special fish called 'Mahasheer'. A few more varieties are also caught namely -*Khenka - Khai, Bansa-patri, Gami, Gargad and Tikili* by the Ujia. After collecting the fish poison they pound them (either the bark, leaf, fruit or stem). Some poisons are also collected from the latex of the plant. The first of all they check the flowing water on both sides by the help of clay/mud weir leaving a narrow flow of water by a side. Then they distribute the pounded poison on the water. After some time the fishes in the water are stunned or stupefied and float in the water. The tribals catch them very easily and bring home. Such poisons usually make the fish blind and senseless. As the people say, it has no adverse effect on human beings.

In addition to the catching of Hill myna, and Horn bill, the tribals (almost all local tribes) catch forest fowls and peacock by using bird traps which is definitely harmful for the forest. The hunting of small and big game animals are mostly practised by all the local tribes through out the year. It has different names and technologies. In Similipal forest six types of hunting expeditions are organized, namely:

1. *Baha Akhan Sendra* (*'Sendra'* in Mundari means hunting)
2. *Dah Sendra*
3. *Panta Sendra*
4. *Pata Akhan Sendra*
5. *Dihirigiri Sendra, and*
6. *Kulei Sendra*

On the basis of the type of the game animal, the group size and weapons are usually decided. The large game animals require more number of people (hunters), about 15 - 20 where as for the small game animals,a small group of 3/5 members are required. But the Hill Khadias and Ujias occasionally or rarely go for hunting in Similipal. As most of them do not carry bow and arrow, they join very rarely other tribal communities for large game hunting. Very rarely also they go for an exclusive hunting expedition. Generally their hunting activities are organized on the spot whenever they encounter any small game animal within their reach while collecting minor forest produces. Such small game animals are mouse deer (*Tragulus meminna*), barking deer (*Muntiacus muntjack*), Indian Percupine (*Hystrix indica*), Indian hare(*Lepus nigricollis*), Belra mouse (*Ratufa indica*) *etc.* Therefore, in large game hunting like the hunting of wild buffalo (bison), Sambar, Chittal or spotted deer *etc.* the neighbouring agriculturist tribes participate and mostly because of them the population of wild animals are severely threatened till the date.

In all large game hunting types which are organized by a large group of tribal hunters, a common procedure is followed. Such hunting expeditions are usually called as *'Panta Sendra'*. The meaning of *'Panta'* is to chase. In a large game hunting group there are four groups of participants.

1. Three groups of chasers or *'Panta'*.
   (a) Middle group of *'Panta'*.
   (b) Left side chasers - *'Kanjen Kaniari'* (in Mundari)
   (c) Right side chasers - *'Etam Kaniari'* (in Mundari)
2. The group of hunters - *'Tanha -Ka'* (in Mundari) or *'Ghatia'* (in Odia). Such hunters are sharp shooters who kill the game finally.

Usually before the starting the hunting process, all the members of the group meet in a particular site, known as *'Takabel'*. The priest (tribal) plays the pivotal role propitiating the hill gods and goddesses to satisfy them for a successful hunting expedition. The most dangerous of all hunting practices is *'Akhanda Shikar'* (in local Odia) or *'Akhani Sendra'* (in Mundari). Almost all Mundari speaking tribes consider *'Akhanda Shikar'* as their annual hunting ritual or festival. Considering it as a ritual practice and essential part of their cultural tradition, they ruthlessly kill thousands of game animals on this occasion. Every year Similipal authority tries their best to check such destructive practices but of no result. The forest officials along with the police force guard all the forest check gates to arrest the hunters. But hunters never pass through such check gates. They have innumerable foot paths connecting their villages to Similipal. Very easily they enter into the dense forest and plan their hunting programme which no forest official or police force can even detect. Only occasionally they caught tribals with bow and arrow which bears no importance at all.

Special Role of Hunter-Gatherers

No doubt, Similipal is a green paradise on Earth. As the controller of climate, provider of food and shelter to the poor tribals and non-tribals, Similipal has played a motherly role for the state as well as country. It is the feeding and breeding ground for innumerable plant and animal species on the soil of Odisha. But the so called dependent tribals by their destructive role through over exploitation have made this heaven to a denuded hell for all time to come. This is not an exaggeration, but a reality. The green Similipal is every moment being robbed of its valuable resources, both biotic and abiotic by the secondary and tertiary dependents. It is losing its energy, life gradually and just breathing to regain its green vigour of the past. Tigers are no more roar, bird no more chatter, riverins no more flow with musical steps and the elephants no more dance with their blowing trumpet here. Why? In order to realise the beauty and youth vigour of Similipal, we have to come to realize the special role of the food gatherers and hunters like Hill Khadia, Ujia and Birhor in the maintenance of Similipal.

Already I have highlighted their role for the conservation of Similipal which have gone un-noticed since long. The so called food-gatherers while making out a living out of the resources of Similipal both edible and inedible, take systematic and decisive steps for the replenishment of such plant species, so that the species in the forest will propagate and continue for the future and the food-gatherers will harvest them every year without fail. Some major resource collections have already been presented in the paper (Table 23.7). But at the same time, few more specialized roles may be discussed here to justify the beneficial roles of Hill Khadia, Ujia and Birhor. Let us start from the beginning with reference to a major seasonal collection of resources.

(i) Paradise Maintained

Not simply by virtue of pertaining to food gathering economy, Hill Khadia, Ujia and Birhor make beneficial contributions both to the forest and their economy, there are several other factors. No body kicks out its own bread and butter at any cost. It is the basic human nature. The food gathering tribes have no land, no cultivation. Therefore, the neighbouring agriculturist tribes do not take care whenever they

Table 23.7. Seasonal Collection of Forest Resources

	Types	*Khadia (Savara) and Ujia (Savara)*	*Birhor*
Plant resources (Edible and Inedible)	1. Major seasonal Collections	1. Honey 2. Sal resin 3. Arrow root	1. Chopped fibres for rope making
	2. Subsidiary seasonal collection	1. Firewood 2. Timber 3. Sal leaves 4. Sal flowers 5. *Giliri* flowers 6. Tubers 7. Fruits 8. Nuts and berries 9. Herbal Medicines 10. Kusum Seeds 11. Greens 12. Mushrooms 13. Chopped fibres 14. Broom Stick	1. Firewood 2. Sal leaves 3. *Giliri* flowers 4. Tubers 5. Fruits 6. Nuts and berries 7. Herbal Medicines 8. Greens 9. Mushrooms 10. Broom Stick 11. Honey
	3. Specific Collection	1. Gum-Khadia only 2. Grasses - Khadia only	
Other resources	4. Animal product	Antler, tusk, porcupine stick, tiger's milk, teeth, skin, bone, hairs, *etc.*	Teeth, bone, skin or hide, placenta, bone, feathers, hair, *etc.*
	5. Other collection	Soil, stone, Hornbill soil, peacock's feathers, *etc.*	Porcupine stick

collect any plant resource from Similipal. It is for them neither the bread nor the butter, but something additional.

(a) The statement of Mr. Saraj Raj Chaudhury (1974), the founder Director of the Similipal Tiger Reserve is very much appropriate here. In the management Plan of Similipal Tiger Reserve, he has rightly maintained -"The Erenga Kharia (Hill Khadia) live in the forest and depend entirely on it. Most of the minor forest produces - honey, wax, resin, arrowroot *etc.* are collected by the Kharias (30: 1). Rice is the staple food of the people except for Erenga Kharias who depend most on roots and tubers from the forest "(30: 2). Elsewhere in the document, Mr. Choudhury has mentioned - "Bulk of the ethnic strata inside and in the peripheral vicinity of Similipal is of aboriginal status, some still in their autochthonal primitiveness as in the case of the Erenga Kharias, the Mankidias (Birhor) and the Saharas (Ujias?) who principally live off the forests. The Erenga Kharias (Hill Khadia) in particular consider Similipal as their Primordial home "(30: 1). He discussed further that "collection of MFP., *i.e.,* honey, arrow-root, hill myna, sal seeds *etc.* is done by the local tribal people for the department or department agents. There are no rights or privileges "(18). Thus, the views Mr. Choudhury, was very scientific and he has the idea that, tiger cannot be conserved in isolation. For this reason perhaps in the core area he, officially made a colony or settlement for few Hill Khadia especially for getting their help and cooperation in various conservation activities. But to our surprise, soon after his death, his followers in the chair couldn't accept this view. They ousted the the Hill Khadias along with other communities from the core area.

(b) Mr. Choudhury has kept two sacrosanct areas out of (10 sq.km each) core area of 303 sq. kms completely a restricted zone by saying that, "No one will enter this area except for authorized scientific studies (35.10.1A-ii)". Similarly, the rest of the core areas - "Minor Forest Produce collection - honey, wax, arrowroot, sal seeds, hill myna will be strictly controlled" (35.10.1B-v.). But in the inner portion of the buffer zone covering over 1200 sq.km, "human occupancy will be regulated to yield highest multiple use of the land without marring the prestine look of the rolling landscape, the interesting fauna and the pockets of tribal villages "(35.10.2A).

(c) When I was working inside Similipal hill range, during the 1980s, at that time, for the outsiders, Similipal was mostly known for four things namely, the exploitation of 'champa'(*Michelia champaca*), nageswar (*Mesua ferrea*), catching of hill myna and rhesus monkeys. They were exported to outside through the National Park authority of Similipal. The timber of 'champa'(white flower) was exported to Japan.

The hill myna and rhesus monkeys, similarly exported to Japan and European countries for research. The flower and branches of nageswar plant were also sent to Kolkata as well as different markets of Odisha for medicinal requirement.

(d) Orchid Wealth of Similipal

For some time, the Orchids were also marketed to Kolkata for its medicinal as well as floral uses. As Mr. Sarat Mishra has pointed out, the Orchids are very much habitat-specific and he mentioned about two types of habitat in Similipal - **open habitats** (large stretches of grassland and meadows in valleys) and **forest habitats** comprised of (i) Open forests, (ii) Montane forests, (3) forests of high hills and (4) vegetation in deep forests along with water-courses. Through very painstaking research in 1985 he has taken up a maiden attempt to document the Orchid flora of Similipal, which includes 65 species (14 terrestrial and 51 epiphytic). Later in 1989 he enumerated 92 species (35 terrestrial + 57 epiphytic). Between 1989 to 1996 he added 6 more species to the list of 92 orchid species. Making a distributional analysis, he pointed out that "Similipal harbours as many as 19 Himalayan elements and 4 South-Indian elements; 12 species are common both to the North and the South. This could be attributed to the Northern location of Similipal. There are as many as 20 orchids of Odisha which are known only from the Similipal forests"(Mishra,

1997: 88). Orchids have very beautiful flowers because of which some species are also popularly denoted as ornamental orchids. In Similipal, he has recorded 14 epiphytic and 7 terrestrial ornamental orchids.

(e) Socio-cultural and Mutual Division of Similipal

Everybody knows, Similipal is a national property, which was declared as National Park, Sanctuary, Tiger Reserve and Biosphere Reserve. How can it be socio-culturally and mutually divided among the tribes? Despite its national status, for the local people especially the food gathering tribes, it is their motherland where they have seen the first light of the day on earth. It is their breeding and feeding ground too. Therefore, the food-gathering tribes Hill Khadia and Ujia both inside Similipal and outside divide the resourceful sites of among themselves on mutual basis. Every forest as well as fringe village maintains a traditional boundary with certain natural land marks. All the village heads sit together, finalise the boundary and set the principles for resource collection. However, all of them very well knew that forest does not belong to them and except on mercy or sympathetic grounds they have no right also to exploit the resources in the reserve forest. Such a tradition of fixing the boundary for resource collection has been continuing since the reign of the kings in the past. From the Table 23.8, the mutual division may be understood. But the case of Birhor is a little different as they move as a nomadic community and change their micro-ecosystem regularly.

Table 23.8: Micro-Ecological (Mutual) Zones of Kharia, Ujia and Birhor

Ethnic Groups	*Target Villages*	*Location*	*Micro-Ecological Zones of Similipal Exploited*	*Major Collection/ Hunting*	*Specific Items of Ecological Zone*
Khadia (Savara)	Baniabasa	South-Eastern Similipal (foothill)	10-15 km. in the North and West, 5km. in East	Honey, Sal, resin, arrow root, fire wood	Same as the major collection and sal leaves
	Khejuri	Central Similipal (Forest)	On all direction within 10-15 km. radius	Honey, Sal, resin, arrow root	Gum (edible)
Ujia (Savara)	Badjunpal	Southern most Similipal (foot hill)	Within 10-15 km. radius towards the North	Honey, Sal, resin, arrow root	Hill myna, Sal leaves and flowers
	Gandidar	Southern most Similipal (foothill)	Within 10-15 km. radius towards the North West Similipal	-do-	-do-
Birhor	Tanda-I	South-Eastern foot-hill plains	Within 15 km. radius	Chopped fibres and monkeys	Same as major items
	Tanda-II	Southern foot hill plains	-do-	-do-	-do-

(f) Socio-Religious and Emotional Shadows of Similipal

This part has already been highlighted in the paper. The purpose of repeating it at the end, has the intention to reveal that Similipal is not simply a geological, botanical and zoological multi-structural forest. With both biotic and abiotic components it is also not a dormant-land. Food gathering tribes like Hill Khadia, Ujia and Birhor have brought life and heart to it, creating a sense of emotion within.

Joining the green ground, its roar, chattering, tumpet with the supernatural entities, the tribals brought the culture of man, the culture of the plants and animals to a common socio-religious and emotional ground. They consider all the hills, rivers and waterfalls as goddesses, each forest a god, the caves as their heavenly abode and so on. They, on the whole, consider the natural entity of Similipal as their mother goddess which only gives, cares, supports and sacrifices. As obligation, Hill Khadia, Ujia and Birhor also take great care for the renewal of majority of useful seasonal resources for the propagation of the concerned species and continuity of their dependence. Whatever the destruction sometimes attributed to them is meagre and not so much accountable in comparison to that of the surrounding agriculturist tribes. Thus, the eco-friendly culture of the food-gathering tribes has been the sole factor responsible for making Similipal a paradise on Earth. Looking at the supernatural entities who are taken to be the guardians and custodians of the hill range, they developed a sense of disciplined behaviour in all forest-based activities. They fear and obey the forest gods and goddesses and never destroy their sacred land, *i.e.* Similipal.

(g) Knowledge is Power

The power of knowledge has been widely recognized in all types of techno-economic activities. In the forest, knowledge is most indispensable. For identifying the types of forest, availability of required resources, knowing the technology of collection, understanding the behaviour of the birds and animals *etc.* one has to learn the lessons through participation. Without knowledge, no food gatherer will be able to survive.

On the other hand such food gatherers are so knowledgeable on the matters related to forest that often the forest authorities consult them for understanding the forest ecosystem. By reading the nature and behaviour of the tigers, bears, elephants *etc.*, they take timely action for self-protection. Perhaps, this is one of the reasons which tempted Mr. S. R. Choudhury to build a colony for the Hill Khadia at the Forest office of Burha Kamuda, so that Khadias will work for the forest as labourers and help at the time of need through their knowledge and experience. There are so many other observations by which one can feel the Paradise status of Similipal and its careful maintenance by the food gatherers like Hill Khadia and Ujia.

(h) Paradise Lost

The overexploitation of resources without caring for their renewability and destruction of plant and animal species by secondary and tertiary dependents have already been discussed earlier in the paper. At present few more little known situations may be presented to realise how the Paradise of Similipal is going to be lost.

Turning Points to Destruction

(a) Displacement of Hill Khadias

The presence of food-gathering tribes have made the green Similipal greener due to the personal, emotion, and religious care of the food-gathering people. Their presence has also helped the forest officials in a great way to expel the poachers and trespassers to some extent.

Soon after the sad demise of Mr. S. R. Choudhury, Hill Khadia and other tribes of the core and protected are displaced outside. Except Burha Kamuda, rest other Khadia settlements are ousted from the forest. They have been resettled in Jashipur and Udala areas of Similipal where there was no forest acquaintance and Khadia started living like the daily wage labourers. The absence of Hill Khadia in core and protected areas resulted in the mere destruction of the renewable resources by outsiders. Though Hill Khadias in buffer and fringe zones remain as before, they were restricted to enter into core and reserve areas. It bears several repercussions on the destruction of the forest.

(b) Coming of SFDC

In 1979 Similipahar Forest Development Corporation Ltd. was established. Its basic purpose was to conserve nature and use its produce most decisively. It banned tree felling from Similipal and tried to procure timber and MFPs on its own for a better harvest and marketing. But the ultimate consequences go in the opposite direction. Before the coming of SFDC, the suitable forest patches were marked by an expert team from the forest department, for tree felling and marketing. Some experienced contractors were given a lease to fell only the marked trees under the supervision of the foresters. Though in the processing of tree felling, few more unmarked trees are exported illegally by paying bribes to the forest officials, in comparison to such loss, the ultimate gain of the forest was much more, which is very interesting. While felling down the trees from the leased area of the forest, the forest contractors were very carefully protecting the forest from the local trespassers. They also check the movements of the timber mafias and hunters from outside. Their watchfulness was saving much more than that of SFDC's restriction.

Earlier when the food gathering tribes were collecting MFPs for the forest department, other tribals were considering Khadias and Ujias as forest workers and fearing them. As SFDC also continued the same work as before, Hill Khadias and Ujias enjoyed a better status which helped in protecting the Similipal from the poachers, trespassers, and mafias. However, the ban of the timber contractors by SFDC and the displacement of food gathering tribes caused severe loss to Similipal. The poachers, trespassers and mafias understood that the contractors, as well as food gathering tribes, are no more having any official right or responsibility. Both by paying bribe and by trespassing they rampantly hunt and steal the wealth of Similipal. The local tribes also started stealing the timber wealth in different inaccessible forest patches. However, with a lot of failure, SFDC was closed in 1990.

(c) Failures on Several Frontiers

The paradise of Similipal has been severely devastated by failures on several frontiers. Late Maharaja Rama Chandra Bhanj Deo started for the first time official human habitation inside the hill range. Declaration of National Park and Wild Life Sanctuary was also designed to protect the biodiversity of Similipal. Further, the declaration of Tiger Reserve and Elephant Reserve has also maintained a similar intention of protecting tiger and elephant population specifically. The establishment of SFDC had also similar intentions in addition to organize the MFP collection and their marketing through official procedures. However, every attempt failed mainly due to the coming of trespassers, poachers and mafias into the hills. The absence of the food gatherers like Hill Khadia and Ujia added fuel to the problem. The forest department simply worked officially, never through personal or social service approach. Besides, few forest officials also promote such destructive activities by taking bribe. All such happenings are not only confined to the Similipal of Odisha. The forest cover of the whole country has been seriously affected through similar activities.

For some time, Mr. Saroj Raj Chaudhury has understood the process of conservation for which he had written that conservation of Tiger is not possible in isolation. He has understood perfectly the subsistence-oriented cultures of food gathering tribes like Hill Khadia and Ujia. But soon after his sad demise his approach also went to the graveyard and the situation in Similipal became so precarious that one cannot even dream or imagine. Since 1977 I have been observing its nook and corner. To my surprise, all the paradise look has been lost forever. The matured Sal forest is thoroughly destroyed; orchids, tigers, leopards, elephants and spotted deer populations are being depopulated every day, every moment in large numbers. Alas! the paradise of Similipal is no more existing. Its pristine beauty is disappearing steadily, though slowly.

Lastly, therefore, I conclude that the displacement of Hill Khadia from the core area resulted in the:

1. Loss of a forest-based knowledge system that no forestry curriculum can develop in the classroom.
2. Death of a forest-based culture which after displacement has been wage-labour oriented. Instead of displacing, their behaviour would have been regulated through proper training to cooperate the forest department for the most efficient management of the Biodiversity and Biosphere reserve.

   As they had enormous knowledge stock about all the aspects, dimensions and areas of Similipal, the forest department would have utilized them for everything best in Similipal. They would have been also utilized for guiding the tourists, watching, and guarding the wealth of Similipal in all respect. With their disappearance from the scene of Similipal, not only Similipal will lose so many things, the anthropologists will also lose such forest-based cultures from the land of Odisha as well as India for all time to come.

3. Thus, the withered wilderness of Similipal will no more be strengthened by rich Sal patches, beautified by parrots, hill mynas, hornbills and so many other members of avifauna, attracted by the trumpets of the elephants and roaring of the tigers. At present the luxuriant growth of white Champak trees are no more visible. The sandalwood patches are also no more existing and so on. The population of plants and animals have been severely depleted by human intervention.

The playful evergreen Similipal, like any small denuded forest, is now bed-ridden, breathing just for existence. I hav seen here the playful tigers cubs and mother, movements of black panther, dancing of elephants with beautiful tusks, the call of the hill myna and the murmuring music of the hill streams and waterfalls during 1980s. Along with Mr. S. R. Choudhury and Mr. B. Nayak, they have disappeared. I simply observe now a green paradise lost forever due to lack of our humanism, humanistic bent of mind and inhuman approach as well as activities.

While concluding, I recollect saying that "forest plays a vital role in reducing carbon in the atmosphere and in any initiative to combat climate change. Both at local and global levels, forests provide essential ecosystem services beyond carbon storage -such as watershed protection, water flow regulation, nutrient recycling, rainfall generation and disease regulation "(Ojha, RCDC). It is also further quoted that in general, forest resources especially the MFPs (Minor Forest Produces) support the livelihoods of 90 per cent of the 1.2 billion people living in extreme poverty and are home to nearly 90 per cent of the world's terrestrial biodiversity. Local communities depend on forests as a source of fuel, food, medicines and shelter. The loss of forests jeopardizes poverty alleviation. As already highlighted, indigenous and forest-dependent people are stewards of their forests, providing the rest of humanity with vital ecosystem services "(ibid.). At the end, for the sake of Similipal, quoting a saying from *Say trees.org* I state that -

"I clean the air you breathe,

I can do that for your children too,

if you let me live".

REFERENCES

Choudhury, S. R. (1974) Management Plan of Similipal Tiger Reserve

Champion, Harry G., and Seth, S. K. (1968) A Revised Survey of the Forest Types of India, New Delhi, Govt. of India.

Dash, Jagannath - Human Ecology of Foragers, Commonwealth Publishers, New Delhi, 1998.

District Gazetteer of Mayurbhanj, Govt. of Odisha, 1967.

Mahalik, N. K. (1997). Geology and Geomorphology of Similipal Biosphere Reserve, in Similipal: A Natural Habitat of Unique Biodiversity. Odisha Environmental Society, Bhubaneswar, Memorandum and Articles of Association of the Similipahar Forest Development Corporation Ltd.

Misra, R. C. (1979). Similipal Biosphere Reserve, Regional Plant Resource Centre, Bhubaneswar, 2006.

Misra, S. (1997). Orchids of Similipal forest. in Similipal: A Natural Habitat of Unique Biodiversity. Odisha Environmental Society, Bhubaneswar.

Ojha, Nabaghan - REDD Governance, in Community Forestry, RCDC, Bhubaneswar.

Rodgers, W. A. and Panwar, S. H. - Biogeographic Zone in India, New Forest, Dehra Dun, 1988.

Vasu, Nagendra Nath (1911). The Archaeological Survey of Mayurbhanj, Vol.1, Published by Mayubhanj State.

Chapter 24

The People and their Livelihood Inside Similipal Forest, Odisha, India

Debabrata Swain

Member, Lokayukta Odisha and
Former Principal Chief Conservator of Forests and Head of Forest Force, Odisha
e-mail: dswain2008@gmail.com

"It is forbidden to decry other sects; the true believer gives honour to whatever in them is worthy of honour."

– Ashoka the Great (264 BC - 227 BC)

Similipal biosphere and tiger reserve is situated in Mayurbhanj district of Odisha, India. The tribes residing in the district are Bathudi, Bhumija, Kharia, Kolha, Mankidia, Munda, Santal, Bindhani and others. There are 62 villages inside Similipal (four in the core and 58 in the buffer zone of the reserve), and about 1200 villages located in the transitional zone. The people in and around Similipal lead a subsistence living. The study reveals that there is extreme poverty, unavailability of schools and remoteness of tribal habitations in the core area. However, relocation and successful rehabilitation of core area families is a win-win situation for management of Similipal reserve. There is an increase in per capita income and a decent living for the families of relocated community. Simultaneously, wild animals found extensive inviolate area for movement and reproduction.

Keywords: *Similipal, Biosphere and tiger reserve, Core area villages, Socio-economic condition, Village relocation, Implementation of forest right act.*

INTRODUCTION

Similipal forest, notified as a tiger reserve and a biosphere reserve, is situated in Mayurbhanj district of Odisha, India (Figure 24.1). I was once trekking from

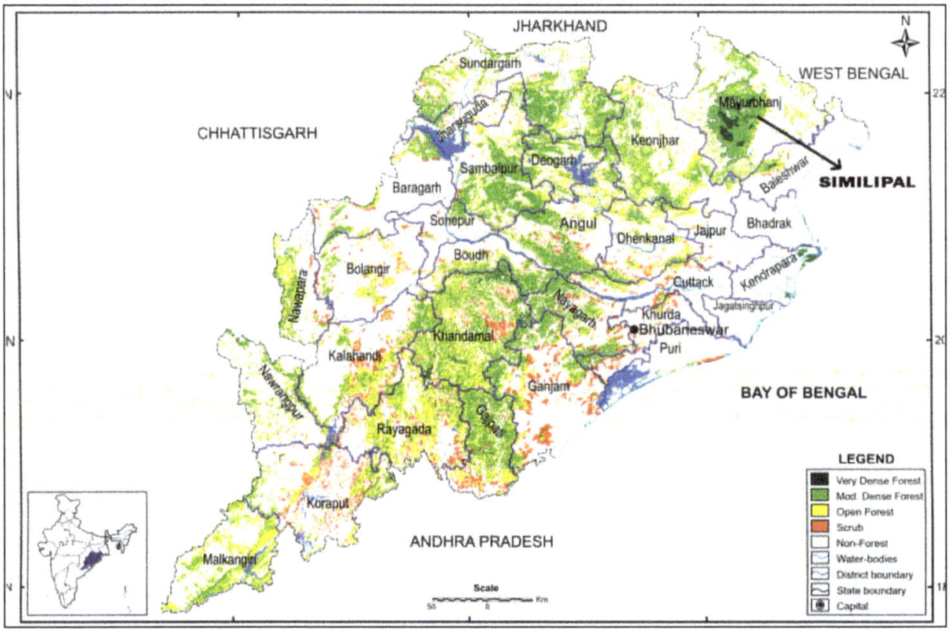

Figure 24.1. Location of Similipal Forest, Odisha, India.

Kaliani to Gudgudia inside Similipal. On the way at the bank of river Khairi, I saw a group of people belonging to Kharia community, a primitive tribe residing in and around Similipal. They apprised me that they were on their way to a place named Mahubhandar, about 10 km from that location deep inside the forest. They would stay there few days and collect honey, arrowroot, resin and other such jungle products. They further told me that entire Similipal had been divided amongst their community for collection of the various forest produce and they used to enjoy this right generation after generation without encroaching on the area of other people from the same community. Therefore, even though they resided at Kapand near Jashipur at the fringe of Similipal, they travelled such a long distance to collect the forest produce. Incidentally, these are the people shifted from Jamunagarh and Kabatghai inside core area of Similipal in 1994. I explained to them that collection of NTFPs (non-timber forest products) from the reserve was illegal and they would be prosecuted. It was not convincing to them as they believed it was their right since time immemorial to collect NTFPs from the assigned area and if they did not obey the law of the community, the forest deity would be annoyed inviting misfortune on them.

There is another group of nomadic tribe called Mankidias (Birhor tribe) who catch monkeys and even eat them. They move inside Similipal collecting Siali fibres and sell them in local market.

According to a local folklore, a group of aboriginal people used to take cattle of the king of Mayurbhanj everyday inside Similipal for grazing. Years passed by and when the king disengaged these people from the practice of grazing his livestock, some of them resorted to settled agriculture outside Similipal and they are now called Bathudi tribe. The people who did not take up agriculture and roamed inside Similipal and depended on roots, tubers, barks and leaves are called Kharia tribe. Another set of people who remained in Similipal and resorted to kill monkey and other small creatures for their consumption are called Mankidia tribe.

History

The earliest evidence of man in Mayurbhanj district, where Similipal is located, is believed to belong to the Lower Palaeolithic or Early Stone Age culture (Senapati and Sahu, 1967). The Early Stone Age man lived here during the Middle Pleistocene time. The sites of this culture are concentrated along the banks of river Budhabalanga and its tributaries, the Sone and the Gangahar. Senapati and Sahu (*supra*) report that Middle Stone Age Culture in Mayurbhanj flourished on the western side of the Similipal hill range along the river Kharkei and its tributaries. After the Middle Stone Age the Microlithic culture of the Late Stone Age also flourished in this region. In comparison with the vast collections of the Early and Middle Stone Age tools, the microlithic collection of this district looks very insignificant and gives indication that this culture did not live long in this district. But the other Late Stone Age culture, *viz.*, the culture of the polished stone celts had a very flourishing time in this district. The long list of sites of this culture in Mayurbhanj, gives this district a unique place of the Neolithic civilization in eastern India. For the first time, evidence of the prehistoric pottery is noticed here. Besides pottery, grinding and polishing of the stone tools and domestication of animals might have been practiced during this period. The people lived in well-built houses and settled as agricultural communities. The shouldered celt culture is a later development of the Neolithic. The people of this culture knew the use of metal. It came to India with the people speaking Austro-asiatic language. The tribal people of the district, namely, the Santals, the Mundas, the Kolhas and the Savaras, now speak this language. There is also evidence of primitive copper celts in this district. Senapati and Sahu (*supra*) further state that the earliest type of celts, the shouldered is dated to the Yin dynasty (about 1300-1028 BC).

The tribes residing in the district are Bathudi, Bhumija, Kharia, Kolha, Mankidia, Munda, Santal, Bindhani and others. Not much study has been made about the tribes of Similipal. One of the earliest studies about the economic condition of the inhabitants of Similipal was reported by Das and Das (1968). Dash (1998) gives an account of the life and culture of the Kharia Savara, Ujia Savara and Birhor (Mankidia Savara) in and around Similipal hills. Barik (2003) gives an account of tribal population and settlement inside Similipal. According to a legend, Bathudis are the earliest settlers of Similipal (Das and Das, 1968; Dash, 1998). The exact time of settlement is not known. Dash (1998) citing Vasu (1911), the time period of early settlement is predicted to be that of King Ashoka's period as the Bathudis of Similipal are said to be descendants of Madan Nayak and his three sons Nanda, Benu and Beni. Madan Nayak, as believed by Bathudis, was either a Sardar in Ashoka's army or simply belonged to Ashoka's army or just belonged to Ashoka's period in Odisha.

Kharias were originally living in Panchpir and were eking out their livelihood by collecting forest produce from Similipal. Their Chief was at Jashipur and Bathudis under the leadership of Nand Das defeated the Kharia Chief and captured

Jashipur fort. Kharias after the defeat of their chief migrated into the deep dense forests of Similipal. Similarly, Bamanghati which was then under the control of a Gond king was captured by Bathudis and Gonds migrated and settled in certain parts of Similipal. But due to depredation of wild animals, diseases like malaria and gastroenteritis, and constitution of Similipal as a reserved forest and Game Reserve for the royals, large-scale settlement was never possible. It is stated that depopulation happened as many as seven times in Similipal and fresh migrations took place (Das and Das, 1968).

After accession of Maharaja ShriRama Chandra Bhanja Deo to the throne of Mayurbhanj state in 1890, he initiated steps to improve the living conditions inside Similipal. Temporary leases for cultivation were granted to a few settlers. One Peter Dubraj, a Kolha from Singbhum who was working as a Postmaster at Chaibasa was granted a lease of 920,642 acres at an annual rent of Rs. 254 and Annas 9 only. He was declared intermediary Sardar for the area and was allowed to establish villages inside Similipal. Similarly, a lease was granted to Durjadhan Mahakud for Kasirapirh, where seven revenue villages were formed in due course. When Dubraj died in 1934, it was found that he was in heavy arrear of rent. His successors were forced by circumstances to sell their lands. Ganesh Ram Ho of Nawana purchased some lands in 1949. In 1955, the Sardari system was abolished. Lands of Peter Dubraj were auctioned towards arrear revenue in 1965 and were purchased by a group of persons belonging to Mahato caste. The Mahatos of Bhanjkia set up a Doharaghar (farm house) in Similipal.

After integration of Mayurbhanj state into the Union of India on 6 November 1948, it became part of Odisha as a district on 1 January 1949 and forced labour and monopoly system of timber extraction were abolished. Forest coupes were auctioned to the highest bidders. People from inside Mayurbhanj district and from outside the district, even far off places of the then Bihar (now Jharkhand) settled inside Similipal for forest operations and collection of minor forest produce. People cleared forests in the valley near the perennial water sources to settle down in protected areas for which they were penalized several times by the Forest Department. Finally in 1955 those lands were declared as 'Nayabadi' and were allotted to the occupants. The availability of Nayabadi lands stimulated others to encroach upon forest land and people from far off places like Ranchi and Singbhum rushed to Similipal to establish villages encroaching upon forest lands and destroying forest wealth. This infiltration of outsiders has considerably increased the population, particularly at Jajdihi, Kukurbhuka, Budhabalang, Saruda and Bunduriabasa. They mostly belong to Munda tribe converted to Christianity (Das and Das, 1968).

Das and Das (*supra*) further state that other processes through which Similipal became populated are matrimonial relations and kinship bond. On many occasions, the son-in-laws are made to live in Similipal after marriage. Many did immigrate on request of their relatives who had settled in Similipal earlier. Many came down to this place to remain as '*barmasia*' or annual agricultural labourer. Subsequently, they acquired patches of land and settled down permanently.

Gudugudia Panchayat is found to be populated by immigrants from other parts of the district of Mayurbhanj, whereas Astakunhar is mainly populated by outsiders. Bamanghati kolhas are found in older settlements like Sanakasiria, Barheipani, Budhabalang, *etc.*, whereas, Kolhan kolhas are noticed in Nawana, Balarampur, and Garh Similipal.

Settlements in the interior of Similipal did not take place during any particular period. It took place over a long period of time by successive flow of immigrants from within Mayurbhanj and from outside. During 1911 to 1931, immigration was casual, but it increased from 1931. While Bathudis and Kharias migrated from within Mayurbhanj, the majority of Kolhas came from the adjoining Kolhan Government Estate of Singbhum which formed an integral part of the territory of Mayurbhanj till 1937, when their administration went over to the British Government and which formed a compact block in the Eastern frontier of Kolhan estate (Das and Das, 1968).

The virgin soil, abundant perennial flow of water, easy access to forest resources and availability of engagement in forest operations, encouraged the people from adjoining places particularly from the then Bihar state (and present Jharkhand) to settle inside Similipal. The old inhabitants were also not hostile to the new comers. Initially, the forest department did not take much notice of it as land resources were extensive and there was easy availability of migrant labourers for forest department work of various types. But when the population started swelling enormously and encroachment of forest land increased the forest department started a drive to evict the unauthorized emigrants.

Those who could produce Amalnamma signed by Dubraj, Mahakud or their successors are considered as bonafide settlers and others are treated as encroachers. There are two types of immigrants, permanent and temporary. Permanent immigrants came down from within and outside the State, encroached on forest lands, destroyed forests, and settled in regular villages. The temporary immigrants were persons who migrated from outside Similipal for some specific purpose and returned after that. The herdsmen who used to come with their cattle for grazing belonged to the temporary category.

Socio-Economic Condition

There are 62 villages inside the Similipal reserve – four in the core and 58 in the buffer zone of the reserve. In 1971, 7,700 people lived in these villages, but according to 2001 census, the population swelled to 12,449 in those thirty years. In 2011, the population became 17,472 inside Similipal. Similarly, about 1200 villages are located within 10 km of the reserve (transitional zone), with a total population of approximately 450,000.

Swain (2009) states that the people in and around Similipal reserve lead a subsistence living. The annual per capita income for people in the core, buffer, and transitional zones are respectively Rs. 10,150, Rs. 13,595 and Rs. 6,950, whereas that of Odisha and India are Rs. 17300 and Rs. 27150 respectively in 2006 price. The situation is similar for the other socio-economic profiles mentioned in Table 23.1. The literacy rate is abysmally low in the core and buffer areas (16.51 per cent in the core, with a female literacy of 3.53 per cent), in comparison to 51.91 per cent of the district of Mayurbhanj (female literacy 37.84 per cent) and 63.08 per cent of the state of Odisha (female literacy 50.51 per cent).

Table 24.1. Comparison of Demography, Socio-economy, and Health and Development in the Core, Buffer and Transitional Area of Similipal Biosphere Reserve (Swain 2009)

Sl.No.	*Criteria*	*Core*	*Buffer*	*Transitional*
1	Sex ratio (F: 1000M)	1076	994	974
2	Average household (Number of members)	7	5	6
3	Social category	Tribe 100 per cent	Tribe 98 per cent	Tribe 90 per cent
4	Housing pattern	Mud 100 per cent	Mud 100 per cent	Mud 92 per cent
5	Literacy per cent (over all)	16.51	12.31	42.77
	Literacy per cent (Male)	34.47	18.39	44.69
	Literacy per cent (Female)	3.53	6.19	40.80
6	Primary education (per cent)	77.77	47.03	53.52
	Secondary education (per cent)	5.50	30.81	29.82
	College education (per cent)	11.00	16.76	12.62
7	Average land holding (in Ac)	8.40	3.0	2.7
8	Productivity (Paddy in qtl/Ac)	3.18	3.01	4.70
9	Land holding pattern (per cent)			
	a. Big >10Ac)	11.11	13.89	4.74
	b. Medium (>5 and <10Ac)	55.55	13.89	9.62
	c. Small (>2 and <5Ac)	33.34	61.11	67.04
	d. Marginal (<2Ac)	0	11.11	9.96
10	Land less (per cent)	0	0	8.64
11	Avg. cattle/household	27.87	16.19	11.19
12	Annual household income (in Indian Rupees)			
	a. Agriculture	7430	5030	1325
	b. Animal husbandry	1120	1350	695
	c. NTFP	800	1775	4515
	d. Wage earning	800	5440	415
	e. Total	10150	13595	6950
13	Chief NTFP (per cent)			
	a. Sal leaves	10	40	70
	b. Kusum seeds	10	25	10
	c. Honey	40	10	02
	d. Mohua flower and seed	0	20	17
14	Irrigation facility (per cent)	90	50	63

F: Female; M: Male; Ac: Acre, 1 Acre = 0.4 Hectare; qtl: quintal, 1qtl = 100kg; NTFP: Non-timber forest product.

The people's livelihoods depend upon the collection of NTFPs (non-timber forest products) (Plate 24.1), agriculture (Plate 24.2) and rearing cattle, goat, and poultry. Almost all the households in the core, buffer, and transitional zones depend upon the Similipal forests for fuel wood and small timber requirements. Some of the landless, marginal and small farmers resort to illegal felling of trees for fuel wood and timber to eke out their living.

Plate 24.1. Collection and Processing of Palua (Arrawroot) by the Core Area People Inside Similipal, Odisha, India.

Plate 24.2. Cultivation of Rice Inside Similipal, Odisha, India.

Therefore, there is extreme poverty, unavailability of schools and remoteness of tribal habitations in the core area of Similipal reserve. Mainly the villagers are small and marginal farmers, and the land is rain-fed. The people are involved in collection of various NTFPs, but these are at subsistence level, and they are facing problems of transporting and marketing of the material. The population of livestock is increasing in the core area, which causes degradation of grassland for herbivorous wildlife. The problem of unemployment is also prevailing. Most of the villagers are involved in wage earning, mainly provided by the forest department. The people in the core area face a number of health hazards due to poor drinking water facilities and unhygienic living conditions. In the buffer zone the pre-dominant tribal communities are in higher sex ratio and enjoy better educational facility. Many features in relation with the socio-economic condition, village institutions and development in the buffer zone are same as in the core area. In comparison with the core area the villagers in the buffer area are in an advantageous position in the collection of NTFPs and medicinal plants, which support the income generation by selling in the local market. In the transitional zone, the villagers belong to the Scheduled Tribes, Scheduled Castes, other Backward Classes and General Castes. The literacy rate in the transitional zone is better than the other two zones. The average income of the households in the transitional zone from all sources is comparatively less, but they earn more by selling of NTFPs. Due to implementation of watershed activities, the irrigation facilities in the transitional zone are better. In the buffer and transitional area some Civil Society Organizations are involved in watershed management, herbal cultivation, and rural employment through SHG (Self Help Group) formation. In the transitional zone, in some villages, the electricity and telephone facilities are available.

Village Relocation

In search of inviolate space for tiger and associated fauna and flora, families living years together in deep forest are displaced out of the forest. This is a painful exercise for people being relocated and people involved in relocation process. Displacement resulting from the establishment and enforcement of protected areas has troubled relationships between conservationists and rural groups in many parts of the world (Brockington and Igoe, 2006). Colcester (2004) describes that conservation through the establishment of 'national parks', pioneered in USA and applied worldwide, has violated the rights of indigenous people causing impoverishment and social problems. But conservation related displacement, revealing the forms of deprivation, social and equity issues have not been studied properly in south Asia as revealed by Agrawal and Redford (2009), Ferraro (2008), and Rangarajan and Shahabuddin (2006). Lasgorceix and Kothari (2009) opine that local communities usually have unclear or unrecognized rights to park resources and, being located away from mainstream society, are already disadvantaged. This contributes towards their poverty, which further worsens when non-consultative relocation occurs and causes conflict with protected area managers. Panwar (2003) citing the case of Kanha tiger reserve and Kuno National Park says that when displacement is consultative and participatory has a beneficial impact on tribals and other forest dwellers.

In 1994, 11 kharia families from Jamunagrh and 30 Kharia families from Kabatghai were shifted to Kapand near Jashipur at the fringe of Similipal. They were provided only residential accommodation and without any provision for earning their livelihood. Kharias are basically nomadic tribes as mentioned earlier and they were relocated outside Similipal under Kharia and Mankdia Development programme of Odisha. The resettled tribes depended on Similipal for collection of NTFPs like honey and resins. Four years later 23 tribal families from Jenabil were relocated to Ambadiha near Udala township (Plate 23.3). This time each family was provided with residential accommodation along with two acres of irrigated land as per the policy of Project Tiger, Government of India. The families were given eight decimal each of homestead land and during shifting the families were provided free rations and transportation cost of their household belongings. They were even provided temporary accommodation at the rehabilitation site, so that staying there, they constructed their own house as per their requirements within the fund provided to them. Such a scheme became attractive to others and in 2003, eight families from Kabatghai opted to come out to Ambadiha rehabilitation colony. Remaining families of Jenabil was also willing to be shifted to Ambadiha. But it could not be possible due to Kalinganagar mishap where 13 tribals were killed in 2006 when they protested against acquisition of their land for Tata Steel Limited. In the mean time, when National Tiger Conservation Authority was constituted in 2006 and village relocation policy from inviolate area of tiger reserves was revisited in 2007, two options were formulated.

Plato 24.3: A Relocated Tribal Family from Jenabil Village Inside Similipal Forest to Ambadiha, Near Udala Township, just Outside of Similipal Forest, Odisha, India.

Option I- Payment of one million rupees per family as one time package in case the family opts so, without involving any rehabilitation/relocation process by the Forest Department.

Option II- Carrying out rehabilitation/relocation of village by the Forest Department at the above rate of one million rupees per family in the following manner:

(a) Agriculture land procurement (two hectares) and development - 35 per cent of the total package
(b) Settlement of rights- 30 per cent of the total package
(c) Homestead land and house construction – 20 per cent of the total package
(d) Incentive – 5 per cent of the total package
(e) Community facilities commuted by the family (access road, irrigation, drinking water, sanitation, electricity, telecommunication, community center, religious places of worship, burial/cremation ground) – 10 per cent of the total package

With the revised package of one million rupees, rest 61 tribal families from Jenabil came out of Similipal and settled at Ambadiha in 2010. There were few Kharia families kept at Upper Barhakamuda and Bahaghar in Upper Barhakamuda Range since tiger reserve was constituted for getting labour to carry out departmental work. With the passage of time, these families grew in size for which they were required to be shifted out of Similipal. In 2013, there were 22 families in Barhakamuda and 10 families in Bahaghar camps of Similipal. They were shifted to Asankudar village near Karanjia.

Further in 2015, 35 families from Jamunagarh were shifted to Nabra near Podadiha. Only three families were left behind at Jamunagarh, who did not give their willingness as per Scheduled Tribes and Other Traditional Forest Dwellers (Recognition of Forest Rights) Act, 2006 to come out of Similipal. They are still continuing at Jamunagarh. In 2016, rest 47 families of village Kabatghai, resettled at Manda near Jashipur. The name of villages and number of families relocated from the core area of Similipal tiger reserve are furnished in Table 24.2.

Besides shifting of villages from core area, a village remotely located in the buffer, namely Kiajhari, was relocated out of Similipal. Entire 79 families of that village were shifted and settled at Khandiadar, near Karanjia in 2018.

Mahapatra *et al.* (2015) examined in detail the income from different sources and livelihood diversification of displaced dwellers from similipal and rehabilitated at Ambadiha and found an increase in per capita income for poorer segments with an overall 8 per cent increase in absolute household income and corresponding improvement in the poverty ratio and the poverty gap index for the relocated community. Mahapatra *et al.* (2015) further state that the tribes at Ambadiha relocation colony received sizable compensation packages and had closer market access. The rehabilitated families find themselves in a new world with little forest resource but better infrastructure. Educational facilities and access to transport and health care are some of the major benefits. Damage to property by the elephants, to crops by herbivores, and fear of Left Wing Extremists attack are the risk factors with which they had lived inside Similipal. Pecuniary benefit has come in the form of compensation to construct houses and create agriculture land, which has helped

Table 24.2. Name of Villages and Number of Families Relocated from the Core Area of Similipal Tiger Reserve

Village	*Family*	*Year*	*Year-wise Break Up of Families Relocated and Rehabilitated in the Rehabilitation Colony*						*Balance Family (as per survey)*
			Kapand	*Ambadiha*	*Asankudar*	*Nabra*	*Manada*	*Total*	
Jamunagarh	49	1994	11	–	–			46	03 (2014- Survey)
		2015	–	–	–	35			
Jenabil	84	1998	–	23	–			84	0
		2010		61	–				
Kabatghai	85	1994	30	–	–			85	0
		2003	–	8	–				
		2016	–	–	–	–	47		
Bakua	61	-	0	0				0	61 (1998-Survey)
Barakamuda	22	2013	0	0	22			22	0
Bahaghara	10	2013	0	0	10			10	0
Total	**311**		**41**	**92**	**32**	**35**	**47**	**247**	**64**

some households to come out of below poverty line status. The social integration with locals, complications in observing tribal rituals-cum-festivals and ethnic discrimination post-displacement considered major social setbacks in conservation-induced displacement are not evident at Ambadiha.

Rehabilitation of families from Similipal is a win-win situation as evident from the case study in (Box 1).

Box 1: Rehabilitation of Families from Similipal at Ambadiha – A Case Study

Sundar Mohan Nayak belongs to Bathudi community and he was resident of Jenabil village inside core area of Similipal tiger reserve. He was the first person opted to receive rehabilitation package and settled at Ambadiha in 1998. He received two acres of irrigated land and eight decimal of homestead land. His father simultaneously had also received two acres of irrigated land and eight decimal of homestead land. Sundar Mohan has a son and two daughters. They are now reading in nearby college. Every year the family gets about 25 quintals of paddy from their two acres of land. Further 25 quintals of paddy are received from two acres of land assigned to Sundar Mohan's father who is now no more and therefore Sundar Mohan enjoys father's property. Sundar Mohan cultivates NTFPs (like arrowroot), vegetables and pulses in his land and earns handsomely. He has a pisciculture pond (Plate 23.4), and on the bank of it he cultivates mango and banana. He also raises goats and poultry. Udala township being nearby he gets better price for his crops and animals. His wife works as a cook in the Anganwadi centre of the village. There are modern amenities like television, smart phones, cooking gas in their home. There are also better health and transport facilities. With four acres of land, cost of each acre being about 10 million of rupees, Sundar Mohan is a millionaire. He is happy here at Ambadiha and never wants to go back to Similipal.

Plate 24.4: Rehabilitated People at Ambadiha Earns Livelihood from Fishery and Horticultural Cultivation besides Rice Cultivation.

Similar is the case with Rabi Nayak, a neighbour of Sunder Mohan Nayak. He has no child. But both Rabi Nayak and his wife are happy at Ambadiha. They have about two lakh of rupees in their Bank Account. Sushil Dehury belonging to Kharia community has a son and a daughter. While he married off his daughter, his son after marriage work at Chennai. Sushil lives happily at Ambadiha with his wife and daughter-in-law.

Relocation of Jenabil village from core area of Similipal to Ambadiha is a win-win situation for communities and wild animals of Similipal. Forest dwellers got a satisfactory life as that of mainstream population. There is now a large inviolate space available for wild animals to roam around inside Similipal. The village Jenabil after shifting of inhabitants has been developed into a sprawling meadow.

Implementation of Forest Right Act

The Scheduled Tribes and Other Traditional Forest Dwellers (Recognition of Forest Rights) Act, in short 'Forest Right Act' was promulgated in December 2006. The preamble of the Act *inter alia* states that it is an Act to recognise and vest the forest rights and occupation in forest land in forest dwelling Scheduled Tribes and other traditional forest dwellers who have been residing in such forests for generations but whose rights could not be recorded; to provide for a framework for recording the forest rights so vested and the nature of evidence required for such recognition and vesting in respect of forest land.

There are 62 villages and hamlets inside Similipal. Four villages are situated in the core area of the tiger reserve and 58 villages are in the buffer zone of the reserve and all these villages lie in the centre of the Similipal sanctuary. All these villages (62 villages) have an area of 111.18 km^2. Besides these villages inside Similipal, most of the Rayati lands of over 250 villages lie close to the boundary of the sanctuary/ reserve. However, so far 2.21 km^2 individual rights have been settled under Forest Right Act, 2006 (Table 23.3).

Table 24.3. individual Rights for Homestead and Agricultural Purposes Settled under Forest Right Act 2006 in Similipal Tiger Reserve, Odisha, India

Sl No.	*Name of the Forest Division*	*Villages in STR*	*No. of FRC Constituted*	*Individual Rights – Titles Issued after Approval from DLC*		*Area In Acres*	
				Homestead	*Agriculture*	*Homestead*	*Agriculure*
1	Baripada	15	1	9	26	7.58	110.85
2	Karanjia	21	4	0	120	0	161.93
3	Rairangapur	29	26	14	116	7.14	261.46
	Total	65	31	23	268	14.72 Acre or 5.93 ha	534.24 Acre or 215.42 ha

STR: Similipal tiger reserve; FRC: Forest Right Committee; DLC: District Level Committee.

Besides individual rights, community rights of 1751.73 ha settled in the buffer area of Similipal tiger reserve for collection of non-timber forest produce.

CONCLUSION

There are 62 villages inside Similipal inhabited mostly by tribals like Bathudi, Bhumija, Kharia, Kolha, Santal, and others. Out of four core area villages of Similipal tiger reserve, namely Jenabil, Jamunagarh, Kabatighai and Bakua, only Jenabil and Kabatghai villages have been relocated outside Similipal successfully. Except three families of Jamunagarh, rest 35 families have also been rehabilitated. The study reveals that there is extreme poverty, unavailability of schools and remoteness of tribal habitations in the core area. In comparison with the core area the villagers in the buffer area are in an advantageous position in the collection of NTFPs that support the income generation by selling in the local market. In the transitional zone, the people belong to the Scheduled Tribes, Scheduled Castes, other Backward Classes and General Castes. The literacy rate in the transitional zone is better than the other two zones. The average income of the households in the transitional zone from all sources is comparatively less, but they earn more by selling of NTFPs.

However, relocation and successful rehabilitation of core area families at Ambadiha, close to Udala municipality, is a win-win situation for management of Similipal tiger reserve. There is an increase in per capita income and a decent living for the families of relocated community. Simultaneously, wild animals found extensive inviolate area for movement and reproduction. The Government of India have promulgated the Scheduled Tribes and Other Traditional Forest Dwellers (Recognition of Forest Rights) Act, 2006. The Act grants legal recognition to the rights of traditional forest dwelling communities. Granting of individual and community rights is in progress. It is therefore necessary to involve local communities in the protection and management of Similipal reserve failing which conservation of tiger and its habitat will remain as a distant dream.

REFERENCES

Agrawal, A. and Redford, K. (2009). Conservation and displacement: An overview. *Conserv Soc* 71: 1-10.

Barik, B. (2003). Tribal population and settlement inside Similipal Biosphere Reserve: Their impact and management. Dissertation submitted in partial fulfillment for the award of Master of Philosophy in Geography, Post Graduate Department of Geography, Utkal University, Bhubaneswar, Odisha, pp. 26.

Brockington, D. and Igoe, J. (2006). Eviction for conservation. A global overview. *Conserv Soc* 43: 424-470.

Colchester, M. (2004). Conservation policy and indigenous peoples. *Cult Surviv Q* 28: 17-22.

Dash, J. (1998). Human ecology of foragers (A study of the Kharia (Savara), Ujia (Savara) and Berhor in Similipal hills). *Commonwealth*, New Delhi, pp.483+xx.

Das, R.N. and Das, N, (1968). Living conditions of tribals of Similipal hills. *Adivasi* (special issue) 10(2): 1-69.

Ferraro, P. J. (2008). Protected areas and human well-being economics and conservation in the tropics: A strategic dialogue. *RFP Conference Paper*, Jan 31.

Lasgorceix, A. and Kothari, A. (2009). Displacement and relocation of protected areas: A synthesis and analysis of case studies. *EPW* 49: 37-49.

Mahapatra, A. K., Tewari, D. D. and Baboo, B. (2015). Displacement, deprivation and development: The impact of relocation on income and livelihood of tribes in Similipal tiger and biosphere reserve, India. *Environ Manage* 56(2): 420-32.

Panwar, H. S. (2003). Historical overview of relocation issues in conservation. In Proceeding of National Seminar-cum-Consultation on Relocation of People from Protected Areas in India: Policy and Process. Council for Social Development, New Delhi, September 18, 2003.

Rangarajan, M. and Shahabuddin, G. (2006). Displacement and relocation from protected areas: Towards a biological and historical synthesis. *Conserv Soc* 43: 359-378.

Senapati, N. and Sahu, N. K. (eds.) (1967). Odisha district gazetteers, Mayurbhanj. Printed by the Superintendent, Odisha Government Press, Cuttack.

Swain, D. (2009). Role of India's tiger reserves in the protection of origin/catchment of water courses: A case study of the Similipal Tiger Reserve. XIII World Forestry Congress, Buenos Aires, Argentina, 18 – 23 October 2009.

Vasu, N. (1911). The archaeological survey of Mayurbhanj, Vol. I, Published by Mayurbhanj state, Odisha, India.

Chapter 25

Similipal Biosphere Reserve and Tribal Livelihood

Sundara Narayana Patro

President, Orissa Environmental Society
e-mail: snpatro11@gmail.com

Similipal Forest complex in Mayurbhanj District of Odisha State under the Indian Union is a pristine heritage of unique biodiversity. It forms the northern extreme of the Eastern Ghats. The Union Ministry of Environment and Forests notified it as the eighth biosphere reserve (BR) of the country in June 1994. The forest complex is rich with myriads of plant and animal species. The population of various tribes is sizable inside and outside it. Their economy, culture, and lifestyle are intricately linked to the habitat and resources of the forest.

The density of floral and faunal species is decreasing fast as an impact of developmental activities like mining, construction of roads, plying of vehicles, annual mass hunting ritual ("akhand shikar"), forest fire, over-grazing, clandestine tree felling and removal of firewood, collection of minor forest products like sal leaf, kendu leaf, mohua flowers, *etc.* beyond the carrying capacity. Eco-development activities like apiculture, floriculture, horticulture, silviculture, sabai grass cultivation, plantation of medicinal and economically beneficial plants, conservation of genetic species, food processing, bamboo and cane crafting, cottage industries, eco-tourism, and various other eco-friendly activities, market linkage, more particularly in the transition zone, need to be undertaken to promote the biodiversity as well as to protect the livelihood security and economy of local people including the tribes dwelling inside and in the periphery of this forest complex.

INTRODUCTION

'Simlipal' derives its name from the abundance of Semul (red silk cotton trees) that bloom in this forest. According to the other school of thought, the name is derived from Seuli (it is the local Odia name of moss) that covers the exposed rocks here. Similipal forest is in the eastern extreme of the Eastern Ghats and is coming under the Chhotanagpur biotic province of the Mahanadian bio-geographical region. It lies between 20°17′ – 22°10′N and 85°57′-86°47′E. It is a complex of thick and green forests, extensive grassy lands, and meadows. It claims to be the pride of Eastern Ghats, because of its varied topography, dense forests, rich biodiversity, and endemism. The Similipal forest complex and the adjoining areas with good forest coverage as an ecological niche play the pivotal role in maintaining the ecological balance, thermal balance, and the water cycle (Patro, 2017)

The average mean elevation of the Similipal plateau is 900 meters. The highest peaks are Khairiburu (1178 meters), and Meghasani (1158 meters). It is the biggest watershed of northern Odisha, giving origin to major rivers like Budhabalanga, Khairi, Salandi, Palpala, and innumerable perennial rivulets and streams meandering through the forest. Amongst several waterfalls Barehipani (217 meters) and Joranda (181 meters) are enchanting. The forest area experiences relatively higher annual precipitation of over 200 cm spread over about 135 days. There is a marked variation of temperature range between the central and southern regions. In winter there is the occurrence of frost valleys in central and south Similipal. Similipal forest complex has a unique mixture of tropical evergreen, moist-deciduous and dry deciduous forest types with numerous sub-types, varying in height from 40 MSL to 1068 MSL, thus providing an unparalleled array of habitat types.

Biodiversity Richness

Interestingly there is a degree of resemblance of floral and faunal composition of Similipal on the Eastern Ghats with those of the Western Ghats and North-East India while being a representative eco-system within the Mahanadian bio-geographic region. About 55 species of mammals, 29 species of reptiles, 231 species of birds, and 1076 plant species are listed in this plateau. Among the fauna tiger, leopard, elephant, gaur, sloth bear, hyena, ratel, sambar, mouse deer, chital, chousingha, deer, hill myna, peacock, giant squirrel, flying squirrels, Indian tree shrew, reptiles like king cobra and other snakes, *etc.* are appealing. Similipal is also the abode of the rare black and melanistic tiger. In its river and stream waters, it has marsh crocodile, mahaseer (*Tor putitora*) fish, and numerous other aquatic species. Among the flora Sal (*Shorea robusta*) is the climax tree species. Champa (*Michelia champaka*), Arjuna (*Terminalia arjuna*), Asana (*Terminalia alata*) are some of its associated species. The rich orchid flora is the indicator of Similipal's eco-health and naturalness.

Conservation and Management

The Similipal forest enjoyed the status of a sanctuary from 1979 with an area of 2,200 sq km. A national park was created in 1980 with an area of 303 sq km only within the sanctuary area, which was later enhanced to 845.70 sq km in 1986. The Similipal Tiger Reserve created in 1956 was included in the first list of 9 Project Tiger areas launched in India in 1973. The Mayurbhanj Elephant Reserve, notified in 2001, is one of the three elephant reserves created in Odisha. The Mugger Crocodile (*Crocodylus palustris*) scheme was started in 1979 at Ramatirtha near Jashipur. The eighth biosphere reserve of the country was declared in Similipal forest on 22.06.1994 taking a vast stretch of forest land measuring 5569 sq km. UNESCO included it in the World Network of biosphere reserves in 2009. As per the provisions of the national park, no human habitation is allowed inside it. The State Government has successfully relocated villagers from Jenabil, Baghagahan, Upper Barhakamara, Kabatghai, and Jamunagarh inside the Similipal Tiger Reserve area between 2010 and 2016. From Bakua 61families have volunteered to move out.

Biosphere Reserve (BR) is an UNESCO program for the conservation of representative parts of natural and cultural landscapes extending over large areas of terrestrial or coastal/marine ecosystems or a combination thereof. These are intended to promote conservation of biodiversity and also to promote alternate livelihood for man and preserve cultural values. In India, the BR Programme was launched in 1986. The BRs are different from wildlife sanctuaries and national parks in the following aspects-its emphasis are on overall biodiversity and landscape rather than some specific flagship species. It takes into account the overall developmental activities with a special focus on conservation. It ensures security to tribal livelihood by sustainable use of natural resources through their participation. Local tribes are the sacred trusts of the natural resources of the forests. Without their association the forest resources cannot be conserved; they have the traditional knowledge and skill about the management of such resources.

The densely forested Similipal Biosphere Reserve area constitutes the core, the buffer, and the transition zones. There are 4 villages in the core and 65 villages in the buffer. The transition zone covers 1200 villages with a population of about 5 lakhs. The core (1194.75 sq km), buffer (1555.25 sq km), and transition (2819 sq km) together compose a vast area of 5569 sq km. The functions of three zones in a biosphere reserve are defined. The **Core zone** allows management and research activities that do not affect natural processes and wildlife. The core zone is to be kept free from all human pressures external to the system. The **Buffer zone** adjoins the core zone. Only those activities which protect and promote the core zone are allowed here. It includes research, education, limited recreation, tourism, controlled grazing, fishing, *etc.* The **Transition zone** is the outermost part of the biosphere reserve. Here conservation knowledge and management skills are applied primarily to foster alternate livelihood and reduce dependence on consumptive use of the forest. Eco-development activities are permitted in this zone to meet local people's needs only but not for commercial purposes. Now there is the inclusion of Hadagarh Wildlife Sanctuary in Similipal Tiger Reserve (STR) for the benefit of better wildlife management. An area covering 25 sq km out of the total area of 179.310 sq km of Hadgarh sanctuary is within the STR. Moreover, there is frequent movement of elephants and tigers from STR to this Hadagarh sanctuary.

The threats to the Simlipal forest are loss of diversity due to the collection of timber and fire-wood; indiscriminate vegetation clearance and tree felling; forest fire; Akhanda shikar (annual mass hunting ritual) and animal hunting; *etc.* Due to loss of density and fragmentation of the forest and increased human pressure, the man-animal conflict is increasing. An account of the death of human beings by elephants is stated to be 93 in 2012-13, 93 in 2013-14, 70 in 2014-15, 89 in 2015-16, 66 in 2016-17, 105 in 2017-18, and 85 in 2018-19. On the other hand, the death of elephants was 82 in 2012-13, 70 in 2013-14, 53 in 2014-15, 85 in 2015-16, 75 in 2016-17, 78 in 2017-18, and 93 in 2018-19.

Eco-development

Similiapal hill complex is a fascinating cultural landscape with numerous tribes co-existing with nature. The tribal population comprises of 73.44 per cent, the scheduled cast population is of 5.21 per cent and the population of other castes is at 21.35 per cent. The tribes of Similipal forest are Kharia, Bhumija, Bathudi, Kolha, Gonda, Santhal, Sabar, Paroja, Kondha, Juanga, Mundari, and Mankadia. The forests and the biodiversity resources to a great extent sustain the livelihood of these people. Emphasis needs to be put on suitable programmes of eco-development in the transition zones and to some extent in the peripheral zone to promote various alternative livelihood options, and thereby reducing people's pressure on the forests.

There is a local committee chaired by the Director, Similipal Biosphere Reserve, and comprising of representatives of local district level officers of various departments, non-government organizations and local communities, which formulates suitable action programmes on eco-development activities and tribal livelihood security. Some of the projects undertaken are the construction of water harvesting structures, improvement of irrigation channels to fields, digging and renovation of ponds and pisciculture, habitat improvement programmes (grassland and wetland improvement, water retention structures, weed eradication, *etc.*), the supply of mosquito nets and solar lamps, health camps, veterinary health camps, anti-malaria campaign, immunisation of people, immunization of domestic animals, vaccination of animals against foot and mouth diseases. The facility of a revolving fund for micro-credit has been extended to selected community groups. Training on sericulture, sabai rope making, improved agricultural technique, bee keeping, orchid collection and cultivation, eco-guide, *etc.* are the other initiatives. Sabai rope making units, sal-leaf plate-making units, food processing units have been supplied to the villagers (Satapathy, 2010). Tasar silk under the royal patronage of Bhanja dynasty had flourished in Mayrbhanj. The salubrious climate of the Similipal forest is conducive to the cultivation of Tasar silkworm. It encompasses the cultivation of tropical silkworm *Antheraea mylitta* Drury and its eco-race Modal which produces the heaviest cocoon among all the sericigenous Lepidoptera inset of the globe (Kar, 2017). Besides Modal, Jata and Nalia are two other eco-races found in Similipal. Later Daba and Sukunda varieties were also introduced. Sal. Asana and Arjuna are the ideal food plants of tasar silkwork plentily grow naturally in Similipal. The tribal in Similipal have the necessary skill of its cultivation. Similipal is known for the quality honey and the Bagua variety is the best one (Tudu and Mohapatra, 2020). The Odisha Forest Development Corporation is marketing honey with the brand name 'Khairi' (named after the famous domesticated tigress Khairi in Similipal). The Corporation was lobbying to get the Fast Moving Consumer Goods (FMCG) tag to Khairi brand honey. A honey production unit is already in operation in the foothills of Similipal at Jashipur. The Corporation had set a target of 250 tonnes honey production by utilizing services of the local honeybee farmers. The Kharia and other tribes have the necessary skill of honey collection from the wild.

The long term programs would comprise of an in depth study of the status of biodiversity resources, eco-systems, and ecological processes in Similipal to establish a bench mark; a critical study of the socio-economic setting in the biosphere area, and exploration of the various alternate livelihood options; and nurturing the process of organizing the local communities into eco-development committees and formation of green brigades to build up proper stake-holding of the natural-biological resource. The short term interventions would consist of measures to build up awareness on the destructive practices which need to be curbed, to explore the potential and the prospects of eco-tourism, to channelize the water flow out of Similipal to agricultural fields, to support suitable micro enterprises and microcredit to self-help groups, and to mount a vigorous program for controlling malaria and cattle disease.

The forest plays the pivotal role of climate change impact reduction as the carbon sink and also is the eternal source of livelihood support of the local people. The local community can be associated with conservation practices based on the local natural resources available. Cottage industries, value addition of products, skill development of local people, development of market net-work, cultivation of local medicinal, fodder and other valuable plants, rearing of domestic animals for cultivation, milk, meat, *etc.* would go a long way in conservation of local resources and their wise use. Nursery raising, plantation and post plantation care, cultivation of market-driven economically beneficial medicinal plants, non-timber forest products (NTFP) collection (Rout *et al.*, 2010), processing and storage-tamarind, mango, turmeric, arrowroot, ginger, mohua, kendu, broom, food and fodder, medicinal plant parts, land use practices-agricultural, horticultural and fodder crops over the hills and plain land, traditional skills and economic activities-leaf plate making, rope making, bamboo and cane crafting, plant sap collection for country liquor, honey collection, food processing, hill agriculture, popularization of organic composting to save fuel wood and fire wood species and cow dung are some of the activities that best suit to conservation practices. Development of education and awareness programs ideally should be launched on continuous basis.

Integrated development of the forest complexes is intricately linked to the economic development and wellbeing of the local population-most of them are tribal poor, who have been historically protecting the resources since time immemorial. Kendu leaf trade sustains lakhs of tribal people who work as pluckers. Similarly, Sal leaf cup and plate making is a low investment project that gives self-employment to lakhs of people in Mayurbhanj and adjoining Kendjhar and Balasore districts. Sal seed collection and mohua flower collection, medicinal plant parts collection is some of other occupations of the tribes. Wild mushrooms like Russula, Termitomyces, Volvarlella, Pleurotus, *etc.* are being used as source of food and medicine by the local tribes (Singdevsachan and Thatoi, 2017) Odisha some medicinal plant species have been identified that are included in the IUCN Red Data Book. The list includes two critically endangered species namely *Saraca asoca* (Ashoka-Odia), and *Symplocos racemosa* (Lodha-Odia). Among endangered ones the species are *Gloriosa superba, Blepharispermum subsessile, Rauvolfia serpentine, Uraria picta, Mucuna gigantean, Piper longum, Cordia macleodii, Oroxylum indicum, Pterocarpus marsupium, Stereospermum chelonoides, Xylocarpus granatum*. Among vulnerable ones the species are *Hedychium coronarium, Thalictrum foliolosum, Caesalpinia digyna, Embelia tsjeriam-cottam, Gardenia gummifera, Celastrus paniculatus, Embelia ribes, Gnetum ula, Operculina turpethum, Paederia foetida, Pueraria tuberose, Scindapsus officinalis, Stemona tuberose, Aphanomixis polystachya, Cerbera odollum, Crataeva magna, Cycas circinalis, Garcinia xanthocymus, Litsea glutinosa, Mesua ferrea, Polyalthia simiarum, Salvadora persica, Schrebera swietenioides, Strychnos potatorum, Zanthoxylum armatum, Zanthoxylum rhetsa* (Anonymous 2007). Most of these species are available in the Smilipal forest. These medicinal plants can be cultivated, collected, and promoted with good market linkage to ayurvedic preparation institutions. Wildlife tourism and tribal tourism are some areas where the local tribes could be associated, but it should be strictly under the guidance of the management authorities so that habitat does not tamper and conservation is not affected. Tribes lead a very simple lifestyle and sustain forest resources for their food and nutrition. In addition to solid food, the tribes drink liquors such as handia, mahuli and eat fermented mahua cakes and other such preparations having anti-oxidant properties. Malnutrition is a chronic problem with them. A study needs to be conducted on food habits of the tribes and their nutrition status so that they can be educated for better nutrition and health.

The sacred groves, Jahira for example, are the traditional cultural entities of micro-ecosystems that address tribal way of worshipping nature. In Jahira Sal and other associated plants are not removed or cut in the name of god. The offerings and celebrations centre on the life cycle of sacred trees-during unfolding of foliages, flowering, fruiting, *etc.* Happily, the government is encouraging with financial support to maintain such cultural entities. This leaves scope to ecologists and anthropologists to understand more deeply about such conservation practices of ecological wisdom.

CONCLUSION

Because of its bio-geographical situation, geological features, the international recognition as one of the first nine prime areas for tiger conservation and for being one of the world network of biosphere reserves of India, the scientific value of Similipal makes it a paradise for scientists pursuing studies on biodiversity, and many features are believed to have remained unexplored yet. The discoveries of new paddy species and the identification of over 500 species of medicinal plants, 94 orchid species out of about 120 in Odisha State are a few of the examples of the 'gene pool reserve' in Similipal and the vast opportunities for biotechnological research and applications in future. *Oryza officinalis*, a species of paddy known from Kerala was first collected in Odisha from Similipal near Khejuri hills in the late 1980s. The species is rare and sporadic, not used for any purpose locally. Another species of paddy, *O. granulata* previously recorded from Jeypore in south Odisha has also been recorded in Similipal.

Organizations and institutions of repute have remained associated with studies on habitat, eco-system and wildlife of Similipal forest complex. More studies to update the biodiversity status and the carrying capacity of the Similipal forest to sustain the wildlife, more particularly the key-stone species, livelihood support of local tribal and non-tribal communities, and the anthropogenic pressure are warranted. Based on the studies appropriate conservation and management strategies have to be worked out. As Similiapl BR is considered one of the rich natural habitats to support tiger and other key-stone species a bold and robust decision have to be taken to declare the entire area inviolable for wildlife. This will also help protect the livelihood of the tribes who are used to live with wildlife from time immemorial.

REFERENCES

Patro S.N. (2017). Similipal Biosphere Reserve: Biodiversity Conservation and Management, in 'Facets of Similipal', North Orissa University, Baripada, Odisha, pp. 180-186.

Rout, S.D., Panda, S.K., Mishra, N. and Panda, T. (2010), Role of Tribals in Collection of Commercial Non-Timber Forest Products in Mayurbhanj District, Odisha. *Studies of Tribes and Tribals,* (8) 1, 21-25.

Satpathy, A.R. and Sahu, U.N. (2010). The Role of Sabai Grass Industry in The Economic Development of Mayurbhanj District of Odisha (India). *International Journal of Applied Agricultural Research,* (5)2,. 221–242.

Singdevsachan S.K. and Thatoi H. (2017). Biodiversity and Biotechnological potential of Wild Mushrooms from Similpal iosphere Reserve, Mayurbhanj, Odisha, in 'Facets of Similipal', North Orissa University, Baripada, Odisha, pp. 1-13.

Tudu, F., and Mohapatra, R. (2020). Forest products collection of hill kharias of mayurbhanj of odishain eastern india: A recent study. *International Journal of Psychosocial Rehabilitation*. 7(6), 327-333.

Chapter 26

Ensuring Biodiversity Conservation and Local Livelihood Outside Protected Areas in Odisha: A Critical Analysis

Sushil K. Dutta[1], Pratyush P. Mohapatra[2], Prasad K. Dash[3], Cuckoo Mahapatra[4] and Arun K. Mishra[5]*

[1]Department of Zoology, Assam Don Bosco University, Tapesia garden, Kamarkuchi, Sonapur, Tepesia – 782 402, Assam
[2]Zoological Survey of India, Central Zone Regional Centre, Vijay Nagar, Jabalpur – 482 002, Madhya Pradesh
[3]Odisha Biodiversity Board, RPRC Campus, Nayapalli, Bhubaneswar – 751 015, Odisha-
[4]P.G. Department of Zoology, North Orissa University, Takatpur, Baripada – 757 003, Odisha
[5]Divisional Forest Officer, Rairangpur Forest Division, Mayurbhanj, Odisha
**e-mail: pratyush.m@zsi.gov.in*

The forest ecosystems of Odisha beyond the Protected Areas are equally biodiversity-rich. Since time immemorial, many such sites are being protected and conserved by the local communities for their sustenance and livelihood. Keeping in mind people's participation in the protection and conservation of forests outside the Protected Areas, there are also various provisions at the state and central levels. The present study throws light on some important aspects of community participation and biodiversity conservation in Odisha and presents a bird's eye view of the existing practices. It was found that there are enough opportunities for the conservationists to explore and prioritize biodiversity-rich areas beyond the Protected Areas. Initiatives taken by the communities need to be encouraged and traditional conservation practices to be given priorities to ensure livelihood security.

INTRODUCTION

Humans have influenced the ecosystem more than any other species on earth by steadily transforming the ecosystems into depauperate systems (Vitousek *et al.*, 1997). The anthropogenic impact leads to an imbalanced biogeochemical process in the form of global warming, climate change, irregular rainfall *etc.* Hence, the relationship between humans, biodiversity and ecosystem functioning has emerged as a central issue in environmental sciences since the past few decades (see Loreau *et al.*, 2001). The relationship between humans and biodiversity is as old as human evolution and with the increasing use of bioresources, conservation has been an integral part of diverse cultures in different ways. The traditional worship practices show the symbiotic relation of human beings and nature. Indigenous communities all over the world lived in harmony with nature and conserved its valuable biodiversity. Whether in the name of sacred groves or reserve forests or in later time as Protected Areas, the goal is to save an entire ecosystem. Even in our sacred literature and folklores, many species-level conservation efforts are seen in the form of species associated with Gods and Goddesses, sacred animals, sacred plants, *etc.* Even before any formal laws to save species and to prevent unnecessary persecution, many animals were designated as dangerous, so that humans do not dare to go near them.

The Indian subcontinent sustains the second largest human population in the world and it is home to one of the richest and most diverse biological communities (Singh 1985). Growth in the human population and changes in socio-economic values triggered a decline in charismatic wildlife (Tigers, Elephants, Gharials, *etc.*) populations and in some instances led to (local) extinction of native flora and fauna. Hence during the latter half of the twentieth century, the Government of India took pro-active measures to restrain further deterioration. The Wild Life (Protection) Act, 1972 was framed and under this Act, Protected Areas (PAs) were made free of human disturbances and were managed to prevent further decline of wildlife. During the twenty-first century, the National Wildlife Action Plan (2017-2031) emphasizes people's participation and their support for wildlife conservation (MoEFCC, 2017). As per the plan, the Biological Diversity Act 2002 and subsequently the Biological Diversity Rules, 2004 were formulated. As per the Act and the Rules, the National Biodiversity Authority and State Biodiversity Boards were formed with a herculean task of conservation of biological diversity, sustainable use of bioresources, and focused on conservation and judicious equitable sharing of benefits out of the commercial utilization of bioresources in their respective jurisdictions. People's participation in biodiversity conservation was promoted and local communities and their perspectives towards saving biodiversity were given priority. Unlike the Wild Life (Protection) Act, 1972, the Biological Diversity Act, 2002 has a vast scope in people's participation and can safely be called a pro-community Act which not only safeguards biodiversity but also has provisions for benefit sharing.

In the global arena, India is a party to a multilateral treaty- the Convention on Biological Diversity (CBD) and the Convention has three main goals; conservation of biological diversity, sustainable use of its components and fair and equitable sharing of benefits arising from genetic resources. Mainstreaming biodiversity in sustainable development can only be achieved by giving due recognition to cultural diversity, and the rights of indigenous peoples and local communities. With the acquired traditional knowledge available with the local community, sustainable customary use and practices, innovations and values, the indigenous peoples and local communities are the key actors in the sustainable management of biodiversity. They are also key players in building resilience in the face of the current social and environmental changes. Action and initiatives at the local and grassroots levels are vital for the achievement of the three objectives of the CBD and the Aichi Targets.

The present study throws light on the community conservation initiative outside the Protected Area network of the state and provides a critical analysis of the existing scenario. Furthermore, a set of recommendations have been suggested for encouraging community conservation efforts.

Study Area

Odisha has rich biological and cultural diversity. The state is home for 103 species of mammals, 473 species of birds, 131 species of reptiles, 29 species of amphibians, 600 species of fishes (both marine and freshwater) (Dash *et al.*, 2015; Mohapatra *et al.*, 2017a and b; Mohapatra *et al.*, 2019). Among the invertebrates, more than 300 species of butterflies, 90 species of odonates, 48 species of marine molluscs, 12 species of blattaria (Arthropoda: insecta), 14 species of Dermaptera (Arthropoda: Insecta), 31 species of Isoptera, 32 species of land Molluscs, 48 species of Nematodes and 46 species of Oligocheat have been recorded from the state (Nair, 2011; Dash *et al.*, 2015). The flora of Odisha consists of 6000 species of plants. This includes 3000 species of angiosperms, 176 species of pteridophytes and 22 species of gymnosperms which includes 313 species of wild relatives of crop plants and 450 species of medicinal plants. Apart from higher plants, the lower group of plants also significantly contributes to the floral richness of the state. This includes 300 species of bryophytes, 270 species of lichens, 1000 species of both freshwater and brackish water algae (Saxena and Brahmam, 1994-1996; Dash, 2012; Upreti, 1996; Dash *et al.*, 2015). The fungal diversity consists of 800 species of macro and microfungi. In total 116 species of plants are covered under threatened and endangered categories and 21 species of plants are endemic to Odisha (Dash *et al.*, 2017).

As per the Rodgers *et al.* (2000) classification, the state falls under four biogeographic provinces and as per vegetation, it is divided into four major forest types, with the most dominant – moist deciduous Sal forest. Similarly, as per global ecoregion classification by Wikramanayak *et al.* (2002), the vegetation diversity of the state can be categorized as Lower Gangetic Plains moist deciduous forests, Chhota-Nagpur dry deciduous forests, Eastern highlands moist deciduous forests, Northern dry deciduous forests, Odisha semi-evergreen forests and Godavari-Krishna mangroves ecoregions. Harboring five forest types out of 16 found in India, the landscape is naturally low in species diversity and endemicity in comparison to the eastern Himalayas and the Western Ghats but the wet zones scattered across the state serve as refuges for many interesting floral and faunal species. Apart from the above four types of forests, some subsidiary and several types of vegetation and special plant groups also occur in the state which is, Shola Forests, Grasslands, Hydrophytic vegetations and Coastal sand-dune vegetation.

The state has a protected area network of 19 wildlife sanctuaries, one national park, one biosphere reserve (Similipal), two tiger reserves (Similipal and Satkosia Tiger Reserves) and three elephant reserves (Mayurbhanj ER, Mahanadi ER, and Sambalpur ER). The protected area spans over 10.37 per cent of the total forest area and 5.36 per cent of the total geographical area of the State. Apart from these, community protected forests are also scattered across the state in the name of community forests, sacred groves and village forests (= gramya jungles).

Results and Discussion

Human species as a part of the ecosystem has left foot-print in all the habitat types for livelihood and day-to-day activities. The communities those are mostly dependent on the local biodiversity have a greater and direct dependency on the forest for traditional medicine, wild food, shelter, and a long-list of ecosystem provisioning they get from the forest. On the other hand, many people are also dependent on the forest for commercial purposes and in some cases, they are perceived as parties exploiting natural resources. However, the community who truly understands the value of biodiversity deliberately conserve it in a sustainable way for the long-term utilization of forest products. Furthermore, there are domesticated and non-domesticated plants and animals endemic to a particular region and hence provide an opportunity for systematic *in situ* maintenance of the genetic resources.

Sacred Groves

Sacred groves are patches of primary and natural forests, protected through socio-cultural laws by the local communities near their villages and are always dedicated to certain deities. This practice of conserving such groves is linked to society since the beginning of the human settlement. There are several hundreds of sacred groves in Odisha, most of which are still not known to the forest managers across the entire state. A perusal of available literature on biodiversity of sacred grove indicated scattered and incomplete information in these SGs of Odisha except for few (Mohanta and Adhikari 2012; Palei *et al.*, 2012, Panda *et al.*, 2014; Pradhan *et al.*, 2016 a, b and c, Mohanty *et al.*, 2016; Pradhan *et al.*, 2017, Mohapatra *et al.*, 2017; De and Palita 2018; Panda *et al.*, 2020). Sacred groves are typically well preserved despite massive land-use changes in the adjacent areas and are protected by the indigenous community for religious and spiritual beliefs (Pradhan *et al.*, 2016c). In India, more than 1,00,000 sacred groves have been documented and it is a fact that many of these groves are losing their sanctum sanctorum due to cultural changes (Ormsby and Bhagwat 2010). In Odisha sacred groves are also known by several colloquial names such as *Jaher, Thakurrama, Pirha, Demul* and Malhotra (1998) reported about 322 sacred groves from the state but as per Pradhan *et al.* (2016c), more than 1500 sacred groves of varying sizes are there only in western Odisha. The authors also emphasized that even these sacred groves are relatively smaller in sizes their contribution towards biodiversity conservation and carbon sequestration potential must be recognized. Biodiversity studies in eighty such sacred groves of

Bonai forest division, Odisha by Mohapatra *et al.* (2017) reported 306 species of plants (107 species of trees, 118 species of herbs, 14 species of shrubs, 47 species of climbers and 10 species of orchids, 4 lichens, 7 bryophytes and 45 species of fungi) and among fauna 25 species of mammals, 133 birds, 22 reptiles, 13 amphibians, 122 butterflies, 26 odonates, 27 arachnids and 31 species of other invertebrates were recorded during the study period. The size of sacred groves ranges from 0.01 Acre to 7.9 Acre, covering about 2.2 per cent of the total geographical area of the division. Ecological services rendered by sacred groves need to be highlighted and community protecting large forest patches needs to be encouraged.

Figure 26.1. Sri Sri Baba Jhadeshwar Pitha, Borigaon, Ganjam. This sacred grove is just 1 ha but is home for an endemic species of lizard (*Hemiphyllodactylus* sp.) along with forty more herpetofaunal species.

Community Conservation Efforts in Odisha

Beyond the PAs, there are many landscapes rich in biodiversity but are either rapidly depleting or are vulnerable due to 'developmental activities. On the other hand, some habitats and species are dedicatedly conserved by local communities through self-initiated institutional arrangements, which are popularly known as Community Conserved Areas (CCAs). In Odisha, 27 such places have been documented spanning across 18 out of 30 districts of the state (Annonymous 2007; Dash *et al.*, 2015). Most of the CCAs are in forest ecosystems except for a few, such as Baghamunda and Praharajpur (wetland), Bhetnoi-Balipadar (blackbuck habitat), Mahalasarua, Mangalajodi and Rugudipali (wetland and waterfowl) and Rushikulya (beach and coastal wildlife). Among these CCAs, Mangalajodi (Khurdha district), Bhetnoi and Pakidi (Ganjam district) received the most prestigious wildlife conservation award of Odisha, the Biju Pattnaik Award for their outstanding efforts in conserving wildlife.

A total of 13,800 Vana Surakshya Samities (VSS) and 568 Eco-Development Committees (EDC) have been formed in the state and these are managed under the Joint Forest Management (JFM) policy. Furthermore, as per Section 28 of the Indian Forest Act 1927, there is provision for declaring village forests (VF) and as per the rule, the villages get powers similar to the forest department. Although forest patches adjacent to the villages are factually known as 'gramya jungle' (= village forests), none of these have got formally declared under the Village Forest Rules of 1985 and hence these are not been formally recognized under the Odisha Forest Act (Mohanty, 2011). Even in the country few states namely Uttarkhand, Karnatak and Mizoram have designated Village Forests. Mohanty (2011) in his report strongly recommended the implementation of Village Forests which can be a strong category to support Community and Conservation Reserves. Many potential sites are there in the state and the local communities are also protecting these forests for several decades, which can be seen as self-motivated efforts. The village forest category entails handing over government-controlled reserve forests to local villagers for conservation and sustainable use and hence suits the purpose well. There are cases where many JFM villages have been demanding that theirs may be declared as village forests. Even the MoEF steering committee for the Eleventh Five Year Plan (2007-2012) had strongly recommended about the village forests for giving legal backing to existing JFM sites as well as other initiatives of the communities towards conservation of forests, not much progress has been seen. The Forest Rights Act (FRA), 2006 has also provision for granting Community Forest Right (CFR) to the tribal and forest-dependent communities but because of several reasons, no such rights have been delegated to any local tribal community yet.

Figure 26.2. The First Biodiversity Heritage Site of Odisha–Mandasaru, a Gorge in Raikia Block of Phulbani.

After the promulgation of Biological Diversity Act, 2002, the Odisha Biodiversity Board with the physical support of Panchayat Raj Department has constituted more than 6798 Biodiversity Management Committees (BMCs) at GP level, 114 BMCs at Urban level, 68 BMCs at the block level and eight BMCs at the district level and for all the BMCs at GP level have prepared 181 People's Biodiversity Registers (PBRs) and 30 district level PBRs are under progress. Furthermore, Mandasaru, a gorge in Raikia block of Phulbani forest division of Odisha was declared as the first Biodiversity Heritage Site of the state by the Odisha Biodiversity Board in 2019. Strengthening and capacity building of BMcs towards self-sustenance, including levy charges against the commercial utilization of biological resources from their respective jurisdictions under Access and Benefit Sharing Agreement (ABS) guidelines, 2014 is far from reality.

Local Community and Biodiversity Conservation

Several tribal and non-tribal communities depend on their local biodiversity for day to day activities, as the source of livelihood and other ecosystem provisioning such as traditional medicines, materials for house construction *etc.* In Odisha, the communities cultivate a diverse crop species with varying seasons to meet their food and economic requirement throughout the year. This was supplemented by wild food collections during lean periods. It has also been observed that based on the traditional knowledge of the community, there are vast scopes for improving food production and resolving health and nutritional insecurity. Hence we recommend *in situ* conservation of these valuable resources through the communities for ensuring food security and germplasm conservation centers can be established at the village level. Conventional plant breeders and biotechnologists should undertake scientific validation of these crops and plants to make their value explicitly known. This knowledge can be utilized for crop improvement and bio-prospecting. To achieve the goal, an effective community-level initiative in curbing challenges like degradation of forests, land-use changes, invasive alien species, and erosion of traditional knowledge that greatly affects availability and access to wild biodiversity is a prerequisite. Similarly as noted, women's knowledge of wild edible plants is exemplary and it increases the habitat value by restoring wild vegetation. Traditional knowledge of the use and management of forests still has a pivotal role to play in forest management today. This is because traditional practices are believed to be often favorable towards conservation and sustainable use. Hence valuation of different ecosystem services provided by a particular landscape should be a prerequisite before carrying out any developmental activities.

Living in Harmony: A Novel Perspective to Deal with Conflict with Wild Animals

Several communities outside Protected Areas are living in harmony with wild animals and there are ample examples where conflict with wild animals is dealt with tolerance. To cite some examples, the case study of Jeypore local communities living with sloth bears, villagers in Buguda, Bhetnoi-Balipadar, and Khallokote areas of Ganjam district living in harmony with blackbucks, villagers in Pakidi living with Peafowls and people of Humma and Maneswar in Sambalpur, Katramal in Kandhamal are conserving Mahaseer fishes in aquatic ecosystems. Similarly, the freshwater turtles and tortoise in many ponds and temples covering different districts of Odisha are being worshiped by the local communities for ages. Although crop depredation due to wildlife is prevalent in these areas, the local communities have a higher level of tolerance towards such loss.

Figure 26.3. Most of the Mountain Ranges like Mahendragiri in Odisha are Potential Sites with many Unique Species of Flora and Fauna, which Need Conservation Intervention by the State Forest Department.

Traditional Knowledge: Challenges and Threats

With a rich diversity of tribal communities in the state, the Traditional knowledge practiced by various indigenous communities of Odisha has enormous potential for future bio-prospective applications. Furthermore, as per the Biological Diversity Act, 2002, fair and equitable sharing of benefits arising from the use of such biological diversity and associated traditional knowledge has provisioning to benefit the local community when such products are used commercially and even for research purposes. However, as per Nair and Harimurthy (2010) in their critical assessment of ABS the authors raised some very important concerns regarding the assumption that monetary incentives will motivate communities to conserve their biological diversity and associated traditional knowledge and practices. While citing India's most famous ABS case relating to the commercialization of the Kani tribe's ethnobotanical knowledge of ArogyaPacha (*Trichopus zeylanicus*), Nair and Harimurthy (2010) stressed on various key challenges which may lead to a serious negative impact on that species. Another crucial issue arises when ABS enters to the legal provision of intellectual property rights, that only recognizes private ownership, and in due course, it may lead to a conflicting situation with other communities.

In Odisha's scenario, the state Biodiversity Board has taken steps to address the issue by communicating to various companies utilizing the biodiversity from different parts of the state and further progress in forming BMCs in such areas has been achieved by the board. It is also pertinent here to mention that the Board is also in the process of listing the exclusive traditional knowledge and practices with local communities. The board also sensitisatizes Ayurvedic drug manufacturers, traders, forest managers, officials of line department of state govt., PRI members, print and electronic media, scientists and academicians of the Universities and research institutes of the state in regular intervals on key provisions of BD Act, 2002 especially on ABS and IPR issues to make the BMC members self-reliant.

Figure 26.4. A Kondh Woman with MFC Collected from Nearby Forests.

The Way Ahead

We argue that the local conservation efforts and sustainable utilization of resources can be achieved by empowering the BMCs at par with other rules implementing institutions and by conferring the power to regulate unsustainable use of natural resources. Information from People's Biodiversity Registers needs to be analyzed to prioritize species and areas needing conservation intervention. In many cases, the local conservation efforts need scientific intervention for better management of species. Local communities should be incentivized for maintaining associated traditional knowledge and such practices can be conserved by integrating modern techniques. Steps should be taken by government agencies for formal collaboration with R and D institutions, academicians, NGOs and other line departments towards safeguarding the biodiversity of the state more transparently. Only the 22 animal breeds of the state have been recognized and registered by the National Bureau of Animal Genetic Resources (NBAGR) and it is high time to register all the local plant verities and wild crop relatives under PPVFRA and the BMCs have to be given the ownership. A large chunk of funds is available under the Gram Panchayat Developmental Plan (2018 guidelines) towards conservation and protection of biodiversity and strengthening of BMCs. The same can be utilized for the financial empowerment of the BMCs. Necessary certificates should be issued to the traditional knowledge holders and practitioners after validations of their claims of curing respective diseases and ailments and their knowledge should be recognized and registered by the National Biodiversity Authority. NTFPs have been collected by the traders and large pharmaceutical companies without giving any collection fees to the stakeholders. Hence ABS issues must be dealt with priority. There are also prevailing threats due to the increasing commercialization of biodiversity products as per the demand of the global market, which has been carried out without sharing the benefits with the local community. Although steps have been taken by the State Government, it needs serious attention and effective implementation of the existing Acts and rules related to biodiversity conservation. Traditional knowledge and management of forests still have a vital role to play in forest management today. Hence valuation of different ecosystem services provided by a particular landscape is a prerequisite before any developmental activities.

ACKNOWLEDGMENTS

We thank Prof. P.K. Chand, the Vice-Chancellor of North Orissa University for giving us the opportunity to write this article. Thanks are also due to Prof. H.N. Thatoi for the necessary editing of the manuscript. PPM would like to thanks Dr. K. Chandra, Director, Zoological Survey of India and Dr. S. Sambath (Officer-in-Charge and Scientist-E) ZSI, CZRC, Jabalpur for encouragement. PKD thanks Sri M. Mohan IFS (Chairman, OBB) and Dr. P.C. Panda (Member Secretary, OBB) of Odisha Biodiversity Board for guidance. CM thanks H.O.D., P. G. Department of Zoology for necessary facilities. Mr. Rabindranath Sahu is acknowledged for providing the photograph of Sri Sri Baba Jhadeshwar Pitha, Borigaon, Ganjam.

REFERENCES

Annonymous, (2007). *Surakshya se Samrakshyan*, Vasundhara, Odisha, 12 pp.

Dash, P.K. (2012). Biodiversity of aquatic plants of Similipal Biosphere Reserve, Odisha, Ph.D. thesis. Utkal University, Bhubaneswar, Odisha.

Dash, P.K., S. Behera, S. Nayak, U.P. Das, H.S. Upadhyay and J.P. Singh (2017). *Biological diversity of Odisha*, Odisha Biodiversity Board, Bhubaneswar, Odisha, 36 pp.

De K. and S.K. Palita (2018).A checklist of spiders from six sacred groves in Southern Odisha, India.*Serket*, 16(1): 30-40.

Loreau, M., S. Naeem, P. Inchausti, J. Bengtsson, J. P. Grime, A. Hector, D.U. Hooper, M.A. Huston, D. Raffaelli, B. Schmid, D. Tilman, and D.A. Wardle (2001). Biodiversity and Ecosystem Functioning: Current Knowledge and Future Challenges. *Science* 294: 804-808.

Malhotra K.C. (1998). Anthropological dimensions of sacred groves in India: an overview. *In: Ramakrishnan PS, Saxena KG and Chandrashekara UM (Eds.) Conserving the Sacred for Biodiversity Management.* UNESCO and Oxford-IBH Publishing, New Delhi: 423-438.

MoEFCC (2017). India's National Wildlife Action Plan (NBAP). Ministry of Environment, Forests and Climate Change, India. 140 pp.

Mohanta, R.K. and B. S. Adhikari (2012). The Biological Values and Conservation Status of Sacred Groves in the Balasore Wildlife Division, Odisha: A Case Study, *International Journal of Conservation Science*, 3 (3): 207-216.

Mohanty, P. (2011). Conservation Reserve and Community Reserve in Odisha: A Study on the Potentials and Initiatives Taken so far. Research report, Regional Centre for Development Cooperation (RCDC), Bhubaneswar, Odisha, India: 47 pp.

Mohanty S, P.K. Das and Kumar S. (2016). Role of sacred groves in the conservation of traditional values of Odisha. *Adv. Plants Agric. Res.*, 3(3): 56 58. DOI: 10.15406/ apar.2016.03.00094

Mohapatra, P.P., C. Mahapatra and S.K. Dutta (2017). Studies on amphibians of Odisha: an overview. (IN) Das, A. (Ed.) *Diversity and Ecology of Amphibians of India.* ENVIS bulletin: Wildlife and Protected Areas. Vol. 19. Wildlife Institute of India, Dehradun, 248001, India, 11-32 pp.

Mohapatra, P.P., P.K. Dash, H.S. Palei, S. Debata, V. Sarkar, A.K. Mishra and S.K. Dutta (2017). Ecological and Social Aspects of Biodiversity Conservation in Sacred Groves of Bonai Forest Division, Odisha, India. (IN): *Biodiversity conservation and Wildlife Management*, Thatoi *et al.*, Eds, Centre for Similipal Studies, North Orissa University, 35-50pp.

Mohapatra, P.P., S.K. Dutta and C. Mahapatra (2017). Herpetofauna of Similipal and their conservation. (IN) *Facets of Similipal.* Thatoi *et al.* (Eds.) North Orissa University, Baripada, Odisha, 757003, India, 164-179 pp., ISBN: 978: 81- 933281-2-5.

Mohapatra, P. P., S.R Behera, S.K. Dutta, C. Mahapatra and H.S. Upadhaya (2019). Mammals of Odisha, India: updated checklist, distribution, and conservation status. *PRANIKEE, Journal of Zoological Society of Orissa*, XXXI: 112 pp. ISSN 0970-4450.

Nair, M.N.B. and G. Hariramamurthy (2010). Access and benefit sharing and the *in situ* protection of traditional knowledge. *Policy Matters*, 17, IUCN: 169-172.

Nair M.V. (2011). *Dragonflies and Damselflies of Orissa and Eastern India*, Wildlife Organisation, Forest and Environment Department, Government of Odisha: 254pp.

Ormsby, A.A. and S.A. Bhagwat (2010). Sacred forests of India: a strong tradition of community-based natural resource management. *Environmental Conservation* 37 (3): 320–326.

Palei, H. S., A. K. Mishra, P. K. Dash and P. Mohapatra (2013). Biodiversity of Sacred Groves of Koliposh Range in Bonai Division, Odisha, India, *e-planet 10 (1): 20 – 30.*

Panda D., S.S. Bisoi and S.K. Palita (2014). Floral Diversity Conservation through Sacred Groves in Koraput District, Odisha, India: A Case Study, *International Research Journal of Environment Sciences,* Vol. 3(9), 80-86.

Panda, T., N. Mishra, B.K. Pradhan, S. Rahimuddin and R. Mohanty (2020). Sacred groves in conservation of biodiversity in Odisha, *The Holistic Approach Environ.* 10 (1), pp. 10 – 15.

Pradhan, A, S.P. Mishra and N. Behera (2016b). Avian diversity in a sacred natural forest site in Odisha, *The Ecoscan*, 10(1 and 2): 91-95.

Pradhan, A., S.P. Mishra and N. Behera (2016a).An ethnomedicinal survey of medicinal plants from a sacred forest of western Odisha, India. *International Journal of Phytomedicine*, 8: 325-332.

Pradhan, A., S.P. Mishra and N. Behera (2016c). Species diversity and biomass carbon analysis of the tree layer in a sacred natural forest patch from Western Odisha, *International journal of environmental sciences*, 7 (2): 113-122.

Pradhan, A., S.P. Mishra and N. Behera (2017).Diversity, population structure and regeneration potential of tree species in a community conserved sacred forest from Western Odisha, India. *Indian Forester*, 143 (6): 566-572

Rodgers W.A., H.S. Panwar, V.B. Mathur (2000). *Wildlife protected area network in India: A review* (Executive Summary). Dehradun: Wildlife Institute of India; p. 53.

Saxena H.O. and Brahmam, M (1994-1996). *Flora of Orissa*. Vol. I-VI. Regional Research laboratory (CSIR), Bhubaneswar.

Singh, S. (1985). An overview of the conservation status of the National Parks and Protected Areas of the Indomalayan realm. In: J.W. Thorsell (ed.). Conserving Asia's Natural Heritage: The Planning and Management of Protected Areas in the Indomalayan Realm. Pp. 1-4. IUCN, Gland.

Upreti, D.K. (1996). Lichen on Shorea robusta in Jharsugada district, Odisha, India. *Flora and Fauna*, Jhansi 2(2): 159-161.

Vitousek, P.M., H.A. Mooney, J. Lubchenco and J.M. Melillo (1997). Human Domination of Earth's Ecosystems, *Science*, 277: 494-1499.

Wikramanayake, E.D., E. Dinerstein, C.J. Loucks, D.M. Olson, J. Morrison, J. Lamoreux, M. McKnight, and P. Hedao (2002). *The terrestrial ecoregions of the Indo-Pacific: a conservation assessment.* Island Press, Washington, D.C. pp. 643.

Chapter 27

Tasar Culture as Befitting Sustainable Livelihood Option for Tribal People of Mayurbhanj District of Odisha

Prasanta Kumar Kar

Regional Sericultural Research Station, Central Silk Board, Prafulla Nagar, Baripada, Mayurbhanj, Odisha
e-mail: prasantakar@rediffmail.com

Tasar culture is a traditional practice of local tribes in Mayurbhanj district of Odisha since time immemorial. Before independence also they had the practice of rearing in forest patches. Over period of time new technologies developed and adopted by farmers with rearing on economic plantations. This led to increase in production and productivity. In Mayurbhanj, 20 Tasar Rearers' Cooperative Societies are functioning. The reared as well as forest collection cocoons account for about 36-40 per cent of raw silk production of state. More and more people are now associated with tasar culture, as it proves to be the best livelihood option. However, the adoption of technology is not uniform among farmers and also the pattern of income generation. So, a study was taken to ascertain the socio-economic conditions *vi a vis* adoption of tasar technologies and income through tasar culture. Farmers from three productive TRCSs were categorized into three groups *viz.* Seed Rearers (SRs), Private Graineurs (PGs) and Commercial Rearers (CRs). On an average 25.67 per cent of farmers are illiterate, 39 per cent are below primary level, 21.67 per cent are non-matriculate, 10 per cent are matriculate and only 3.67 per cent have studied beyond matriculation level. Family size is almost identical in all groups. The PGs done exceptionally well as 40 per cent of them earned between Rs 40 – 50,000 and even 35 per cent earned more than Rs50,000 from a single season. Highest productivity (41 to 50 cocoons/dfl) was recorded by 35 per cent of PGs and 31.82 per cent of CRs whereas 31-40 cocoons/dfl by 51.67 per cent of SRs. The adoption level of plantation management as well as tasar silkworm rearing and grainage technologies was also highest in GRs and SRs. There has been increased awareness for silkworm disease control and pest, predator management. Thus tasar culture in Mayurbhanj is proved to be the best livelihood option as farmers get maximum return in less time period with minimum investment. In order to increase the income of poor tasar farmers importance of dissemination of technologies as well as extending training support are discussed.

Keywords: *Antheraea mylitta, Socio-economic status, Tasar silkworm, Rearing, Plantation maintenance.*

INTRODUCTION

Tasar is a variety of silk produced from tropical tasar silkworm *Antheraea mylitta* which feeds primarily on Asan (*T. tomentosa*), Arjun (*Terminalia arjuna*) and Sal (*Shorea robusta*) and secondarily on a dozen of plants. Tasar silk is copperish in colour with unique coarse filament, has its unique feel and appeal. This silk is specific to India and production is a forest based agro-industry being practiced by the tribal residents in forests and areas nearby forests. Earlier, tasar cocoons were collected from forests and sold to middlemen and then they were reaching at reelers and weavers. The greatest boon of tasar culture is that it does not have any investment on electricity, plantation, rearing appliances and extra essentials as tasar silkworms are reared on the forest trees in the untamed conditions. Perhaps, Tasar culture not only prevent the destruction of forest but also endow with a productive exploitation of cosmic natural resources with awfully minimum investment along with massive employment generation in rural areas and progress of downtrodden (Mohanty, 1998) It holds a great promise for the rural populace, especially for tribal people in providing their livelihood. Tasar Sericulture industry is gaining momentum for the reasons that this industry creates more employment avenues up to 2.5 lakh rural aboriginal families of the tropical belt of India and gives high returns with low investments for sustainable development. It also grants persistent lofty economic returns, engenders employment opportunities and has prospective to pave a strong basis for tasar silk based rural cottage industries. It is suggested from the diverse studies of global description that endorsing and encouraging sericulture will help to rebuild the ecological balance and not only check environmental degradation but also conserve endangered silk insects, associated flora and fauna for posterity (Dewangan, 2011).

Tasar culture is a major source of livelihood for rural poor, mostly in the tribal dominated districts of Mayurbhanj, Keonjhar, Sundargarh, Jajpur, Deogarh, Angul, Dhenkanal, Boudh and Nawarangpur (Sahu, 2015). Tasar culture in Odisha was first organized by princely state of Mayurbhanj under the Royal patronage of Bhanja Dynasty. Odisha is presently positioned 3rd in tasar cocoon production in India next to Jharkhand and Chhattisgarh. Tasar culture in Odisha is a traditional occupation of tribal people who are utilizing about 12,450 hectares of Arjun and Asan plantation. There are more than 46,828 families enrolled as members of primary societies *i.e.* Tasar Rearers' Cooperative Societies (TRCS) in tasar producing districts of which nearly 15000 families are actively engaged in tasar culture. Most of such people engaged in tasar culture come under BPL group. Out of the registered tasar farmers, about 68 per cent rear single crop, 26 per cent rear 2 crops and only 1 per cent rear 3 crops in a year. Annually more than 60,000 kahans of tasar cocoons are produced and marketed in the state. The tribal farmers earn about 18 crore rupees from sale of cocoons with a very nominal investment on seed. The production of tasar cocoons has increased over the years. Considering the vast forest wealth, availability of skilled weavers and human resources in the state, tasar industry has a bright future. The state has four types of forests of which tropical humid forest constitutes 80 per cent of the total forest area where primary tasar food plants are available in abundance. The semi-domestic race Daba constitutes 66.62 per cent of the total production, where as wild cocoon collection accounts for about 33.38 per cent during the last five years. The state has contributes about 130 MT of tasar raw silk during 2018-19.

Mayurbhanj is the lead district for tasar cocoon production in Odisha with share of about 36 – 40 per cent. Earlier, the wild genetic resources of tasar silkworm *viz.* Modal, Nalia, Jata ecoraces available in Similipal Biosphere Reserve mainly contributed to the bulk of cocoon production. In some of the areas farmers still collect Modal cocoons during the month of August which are used for traditional egg production followed by rearing of Bogai crop. During eighties, Daba and Sukinda ecoraces were introduced in the state. In recent years, Daba (BV) and Daba (TV) are the two ecoraces reared by farmers. There are 20 TRCSs functioning in the districts, notable among them are Bhuasuni, Mudrajodi, Chandua, Kansapal, Thakurmunda, Kendujuani, Bisoi, Dantiamuhan and Nuagaon. In recent years five Tasar Silkworm Seed Centres have been established at Chandua, Bangiriposi, Bisoi, Thakurmunda and Kaptipada to cater the need of basic tasar seeds.

If one considers the geographical features of Mayurbhanj district, it is found that the vast landscapes in the eastern part of Similipal Biosphere Reserve (SBR) fall below 50m MASL whereas the altitude of the tasar rearing areas in the western part of SBR ranges from 121 to 530m MASL. The weather condition is also influenced by the topography of the region. The eastern region is suitable for Daba (trivoltine) and the western part for Daba (bivoltine). Thus the production and productivity also vary. Earlier, tasar silkworm was traditionally reared on forest trees where production level was low. Gradually, with the development of different technologies people started adopting them. However, the adoption level also varies from farmer to farmer where socio-economic conditions play a major role. Studies on socio-economic conditions and adoption level of technologies in tasar culture are scanty in tropical tasar sector baring a few citations (Goel *et al.*, 1993; Suryanarayana, 2007; Sinha *et al.*, 2008).

Thereafter, several technologies are developed by CTR and TI, Ranchi, test verified in regional centres and also at selected farmers' level (Sinha *et al.*, 2018). However, till date the adoption level at farmer field in relation to their socioeconomic conditions has not been studied so far in recent years. Hence the present study was conducted with the objectives to assess the socio-economic status of tasar farmers of Mayurbhanj, their livelihood generation through tasar culture and also adoption level of different technologies.

Materials and Methods

Study Places

The present study was conducted in Mayurbhanj district of Odisha. Villages were selected under the command areas of Chandua Tasar Rearers' Cooperative Society (TRCS) which lies in the eastern Mayurbhanj and Thakurmunda and Kendujuani TRCS which lie in the western part of Mayurbhanj.

Data Collection Method

A set of questionnaire was prepared to evaluate the socio-economic status. A comprehensive questionnaire was also prepared for collection of data taking into consideration the technologies prevalent in tasar culture. The total sample size was 300, covering Seed Rearers (SRs, 60 nos.), Private Graineurs (PGs, 20) and Commercial Rearers (CR, 220) and the nomenclature is described below. The data were collected through personal interview method at the farmers' residences and work places. The contents of the questionnaire were explained to farmers and the response was recorded meticulously. The information on tasar seed production, collection, cocoon production, marketing and other related aspects was also collected.

Seed Rearers (SRs): These farmers conduct their rearing for the production of cocoons which are utilized for egg production at their own level or at TRCS/Seed Production Centre. Thus, for Daba (bivoltine), the first crop during rainy season (June – August) is the seed crop whereas for Daba (trivoltine), the second crop (August-September) is seed crop.

Private Graineurs (PGs): These are the farmers who conduct tasar silkworm egg production activities called as grainage out of the cocoons produced during seed crop. They collect or purchase seed cocoons from SRs for this purpose. Some of them also conduct rearing of seed crop. Further, they sale disease free layings (dfls) to the CRs. They also conduct commercial crop rearing.

Commercial Rearers (CRs): These are the farmers who rear the tasar silkworm for production of commercial cocoons meant for silk reeling purpose. They purchase dfls from TRCSs, PPCs and other private grainages.

Data Analysis

The cultivation practices followed by the farmers/rearers are presented as percentage to make simple comparisons. The per cent adoption of recommended package and practices was calculated.

Results

Tasar culture in Mayurbhanj is an important livelihood activity of the tribal people who also inherit from their forefathers. They have their own perception of seed cocoon collection, egg production and rearing but over time they have modified the practice to suit the conditions. With the advent of technologies, many of them have been adopting the modern system. Thus, it is necessary to assess the socio-economic conditions of tasar farmers in order to know adoption level of technologies.

Status of Education Level in Tasar Farmers

Education status of farmers is presented in Figure 26.1. Most of them are either below matriculation or uneducated. Among the illiterate group some know their signature only. Only 8.34 per cent of SRs and 11.37 per cent CRs were found to be matriculate and above matriculation level. Contrastingly 55 per cent of GRs were either matriculate or above. The illiteracy rate was also bare minimum in GRs (5 per cent) where it was 28.33 and 26.82 per cent in SRs and CRs respectively. If we take all the tasar rearers into account it is found that 25.67 per cent are illiterate, 39 per cent are below primary level, 21.67 per cent are non-matriculate, 10 per cent are matriculate and only 3.67 per cent have studied beyond matriculation level. Literacy rate is comparatively very low in case of CRs.

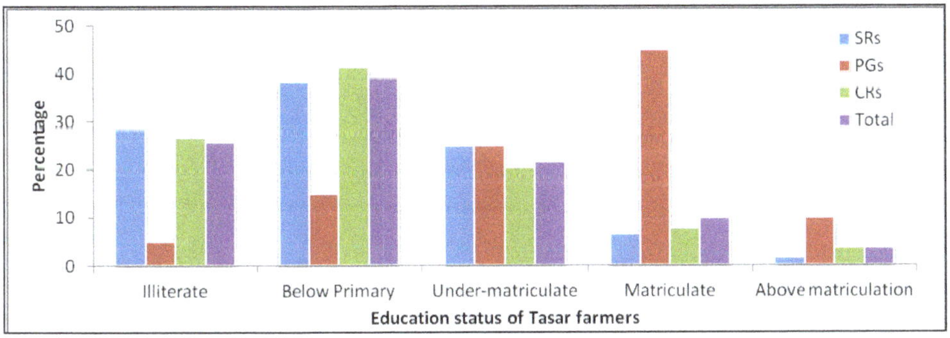

Figure 27.1. Educational Status of Tasar Farmers in the Study Area of Mayurbhanj.

Family Size

Generally, a tasar farmer's family can rear 100 to 600 dfls in a crop. This depends on their family size because during the rearing season almost all of their family members are engaged by one way or another. The distribution pattern of family members is presented in Figure 26.2. Minimum family size with 3 to 5 members in a family was most prevalent in PGs (40 per cent) whereas 6 to 8 members family

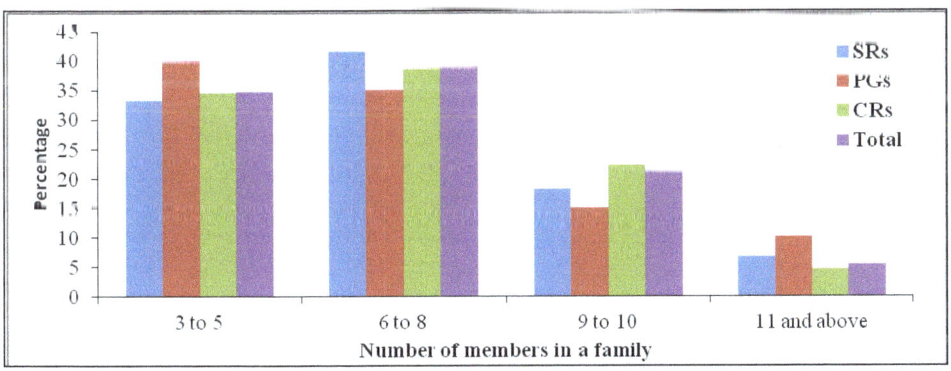

Figure 27.2: Distribution of Tasar Farmers According to their Family Size.

size was more prevalent in SRs (41.67 per cent) and CRs (38.645). If we consider overall distribution of the family size 34.67 per cent had 3 to 5 members, 39 per cent had 6 to 8 members, 21 per cent had 9 to 10 members and only 5.33 per cent of them comprised of 11 or more members in their families.

Annual Income from Tasar

The farmers usually take one crop in a year, may be the seed crop or commercial crop. Thus, the crop is seasonal and within 30 to 45 days of rearing period the cocoons are produced and sold. According to the income from tasar culture, the farmers were distributed in four groups (Figure 26.3). In case of SRs, majority (51.67 per cent) earned Rs 30 – 40,000 from a single crop. The CRs maximum frequency (35.45 per cent) fell in the category of earning Rs 20 – 30,000 from crop. The PGs done exceptionally well as 40 per cent of them earned between Rs 40 – 50,000 and even 35 per cent earned more than Rs50,000 from a single season. It may be noted here that this group takes the lead in silkworm seed preparation at their own level and sale to other CRs. Besides, they also conduct commercial crop rearing where sale of cocoons also add to their income. Overall 15 per cent of farmers earned less that Rs20000/-, 31 per cent earned Rs20–30000/-, 32.33 per cent earned Rs 30 to 40000/-, 15.33 per cent earned Rs40-50000/- and 6.33 per cent earned more than Rs50000/- per season.

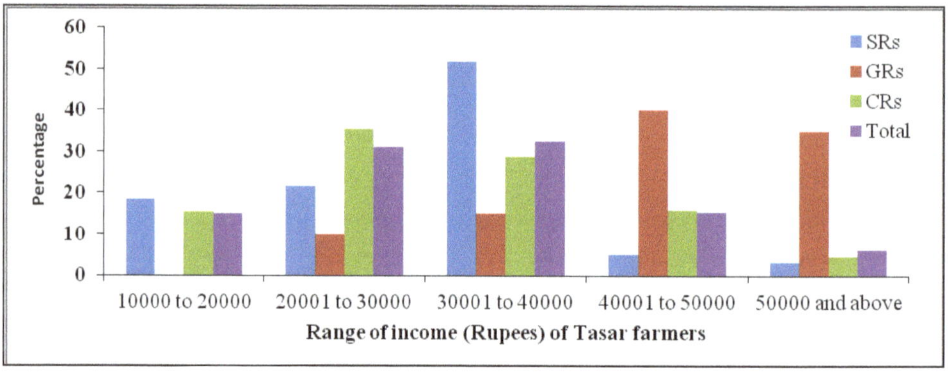

Figure 27.3. Distribution of Tasar Farmers According to their Annual Income.

Tasar Cocoon Productivity

The tasar cocoon productivity assessed through production of cocoons per disease free laying (dfl) of a female moth (200 eggs) is presented in Figure 4. According to the availability of host plants and family size they conduct rearing of 150 to 500 dfls of tasar silkworm. In earlier days cocoon productivity was very low *i.e.* < 20 cocoons/dfl. Due to technology intervention and creation of awareness among farmers, there has been substantial increase in cocoon production. Availability of good quality dfls after moth testing and adoption of technology by farmers have led to this increase, thus making tasar crop remunerative. Figure 26.4 depicts that only 4 per cent of total farmers harvested < 20 cocoons/dfl. Highest productivity

(41 to 50 cocoons/dfl) was recorded by 35 per cent of PGs and 31.82 per cent of CRs whereas 31-40 cocoons/dfl by 51.67 per cent of SRs.

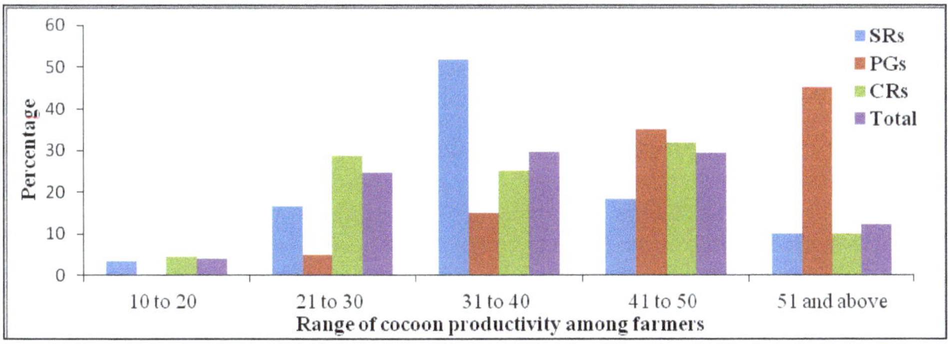

Figure 27.4. Cocoon Productivity Status (Cocoons Per Dfl) of Tasar Farmers.

Adoption Level of Technologies

Technologies Related to Plantation Maintenance

The details of adoption level of technologies related to plantation maintenance are presented in Table 27.1. Perusal of Table 27.1 indicates that most of the technologies related to plantation maintenance are adopted by PGs followed by SRs whereas adoption level is less in CRs. About 66.67 per cent of the SRs and 90 per cent of PGs have the opportunity to utilize the economic plantations of *T arjuna* developed through state government agencies. About 68.18 per cent of CRs depend on the forest patches of Asan or isolated trees in the bunds of paddy fields. Cultural operations and leaf enrichment related technologies are adopted completely by PGs

Table 27.1. Adoption Level of Major Plantation Maintenance Technologies

Sl.No.	*Technology*	*SRs (60)*	*PGs (20)*	*CR (220)*
1.	Adoption of rearing in economic plantation	40 (66.67)	18 (90.00)	70 (31.81)
2.	Adoption of rearing in forest plantation	20 (33.33)	2 (10.00)	150 (68.18)
3.	Food plant used for rearing			
	Only Asan	5 (8.33)	1 (5.00)	168 (76.36)
	Only Arjun	37 (61.67)	15 (75.00)	11 (5.00)
	Both Asan and Arjun	18 (30.00)	4 (20.00)	41 (18.64)
4.	Organic manure application (chawki plot)	24 (40.00)	12 (60.00)	10 (4.55)
5.	Chemical fertilizer application (chawki plot)	6 (10.00)	2 (4.00)	3 (1.36)
6.	Pruning and pollarding method adopted	56 (93.33)	20 (100.00)	105 (47.73)
7.	Protection against gall insect	56 (03.33)	20 (100.00)	75 (34.09)
8.	Leaf enrichment measures for leaf yield (Foliar spray of urea)	35 (58.33)	20 (100.00)	40 (18.18)

Figures in parentheses indicate percentage.

followed by SRs and least in CRs. The application manure and fertilizer in their rearing field are adopted by about 60 per cent of PGs, but in other cases it is meager. The CRs pay least attention for plantation maintenance as most of the rearing is conducted in the Adapahi located in forests.

Technologies on Silkworm Rearing and Egg Production

The rearing practices followed by farmers in the study area are presented in Table 27.2. All farmers preferred Daba race (bivoltine and trivoltine in respective zones) of tasar, but for seeds (eggs) they depend on government agencies and associated private graineurs (egg producers).All the group of farmers are aware about brushing period as well as brushing technique. Farmers keep eggs in plastic

Table 27.2. Adoption Level of Tasar Silkworm Rearing and Grainage Technologies

Sl.No.	*Tasar Silkworm Rearing Technology*	*SRs (60)*	*PGs (20)*	*CR(220)*
1.	Awareness on brushing period	60 (100)	20 (100)	200 (90.91)
2.	Silkworm brushing technology	58 (96.67)	20 (100)	190 (86.36)
3.	Chawki rearing adoption	40 (66.67)	18 (90)	5 (2.27)
4.	Awareness about diseases	60 (100)	20 (100)	150 (68.18)
5.	Use of Jeevan Sudha for control of virosis	60 (100)	20 (100)	50 (22.73)
6.	Spray of Sodium hypochlorite	50 (83.33)	20 (100)	10 (4.55)
7.	Disposal of diseased and infected silkworm	54 (90)	20 (100)	132 (60)
8.	Care for silkworm transfer when leaves are consumed	54 (90)	20 (100)	180 (81.82)
9.	Care during moulting period	60 (100)	20 (100)	200 (90.91)
10.	Maintenance of hygienic conditions of field by using Bleaching powder and lime dusting	50 (83.33)	20 (100)	35 (15.91)
11.	Care during spinning (not to disturb larvae)	60 (100)	20 (100)	220 (100)
12.	Foliage left on plant for hammock formation	50 (83.33)	19 (95)	205 (93.18)
13.	Checking of crawling down of silkworms by	55 (91.67)	20 (100)	176 (80)
14.	Use of sticky trap for pest and predator	60 (100)	20 (100)	200 (90.91)
15.	Care for cocoon harvesting	60 (100)	20 (100)	200 (90.91)
16.	Watch and ward/crop monitoring	60 (100)	20 (100)	180 (81.82)
17.	Grainage/Egg production technology			
i	Sorting and garlanding of seed cocoons	40 (66.67)	20 (100)	-
ii	Disinfection of grainage house	5 (8.33)	20 (100)	-
iii	Care during coupling	8 (13.33)	20 (100)	-
iv	Care for oviposition	5 (8.33)	20 (100)	-
v	Mother moth testing for pebrine	10 (16.67)	20 (100)	-
vi	Disinfection of eggs	12 (20.0)	20 (100)	-
vii	Packaging of dfls	15 (25.0)	20 (100)	-
viii	Incubation of eggs	15 (25.0)	20 (100)	-

Figures in parentheses indicate percentage.

or bamboo trays and small Asan/Arjun twig is kept on the eggs of silkworm. The newly hatched worms crawl on the twig and then the twig is tied to food plants to facilitate crawling of worms to leaves of the food plants. The age old method of tying leaf cup containing tasar eggs on the twigs of food plants is now obsolete.

Young stage rearing or chawki rearing with nutrient enriched leaves is conducted by PG and SRs, but least in CRs. These rearers use nylon net for rearing of early age silkworms in chawki garden and net is provided by state government. Majority of CRs did not know about the importance of young age rearing in chawki garden.

Awareness of tasar silkworm diseases is mostly found in SRsand PGs while 68.18 per cent have it. They have gathered knowledge of silkworm diseases and control measure through training by government agencies. The SRs and PGs are also aware of using "Jeevan Sudha" for the control of virosis of tasar silkworm as they use it for during rearing period. Similarly, they are also ahead in the use of sodium hypochlorite as prophylactic measure. All the ASRs and CRs use to collect and bury or burn the diseased silkworms whereas CRs do not care about the disposal of diseased and dead silkworms. All SRs and PGs take prophylactic measures like sprinkling of bleaching powder and lime dusting provided by government agencies.

The awareness on the moulting period of the worms was 100 per cent but 90.91 per cent in CRs. During this period they do not transfer or disturb the silkworms. Care for silkworm till spinning and harvesting of cocoons is mostly common in SRs and CRs and to a larger extent in CRs also. To protect the crop from predators/ birds all framers resorted to watch and ward where their family members also help. They monitored their crops in morning and evening hours when the birds attack are at the maximum. They used catapult to protect the crop from birds. Nearly all farmers know about the sticky trap and use it to capture enemies like yellow fly, uzi fly, praying mantis and wasps. Generally farmers sell their cocoons to TRCSs. Some farmers sell their cocoons to Mahajans also when they get more rate but this is at a very low level in Odisha. SRs and GRs are in the habit of garlanding of cocoons as 44.21 per cent and 38.82 per cent of them are habituated to do so. Majority farmers transport the cocoons transport cocoons in plastic crates provided by state government units.

The cocoons produced from seed crop are utilized for egg production or grainage activities taken up by PGs. Grainage activities completes within a span of 20 to 25 days and the selling of dfls is quite remunerative. So, PGs adopt almost all technologies (Table 26.2). Since seed rearers are linked to PGs, many of the SRs also are associated with grainage activities as depicted in the table.

DISCUSSION

Odisha has a unique place in the tasar silk production in the country. Tasar culture has been kept alive by the people who are economically and socially backward. The tribes gathered knowledge of tasar crops simply by trial and error from generation to generation. Even agriculture is not a full time job for them because villagers of the area use to grow mostly paddy and they are mostly dependent on forest resources. The tasar culture is largely practiced in the nearby forests on Asan plants and in some cases on developed Arjun plantations. The average production under such condition comes to very low with 15-20 cocoons per dfl. In this regard the role of improved technologies of tasar holds promises. The technology intervention hasled to the increase of cocoon productivity.

Farmers do not raise any economic plantation but they conduct rearing of tasar silkworms in economic plantations developed by state government. Many of them also conduct rearing in forest patches, mostly dominated by Asan called as *Adapahi*. In recent years about 4000 hectares of new plantations are developed in Mayurbhanj under different schemes like MGNREGS and RKVY. Now farmers are willing to develop plantation in their own land also. The farmers are also engaged for seed collection, nursery preparation, seedling transplantation, digging of pits, transplantation to field, manuring, maintenance for four years. It is a nice system that tasar farmers from the respective areas are engaged and they get wages under the rates specified in the scheme and this is the income additional to tasar cocoons.

From the present study it was clear that the adoption level of the improved technologies among the farmers varied considerably. The private graineurs (PGs) were at the top followed by seed rearers (SRs), they also obtained training and facilities provided by government agencies. In general, the traditional farmers (CRs) follow the age old method and only a few recommended technologies bearing low or no cost in nature were adopted. On the method of brushing silkworms, transfer and disposal of dead worms *etc.* are practiced. But regarding other improved technologies, in most of the cases of CRs, the study revealed little or least awareness of the technologies. Similar observation of importance of adoption of technologies was reported in mulberry silkworm cultivation (Singhvi at al., 2002; Mishra *et al.*, 1999).

Earlier studies on socio-economic conditions of tasar farmers revealed the poor economic status of tasar farmers resulting in lesser adoption of new technologies (Goel *et al.*, 1993; Surynarayana, 2007; Sinha *et al.*, 2008). A number of workers studied socio-economic conditions of mulberry silkworm farmers, adoption of technologies by the sericulturists and impact of the technologies on economy of the farmers (Mishra *et al.*, 1999; Subramanyan, 1976; Prakash Kumar, 1986; Srinivasa G *et al.*, 1990; Iyenger and Datta, 1994; Doddagadad, 1996; Desai at al., 1997; Das *et al.*, 1999; Gulta *et al.*, 2001; Vijayaprakash and Dandin, 2005). It appears that the illiteracy prevailing in the community may be one of the factors responsible for the poor awareness and poor adoption level which was also advocated in mulberry sericulture.

As tasar culture is linked with the age old tradition of the tribes, they have developed their own traditional methods of tasar rearing and it is sometimes difficult to motivate them for new technologies. Therefore extension efforts need to be taken to realize the farmers that improve technologies of tasar culture if adopted by them will lead to increase their economic gain. Due to poor economic condition, they are unable to spend any money on purchase of materials related to tasar culture. They expect the materials like bleaching powder, lime and sodium hypochlorite *etc.* to be provided by government. However, in recent time more and more young people are associated with tasar culture, because of the good return from sale of cocoons and eggs with bare minimum investment within short time span. In Mayurbhanj, value addition and women employment generation through post cocoon activities like reeling, spinning of silk yarn and weaving are lacking which must be looked into.

Tasar culture has given a boost to the economy of farmers in Chhattisgarh (Dewangan, 2012, 2016). In addition to income generation and checking urban migration, sericulture activities also contribute to conservation of environment. CTR and TI, Ranchi has developed a number of technologies for tasar culture but there is the need for large scale dissemination leading to adoption at farmers' level (Sinha *et al.*, 2018). So, along with emphasis on adoption of modern technologies, tasar culture should be promoted on a larger way as one of the best livelihood options.

REFERENCES

Das, K.K., Sahu, P.K., and Das, N.K. (1999). Identification of factors influencing sericultural productivity in Malda district of West Bengal. *Indian Journal of Sericulture*, 38: 53.

Desai, B.R., Girish, K.A. and Patil, R.P. (1997). Constraints faced by contact farmers in adoption of new technology. *Agricultural Entomology Review*, 9(2): 14-16.

Dewangan, S.K. (2011). Sericulture - A tool of ecosystem checking through tribal. *Journal of Environmental Research and Development*, 6(1): 165-173.

Dewangan, S.K. (2016). Livelihood Opportunities for Tribe through Sericulture in Raigarh, Chhattisgarh, India. *Nature and Environment*, 21 (2): 1-12.

Dewangan, S.K., Sahu, K.R. and Soni, S.K. (2012). Breaking of poverty through sericulture among the tribe- A socio-economic study of Dharamjaigarh block of Raigarh Dist, CG, India. Research. *Journal of Recent Sciences*, 1: 371-374.

Doddagadad, C.B. (1996). Adoption of behaviour of sericulturists and constraints of extension personnels – A study in Dharwad district. Karnataka. PGDS Dissertation, CSR and TI, Mysore.

Goel AK, Brahmachari BN, Thandapani and Thangavelu K (1993). Socio-economic study of tasar culture. *Indian Silk*, 31(12): 38 - 41.

Gulta, G.S., Srinivasa, G., Jayaram, H., Iyengar, M.K.S.and Vijayaprakash, N.B. (2001). Socio-economic determinants of farmer oriented technology package for sericulture – a field study. *Indian Journal of Sericulture*, 40(1): 96-99.

Iyenger, M.N.S. and Datta, R.K. (1994). Knowledge level and adoption of new technology by farmers in Hunser taluk, Mysore district, Karnataka- an evaluation. *Indian Journal of Sericulture*, 33(11): 48-55.

Mishra, R.K., Jack, K.S., Chaudhary, P.C. and Rao, Y.R.M. (1999). Studies on the extent of application of input of mulberry and their role in production in rain fed tracts of Chamrajnagar, Karnataka. *Indian Journal of Sericulture*, 44: 125.

Mohanty, P.K. Tropical tasar culture in India, 1998, 186.

Prakash, K (1986). A study on adoption of improved sericultural practices and labour utilization among big, small and tenant farmers of Ramnagar taluk,

Bangalore district, M. Sc. (Agri.) Thesis, University of Agricultural Sciences, Bangalore, India.

Sahu, K. K. (2015). Status and performance of sericulture in Odisha, *Odisha Review,* LXXII (4): 37-42.

Singhvi, N.R., Rao, M.K.S., Rao, Y.R.M., Greiss, H. and Tzenov, P. (2002). Effect of improved technology on the main economic cocoon parameters of silkworm, *Bombyx mori* L. *Bulletin of Indian Academy of Sericulture*, 6: 27-33.

Sinha, A.K., Mittal, V., Pandey, J.P. and Singh, G.P. (2018). Tasar Technology Compendium. Central Tasar Research and Training Institute, Ranchi, India, 1-28.

Sinha, M.K., Mishra, R.K., Sharma, K.K., Singh, B.M.K. and Suryanarayana, N. (2008). Adoption of improved technologies by tasar farmers in Shikaripara block of Dumka district: a case study. *Biospectra*. 3(1): 163-166.

Srinivasa, G., Dolly, S.S., Raveendra, M. and Iyenger, M.N.S. (1990). Socio-economic factors and their relation to adoption of improved sericultural practices. *Indian Journal of Sericulture*, 35(1): 43.

Subramanyan, V.S. (1976). Relative knowledge gained through selected extension methods and their combinations. *Journal of Extension Education*, 12: 20.

Surynarayana, N. (2007). Contributions of R and D in Indian Tasar Industry, Souvenir Rashtriya Tasar Krishi Mela, BTSSO, Bilaspur.

Vijayaprakash NB and Dandin SB (2005). Factors influencing the adoption of bivoltine sericultural practices in Mandya district of Karnataka. *Indian Journal of Sericulture*, 44(1): 55-58.

Chapter 28

Wild Edible Mushrooms as Livelihood Support for Rural Tribal People in Similipal Biosphere Reserve, India

Hrudayanath Thatoi[1], Mrunmaya Kumar Panda[1] and Kumananda Tayung[2]*

[1]Department of Biotechnology, North Orissa University, Baripada – 757 003, Odisha
[2]Department of Botany, Guwahati University, Guwahati – 781 014, Assam
**e-mail: hn_thatoi@rediffmail.com*

The tribal's are exclusively depending on forest resources for their livelihood by practicing hunting, gathering, and collecting forest products including mushrooms. They have extensive ethno-mycological knowledge based on which they properly identify, collect and sell the wild edible mushrooms. In this study, the wild edible mushrooms collected from the forests and sold in the local market of northern Odisha by the ethnic tribes namely Kharia, Mankidi, Santal, Kolha, Munda, Bhumija, Bhuyan, Bathudi, Kudumi, Ho, Mankidia, and Birhor have been documented. The prime mushroom species that sold by the tribal peoples of northern Odisha belongs to the genus *Russula, Termitomyces, Amanita, Astraeus* and *Volvariella*. This study also highlights the ethnomedicinal uses of wild edible mushrooms of northern Odisha. About 14 wild edible mushroom species have been reported to have some kinds of ethno-medicinal potential used by the tribal's. Thus, the present study suggests that as a source of income and ethnomedicinal properties these mushrooms could play a significant role in the scocio-economic conditions of tribal peoples of northern Odisha.

Keywords: *Edible mushrooms, Livelihood, Triabal people, Ethnomedicine, Similipal Biosphere Reserve.*

INTRODUCTION

Mushrooms have been attracting much attention to the mankind since ancient times, for its nutritional, medical, biotechnological properties and environmental applications. Today, approximately 14000 described mushroom species from 1.5 million fungi estimated, out of which only about 100 edible mushroom have so far been artificially cultivated and commercially exploited all over the world addressing food security and medicinal application (Moore and Chi, 2005). Wild Edible Mushrooms are one of the important food sources of rural/tribal peoples. Mushroom collection is a means of generating additional income and improving nutritional security for many ethnic communities. In India the total recorded mushrooms are approximately 850 species (Deshmukh, 2004). There are references to the use of mushrooms as food and medicine in India in the ancient medical treatise, Charaka Samhita (3000±500 BC). However, the scientific study of mushrooms in India started with the identification and description of *Podaxis pistillaris* (L.: Pers.) by Linnaeus in the 18th century which was collected and sent by Koening from Tamil Nadu State. Later, Sir J.D. Hooker made extensive collection mostly from Assam, Darjeeling, Sikkim and Khasi hills which led to the publication of a series of papers by an English mycologist, Revd M.J. Berkeley between 1850 and 1882 (Natarajan, 1995). Wild edible mushrooms and their nutritional value have been reported from northern India which includes North Western region, Eastern Himalaya proper and North-Eastern hilly areas. North-Western Himalaya has been the center of intensive research on higher fungi since the 1950s. An edible species of *Agaricus*, namely *Kbasianulosus* was first reported from Punjab of North-West Himalaya by Paracer and Chahal (1962). Now there are two active centers (Department of Botany, Punjab University, Chandigarh and Punjabi University, Patiala) of research on macrofungi in this region and collections have mostly been made from North-West Himalaya. A number of reports on mushroom from North Western Himalaya have been provided by Atri and Saini since 1988 to till date from Department of Botany, Punjab University, Patiala (Atri and Saini, 1989). Atri *et al.* (1997) also studied the taxonomy, distribution, ecology and edibility of 30 taxa of genus *Russula* which are new records from India. Study on mushrooms in South India such as Tamil Nadu, Kerala, Karnataka and Andhra Pradesh was neglected as regards to studies on agarics until 1975 (Natarajan, 1995). Mushroom research of central part of India covers Maharashtra, Madhya Pradesh, West Bengal and Odisha. Recently, 24 fleshy mushroom species were reported by the authors from forest areas of northern Odisha, out of which 19 species were found to be edible and were represented by 11 families and 11 genera (Panda and Tayung, 2014). In another study, Sachan *et al.* (2013) reported a total of 14 species of fleshy mushrooms belonging to 8 genera and 6 families from Similipal Biosphere Reserve, Odisha, India.

Similipal Biosphere reserve (SBR) in the Mayurbhanj district of Odisha (India) is one of the tropical forest ecosystems rich in diversity of flora including mushroom. Some of the ethnic tribe *viz.* Santal, Kolha, Munda, Khadia, Bhumija, Bhuyan, Bathudi, Kudumi, Ho and Mankdias are residing in the Similipal forests area (Panda *et al.*, 2011). The origin and distribution of ethnic castes are found to localize in particular zone or area. These ethnic groups are purely depends on forest

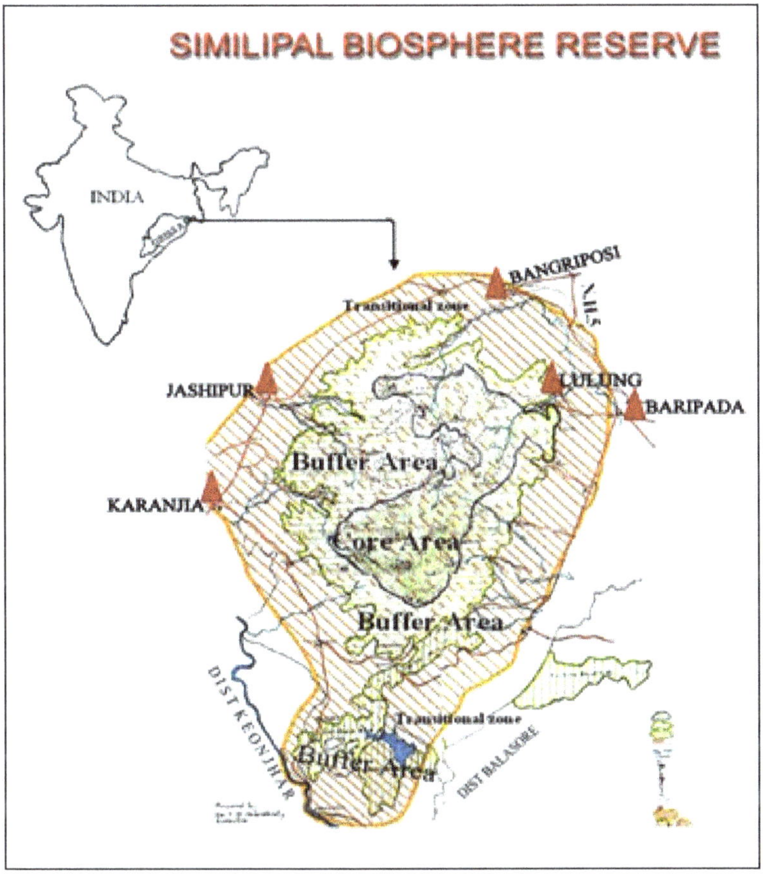

Figure 28.1. Study Area of Similipal Biosphere Reserve.

resources by hunting, gathering and collecting forest products for their livelihood. They are the traditional collectors and consumers of varieties of wild mushrooms. Even though, wild edible mushrooms contribute towards livelihood and economy of the rural and tribal folks, information on their diversity and demand are very limited in Similipal regions except some few sporadic studies on their occurrence and ethnomedicinal uses (Sachan *et al.*, 2013; Panda and Tayung, 2014).

Mushroom Diversity in Similipal Biosphere Reserve

A total of 20 wild edible mushroom species were recorded during the study period from different places of Similipal, Odisha, India (Sachan *et al.*, 2013; Panda and Tayung, 2014). Out of 20 wild edible mushroom species 17 species were found in forest soil substratum followed by two species from decaying wood logs and only one species from the decaying paddy straw. The dominant mushrooms species belong to genera *Russala* (seven species) followed by *Termitomyces* (three species), *Amanita* (two species) (Table 28.1). Species of the genus *Russula* were found abundantly in forest soil substratum during the peak growing periods and *Geastrum*

Table 28.1. Habitat and Period of Availability of Wild Edible Mushrooms of SBR, Odisha

Sl. No.	*Species*	*Habitat*	*Host/ Substratum*	*Population*	*Period of Availability*	*Edibility*	*Toxic/Non Toxic*
1	Russula rosea	Decaying Sal leaf mixed soil	Soil	Abundant	June-September	Edible	Non toxic
2	Russula delica	Decaying Sal leaf mixed soil	Soil	Abundant	June-September	Edible	Non toxic
3	Russula densifolia	Decaying Sal leaf mixed soil	Soil	Abundant	June-September	Edible	Non toxic
4	*Russula violeipes*	Decaying Sal leaf mixed soil	Soil	Abundant	June-September	Edible	Non toxic
5	*Russula cyanoxantha*	Decaying Sal leaf mixed soil	Soil	Abundant	June-September	Edible	Non toxic
6	*Russula virescens*	Decaying Sal leaf mixed soil	Soil	Abundant	June-September	Edible	Non toxic
7	*Russula emetica*	Decaying Sal leaf mixed soil	Soil	Abundant	June-September	Edible	Non toxic
8	*Termitomyces heimii*	Soil with underground termite nest	Soil	Moderate	August-November	Edible	Non toxic
9	*Termitomyces microcarpus*	Sandy Soil	Soil	Moderate	May-September	Edible	Non toxic
10	*Termitomyces eurhizus*	Sandy Soil	Soil	Moderate	June-September	Edible	Non toxic
11	*Termitomyces* sp.	Soil with underground termite nest	Soil	Moderate	August-November	Edible	Non toxic
12	*Amanita hemibapha*	Decaying Sal leaf mixed soil	Soil	Abundant	June-September	Edible	Non toxic
13	*Amanita egregia*	Decaying Sal leaf mixed soil	Soil	Abundant	June-September	Edible	Non toxic
14	*Tuber* sp.	Decaying Sal leaf mixed soil	Soil	Abundant	June-September	Edible	Non toxic
15	*Geastrum* sp.	Decaying Sal leaf mixed soil	Soil	Abundant	June-October	Edible	Non toxic
16	*Astraeus hygrometricus*	Decaying Sal leaf mixed soil	Soil	Abundant	June-September	Edible	Non toxic
17	*Lycoperdon pyriforme*	Decaying Sal leaf mixed soil	Soil	Abundant	June-October	Edible	Non toxic
18	*Volvariella volvacea*	Decaying Paddy straw	Paddy straw	Abundant	June-November	Edible	Non toxic
19	*Lentinus sajor-caju*	Decaying trunk of plant	Plant	Moderate	July-November	Edible	Non toxic
20	*Pleurotus florida*	Decaying trunk of plant	Plant	Moderate	July-November	Edible	Non toxic

species were rare and found only in Sal forest under decomposed leaves, while *Termitomyces* species were found abundantly in the sandy forest soil around termite nests. The favorable time for collection of wild edible mushrooms in the study area was found during the onset of rains, the period when the condition was conducive for the mushroom growth. The wild edible mushroom species started growing on the month of May with the oncoming of rain and the occurrence gradually increases and found to be high in August and then after decreases gradually at the end of October (Figure 28.2).

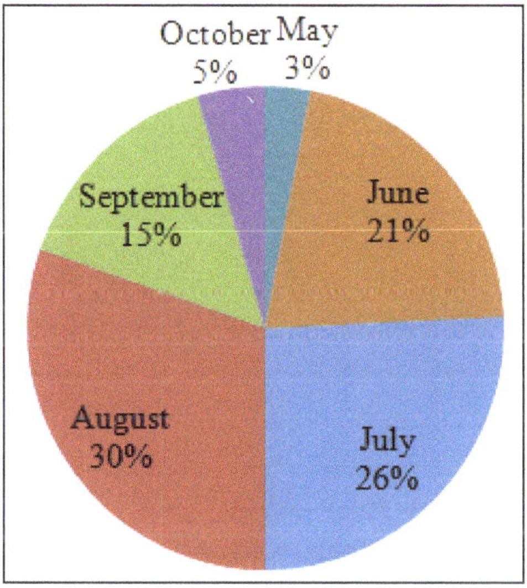

Figure 28.2. Occurrence of Wild Edible Mushroom Species with Respect to Months in Similipal.

Survey on Market Value of Wild Mushrooms in SBR

During the study, it was observed that, the rural tribal peoples in the Similipal Biosphere Reserve of Odisha collects mushrooms from the wild and sells in the market for their livelihood (Figure 28.3). Species belongs to genus *Termitomyces* were found to have the highest market value in rural (30-120 rupees/kg) as well as in urban markets (150-250 rupees/kg). Another wild mushrooms belonging under the genus *Russula* was found to be lower market prices both in rural (10 30 rupees/kg) and Urban markets (60-90 rupees/kg) as compared to other wild edible mushrooms (Table 28.2). Thus the study revealed that the economic value of wild edible mushrooms varied remarkably from rural to urban markets. This might be because of the ignorance of the rural tribal community about the palatability and health benefits of wild edible mushrooms while urban community were too choosy on those wild mushrooms, because they have insufficient knowledge about and the edibility as they considered many of the mushroom species to be poisonous. Further, vendors play an important role on the economy of rural tribal people who depends upon the forest product such as mushrooms for their living. They used

Figure 28.2. Wild Edible Mushrooms of the Study Area.

a and b: Tribal women selling wild edible mushrooms in road side as well as in local market; c and d: Local vendors selling the wild edible mushrooms in urban market.

to purchase the mushrooms at cheaper price from the rural folk and sell in mush higher price in urban markets.

Table 28.2. Market Values of some Wild Edible Mushrooms Sold in Local Markets of Similipal

Mushroom Species	*Local Name*	*Price in Rural Tribal Villages/kg (Rs)*	*Price in city Markets/kg (Rs)*	*Period of Availability*
Russula delica Fr.	Patra chatu	10-30	60-90	Fresh sold during the rainy season (June-Aug)
Russula rosea Quél.	Kukuda chatu, Nali patra chatu	10-30	60-90	
Russula virescens (Schaeff.) Fr.	Patra chatu, Kali kukuda chatu	10-30	60-90	
Russula cyanoxantha (Schaeff.) Fr.	Jamu chatu	10-30	60-90	
Russula densifolia (Secr.) Gillet.	Angarani, Angar Chatu	10-30	60-90	
Russula violeipes Quel.	Kukuda chatu	10-30	60-90	

Mushroom Species	Local Name	Price in Rural Tribal Villages/kg (Rs)	Price in city Markets/kg (Rs)	Period of Availability
Termitomyces microcarpus (Berk. and Broome) R. Heim	Bada bali chatu	30-60	150-240	Fresh sold during the rainy season (June-Aug)
Termitomyces eurhizus(Berk.) R. Heim	Chota bali chatu	30-60	150-240	
Termitomyces heimii Natarajan	Parbana chatu, Ind chatu, Nada chatu	60-120	150-250	
Termitomyces sp.	Bihida chatu	60-100	100- 180	
Amanita egregia D.A. Reid	Mahudhal chatu, Dhala manda chatu	10-30	60-90	
Amanita hemibapha (Berk. and Broome) Sacc.	Haladia Manda	10-30	60-90	
Astrareus hygrometricus (Pers.) Morgan	Rutka chatu	30-80	120-250	
Volvariella volvacea (Bull.)Singer	Kuta chatu	80-120	120-200	

Future Prospective

This study provides the baseline information on wild edible mushroom diversity of Similipal biosphere reserve of Odisha and stresses upon the socio-economic importance of the indigenous edible mushroom species among rural/tribal peoples. The study also recommends regular surveys over an extended period in order to assess the patterns of abundance of mushrooms in different seasons. From such information, harvesting strategies and management plans can be formulated and implemented to ensure the lasting presence of these socially and economically important species. In view of the increasing commercialization of the wild edible mushrooms, more studies on the production and cultivation of some of these socio-economically important wild species should be conducted in this region by introducing simple and appropriate low cost technology. Further, it will be worthwhile to gather different views of the local populace about the value of mushrooms, which would pave a way for the introduction of some known wild edible mushrooms in the diet of rural population. In addition wild mushrooms are known to exhibit novel bioactive compounds and bioactivities. Therefore, further studies need to be carried out in order to assess the mushroom diversity of Similipal, Odisha in a view to highlight their socio-economic importance and ethno-medicinal potentials for discovery of novel compounds for their pharmaceutical applications.

ACKNOWLEDGEMENT

Authors are grateful to the authorities of North Orissa University for providing laboratory facilities to carry out this research programme **and the local inhabitants of Northern Odisha for assisting during the field studies.**

REFERENCES

Atri, N.S., Saini, S.S. (1989). Family Russulaceae Roze-a review. In: Trivedi, M.L., Gill, G.S., Saini, S.S. (Eds). *Plant Science Research in India*. New Today and Tomorrow's Printers and Publishers, Delhi, India, pp. 115-125.

Atri, N.S., Saini, S.S., Saini, M.K. (1997). Studies on genus Russula Pers. from North Western Himalayas. *Mushr. Res*. 6: 1-6.

Deshmukh, S.K. (2004). Biodiversity of tropical basidiomycetes as sources of novel secondary metabolites. In: Jain PC (ed). *Microbiology and Biotechnology for Sustainable Development*. CBS Publishers and Distributors, New Delhi, India. pp. 121-140.

Moore, D. and Chi, S.W. (2005). Fungi products as food. In: Pointing SB, Hyde KO (Eds.), Bio- Explotation of filamentous fungi. *Fungi Diversity Res Lenis* 6: 223-251.

Natarajan, K. (1995). Mushroom flora of South India (except Kerala). In: Chadha KL, Sharma SR (eds). *Advances in Horticulture Vol. 13: Mushroom*. Malhotra Publishing House, New Delhi, India. pp. 381-397.

Panda, A.K, and Swain, K.C. (2011). Traditional uses and medicinal potential of *Cordyceps sinensis* of Sikkim. *J. Ayurveda Integr. Med*. 2(1):9-13.

Panda, M. and Tayung, K. (2015). Documentation and ethnomedicinal knowledge on wild edible mushrooms among ethnic tribes of northern Odisha, India. *Asian J Pharm Clin Res*. 8:139-143.

Paracer, C.S., Chahal, D.S. (1962). A new edible species of genus *Agaricus* in the Punjab (India). *Mycopathol. Mycol. Appl*. 18: 267-270.

Sachan, S.K.S., Patra, J.K. and Thatoi, H.N. (2013). Indigenous knowledge of ethnic tribes for utilization of wild mushrooms as food and medicine in Simlipal biosphere reserve, Odisha, India. *Journal of Agriculture Technology*, 9: 335-348.

Chapter 29

Resource Conservation and Livelihood Generation through Sustainable Agroforestry Systems in the Plateau Zone of Odisha: A review

*Gayadhar Shial[1] * and Umasankar Nayak[2]*

[1]Programme Assistant (Forestry), Krishi Vigyan Kendra, Bhadrak, OUAT, Odisha

[2]Senior Scientist, Regional Research and Technology Transfer Station, Ranital, OUAT, Odisha

**e-mail: gayadharbalasore@gmail.com*

During the last few decades forests in the plateau zone of Odisha have been denuded very fast because almost all the requirements of timber, fuel, fodder, NTFPs (Non Timber Forest Produces) *etc.* are met from them. The area under forest is degrading due to demographic pressure and infrastructure growth needs. To save forests and meet the growing demands of forest dependants there is an urgent need for large scale adoption of agroforestry to make them self-reliant for the innumerable benefits. Agroforestry can play a significant role in such course of action by meeting the diverse needs of people as inter-dependent benefits of the three components, *viz.* trees, crops and livestock in addition to the 6Fs, *i.e.* food, fruit, fodder, fuel, fertiliser and fibre from limited land resources. Livelihood security in rural areas are to be ensured through implementation of many alternatives such as apiculture, sericulture, lac cultivation, gum and resin, medicinal and aromatic plants *etc.* can be integrated with agroforestry. The present work was taken up in Mayurbhanj and Keonjhar districts in state of Odisha which are forested area with a high occupancy by tribal communities, whose live and livelihood closely integrate and interact with forest and agriculture communities. The study aims at suitable agroforestry systems and name of the plants used by the local communities for providing livelihood and environmental securities in the hilly areas of

Odisha. During the investigation 3 types of traditional and 4 types of improved agroforestry systems comprised of 67 species of 54 genera representing 25 families are found to be used by the tribal communities in the plateau zone of Odisha.

Keywords: *Plateau zone, Demographic pressure, Forest dependants, Agroforestry, Livelihood security.*

INTRODUCTION

In India, out of 329 M ha hectare of geographical area, 145 M ha is cultivable, with per capita cultivated land availability coming down from 0.48 ha in 1952 to less than 0.15 ha at present posing a threat to the security of food and livelihood. Odisha is an agrarian state with Agriculture and Animal Husbandry sectors are mainstay of livelihood and providing employment directly or indirectly to 70 per cent of total work force. The per capita availability of cultivated land was 0.39 hectares in 1950-51, which has declined to less than 0.13 hectares. This calls for judicious management of land through adoption of agroforestry practices in this computer era. Agroforestry offer a great scope to stabilize and increase the productivity of the land as a whole. Now-a-days agroforestry gaining importance in the recent years for meeting the manifold objectives of providing opportunities for employment generation in rural areas of achieving food security, economic stability, social and environmental sustainability that provide life supporting systems for human beings, plants and animals. It not only improves the farm production and income but also conserves the natural resources and ensures their optimum utilization. Agroforestry has been defined as a sustainable land management system which increases the overall yield of the land, combines the production of crops (including tree crops) and forest plants and/or animals simultaneously or sequentially, on the same unit of land, and applies management practices that are compatible with the cultural practices of the local population (King and Chandler, 1978). Agroforestry systems based on lac, gum, resin, and apiculture and sericulture cultivation have been developed for providing livelihood opportunities to the stakeholders. Dhyani *et al.* (1996) successfully developed and demonstrated mulberry and muga sericulture based agroforestry systems for the north-eastern hill region. Realizing the importance of this type of land use, emphasis is being given to promote agroforestry in agricultural lands and permanent fallow lands for stabilizing food production and extending the environmental services. Agroforestry promotes economical promotion and environmental amelioration. Timber tree plantations can also benefit from the global environmental facilities like carbon trading (Pandey, 2007 and Dogra, 2007). Cultivation of the suitable agricultural crops/intercrops in the inter spaces of woody components are selected with utmost care. The choice of intercrop also depends on characteristics of particular tree species *e.g.* root system, canopy, allelopathic effect of litter and technical factors *viz.* agro-climatic and edaphic conditions (Batish *et al.*, 2007).

Methodology

This study was conducted in ten *v*illages with five villages each in Mayurbhanj (Dhalabani, Kaliani, Gudugudia, Jashipur and Durdura under Similpal Biosphere reserve) and Keonjhar districts (Sitarampur, Arsala, Gonasika, Saharpur and Godipatana). The study area was purposefully selected as the local community mostly adopting different indigenous and improved agroforestry systems. They also collect and sell the farm produces either in the local market or to the middleman as a source of income for their livelihood. The villagers were mostly tribal group and belong to Bathudi, Ho, Munda,Kolha and Santal communities and lived in compactly assemblage houses. The study area is situated between 21°-01″ and 22°-34″N latitude and 85°-11″ and 87°-11″E longitude. Physiographically, this agro-climatic zone is a part of the Chhotanagpur plateau of Jharkahand and forests cover nearly 50 per cent of the total geographical area of the zone. The climate is hot, moist and sub-humid with mean summer maximum temperature of 36.9°C and means winter minimum temperature of 11.9°C. The mean annual rainfall is 1575.4 mm of which about 86.1 per cent is received during the monsoon season (June to October). Investigation was carried out during the years 2009-10 and 2010-11 to collect information about the adoption different agroforestry systems in hills, home garden and village commons, indigenous management practices, present status and strategies required for their conservation. Techniques like field visits, questionnaire survey, personal discussion and focus group discussion with tribal community, VSS (Van Surakshya Samiti), forest and environment department, Odisha were employed during the survey. The information was also collected through a field survey using semi structured interview schedules (Bargali *et al.*, 2007 and Pandey *et al.*, 2011) with adult members or head of the family. The study involves intensive field visits at regular intervals and careful observation of the floral resources during all the seasons of the year *viz. kharif, rabi* and *summer*.The explorative study was undertaken in different landscape like forest, periphery area, degraded open forest/village forest, waste lands, pasture lands, agricultural field and other village commons by a multi-disciplinary team comprising of professionals from forestry, agriculture, botany specialisation and NGO personnel working in the conservation of natural resources.

Suitability of Potential Agroforestry Systems

For a particular region selection of proper agroforestry system depends on the success of system. Proposed land to be utilised cautiously as per based on land capability classifications. Relevant agroforestry systems based on the local farming situations and needs. In valleys of the plateau zone multipurpose trees/shrubs to be planted in boundary/block plantations with suitable silvicultural operations. On terraced fields of hills trees/shrubs, grasses and legumes could be established either in pure or multiple cropping pattern. The choice of the plant species should be based as per farmer's need, preference, market demand and system followed. The Following major factors determine suitability of the system.

1. Land capability of the area
2. Agro-climatic conditions
3. Needs of the farmers
4. Market demand

Traditional Agroforestry Systems in the Plateau Zone of Odisha

Odisha is traditionally an agrarian state. The traditional agroforestry systems which are prevailing in the state since time immemorial are broadly agrisilvicultural, silvipastoral and agrisilvipastoral systems. The prominent agrisilvicultural systems include podu/jhum cultivation, multipurpose trees on farm lands, multiple species tree gardens and agroforestry for fuel wood production. The important silvipastoral systems are trees on pasture, fodder trees and shrubs on forest land and live fence of fodder trees and shrubs. The very old agrihortisilvicultural system prevailing in the state is home garden. These indigenous agroforestry systems not only support the livelihood through production of food, fodder and fuel wood, but also mitigate the impact of climate change through carbon sequestration (Arora *et al.*, 2011, Singh *et al.*, 2008a, Bargali *et al.*, 2009).

1. Agrisilvicultural System (Tree Crops + Grain crops)

In this system, cultivation of annual food crops and woody perennials enhance productivity and sustainability of land. It helps to stabilize slopes, minimize run off and provides fuel wood, poles, small timbers, fruits, fodder, nuts and organic manures.

(i) Shifting Cultivation

Shifting cultivation (locally Podu) is a primitive agricultural system and involves illicit felling and burning of the existing vegetation to generate nutrients in situ to support the crop cultivation for few years before the area is abandoned. It is also known as slash and burn method prevalent in almost all the tribal areas of Odisha covering 119 blocks. About 5298 sq. km area annually is under this primitive agriculture practice. It is not only for sustenance of their livelihood but also as a way of life for their rituals and festivals. Shifting cultivation is considered to be the evillest practice of ancient agriculture. The effects of shifting cultivations are harmful on environment as well as on biodiversity. Agroforestry systems have been developed as alternative/or improvement of shifting cultivation using local resources and conservation-based measures in the NEH region (Dhyani and Chauhan, 1995).

(ii) Scattered Trees in Crop Land

The practice of growing agricultural crops under scattered trees in farm lands is an age old practice and does not seem to have changed for centuries. Growing multipurpose trees in paddy field bunds while crops are grown alongside or underneath has been adopted by farmers either by protecting and managing the trees that are already there or by planting new trees. There are different spacing pattern and densities of plantation depending on the type of tree selected, soil, agro-climatic condition, requirement of farmer's/market demand of woody crops. A variety of locally-adapted multipurpose trees were widely scattered over farmlands Among these are *Acacia nilotica, Azadirachta indica, Borassus flabellifer, Terminalia arjuna, Butea monosperma, Madhuca latifolia, Ziziphus mauritiana etc.* By growing only a few species of trees the farmers reduced the risk of crop competition with trees.

(iii) Trees in and Around Farm Boundaries

Eucalyptus spp., *Acacia nilotica, Terminalia arjuna, Tectona grandis, Madhuca latifolia, Bamboo* spp. *etc.* are grown in and around agricultural farm boundaries. These trees which are around the farm demarcate one farm from another. This type of agroforestry is being practiced by farmers in Odisha since very long. These trees are grown for shade, fodder, fuel wood, fruit, small timber, agricultural implements and other house hold needs.

(iv) Farm Forestry for Fuel Wood Production

In this system, mostly fast growing fuel wood tree species are planted in/around farm lands to provide fuel wood needs of rural people and also protect agricultural crops adverse climatic hazards. This system is more prevalent in areas of scarcity of fuel wood. Tree species commonly grown *are Acacia nilotica, A. auriculiformis, Pongamia piñnata, Casia siamea, Samanea saman, Eucalyptus* spp., *Ziziphus mauritiana etc.*

2. Silvi-pastoral System (Tree+ pasture/animal)

Silvi-pastoral systems are suited to marginal degraded lands for forage, livestock and wood production. This system comprised of top feed tree species carrying grasses or legumes as under storey crops (Nimbole, 1992). Farmers having unutilised dry land/waste lands holdings adopt this system which provide fodder, fuel and timber. The tree species are Subabul (*Leucaena* spp.), Shisham (*Dalbergia* spp.), Babul (*Acacia nilotica*), Kanchan (*Bauhinia purpurea*) *etc.* can be planted, whereas grasses are Napier hyb. grass (*Pennisetum purpureum*), Sudan grass (*Sorghum sudanense), Deenanath grass (Pennisetum pedicellatum*) and legumes like *Stylo* (*Stylosanthes hamata*), *cowpea* (*Vigna unguiculata*), Country Bean (*Lablab purpureus*) *etc.*can be grown as pasture components.

(i) Fodder Llants for Green Fences/Living Fences

Green fences/living fences are lines of trees or shrubs planted on farm boundaries or on the borders of farmyards and comprises of rows of single or multiple densely planted a mixture of fodder plant species Fodder trees are hardy and can provide year round fodder to be used as a supplement in lean periods and as a viable feed resource to supplement the income of farmers in rural areas. In this system various fodder trees and shrubs are planted as live fence at closer Spacing to protect the crops from stray animals. They are maintained as hedges by regular pruning and the pruned materials are fed to animals. Several fodder species are grown along plot boundaries; the foliage lopped periodically and fed to animals that are kept in stalls. The plant species grown in this system are *Erythrina indica, Gliricidia sepium, Acacia nilotica, Ziziphus mauritiana, Bauhinia purpurea, Moringa oleifera etc.* Trees are also planted on farmland and farm boundaries as a source of edible fruits, fodder, fuel and shade.

Both woody and non woody components were grown in irregular manner. Natural regenerations, plantation of different fodder trees without long term plan and to accommodate more plants in limited space were main causes of irregular arrangements. Fodder shrubs also conserve the soil, supply fuel wood and provide bee forage for honey production. Fodder trees for green fencing of different sizes provides aesthetics, ornamental, improved fodder quality and nutritional security to the animals. Additionally, pod-producing trees may become a very useful source of energy and protein concentrate. Foliage of fodder trees retain sufficient crude protein, minerals and energy due to the deep root systems of these species. Subsistence of farmers usually include trees as an output product of their farm, whether it is for wood, shade, soil conservation, or fodder.

(ii) Scattered Trees on Pasture Lands

In this system trees having fodder values are seen on pasture lands. These trees are scattered and irregular. This system provides fodder to animals from the grasses grown as well as from tree leaves and fruit pods. The trees provide shelter to animals from sun and rain. This system is found throughout the state. The common tree species associated are *Acacia leococephala, Acacia nilotica, Albizia lebbeck, Ficus religiosa, Terminalia arjuna, Leucaena leucocephala, Feronia elephantum, Terminalia tomentosa, Samanea saman, Azadirachta indica, Tamarindus indica etc.*

(iii) Contour Vegetation Strips

This is a traditional system in which rows of grasses, stones/debrises, crop residues and other organic debris are placed across hill slopes to minimise soil erosion. These living barriers of trees and shrubs provide a large numbers of useful products of food, fuel, poles, fodder *etc.* The effectiveness of the strip depends on so many factors such as the type of tree planted, the spacing of the trees and the width of the strip, the steepness of the slope, the amount of rainfall and the type of soil.

3. Agrisilvipastoral System (Tree Crops + Grain crops + animals)

The production of woody perennials (trees or shrubs) with cattle in the same site combined with annuals and pastures are termed as Agrisilvipastoral System. The objective of agrosilvopastoral (ASPS) systems is to provide forest products such as fuel wood, poles, and timber which meet the needs of livestock, domestic needs and also to alleviate the pressure on natural forests.

(i) Home Garden

It is the most ancient form of agroforestry system found in all districts of Odisha. It is also called Homestead Agroforestry/Gharbari Bagicha, Kitchen garden and Household garden. More intensive forms of this practice are seen commonly in rural areas. Home garden consists of on growing trees, shrubs, vines and herbaceous plants in or around the homestead, aiming mostly food production for house hold consumption. Coconut, arecanut, guava, mango, citrus, tamarind, jackfruit, papaya, banana, moringa, custard apple and many multipurpose trees are the major trees found grown in home gardens. It is a system in which an operational farm unit grows a number of tree species along with livestock, poultry and fish for the purpose of satisfying the household needs.

'Bari' (backyard) farming is a common practice in Odisha. In Odisha 'Ghar' (house) is incomplete without 'bari'(backyard). Like other areas, a common backyard or housestead in Odisha consists of trees, herbaceous crops, livestock sheds along with additional features like threshing yard,fuel wood stack,toilet *etc.* In coastal Odisha, pond is a common feature. Among trees *Acacia,* Mangium, Neem, Teak, Tamarind, Mango, Bael *etc.* (mostly creepers) among vegetables, chicken and goat among livestock and fencing (mostly brushwood) are characteristic features of Odisha backyards. Most of these backyards also have some medicinal herbs planted to take care of household ailments.

A home garden is a piece of land with a definite boundary surrounding a homestead,being cultivated with a diverse mixture of perennial and annual plant species,arranged in a multilayered vertical structure,often in combination with raising livestock,and managed mainly by household members for subsistence production. Next to natural forests,home gardens probably are most intensive and optimal biomass production system. In India the home gardens are generally found in tropical and sub-tropical areas and characterized by high specices diversity and usually three to four vertical canopy strata. These consist of an herbaceous layer near ground, a tree layer at the upper level and one to two intermediate layers. The lower most being at less than 1.0 m in height dominated by different vegetables like okra,chilli,tomato,cabbage,dioscorea *etc.* and the second layer is 1-4 m height comprising food plants such as banana, papaya, lemon *etc.* The third layer is at 4-10m height occupied by fruit trees like guava, drumstick,custard apple *etc.* The upper layer is the tree layer which can be divided into two consisting of the emergent full grown timber and fruit trees having height more than 20m and medium size trees of 10-20m. In the upper layer species like *Magnifera indica, Samanea samam,Bambusa vulgaries, B.tulda,Cocos nucifera, Artocarpus heterophyllus, Aegle marmelos, Tamarindus indica,Azadirachta indica, Leucaena leucocephala etc.* are grown. In addition to these components, livestock also contributes significantly to the household income of small-scale home gardens in many developing countries, while fulfilling many social and cultural needs. In some of the very small gardens, where land is a constraint to crop production, livestock are sometimes the main income generators, serving as cash buffers and capital reserves. In raising a home garden; farmers follow the multistoreyed, diversified nature of the forest. Large homesteads maintain fish ponds, fruit trees, flower plants and bamboo grooves along with an area for keeping milk cows, bullocks, goats and ducks (whose dung/faecal matter provide excellent source of organic manure).

(ii) *Bauhinia purpurea* Based System

It is also known as Barada saga by tribals of Mayurbhanj. It is planted surroundings of homestead land and farm boundaries along with sorghum, maize, cowpea *etc.* Tender leaves, flowers and flower buds eaten as vegetables and fodder for cattle, goats. It is an ornamental tree with beautiful and scented flowers.

Improved Agroforestry Systems in the Plateau Zone of Odisha

Keeping in view the constraints and limitations of hill region, the following improved agroforestry measures are suggested to prevent further deterioration of the land under different uses.

1. Agrisilvicultural System

In this system dense planting of trees and shrubs in the form of hedges helps soil conservation. In agrisilviculture system nitrogen-fixing trees (NFTs) meet the nitrogen requirement of crops grown in the inter spaces of trees besides providing organic matter to the soil. Most of the common nitrogen fixing trees species are *Acacia, Albizia, Dalbergia, Gliricidia, Leucaena, Pongamia, Sesbania etc. Leucaena* species are the best suitable hedge row component whereas field crops like Sorghum, Bajra, Arhar, Cowpea, Maize *etc.* can be grown as understorey crops in the alleys formed by hedges.

(i) *Acacia auriculiformis* based Agroforestry System

*Acacia auriculiformis*is grown extensively with commercial production concentrated mainly in the plateau zone of Odisha. Suitability of Acacia trees in coastal region in terms of biophysical and environmental considerations is very good. Generally, Acacia plantation in normal rice field bunds is found to be grown well.

Acacia auriculiformis based agroforestry practices in improved agroforestry practices of different sizes were found playing various roles. They provide aesthetic, ornamental values. They made a significant contribution to the fuel wood and timber. *Acacia auriculiformis* is a useful multi-purpose tree species, being fast-growing and suitable for timber and pulp production. The species is valued also for its timber and high quality chemical pulp. The tree also contains nitrogen fixing bacteria which can help rejuvenate these poor soils. *Acacia auriculiformis* is a useful multi-purpose tree species, being fast-growing and suitable for timber and pulp production. Harsh *et al.* (2000) reported that the returns from this Acacia based agrisilvicultural systems could be improved by extracting gum in improved methods from the trees.

(ii) *Acacia mangium* based Agroforestry System

Acacia mangium in the recent years. has achieved wide popularity in Odisha, as a multipurpose fast growing species. It is fast growing in nature, transit free permit and huge demand of pulp and paper industries in the irrigated as well non-irrigated areas, the farmers have the tendency to grow *Acacia mangium* trees on the paddy bunds. These plants are pruned regularly so that its shade does not affect the yield of rice. Again the pruned branches are usually used as staking materials for vegetable crops and also for fuel purposes. This is an evergreen leguminous tropical tree popular for its rapid early growth.

(iii) Wind Breaks

Wind break is a protective planting round a farm or a field to protect it from strong winds. It is usually composed of one or two rows of trees or shrubs, spaced at 0.5m to 2.5 m apart depending on the nature of the species. Suitable tree species are *Acacia mangium, Acacia auriculiformis, Azadirachta indica, Leucaena leucocephala,Casuarina equisetifolia, Anacardium occidentale* **etc.**

(iv) SALT

The SALT (Sloping Agricultural Land Technology) is the modified model of agroforestry system introduced in Odisha in recent past by the state forest department, Kutia Kandha Development Agency (KKDA). NGOs. In this contour hedgerow intercropping system nitrogen fixing plants are planted along contour with desired food crops and other useful plants in the alleys. *Desmodium rensonni, Indigofera anil,* and *Leucaena leucocephala* are the suitable hedgerow species. Similarly, for SALT, contour hedgerows are prepared and the space between two rows of hedgerow areas is called a strip. Alternate strips are selected for cultivation during the year and the strips left uncultivated area planted with permanent crops of horticultural and wood value.

(v) Planting on Terraces

Terraces are usually constructed across hill terrains can be planted with grasses, trees and shrubs on the ridges of the terrace or at the back of the terrace to stabilise the soil, add leaf manures and provides useful products such as food, fuel, fodder. Suitable plant species are Guava, Amla, Custard Apple, Lemon, Mango, Drumstick, Pomegranate, Papaya, Subabul, Kachnar *etc.*

2. Silvipastoral System

(i) *Leucaena leucocephala* (Subabul) based System

In the plateau zone of Odisha Subabul plants are planted surroundings of homestead land along with sorghum, maize *etc.* to meet the fodder requirements of goat and cattle population, the boundary demarcation and small timber requirements.

(ii) *Acacia nilotica* (Babul) based System

Babul is a multipurpose nitrogen fixing tree that is widespread in wastelands of Nuapada, Kalahandi, Nayagarha. It provides high quality fodder for cattle, sheep, and cattle. fuel wood apart from enriching the soil with nitrogen.Leaves and pods contain 8 per cent digestive protein.

3. Agrisilvipastoral System

In this type of system farmers of Odisha cultivate agricultural crops along with fodder trees/rearing of livestock and the forest trees to meet their food, fodder and other tree based needs. This system has indicated that combination of multipurpose trees are planted with grasses and legumes in an integrated way and their forage is utilised through cut and carry system.

(i) *Gmelina arborea* (Gamhar) based System

In this system farmers of Keonjhar, Mayurbjanj district are planting gamhar plants in compact patches at a spacing of 2.5mx2.5m. Crops like beans, cowpea, brinjal, tomato *etc.* are produced upto 3 years of plantation and after that shade loving crops like colocasia, turmeric, arrowroot, yam, elephant foot yam *etc.* are cultivated. Gamhar is a fast growing fodder tree. In dry season and to some extent during rainy season farmers mostly cut and carry leaves and tender twigs for their animals. Pods are not palatable. Leaves are also used for green manuring and mulching of turmeric, ginger *etc.* Again indigenous fast growing tree species *G. arborea* (gamhar) wood is even-grained, soft, light and strong for which it is demanded for panelling, carriages, furniture, boxes and carpentry of all kinds.

(ii) Improved Home Gardens

With proper layout and design farmers are introducing improved varieties of crops, plants, animals and adopting better management practices in home stead lands. They are cultivating various types of leafy vegetables, fruit crops, tuber crops, medicinal plants. The fast growing trees like Subabul, Mahogany, Gamhar *etc.* are grown in periphery of the boundary. Bamboos are also one of the important non timber forest produce cultivated in the improved home gardens. These homesteads are prevalent in all agro-climatic zones of Odisha. The crop diversity is in such home gardens are high.

4. Other Systems

(i) Aquaforestry

Now-a-days farmers are diverting some portion of land for rearing fish in ponds, which is surrounded by different species of trees like *Mangifera indica, Moringa oleifera, Cocus nucifera, Gliricidia sepium etc.* A well-managed integrated farming including livestock with fish in ponds and fruit trees or forest trees on farm boundary make an appropriate technology for sustainable income to the farmers. The leaves of the many leguminous trees such as *Leucaena leucocephala, Gliricidia sepium, Moringa oleifera etc.* have been found to serve as good fish feed.

(ii) Entomoforestry

Lac is being cultivated in tribal areas of Bolangir,Kalahandi, Balasore, Keonjhar, Mayurbhanj, Sundargarh, Nawarangpur, Sambalpur and Koraput districts of Odisha. Mostly farmers are rearing lac insects on *Butea monosperma, Zizyphus mauritiana* and *Schleichera oleosa* on field bunds and nearby forests. Lac cultivation has very high potential for generating employment opportunity for both men and women. The climatic condition of our country as well as in our state is too congenial for growth and development of this Indian lac insect, *Kerria lacca.* It is a good source of livelihood for resource poor farmers and also a good crop for marginal and degraded lands. Lac cultivation serves as both a complementary and supplementary source of income to the existing livelihood options for the farmers. It is also an assured source of income during lean period of agricultural activities.

Mostly in the tribal dominated districts of Mayurbjanja and Keonjhar farmers reared silk worms. In Odisha, three types of silk worms such as Mulberry, Tassar and Eri are cultivated. Mulberry, Tassar and Eri silk are obtained from trees of *Morus alba, Terminalia tomentosa* and *T. arjuna,* and *Ricinus communis* respectively.

Farmers also rearing tree crop combinations including floriculture. They usually go for tree crops like Neem, Mango, Guava, Pongam *etc.* with crop combinations like mustard, sun flower *etc.*

Results and Discussion

In the present investigation, it was observed that the most common and generally accepted traditional and improved agroforestry systems based on nature of components in the study sites are of 3 types and 4 types respectively which is coincides with the opinion of Nair (1985 and 1993) stated 4 types of agroforestry systems. Agroforestry practices are classified according to the components involved. Combe (1982) classified into 24 systems whereas, Deb Roy (1989) into 6 systems. In the plateau zone of Odisha 3 types of traditional agroforestry systems are Agrisilvicultural System, Silvipastoral system and Agrisilvipastoral system are widespread. There are also 4 types of improved agroforestry systems such as Agrisilvicultural System, Silvipastoral system, Agrisilvipastoral system and Other Systems (Aquaforestry and Entomoforestry) are prevalent. These systems are best known practices for livelihood, suitable land management and sustainable development (Kittur and Bargali 2013, Parihaar *et al.*, 2014).

Agroforestry systems are very specific to site and climate. For the better lands like II and III classes of soil groups, the food based systems like Agrisilvicultural, Agrisilvipastoral system intercropping with multipurpose nitrogen fixing trees/ shrubs are more suitable and give higher returns (Table 29.1). For the resource poor regions like IV and V classes of soil groups the fodder based systems like Silvipastoral, pasture management *etc.* are suitable. These systems will provide assured production of fodder even in the lean months. Where as in marginal and degraded lands like VI and VII classes of soil groups where crop production is not possible can be used for fuel, timber and fibre based systems such as tree farming, timber cum fibre production systems *etc.* (Table 29.1).

Table 29.1. Suitable Agroforestry Systems in the Plateau Zone of Odisha

Need	*Food Production*	*Fodder Production*	*Fuel/Timber/ Fibre Production*
System	Agrisilvicultural, Agrisilvipastoral, Other Systems (Aquaforestry and Entomoforestry) Intercropping with Nitrogen fixing trees (NFTs) *etc.*	Silvipastoral, System,Pasture management *etc.*	Tree farming, Timber cum fibre system *etc.*
Land	Arable lands	Non-arable lands	Marginal degraded lands
Suitable soil classes	II and III	IV and V	VI and VII

The Study reveals that a total 67 plants of 54 genera representing 25 families (Table 29.2) were found in agroforestry systems from the selected villages. An analysis of habits of the species manifest that out of 67 species 46 are trees, 12 shrubs, 03 herbs, 03 palms and 03 grasses (Figure 28.1). Thus, the number of tree species are raised more than others (Table 29.2). The largest number of plant species are distributed under family Fabaceae (20), followed by Poaceae (8) and Rutaceae (6) (Figure 28.2). Out of 67 species, 40 species are used for Miscellaneous uses (fibre, medicinal, oil, green manure, lac *etc.*), 32 for Edible, 23 for Fodder, Fuel 22 and 12 for

Timber (Figure 28.3). Hence there is an urgent need to cultivate suitable plant species that are not only economically important but also conserve natural environment.

Table 29.2. Species-wise Contribution in different Agroforestry Systems in the Plateau Zone

Sl.No.	Scientific Name	Local Name	Family	Habit	Uses
1	*Acacia auriculiformis*	Acacia	Fabaceae	Tree	T, Fu
2	*Acacia catechu*	Khaira	Fabaceae	Tree	Fu, M
3	*Acacia mangium*	Mangium	Fabaceae	Tree	T, Fu
4	*Acacia nilotica*	Babul	Fabaceae	Tree	Fo, Fu
5	*Aegle marmelos*	Bela	Rutaceae	Tree	Ed
6	*Albizia lebbeck*	Sirisha	Fabaceae	Tree	M, Fo, Fu
7	*Albizia procera*	Dhalasiris	Fabaceae	Tree	M, Fo, Fu
8	*Ananus comosus*	Sapuri	Bromeliaceae	Herb	Ed
9	*Andrographis panicuiatus*	Bhuinnimba	Acanthaceae	Herb	M
10	*Anthocephalus chinensis*	Kadambo	Rubiaceae	Tree	M, T
11	*Annona squamosa*	Ata	Annonaceae	Tree	Ed
12	*Artocarpus heterophyllis*	Panas	Moraceae	Tree	Ed, Fo
13	*Azadirachta indica*	Nimba	Meliaceae	Tree	M, T
14	*Bambusa bambos*	Kantabanso	Poaceae	Shrub	M, Fo
15	*Bambusa nutans*	Dababanso	Poaceae	Shrub	M, Fo
16	*Bauhinia acuminata*	Kanchan	Fabaceae	Tree	Ed, Fo, M
17	*Bauhinia purpurea*	Kuliar	Fabaceae	Tree	Ed, Fo, M
18	*Bombax ceiba*	Simili	Bombacaceae	Tree	M
19	*Boraassus flabellifer*	Tala	Arecaceae	Palm	Ed
20	*Buchanania lanzan*	Char	Anacardiaceae	Tree	M
21	*Butea monosperma*	Palasa	Fabaceae	Tree	M, Fu
22	*Cassia fistula*	Sunari	Fabaceae	Tree	M
23	*Cassia siamea*	Sana chakunda	Fabaceae	Tree	Fu, M
24	*Ceiba pentandra*	Amerika tula	Bombacaceae	Tree	M
25	*Cicca disticha*	Narakoli	Euphobiaceae	Shrub	Ed, M
26	*Citrus aurantifolia*	Lembu	Rutaceae	Shrub	Ed, M
27	*Citrus aurantium*	Kamala (Khata)	Rutaceae	Shrub	Ed
28	*Cocos nucifera*	Nadia	Arecaceae	Palm	Ed
29	*Dalbergia latifolia*	Kala Sishu	Fabaceae	Tree	Fo, T, Fu, M
30	*Dalbergia sissoo*	Sishu	Fabaceae	Tree	Fo, T, Fu, M
31	*Dillenia indica*	Oaau	Dilleniaceae	Tree	Ed
32	*Diospyros malabarica*	Kala Kendu/ Maakada kendu	Ebenaceae	Tree	Ed, Fu
33	*Diospyros melanoxylon*	Kendu	Ebenaceae	Tree	Ed, Fu
34	*Embiica officinalis*	Aonla	Euphorbiaceae	Tree	Ed, M
35	*Eucalyptus tereticornis*	Eucalyptus	Myrtaceae	Tree	T, Fu

Sl.No.	Scientific Name	Local Name	Family	Habit	Uses
36	*Feronia elephantum*	Kaitha	Rutaceae	Tree	Ed, M
37	*Ficus hispida*	Baidimiri	Moraceae	Tree	Ed
38	*Ficus racemosa*	Dimiri/Lowa	Moraceae	Tree	Ed
39	*Gliricidia sepium*	*Gliricidia*	Fabaceae	Shrub	M
40	*Gmelina arborea*	Gambari	Verbenaceae	Tree	T, Fo, M
41	*Lawsonia inermis*	Manjuati	Lythraceae	Shrub	M
42	*Leucaena diversifolia*	Diversifolia	Fabaceae	Tree	Fo, Fu, M
43	*Leucaena leucocephala*	Subabul	Fabaceae	Tree	Fo, Fu, M
44	*Madhuca indica*	Mahula	Sapotaceae	Tree	Ed, M
45	*Mangifera indica*	Ambo	Anacardiaceae	Tree	Ed, T, Fu
46	*Moringa oleifera*	Sajana	Moringaceae	Tree	Ed, Fo
47	*Murraya koenigii*	Versunga	Rutaceae	Shrub	Ed
48	*Musa* spp.	Kadali	Musaceae	Shrub	Ed
49	*Pennisetum pedicellatum*	*Deenanath grass*	Poaceae	Grass	Fo, M
50	*Pennisetum purpureum*	Napier hyb. grass	Poaceae	Grass	Fo, M
51	*Phoenix sylvestris*	Khajuri	Arecaceae	Palm	Ed, M
52	*Pongamia pinnata*	Karanja	Fabaceae	Tree	Fo, Fu, M
53	*Psidium guajava*	Pijuli	Myrtaceae	Shrub	Ed
54	*Punica granatum*	Dalimba	Punicaceae	Shrub	Ed
55	*Rauwolfia serpentina*	Patalgaruda	Apocynaceae	Shrub	M
56	*Samanea saman*	Nidrabati	Fabaceae	Tree	Fo, Fu, M
57	*Sesbania grandiflora*	Agasti	Fabaceae	Tree	Fo, Ed, M
58	*Schleichera oleosa*	Kusum	Sapindaceae	Tree	Ed, M
59	*Spondias mangifera*	Ambada	Anacardiaceae	Tree	Ed
60	*Sorghum sudanense*	Sudan grass	Poaceae	Grass	M
61	*Stylosanthes hamata*	*Stylo*	Poaceae	Grass	M
62	*Swietenia macrophylla*	Mahogany	Meliaceae	Tree	T, Fu
63	*Syzygium cumini*	Jamu	Myrtaceae	Tree	Ed, Fu, Fo, T
64	*Tamarindus indica*	Tentuli	Fabaceae	Tree	Ed, Fu, Fo
65	*Tectona grandis*	Saguan	Verbenaceae	Tree	T
66	*Thysanolaena maxima*	Phulochanchuni	Poaceae	Herb	Fo, M
67	*Ziziphus mauritiana*	Barakoli	Rhamnaceae	Tree	Ed, Fo, Fu, M

T: Timber; Fo: Fodder; F: Fuel wood; Ed: Edible fruits/leaves; M: Miscellaneous uses as fibre, medicinal, oil, green manure, lac *etc.*

Criteria for Selection of the Tree Component

The following factors should be kept in mind when selecting tree species. The tree component should be:

(i) **Fast growing:** Fast growing in nature and farmers can get benefit in a short period.

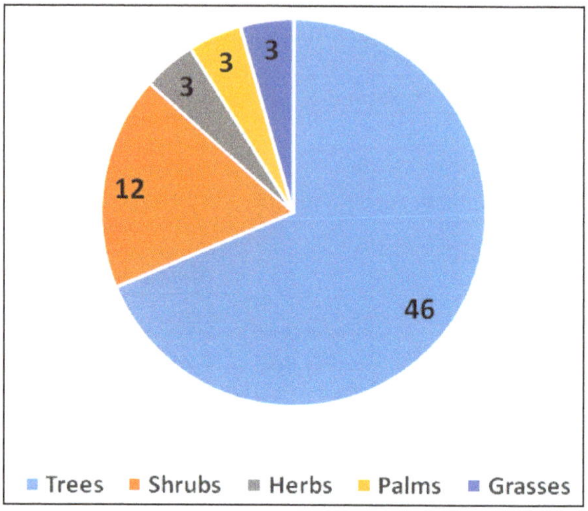

Figure 29.1. Diversity of Plants in different Agroforestry Systems.

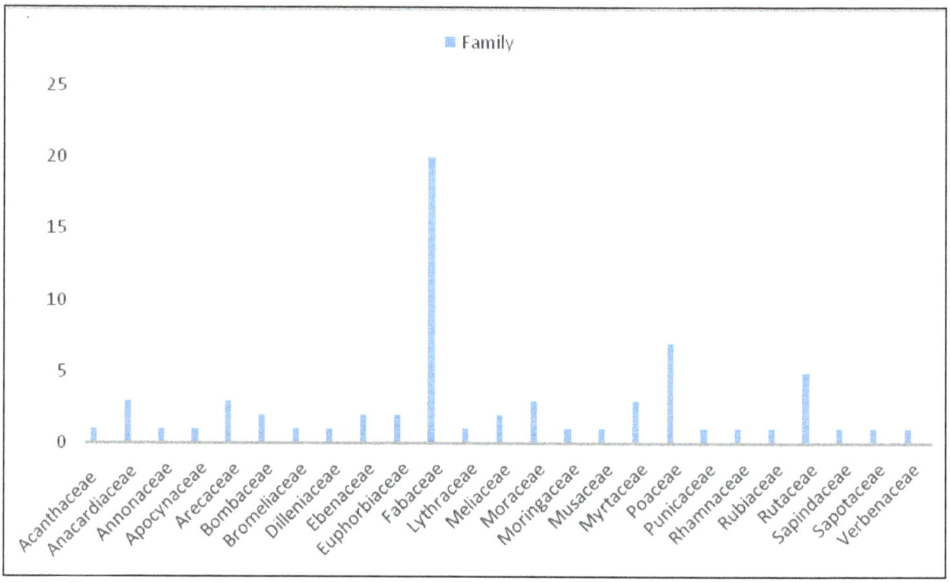

Figure 29.2. Diversity of Plants used in different Agroforestry Systems According to Family.

(ii) **Multipurpose**: Having multipurpose uses, *e.g.* food,fodder, fuel, timber, fibre, green manure *etc.*

(iii) **Heavy and palatable foliage**: High palatable and relished of foliage by livestock.

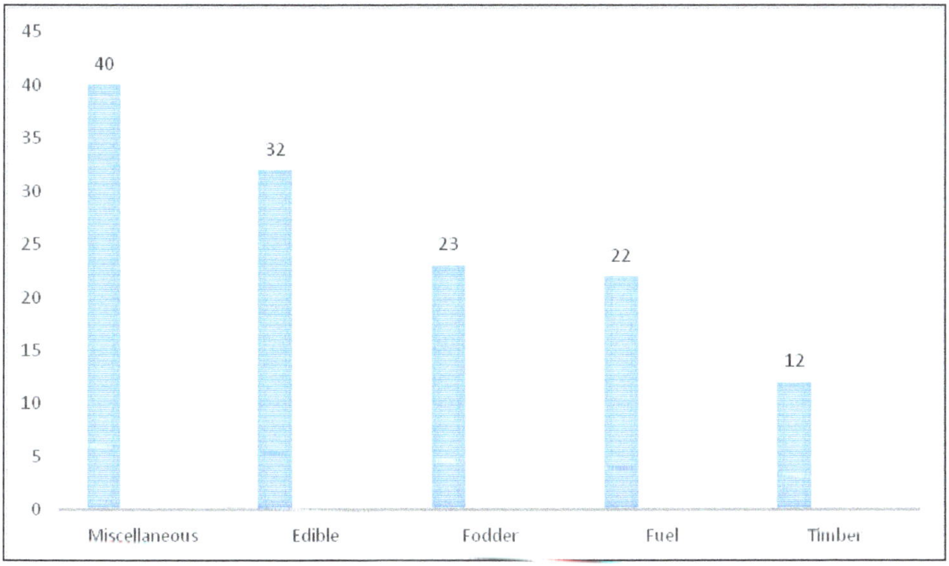

Figure 29.3. Utilisation Pattern of different Plants in the Selected Agroforestry Systems.

(iv) **High coppicing ability**: Having good coppicing/sprouting ability (ability to regenerate by shoot or root suckers).

(v) **Ability to withstand browsing, trampling and lopping**: Having ability to withstand browsing, trampling and intensive lopping.

(vi) **Resistant to climatic hazards:** Resistant or tolerant to drought and extremes of temperature, flood, soil variability and other climatic hazards.

(vii) **Easily established:** Required minimum care for plantation, establishment and maintenance.

(viii) **Cast the light amount of shade:** The crown should be high, small, open, and foliage sparse so that understory crops get the right amount of light for optimum growth and development.

(ix) **Deep rooted:** They can draw nutrients and water from deeper soil layers and will not compete with shallow-rooted crops.

(x) **Nitrogen fixing ability:** Plants should have Nitrogen fixing ability to improve soil fertility and productivity of soil.

(xi) **Compatible with annual and perennial crops:** Plants should not have adverse effects on other agricultural crops they are associated with.

(xii) **Environmental adaptation:** Adapted to the agro climatic condition, soil, topography, plant and animal life of the area.

(xiii) **High income**: A well-known tree is better than an unknown tree but when a new species is introduced, it is important to select the most suitable trees since it is not easy to replace them once they have been planted.

Criteria for Selection of the Pasture Component

To be compatible with tree species, the pasture component should fulfill the following requirements.

It should be able to grow well and be compatible with other forage crops.

(i) Tolerant to drought and extremes to climatic hazards.
(ii) Must be prolific seeded or able to self-propagate vegetative.
(iii) It should have high palatability and good nutritive value.
(iv) It should have the ability to conserve the soil, water and nutrients.
(v) It should be able to withstand over grazing and trampling.

Criteria for Selection of the Intercrop Component

(i) Height of intercrops should be less than woody components.
(ii) Tolerant to partial shade of trees.
(iii) Economic life of inter crops should not more than that of the trees.
(iv) They should have adapted in the locality with reference to agro-climatic, edaphic and socio-ecological conditions.
(v) Intercrops should not have any inimical effect on growth of trees.
(vi) They should not more susceptible to diseases and pests which effect trees.

CONCLUSION

Agro- climatic situations in the plateau zone of Odisha are very unique with respect to soil and climate, which is most suitable for different agroforestry systems and are gaining popularity among farmers. Indigenous and improved agroforestry systems in Odisha found to be immense importance in supporting livelihood of rural poor both for food and for cash. They made a significant contribution of food, fuel wood, fodder, fertilizer, fibre and timber. The major factors for adoption of improved agroforestry systems provide assured income and biodiversity conservation. Over the times indigenous agroforestry systems, have not been static but rather have been constantly modified by farmers to meet demands of society at each point in time. In recent times however, the loss of soil fertility and biodiversity, frequent occurrence of natural calamities among others threaten the farming communities and leads to commercialization of improved agroforestry systems. Improved agroforestry systems are comparatively profitable than the indigenous/conventional agroforestry systems. The contribution of the trees in the farming systems certainly added to the diversity dimension by way of income and employment to the farm households besides fulfilling the household requirements of farming communities. Both forms of agroforestry systems have specific roles to play in the livelihoods and environmental development, which have to be carefully implemented in rapidly changing scenario.

Now-a-days there is much scope to cultivate many other multipurpose trees (MPTs) suitable for the agroforestry system in this agro-climatic region. The most decisive factor for the success of agroforestry technologies is the choice of suitable plant species. To enhance the economic condition of rural poor and progressive farmers, it is essential to impart necessary training to them in systematic and scientific cultivation methods. Adoption of improved agroforestry systems is yet to become widespread. Now farmers have recognized that planting multipurpose plants in the agroforestry practices will give sustainable income.

ACKNOWLEDGEMENTS

The authors are grateful to the Director, Similipal Tiger Reserve, Baripada and Managing Director, SAMBANDH, Jashipur, Mayurbhanj, WOSCA, Keonjhar for providing opportunity and constant support. The authors are also thankful to members of Baidya Sanghas, and VSS of Keonjhar and Mayurbhanj district.

REFERENCES

Ahmed, P. 1991. Agroforestry: A viable land use of alkali soils. *Agroforestry Systems* 14: 23-37.

Arora, V. P. S; Bargali,S. S. and Rawat, J. S. 2011. Climate change: challenges, impacts, and role of biotechnology in mitigation and adaptation. *Progressive Agriculture* 11:8–15.

Bargali, S. S.; Bargali, K.; Singh, L.; Ghosh, L. and Lakhera, M. L. 2009. Acacia nilotica based traditional agroforestry system: effect 154 Parihaar *et al.*, March 2015] 447 on paddy crop and management. *Current Science* 96 (4): 581–587.

Bargali, S. S; Singh, S. P; Shrivastava, S. K. and Kolhe, S. S. 2007. Forestry plantations on rice bunds: Farmers' perceptions and technology adoption. *International Rice Research Notes* 32 (2): 40–1.

Batish, D.R., Kohli,R.K., Jose, S. and Singh, H.P (2007). *Ecological Basis of Agroforestry*. CRC Press, New York.

Combe, J.1982. Agroforestry techniques in tropical countries: Potentials and limitations. *Agroforestry Systems*, 1:13-28.

Dhyani, S. K. 1996. Agro-hortiaquaculture system- an integrated farming system for North- Eastern Hills of India. *J North Eastern Council* 16(1): 18-21.

Deb Roy. 1989. Basic research needs in Agroforestry Indian context. National symposium, Agroforestry systems in India, CRIDA, Hyderabad, Jan.11-13.

Dhyani, S. K. and Chauhan, D. S. 1995. Agroforestry interventions for sustained productivity in NE Region of India. *Range Mgmt and Agroforestry* 16(1): 79-85.

Fonzen, P.F. and Oberholzer, 0. 1984. Use of multipurpose trees in hill farming systems in Western Nepal. *Agroforestry Systems* 2 :187-197.

Getahun, A., Wilson, G.F., and Kang, B.T. 1982. The role of trees in farming systems in the humid tropics. In: MacDonald, L.H. (ed.), *Agroforestry in the African Humid Tropics*, pp. 28-35. United Nations University Press, Tokyo, Japan.

Grainger, A. 1980. The development of tree crops and agroforestry systems. *International Tree Crops Journal* 1: 3-14.

Harsh, L.N; Tewari, J.C. and Bohra, M.D. 2000. Traditional and improved agroforestry systems for sustainable agriculture in the arid region of India. Pages 1035–1037 in Extended Summaries, Vol. 3, International Conference on Managing Natural Resources for Sustainable Agricultural Production in the 21st Century. New Delhi, India: Water Technology Centre, Indian Agricultural Research Institute.

Jambulingam, R. and Fernandes, E.C.M. 1986. Multipurpose trees and shrubs on farmlands in Tamil Nadu State, India. *Agroforestry Systems* 4: 17-23.

Joshi. N.P. and Singh, S.B. 1990. Availability and use of shrubs and tree fodder in Nepal. In: Devendra, C., cd., Shrubs and tree fodder for farm animals. *Ottowa, IDRC.* 211-220.

King, K.F.S and Chandler, M.T 1978: The wasted lands, ICRAF, Nairobi, Kenya, 85 pages.

Kittur, B. and Bargali, S. S. 2013. Perspectives of agroforestry: Present and future facets. *Journal of Progressive Agriculture* 4 (2): 91–4.

Nair, P.K.R. 1985. Classification of agroforestry systems, *Agroforestry Systems,* 3:97-128.

Nair, P.K.R. 1993. An introduction to agroforestry, Kluwer Academic Publishers, Dordrecht, The Netherlands, 499 pages.

Nimbole, N. N. 1992. Performance of multipurpose tree species with or without pruning and understorey pasture). *Stylosanthes hamate. Indian Journal of Dryland Agriculture Research and Development* 7(2): 109-111.

Pandey, D.N. 2007. Multifunctional agroforestry in India. Dogra, A.S. (2007). Contribution of trees outside forests toward wood production and environmental amelioration. *Indian Journal of Ecology,* 38: 1-5.

Pandey, K.; Bargali, S. S. and Kolhe, S. S. 2011. Adoption of technology by rural women in rice based agroecosystem. *International Rice Research Notes* 36: 1–4.

Parihaar, R. S, Bargali, K. and Bargali, S. S. 2014. Diversity and uses of ethno-medicinal plants associated with traditional agroforestry systems in Kumaun Himalaya. *Indian Journal of Agricultural Sciences* 84 (12): (In Press).

Singh, L.; Ghosh. L.; Bargali, S. S. and Saxena, R. R. 2008a. Response of paddy planted under Butea monosperma trees in Chhattisgarh. *International Journal of Ecology and Environmental Sciences,* 34(1): 63–6.

Singh, L.; Ghosh, L; Bargali,S. S. and Saxena, R. R. 2008b. Influence of naturally occurring trees on field bunds and their impact on yield parameters of paddy crop. *Range Management and Agroforestry* 29 (2): 134–7.

Chapter 30

Biodiversity of Underutilised Fruits in the Plateau Zone of Odisha and their Role in Food, Nutrition and Livelihood Security

Umasankar Nayak[1], Gayadhar Shial[2], Madhab Chandra Behera[3] and Ranjan Kumar Tarai[4]*

[1]*Regional Research and Technology Transfer Station (RRTTS), Ranital – 756 111*
[2]*Krishi Vigyan Kendra, Ranital – 756 111, Bhadrak*
[3]*College of Forestry, Siripur, Bhubaneswar – 751 003, Odisha*
[4]*College of Horticulture, Orissa University of Agriculture and Technology (OUAT), Chiplima, Sambalpur, Odisha*
**e-mail: usnayak74@gmail.com*

An investigation undertaken to study the diversity of underutilised minor fruit plants in the North Central Plateau Agro climatic zone located in the northern part of the state of Odisha indicated the presence of 32 indigenous fruit plants under 18 families across the diverse habitats in the study area and hadimmense significance on the food, nutrition, health and livelihoods of the community. Rutaceae and Moraceae are the dominant families with 5 and 3 underutilised fruit plant species, respectively and the presence of such a large number of fruit plant families indicated the rich diversity of the region. Home garden and forest harbour the maximum diversity of the minor fruit plants. These are available throughout the year and mostly consumed by the local community, while a sizable surplus quantity goes to the market either directly or through a local middleman to meet the immediate cash needs during the lean season of the year. During the peak availability period, some of the fruits are preserved as pickles mostly for future use and the preservation techniques and the number of ingredients vary within the study area. However, the population density of fruit plants and their productivity has been drastically declined over the years mostly due to the change in food habits and changing climatic situations.

Keywords: *Underutilised fruits, Biodiversity, Food, Nutrition and livelihoods.*

INTRODUCTION

Plants offer a wide range of ecosystem benefits and play a critical role in the sustenance of human life on the earth. Besides, being the essential constituent of the food chain in all the ecosystems, they regulate some vital ecosystem functions like carbon sequestration, climate control, air purification, soil water conservation, provision of habitat to diverse flora and fauna. Though plant biodiversity is fundamental for ecosystem functioning and sustainable agricultural production yet only a few crop species are utilised for food production throughout the world (Padulosi *et al.*, 2001). Out of the estimated 300000 to 500000 plant species exist in the earth, 30000 are thought to be edible and only 7000 have been either cultivated or collected as food (Garn and Leonard, 1989). Even a greater concern is the fact that only 20 species have provided for 90 per cent of the world's food requirements with wheat, maize, and rice accounting for 60 per cent of the human diet (Collins and Hawtin, 1999). Thus, tens of thousands of edible plant species remain relatively "underutilized", for their ability to contribute to the world's increasing food requirements (Chivenge*et al.*, 2015). The crops, which are neither grown commercially on large scale nor traded widely, may be termed as underutilized crops and these crops are cultivated, traded, and consumed locally (Rai *et al.*, 2005). Underutilized, minor, orphan or neglected crops are labels often applied to plant species that are indigenous, rather than non-native or adapted introductions, and which often form a complex part of the culture and diets of the people who grow them (Padulosi *et al.*, 2002). The underutilized foods can be defined as "the foods which are less available, less utilized or rarely used or region-specific" (William and Haq, 2002). Having higher adaptation to the local environment and greater resistance against the pest and diseases, these crops are less vulnerable to different biotic and abiotic stresses and can improve farm resilience in the fragile ecosystems. Underutilised crops with little requirement of external inputs like fertilisers, pesticides, and less water requirement have minimum ecological footprints in comparison to the high-value crops. According to Salvi, (2016) the use of underutilized plants has long been an intimate part of local cultures and traditions as many neglected and underutilized species play a role in keeping alive cultural diversity associated with food habits, health practices, religious rituals, and social exchanges. Underutilized crops have great potential to alleviate hunger directly, through increasing food production in challenging environments where major crops are severely limited (Mayes *et al.*, 2012). Kunkel (1984) that Once underutilized food crops are properly utilized, they may help to contribute to food security, nutrition, health, income generation, and environmental services when properly utilised (Kunkel,1984). Unfortunately, the lack of attention and authenticated data claim their potential value as under-exploited, and they are in danger of continued genetic erosion, ultimately leading to disappearance (ICUC, 2006).

Minor underutilised fruits are a group of fruits that are growing in a scattered and unattended way on roadsides, homestead land, wasteland, *etc.* that are consumable to human beings but relatively less palatable than other major fruits, which have lesser demand in the market, grown to a limited extent only and are not usually cropped in organized plantations with the application of artificial agro-inputs (Roy and Bauri2019). Wild and underutilised fruits are not only known for their excellent nutritional values but only have been playing a significant role in the Indian system of traditional medicine. Most of them are the rich sources of essential nutrients, vitamins, minerals, dietary fibers and contain a wide range of phytochemicals like flavonoids, glucosinolates, and isothiocyanates that are essential for a healthy and active living. The diversity of minor fruit plant species not only offers varieties in the local food system and tremendously contributes to household food, nutrition, and health security but also provides income generation opportunity by selling the excess quantities in the local markets particularly during the lean season of agricultural activity. The underutilized fruits like Aonla, bael, Jamun, Karonda, Passion fruit, Phalsa, Tamarind, Wood apple, *etc.*are the main sources of livelihood for the poor and play an important role in overcoming the problem of malnutrition (Gajanana*et al.*, 2010). Despite their huge potential for dietary diversification and nutrition security, they are yet to attract research, development, and promotional attention. Most indigenous fruit trees are not commonly cultivated in farms and there is scant and dispersed knowledge about their food and nutritive values but they can contribute to food security, health, or energy needs of local people (Vino, 2016). Their population is very much restricted to home gardens, roadside, village commons, and forest and mostly they grow as wild. They remain underutilised largely due to underestimation of their potential use, little research and extension endeavors on their commercial exploitation and lack of knowledge on their potential food and nutrition value, habitat destruction and focus on few commercial fruit crops. The diversity and abundance of these valuable plant species are declining rapidly due to the increasing emphasis of high-value fruit crops, destructive wild harvesting, and habitat destruction through deforestation. Conservation of this precious plant genetic resource base has been overlooked for a long time particularly with the advent of high value temperate and exotic fruit under the liberalized economy. Widespread adoption of intensive agriculture practices, intensification of a limited number of species, habitat loss through deforestation and uncontrolled grazing, changes in food habit, and unsustainable harvesting of crops has led to a rapid decline of plant genetic resources across the different landscape. Realizing the immense importance of these neglected floral resource bases, renewed emphasis has been given on their conservation and sustainable utilisation through a community-driven, self-sustaining, and holistic approach.

Study Area and Methodology

The investigation was carried out in the 15 selected villages in the Keonjahr and Mayurbhnaj districts of Odisha, India during the year 2018-19 to study the diversity of underutilised minor fruit plants, their present status, and future potential ina different landscape like the forest, backyard home garden, roadside, and farm bunds. The study area comes under the North Central Plateau Agro-climatic zone of the state which is a contiguous patch of Chotanagpur Plateau and is known for its rich biological diversity with a vast repository of neglected or underutilized crop species. Around 50 per cent of the total geographical area comes under forest and forest plays a significant role in the lives and livelihoods of the local community. The climate is hot, moist, and sub-humid with a mean summer maximum temperature of 36.9°C and means winter minimum temperature of 11.9°C. The mean annual rainfall is 1575.4 mm of which about 86.1 per cent is received during the monsoon season (June to October). The region is inhabited by a large group of tribal population and ethnic communities living nearby of forests and they largely depend on the forest for deriving their daily sustenance. Agriculture is the principal source of livelihood, which is characterised by a subsistence rice-based farming system. Collection of forest produces and small scale livestock farming are the important means of living of the tribal communities particularly during the lean season of agricultural income.

The study was conducted from November 2018 to April 2019 to collect information about the diversity of traditional fruit plants in the forest, home garden, roadside, and village commons, present use, propagation techniques, indigenous management practices, potential use, constraints and strategies required for their conservation. The study was based on the primary data collected from the participatory tools like transect walk, field visit, semi-structured interview, and focus group discussion. Semi-structured interviews and discussion with key-research participants such as elderly people, SHG members, local traders, members of VSS (Van Sanrakshyan Samiti), officials of forest and horticulture department, local NGOs were conducted to document the distribution pattern, abundance, existing market, and value chain of the underutilised fruit crops. Frequent field trips were made by a multi-disciplinary team comprising of professionals from agriculture, horticulture, forestry, and botany background along with local NGO personnel working in biodiversity conservation to the study area to document the diverse minor fruit plant and their seasonal availability. The indigenous fruit plants were collected, properly identified and a herbarium was maintained till the completion of the study.

Results and Discussion

A total of 32 numbers of underutilised fruit plants belonging to as high as eighteen families were identified indicating an enormous diversity of traditional fruit plants in the region. These plants along with their habitat, season of availability, medicinal, and nutritional value were documented (Table 30.1). Among the families, Rutaceae (5) has the maximum numbers of traditional fruit plants (Figure 29.1) followed by Moraceae (3). Among the habitats home garden and forest harbours the maximum diversity of fruit plants. These traditional fruits are available throughout the year and mostly consumed by the local community, while a sizable surplus quantity goes to the market either directly or through a local middleman to meet the immediate cash needs during the lean season of the year. Most of these fruits and their nuts or seeds are used for treating various diseases of human and domestic animals, while some are used in agriculture for controlling insect pests and crop

Table 30.1. Diversity of Underutilized Fruit Plants in the North Central Plateau Agro-climatic Zone of Odisha

Sl. No.	*Common Name*	*Botanical Name*	*Family*	*Habitat*				*Season of Availability*	*Medicinal Value*	*Nutritional Value*
				Forest	*Home Gardens*	*Road Sides*	*Village Commons*			
1	Custard Apple	*Annona squamosa*	Annonaceae					August-Sept	Anti-viral, Anti-diabetic, Anti-inflammatory and Analgesic activity. Seeds have insecticidal activity	Rich in total sugars, vitamin C and vitamin B complex, and some minerals.
2	Ramphal	*Annona reticulata*	Annonaceae					December - May	boils, abscesses and ulcers	Rich in total sugars, vitamin C and vitamin B complex, and some minerals.
3	Amla	*Emblica officinalis*	Euphorbiaceae					Oct-Dec	Cold cough, jaundice, haemorrhage	Very rich in vitamin C.
4	Bela	*Aegle marmelos*	Rutaceae					Mar-May	Diarrhoea, dysentery, diabetes, heart diseases	The ripe fruits contain protein, Fat, Carbohydrates with vitamins like Carotene Thiamine, Riboflavin, Niacin, Ascorbic Acid Tartaric Acid
5	Jamu	*Syzygium cumini*	Myrtaceae					June-July	Digestive ailments and for controlling diabetes, Blood purifier	

Sl. No.	Common Name	Botanical Name	Family	Habitat				Season of Availability	Medicinal Value	Nutritional Value
				Forest	Home Gardens	Road Sides	Village Commons			
6	Karunda	*Carissa carandas*	Apocynaceae					June-July	Treatments of malaria, epilepsy, nerve disorder, relieve of pain and headache, fever, blood purifier, itches and leprosy	Rich in iron, contains a good amount of vitamin C, protein, carbohydrates, fat, fibre and calcium. Fruits have been reported to contain carisol, β-caryophyllene, carissone, carissic acid, carindone, carinol, ascorbic acid, lupeol, and β-sitosterol.
7	Kendu	*Diospyrus melanoxylon*	Ebenaceae					May-June	Astringent, diuretic, laxative, urinary problems	
8	Khajuri	*Phoenix sylvestris*	Arecaceae					June -July.	Aphrodisiac, sweetener and diuretic and in the treatment of vomiting, vertigo and unconsciousness.	Vitamin A, vitamin B1, vitamin B2, vitamin B3, vitamin B6 and vitamin K with concentration 0.0039, 0.067, 0.70, 1.34, 0.18 and 0.0042 mg/100 g, respectively. Vitamin A, vitamin B1, vitamin B2, vitamin B3, vitamin B6 and vitamin K
9	Khirkoli	*Manilkara hexandra*	Mimosaceae					Mar-April	Urogenital disorder, Piles, dysentery, wounds	

Sl. No.	Common Name	Botanical Name	Family	Habitat				Season of Availability	Medicinal Value	Nutritional Value
				Forest	Home Gardens	Road Sides	Village Commons			
10	Tentuli	*Tamarindus indica*	Caesalpinia-ceae					Feb-March	Ulcers, constipation	Contain minerals especially sodium, zinc and iron.
11	Kaitha	*Feronia limonia*	Rutaceae					Jan-March	Astringent	Source of essential phytochemicalslike polyphenols, phytosterols, saponins, tannins, coumarins, and triterpenoids, vitamins and amino acids.
12	Ber	*Ziziphus mauritiana*	Rhamnaceae					Dec-Feb	Dysentery and diarrhoea, curing ulcer, fever and wounds.	Vitamin C, A, and B complex, and contain minerals like Ca, K, Br, Rb, and La. - iron, carotene, thiamine, riboflavin, niacin and ascorbic acid
13	Ambada	*Spondias pinnata*	Anacardiaceae					August–September.	Diarrhoea, dysentery.	Sodium, calcium, iron and copper
14	Asadhua	*Capparis zeylanica*	Capparaceae					Sept-Oct	Rheumatism, Swelling, improving appetite	Source of β-carotene, thioglycoside, glycocapparin, n-tricortane, dinna-amyrin and fixed oil.
15	Nara Koli	*Cicca disticha*	Euphobiaceae					June- Dec	Bronchitis	Calcium, Iron, Phosphorous, Ascorbic acid Thiamine, Riboflavin

Sl. No.	Common Name	Botanical Name	Family	Habitat				Season of Availability	Medicinal Value	Nutritional Value
				Forest	Home Gardens	Road Sides	Village Commons			
16	Chironji	*Buchanania lanzan*	Anacardiaceae					March-April	antioxidant activities	A good source of Phosphorus, Calcium, Magnesium and Iron. Also contain thiamine, riboflavin, niacin and vitamin C
17	Pomegra-nate/ Dalimba	*Punica granatum*	Punicaceae					Dec- Jan	Medicine for therapeutic purposes like colic, colitis-diarrhoea, dysentery, leucorrhoea, paralysis and headache Blood purifier and cholesterol control	Ascorbic acid, Calcium, Phosphorous Magnesium, Potassium, Sodium, Iron, Zinc Manganese, Copper
18	Kara-manga	*Averrhoa carambola*	Geraniaceae					June-August	**Anti-microbial and antifungal,antiulcer**	A good source of natural antioxidants like L-ascorbic acid, epicatechin and gallic acid. Also contain calcium, iron, phosphorus, potassium. And various amino acids like tryptophan, methionine and lysine.
19	Tala	*Borassus flabellifer*	Arecaceae					March-Aug	Skin inflammations, nausea and vomiting as well as worm infestation.	Contains gums, albuminoids, fats and the fresh pulp is reportedly rich in vitamins A and C, B-complex.

Sl. No.	Common Name	Botanical Name	Family	Habitat				Season of Availability	Medicinal Value	Nutritional Value
				Forest	Home Gardens	Road Sides	Village Commons			
20	Panas	*Artocarpus heterophyllus*	Moraceae					Jan-July	Antibacterial, antifungal, antidiabetic, anti-inflammatory, and antioxidant activities	Contains amino acids like arginine, cystine, histidine, leucine, lysine, methionine, threonine, and tryptophan. Rich in vitamin C, B-complex group of vitamins.
21	Kandhia	*Citrus aurantium*	Rutaceae					Aug-Dec	Antifungal and antibacterial.	Rich in vitamin C, flavonoids, acids.
22	Santara	*Citrus reticulata*	Rutaceae					Nov-Feb	Analgesic, antiasthmatic, anticholesterolemic, antiscorbutic, antiseptic, antitussive, carminative, expectorant, stomachic.	Rich in Vitamin C and A, potassium ad calcium. It is also a great source of folate and beta-carotene and contains magnesium, potassium, vitamins B2, B1 and B3,compounds such as zeaxanthin and lutein.
23	Jambhira	*Citrus limon*	Rutaceae					November-January.	Improve Digestive Health, throat Infections, prevention of Anemia.	Good source of Vitamin C, Iron, Vitamin B6, Copper, Vitamin B5, Vitamin B1, Potassium, Vitamin B9, Calcium, Phosphorus, of Magnesium.
24	Kusum	*Schleichera oleosa*	Sapindaceae					Apri -May.	Itching, menorrhea, malaria, dysentery, rheumatism, hair loss.	Phosphoric acid and potash.

Sl. No.	Common Name	Botanical Name	Family	Habitat				Season of Availability	Medicinal Value	Nutritional Value
				Forest	Home Gardens	Road Sides	Village Commons			
25	Jamrula	*Syzygium samarangense*	Myrtaceae					May- July	Flowers are astringent and are used to treat fevers and halt diarrhoea. Seeds are used against diarrhoea, dysentery and catarrh.	Rich in Vitamin C, Vitamin A, Calcium, Potassium
26	Dimiri	*Ficus racemosa*	Moraceae					January-March	Diabetes, liver dis-orders, diarrhoea, inflammatory condi-tions, haemorrhoids, respiratory, and urinary diseases.	Vitamin B2, Iron, Copper, Potassium, Magnesium, Calcium and Phosphorus.
27	Ou	*Dilleniai ndica*	Dilleniaceae					July-Sept	Dysentery, jaundice, weakness and rheumatic pain.	Fruits contain calcium and phosphorus
28	Rai	*Dillenia pentagyna*	Dilleniaceae					June-Aug.	Diabetes, diarrhoea and dysentery.	Iron, Manganese, Copper, Calcium
29	Kankoli	*Ziziphus oenoplia*	Rhamnaceae					October-January.	Sore throats, dysentery, and for inflammation of the uterus	
30	Bain-chakoli	*Flacourtia jangomas*	Salicaceae					August-Octobe		Sodium, Potassium, Calcium, Magnesium
31	Sahadakoli	*Streblus asper*	Moraceae					April-June	Ulcers, sinusitis, elephantiasis, boils, haemorrhage, bronchitis, diarrhoea, dysentery, syphilis and haemorrhoids.	
32	Baulakoli	*Mimus opselengi*	Sapotaceae					June-July	Chronic dysentery, astringent and to reduce fever.	Vitamin C, Minerals like Iron, Sodium, Potassium

diseases. During the peak availability period, some of the fruits are preserved as pickles mostly for future use and the preservation techniques and the number of ingredients varies from villages to villages.

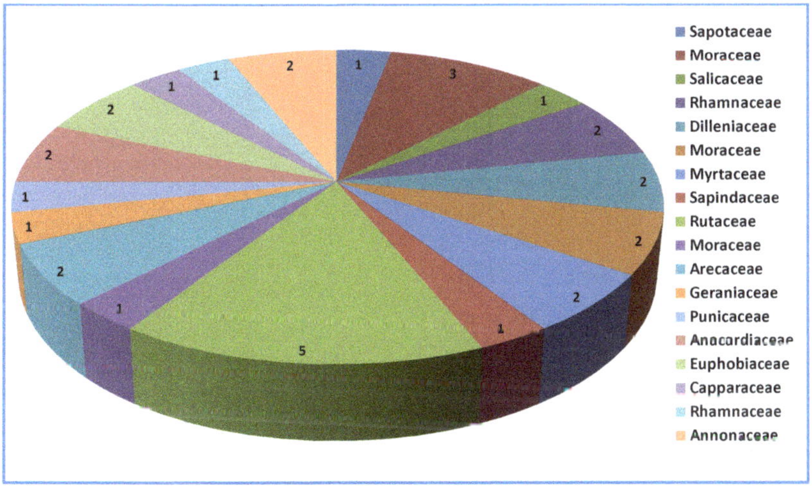

Figure 30.1. Family-wise Distribution of Underutilized Fruits.

However, most of the villagers opined that the population density of fruit plants and their productivity has been drastically declined over the years. Change in food habit and loss of traditional knowledge system that depicts the diverse use and importance of underutilized fruits largely contribute to the decline in their density and diversity. Lack of awareness among the present generation about the nutritional and medicinal value of underutilized fruit crops, less market demand and economic return, inadequate marketing supports and infrastructure facilities for transportation, storage, and processing, poor recognition of these crops in the government and non-government promotion programmes, lack of technology for preservation and value addition further compounded this problem.

Strategy for the Conservation and Promotion of Underutilized Fruit Plants

(i) The Popularization of Fruit-based Farming System

Uplands are the predominant land type in the plateau zone and are always subjected to various forms of erosion and degradation and around 25-30 per cent of the total upland areas of the study sites remained as either wasteland and cultivable fallow lands. Fruit-based farming systems with particular emphasis on the underutilized species along with different suitable intercrops can not only arrest the degradation process of the uplands but also improves the soil productivity. Most of the indigenous underutilized fruit crops are tolerant of adverse agro-climatic situations and can be grown with fewer inputs and little care and therefore have tremendous potential for the rehabilitation of marginal wastelands.

Figure 30.3. Underutilised Fruits in the Local Market to Secure Livelihoods of the Tribals during the Lean Season.

(ii) Crop Improvement and Development of Ideal Propagation Methods

Most of the indigenous fruits have only a few specified local varieties with adaptation to specific agro-ecological situations. Crop improvement through selection and breeding initiatives can increase the yield potential, reduce the high gestation period, and improves the fruit quality which ultimately can attract the farmers for scientific plantation.

(iii) Productivity Enhancement through Better Agronomic Practices

Most of these less known fruit trees establish through natural regeneration of seeds, grow slowly without any nutrition, start bearing fruits after a long period, and produce fruits of inferior quality (Hegde, 2002). Development of improved agro techniques like nursery management, nutrient and pest management, weed control, plastic mulching, water management through drip irrigation can substantially increase the yield and reduce the gestation period. Assessment, refinement, and validation of the developed agro-techniques through farmers' participatory research and demonstration in farmers' fields will increase the adoption level.

(iv) Documentation of ITKs

The traditional knowledge system associated with the underutilized crops

Figure 30.4. Diverse Underutilised Fruits in the Study Area.

is critically essential for their conservation as they possess valuable information on their propagation and production technology, diverse utility and utilization techniques, and sustainable harvesting methods that passed from generation to generation. Documentation of these precious knowledge systems and their scientific validation is the need of the hour before their complete disappearance.

v) Improving the Market Potential of Underutilized Fruits by Preservation and Value Addition

Post-harvest losses in fruits are very high (20-40 per cent), about 10-15 per cent of fresh fruits shrivel and stale, lowering their market value and consumer acceptability (Kumar *et al.*, 2016). Traditional fruit crops are not popular mainly due to their unappealing taste and flavor and adoption of modern value addition techniques can produce many value-added products with superior taste and quality. Besides, the processing and preservation techniques can reduce huge post-harvest. Processing of these underutilized crops mainly helps to increase nutritional security and also for utilization of waste in a better way (Kumar *et al.*, 2016). Processed products from underutilized fruits like pickle, jam, squash and nectar, moraba, kandy, ready to serve drinks, paste, and some dehydrated products will have greater consumer acceptability. The low external input requirements of underutilized fruits offer a tremendous opportunity for their organic production and the typical nutrition and health benefits tagged with these fruits can attract huge national and international consumer demands.

CONCLUSION

Biodiversity is the basis of sustainability in all the ecosystems as it not only maintains variability but also improves resilience. Conservation of valuable underutilized plant species will provide nutritional security; health benefits to the local community and can play a vital role in improving the local economy through value addition, and proper market interventions. Systematic research and extension efforts with a primary focus on productivity enhancement, processing and value addition, and market intervention are the key to the conservation, sustainable utilization, and area expansion of underutilized fruit crops. Policy intervention through the inclusion of some potential underutilized fruit plants in the government plantation programmes, agroforestry initiatives, watershed rehabilitation, and MGNREGA programmes will further provide a fillip to these largely neglected species. A holistic approach of both *in situ* and *ex situ* conservation renewed research and extension focus and a community-based conservation strategy are extremely important to protect these precious plant genetic resource bases before they become extinct forever.

REFERENCES

Chivenge, P., Mabhaudhi, T., Modi, A. T. and Mafongoya, P. 2015. The Potential Role of Neglected and Underutilised Crop Species as Future Crops under Water Scarce Conditions in sub-Saharan Africa, *International Journal of Environmental Research and Public Health*, 12: 5685-5711.

Collins, W.W.; Hawtin, G.C.1999. Conserving and using crop plant biodiversity in agro-ecosystems. *biodiversity in Agroecosystems*; CRC Press: London, UK; pp. 215–236.

Gajanana T.M., Gowda IND and Reddy B.M.C.2010.Exploring market potential and developing linkages–A case of underutilized fruit products in India. *Agricultural Economics Research Review*, 23: 437-43.

Garn, S.M.; Leonard, W.R. 1989. What did our ancestors eat? *Nutrition Review*, 47, 337–345.

Hegde, N.G. 2002. Promotion of Underutilised Fruit Crops Published, In: Fruits for the Future in Asia. Ed. By N. Haq and A. Hughes. International Centre for Underutilised Crops. University of Southampton, U.K.: 45-53.

ICUC.2006.Annual Report 2005-2006 Colombo, Sri Lanka: International Centre for Underutilised Crops. ISBN 955-1560-03-5. ISSN 1800-2315.

Kiran Kumar, G.N., Sudha Vani, V., Dorajee Rao, A.V.D and Viji. C.P. 2016. Value addition of some underutilized fruit crops, *Progressive Research – An International Journal*, 11 (III): 1890-1892

Kunkel, G. 1984.*Plants for human consumption*. Koeltz Scientific Books, Koenigstein, Germany.

Mayes, S., Massawe, F. J., Alderson, P. G., Roberts, J. A., Azam-Ali, S. N. and Hermann, M. 2012. The potential for underutilized crops to improve security of food production. *Journal of Experimental Botany*, 63 (3): 1075–1079.

Padulosi, S., Hodgkin. T., Williams, J.T. and Haq, N. 2002. Underutilised crops: trends, challenges and opportunities in the 21st century. In: Engels JMM, Ramanatha Rao V, Brown AHD, Jackson MT, eds. Managing plant genetic diversity, Vol. 30. Rome: IPGRI, 323–338.

Padulosi, S.; Hodgkin, T.; Williams, J.T.; Haq, N. Underutilized crops: Trends, challenges and opportunities in the 21st century. In *Managing Plant Genetic Diversity*; Engels, J.M.M., Ramanatha Rao, V., Brown, A.H.D., Jackson, M.T., Eds.; Bioversity International: Maccarese, Italy, 2001; Chapter 30, pp. 323–338.

Rai, N., Asati, B.S., Patel, R.K., Patel, K.K. and Yadav, D.S. 2005. Underutilized Horticultural Crops in North Eastern Region. *ENVIS Bulletin: Himalayan Ecology*, 13(1).

Roy, A. and Bauri, F.K. 2019. Scope of minor fruit production in India. Acta Hortic. 1241, 43-50

Salvi, J. and Katewa, S.S. 2016. Underutilized wild edible plants as a potential source of alternative nutrition. *International Journal of Botany Studies*, 1(4): 32-36.

Vino, S.A., Harshita and Sinija, V. R. 2016. Underutilized fruits in India.*Indian Food Industry Magazine*, 35(2): 45-46.

Williams, J.T. and Haq, N. 2002.Global research on underutilised crops - an assessment of current activities and proposals for enhanced cooperation. Southampton, UK: International Centre for Underutilized Crops. ISBN 92-9043-545-3. Accessed on 21st August 2013.

Chapter 31

Wild Foods Consumed by Ethnic and Local Communities of Imphal, Manipur State of India

*Rajkumari Supriya Devi and Sanjeet Kumar**

Biodiversity and Conservation Division, Regional Centre,
Ambika Prasad Research Foundation, Imphal, Manipur
**e-mail: sanjeet.biotech@gmail.com*

Imphal, the capital city of Manipur is known for its rich diversity of wild food plants. The food plants are mostly consumed by the local inhabitants and also sold in the market for earning livelihood. In the present paper, an attempt has been made to survey and enumerate the wild foods consumed by local community. Results revealed that 40 most common wild foods are consumed by the local community of Imphal. The enumerated species belong to 36 genera and 30 families. It was observed that the local people collect food species from nearby forest and hill areas and sell them in the local markets for their livelihood too. The present study emphasis on appropriate awareness to the society considering the socio-economic implications for the conservation and sustainable uses of wild foods.

INTRODUCTION

Urbanisation is spreading at alarming rate specifically in the developing countries like India (Imam and Banerjee, 2016). About 31 per cent of total population of India is residing in urban areas as per 2011 census (Kavita and Gayathri, 2017). The most urbanised states of the country are Maharashtra followed by Tamil Nadu and Kerala (Kadi and Nelavigi, 2015). Urbanisation not only contributes towards the development and economy of the country but also leads to hazardous issues like water pollution, air pollution and biodiversity loss (Kelly and Fussell, 2015). At the present scenario, most part of the country is facing the negative impacts of urbanization, including the North-Eastern regions of country (Yadav and Sharma,

2013; Singh, 2016). The North-East part of India hosts about 13,500 plant species in the Indo-Burma biodiversity hotspot region of India and about 10,000 species in the Eastern Himalayas biodiversity hotspot of which many are endemic in both areas (Mahanta, 2015). The community of this region mostly use foods from the wild plant and animal wealth (Mao *et al.*, 2009). The traditional foods are obtained by nearby forest areas and used in various forms (Patra *et al.*, 2020). The rural and tribal communities of the regions used to sell them for their livelihood (Figure 30.3). These foods are good for health as per the climatic conditions (Debbarma *et al.*, 2017). The ethnic people of North-Eastern part of country mostly use wild fruits and tuberous part of the edible plants. Wild fruits are rich with phenolic compounds and act as sound nutraceutical which is best for good health whereas tuberous part of edible plants provide adequate amount of starch and carbohydrate. Hence, they provide primary and secondary metabolites at a time in the form of organic nutraceutical. Researchers throughout the world, documented the wild foods consumed by different tribal communities and in different landscapes of the world (Ahmad and Pieroni, 2016; Thakur *et al.*, 2017; Thakur *et al.*, 2020). The rational behind their works was to provide optional food for increasing population without any negative impacts on health and environment. The rapid urbanisation in this part of the country is going to vanish the traditional foods and medicines which are intimately connected to their socio-cultural life (Satterthwaite *et al.*, 2010). Among the eight states of this region of the country, Manipur is well known for its traditional foods (Lokho, 2012). The state is bounded by Nagaland, Mizoram, Assam and Myanmar having a total area of 22, 327 km^2 with 3 million population that comprises Meitei, Loi, Yaithibi, Kuki and Naga communities (Mathur, 2011; Patra *et al.*, 2020). The ethnic group, Meitei represents 53 per cent of the total population of state and most of them are settled in the city of Imphal (Ningombam and Singh, 2014; Devi *et al.*, 2010). Imphal city is capital of Manipur and enjoys the humid sub-tropical climate to nurture the rich bio-resources (Devi *et al.*, 2010). At present scenario, city is growing and facing the problems of urbanization (Dociu and Dunarintu, 2012). The extension of city, urbanisation, anthropogenic activities, climate changes, anthropogenic activities and migration of the local people leading the loss of traditional food system practices, loss of wild food plants causing imbalance of ecological systems of valley resulting poor health. Unsustainable harvesting is also a major cause of above problems. Hence, documentation of ethnic food systems is very important to create awareness among the locals, bring attention of young generations, value additions of the available wild foods for providing livelihood to local community, mitigation the negative impacts of non-organic foods and future optional environmentally friendly nutraceutical. Keeping the above aspects for better health, better society, authors made a survey in Imphal city to enumerate the wild foods consumed by local communities.

Materials and Methods

For the enumeration of wild foods consumed by Meitei community of Imphal, a survey was made during 2018-2019. Morning 5: 30 AM to 9: 30 AM is selected for survey and authors visited different Keithels (markets) run in Imphal city (IMA Keithel, Lamlong Keithel, Nambol Keithel) (Figure 31.1) in different seasons Photographs of wild foods are taken and identified by authors using available literature and morphological characters (Kanjilal *et al.*, 1934-40; Chauhan *et al.*, 2000). A questionnaire survey was also made for collection of information from Keithals using standard techniques of ethno-biological approaches (Christian and Brigitte, 2004). For cross questionnaire, authors also discussed with local people who has more than 40 years old.

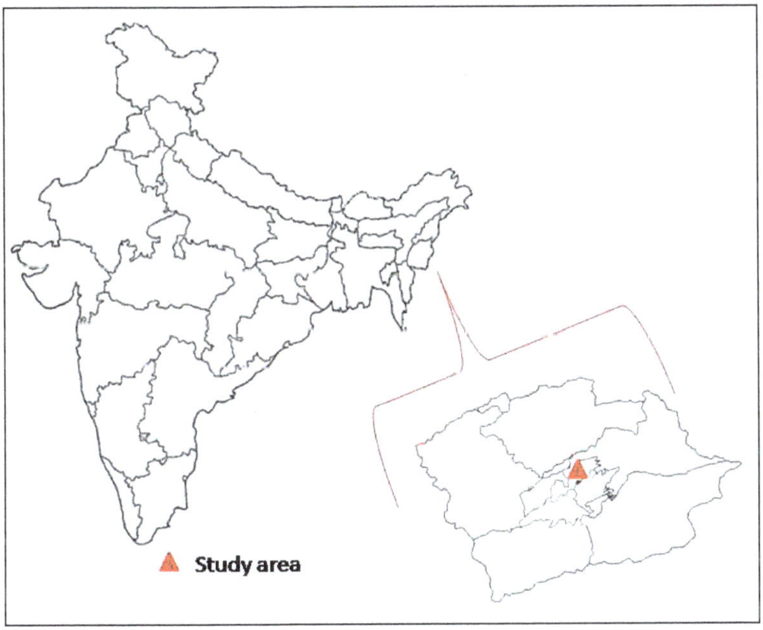

Figure 31.1. Geographical Distribution of Study Area.

Results and Discussion

During the field survey 40 wild foods were recorded from the City. Result revealed that the rich diversity of wild food consumed by local community along with wild mushrooms (Figure 31.2). It was observed that most common wild foods are also commonly sold in local markets. The most common are *Saccharum officinarum, Parkia javanica, Hedychium flavum, Alocasia cucullata, Neptunia oleracea, Capsicum chinense* (Plate 30.1.1), *Oryza sativa, Sechium edule* (Plate 31.1.2), *Nelumbo nucifera, Psophocarpus tetragonolobus, Allium odorum, Phyllantus embelica, Ananas comosus, Elaeocarpus floribundus* (Plate 31.1.3), *Manihot escalenta, Eryngium foetidum* (Plate 31.1.4) *etc.* Very less works has done on Meitei community. They are indigenous people of Manipur and closely associated with their landscapes and bio-wealth. They are not influenced with rapid urbanisation and still consuming local wild seasonal foods. The capital city Imphal facing the urbanisation and their impacts and very slowly losing their traditional skills and ethnic practices to consume wild foods. Hence, documentation is very important. Some researchers have done work on wild food plants of Manipur. In the year 2010, Yumnam and Tripathi reported 64 plant species which were used as medicines whereas Devi *et al.*, 2010 reported 71 plant

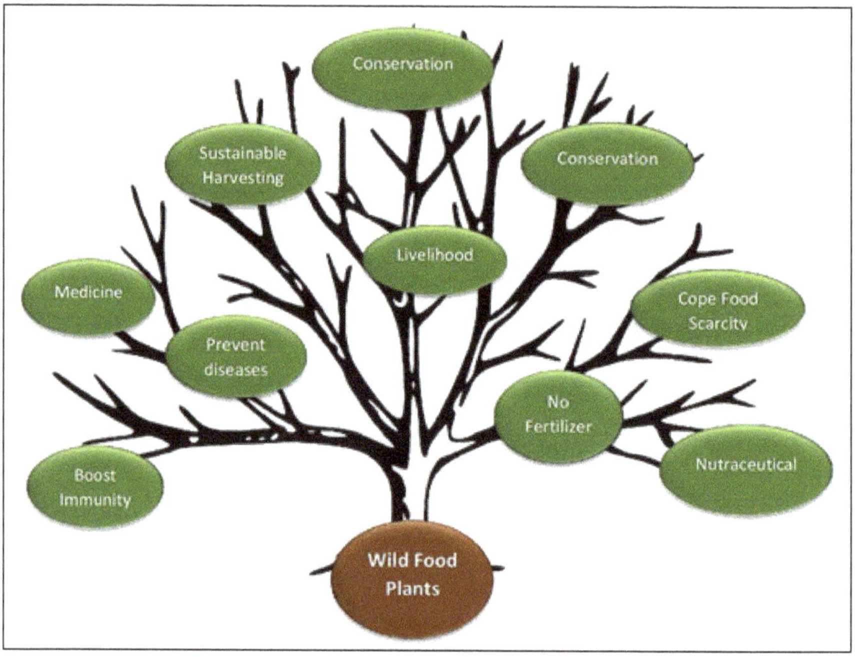

Figure 31.2. Importance of Wild Food Plants.

Figure 31.3. Local People Selling the Mushroom in Markets of Imphal, Manipur.

species used for medicinal and food purposes. They reported 45 wild edible plants, two wild mushrooms, 26 wild fruits, 25 from animal wealth and 18 indigenous fish species. They also documented that among 45 wild edible, 8 are rare species. They reported their demands season wise and it was observed that a mushroom (*Pleurotus ostreatus*) locally known as Uyen costs about Rs. 60-80/kg and most sold plant species is *Oenanthe javanica* (Komprek) costs about Rs. 10-20/kg. The most

Plate 31.1. Wild Foods Consumed by Local Community of Imphal, Manipur.

1: *Capsicum chinense*; 2: *Sechium edule*; 3: *Elaeocarpus floribundus*; 4: *Eryngium foetidum*; 5: *Musa* spp.; 6: *Curcuma angustifolia*; 7: *Dioscorea alata* and *Manihot esculenta*; 8: *Phlogacanthus thyrsiflorus*; 9: *Nelumbo nucifera*; 10: *Parkia javanica*; 11: *Neptunia oleracea.*

common edible plants reported by them are *Curcuma angustifolia* (Yaipal), *Alpinia nigra* (Pullei), *Zizania latifolia* (Eshing Kambong), *Oenanthe javanica* (Komprek), *Ipomea aquatica* (Kollamni), *Trapa natans* (Heikak Yeli), *Euryale ferox* (Thanging), *Nelumbo nucifera* (Thambou), *Neptunia oleracea* (Ekai thibi), *Ficus palmata* (Heiba), *Amaranthus viridis* (Chengkruk), *Ustilago consimitis* (Kambong), *Colocasia antiquorum*

(Lam Pal), *Hibiscus sabdariffa* (Shilhot shougree), *Meriandra benghalensis* (Kanghoo-man), *Enhydra fluctuans* (Komprek tujombi), *Chenopodium album* (Monsaobi), *Oxalis corniculata* (Yensil), *Rumex maritimus* (Torong Khongchak), *Rhus hookeri* (Heimang), *Prunus armeniaca* (Malhei), *Elaeagnus umbellata* (Heiyai), *Morus nigra* (Heijam-Pet), *Syzygium cumini* (Jam), *Elaeocarpus floribundus* (Chorphon), *Ziziphus mauritiana* (Boroi), *Phyllanthus emblica* (Heikru), *Prunus cerasoides* (Chumbrei), *Spondias mangifera* (Heining), *Flacourtia jangomas* (Heitroi) *etc.* In 2013, Gangte *et al.*, reported 84 species used as food by the Zou tribe of Manipur and in the year 2014, Leisangthem and Sharma reported 50 wild food plants belonging to 26 families from Imphal East. They reported *Acorus calamus, Cassia alata, Andrographis paniculata, Kaempferia alangal, Zanthoxylum acanthopodium, Eryngium foetidum, Eupatorium birmanicum, Adhatoda vasica etc.* as a common wild food plants available in this region. A detail study on indigenous edible plants consumed by Senapati district of Manipur was done

Table 31.1. Wild Foods used Among the Community of Imphal, Manipur

Name	*Family*	*Local Name*
Allium ramosum	Amaryllidaceae	Maroi nakupi
Allium hookeri	Amaryllidaceae	Maroi napakpi
Alocasia cucullata	Araceae	Palukabi/Singju-pan
Alocasia indica	Araceae	Pan
Alocasia macrorrhiza	Araceae	Yendem
Ananas comosus	Bromeliaceae	Kihom
Artocarpus heterophyllus	Moraceae	Theibong
Bamboo spp.	Poaceae	Wa
Capsicum chinense	Solanaceae	U- morok
Centella asiatica	Umbelliferae	Peruk
Cinnamomum tamala	Lauraceae	Tejpata
Citrus aurantiifolia	Rutaceae	Champra
Cucurma angustifolia (Plate 30.1-6)	Zingiberaceae	Yaipan
Dioscorea alata (Plate 30.1-7)	Dioscoraceae	Haa
Elaeocarpus floribundus	Elaeocarpaceae	Chorphon
Elaeagnus umbellata	Elaegnaceae	Heiyai
Elsholtzia blanda	Lamiaceae	Lomba
Eryngium foetidum	Apiaceae	Awa-phadigom
Euryale ferox	Nympheaceae	Thangjing
Hedychium coronarium	Zingiberaceae	Takhellei-angouba
Hedychium flavum	Zingiberaceae	Loklei
Houttuynia cordata	Saururaceae	Toningkok
Leucas aspera	Lamiaceae	Mayanglambum
Manihot esculenta (Plate 30.1-7)	Euphorbiaceae	U- mangra
Mukia maderaspatana	Cucurbitaceae	Nom Bil

Name	Family	Local Name
Musa spp. (Plate 30.1-5)	Musaceae	Laphu
Nelumbo nucifera	Neloumbonaceae	Thamchet (Seed)
Nelumbo nucifera	Neloumbonaceae	Thambou (Stem)
Neptunia oleracea	Mimosaceae	Esing ekaithabi
Oryza sativa	Poaceae	Chahou
Parkia javanica	Fabaceae	Yongchak
Phlogacanthus thyrsiflorus (Plate 30.1-8)	Acanthaceae	Nongmankha angangba
Phyllantus embelica	Phyllanthaceae	Heikru
Psophocarpus tetragonolobus	Papilionaceae	Tengnou-manbi
Rhus semialata	Anacardiaceae	Heimang
Sechium edule	Cucurbitaceae	Dashkush
Sesbania sesban	Papiplionaceae	Chuchu-rangmei
Solanum nigrum	Solanaceae	Leipungkhang
Solanum xanthocarpum	Solanaceae	Singkhangkha
Syzygium cumini	Myrtaceae	Jam
Zanthoxylum acanthopodium	Rutaceae	Mukthrubi
Zizania latifolia	Poaceae	Kambong

by Khan *et al.* (2015). They have reported 53 edible plants belonging to 47 genera and 38 families. They reported *Agaricus campestris* (Chengum), *Auricularia delicata* (Uchina), *Schizophyllum commune* (Kanglayen), *Tricholoma giganteum* (Thangii-Uyen) as edible mushrooms whereas *Alpinia galanga* (Kanghoo), *Dioscorea pentaphylla* (Ha Angangba), *Dioscorea alata* (Haa Angouba) as an edible tuberous plant. They reported *Cissus adnata* (Kokngouyen), *Eurya japonica* (Uyangan), *Rhynchotechum ellipticum* (Yembum), *Houttuynia cordata* (Tokningkok) as edible leaves whereas *Litsaea citrata* (Usingsha laba), *Rhus semialata* (Heimang), *Myrica esculenta* (Nonganghei), *Ficus auriculata* (Heirit), *Phyllanthus emblica* (Heikru), *Docynia indica* (Heitup), *Antidesma acidum* (Ching Heiyen) *etc.* as edible fruits whereas present study showed that *Alocasia cucullata* (Palukabi), *Ananas comosus* (Kihom), *Bamboo* spp. (Wa), *Capsicum chinense* (U-morok), *Centella asiatica* (Peruk), *Curcuma angustifolia* (Yaipan), *Dioscorea alata* (Haa), *Elaeocarpus floribundus* (Chorphon), *Euryale ferox* (Thangjing), *Manihot esculenta* (U-mangra), *Musa* Spp. (Laphu), *Nelumbo nucifera* (Thambou; Plate 1.9), *Neptunia oleracea* (Esing, Plate 1.11), *Parkia javanica* (Yongchak, Plate 1.10), *Phyllanthus emblica* (Heikru), *Sechium edule* (dashkush), *Syzygium cumini* (Jam), *Zizania latifolia* (Kambong) are most commonly consumed wild foods by Meitei community of Imphal. In the year 2016, Singh and Binu reported 105 wild edible plants used by Meitei community whereas Konsam *et al.* (2016) reported 68 wild edible vegetables belonging to 42 families consumed by the ethnic communities of Manipur. In the year, 2018, Hazarika and Singh reported 49 species of wild edible fruits of Manipur. The present study and other previous work revealed that local community of Manipur state use diverse wild plants as a food and medicines in general and particular in valley of Imphal. The all reported wild plants not only provide the organic food to fight against diseases and disorders but also give strength in their economic condition. Most of community in and around Imphal, sell the local foods in local markets and fulfil their day-to-day need. Whole world searching nutraceutical from wild and in this aspect, their ethnic food practices provide a platform to screen optional food for increasing population of the world

CONCLUSION

The study highlights the importance of wild foods as folk foods of Imphal. The local inhabitants of Imphal have a wide knowledge about the uses of traditional foods in their day to day life. The local people collect food species from forest and hill areas and sell them in the market for their livelihood too. The present study emphasis on appropriate awareness to the society considering the socio-economic implications for the conservation and sustainable uses of wild foods.

REFERENCES

Ahmad, K., and Pieroni, A. (2016). Folk knowledge of wild food plants among the tribal communities of Thakht-e-Sulaiman Hills, North-West Pakistan. *Journal of Ethnobiology and Ethnomedicine,* DOI: 10.1186/s13002-016-0090-2.

Christian, R.V., and Brigitte, V.L. (2014). Tool and method for data collection in ethnobotanical studies of hommegardens. *Field Methodo*logy, 16(3), 285-306.

Debbarma, M., Pala, N.A., Kumar, M., and Bussmann, R.W. (2017). Traditional Knowledge of Wild food plants in Tribes of Tripura in Noetheast India. *African Journal of Traditional, Complementary and Alternative Medicines,* 14 (4), 156 – 168.

Devi, O.S., Komar, P., and Das, D. (2010). A checklist of traditional edible bio-resources from IMA markets of Imphal valley, Manipur, India. *Journal of Threatened Taxa,* 2(11),1291-1296.

Devi, O.S., Komor, P., and Das, D. (2010). A Checklist of traditional edible boi-resources from Ima market of Imphal Valley, Manipur, India. *Journal of Threatened Taxa,*2(11),1291-1296.

Dociu, M., Dunarintu, A. (2012). The Socio- economic Impact of Urbanization. *International Journal of Academic Research in Accounting, Finance and Management Science,* 2(1),47-52.

Gangte, H.E., Thoudam, N.S., and Zomi, G.T. (2013). Wild edible plants used by the Zou tribe in Manipur, India. *International Journal of Sci Res Pub,* 3(5), 1 8.

Hazarika, T.H., and Sing, T.S. (2018). Wild edible fruits of Manipur, India: associated traditional knowledge and implications to sustainable livelihood. *Genetic Resources and Crop Evolution,* 65: 319-332.

Imam, A. U. K., and Banerjee, U. K. (2016). Urbanisation and greening of Indian cities: problems, practices, and policies. *Ambio,* 45(4), 442-457.

Kadi, A.S., and Nelavigi, K.P. (2015). Growth of Urbanization in India. *The International Journal of Science and Technology,* 3 (7), 2321- 919X.

Kavita, B.D., and Gayathri, N.K. (2017). Urbanization in India. *International Journal of Scientific Research And Education,* 5 (1), 6166-6168.

Kelly, F.J., and Fussell, J.C. (2015). Air pollution and public health: emerging hazards and improved understanding of risk. *Environ Geochem Health*, 37(4), 631- 649.

Konsam, S., Thongam, B., and Handique, A K. (2016). Assessment of wild leafy vegetables traditionally consumed by the ethnic communities of Manipur, northeast India. *Journal of Ethnobiology and Ethnomedicine*, DOI: 10.1186/s13002-016-0080-4.

Leishangthem, S., and Sharma, L.D. (2014). Study of some important wild food plants found in Imphal-East district, Manipur, India. *International Journal of Scientific and Research Publication*,4(9),1-5.

Lokho, A. (2012).The Folk Medicinal Plant of the Mao Naga in Manipur North East India. *International Journal of Scientific and Research Publications*, 2(6) 2250 – 3153.

Mahanta, P.P. (2015). Biodiversity Hotspots. *Exotic Northeast*, 1(2), 2455 – 2135.

Mao, A.A., Hynniewta, T.H. and Sanjappa, M. (2009). Plant wealth of Northeast India with reference to ethnobotany. *Indian Journal of Traditional Knowledge*, 8(1), 96 – 103.

Mathur, A. (2011). A winning strategy for India's North-east. *Jindal Journal of International Affairs*,1(1), 269 – 298.

Ningombam, D.S., and Singh, P.K. (2014). Ethnobotanical Study of *Phologacanthus thyrsformis* Need: A conserved Medicinal Plant of Manipur, Northeast India. *International Journal of Herbal Medicine*, 1(5),10-14.

Satterthwaite, D., Granahan, G.M., and Tacoli, C. (2010). Urbanization and its implications for food and ferming. *Philos Trans R Soc Lond B Biological Sci*ences, 365 (1554), 2809 – 2820.

Singh, A.S. (2016). Conservation of Biodiversity of Manipur: Sustainable Management Strategies. *International Journal of trend in Research and Development*, 3(3), 2394 – 9333.

Singh, D. and Binu, M. (2016). Wild edible plants used by Meitei community of Eastern Himalayas, India. *International Journal of Agricultural Sciences*, 8(53), 2699-2702.

Thakur, A., Somvir, S., and Puri, S. (2020). Exploration of wild edible plants used as food by Gaddis-A tribal community of the Western Himalaya. *Scientific World Journal*, DOI: 10.1155/2020/6280153.

Thakur, D., Sharma, A., and Uniyal, S.K. (2017). Why they eat? What they eat: patterns of wild edible plants consumption in a tribal area of Western Himalaya. *Journal of Ethnobiology and Ethnomedicine*, DOI: 10.1186/s13002-017-0198-2.

Yadav, P.K., Sharma, K., and Kumar, R. (2013). A Framework for Assessing The Impact of Urbanization And Population On Garo Hills Landscape Of North – East India. *International Journal of Conservation Science*, 4 (2), 212 – 222.

Yumnam, J.Y. and Tripathi, O.P. (2012). Traditional knowledge of eating raw plants by the Meitei of Manipur as medicine/nutrient supplement in their diet. *Indian Journal of Traditional Knowledge*, 11(1), 45-50.

Chapter 32

Microbial Biopesticides an Opportunity for Improvement of Livelihood Generation

Kumar Sambhav Giri[1] *and Gargee Mohanty*[2]*

[1]*Department of Agronomy, College of Agriculture, Central University, Imphal – 795 004, Manipur*

[2]*Department of Zoology, North Orissa University, Baripada – 757 003, Odisha*

**e-mail: gargee.mohanty@gmail.com*

Agriculture holds the prime importance of being the leading source of livelihood options in rural India and for ensuring the food security of the nation. Around 58 per cent of the rural population in India are engaged in agriculture as their livelihood. This sector has faced several challenges and proper attention is required for its sustainability. One of such critical challenges is the increasing pest attacks. Pest attacks pose a serious threat to crops and thus affect the food security of the nation. Pesticides are the chemicals used to kill pests in agriculture. Although pesticides have benefits, most of them also have drawbacks. They have been proven to be toxic to humans and other species, and toss a negative impact on the environment. So, alternate solutions to replace harmful chemical pesticides are 'Biopesticides'. Biopesticides are prepared from naturally occurring materials that control pests through non-toxic mechanisms and in an eco-friendly approach. They can be classified as biochemical, plant-incorporated protectants and microbial pesticides depending upon the source of extraction and processing. This chapter focuses on different microbial biopesticides with their mode of action and mechanism of pest killing activities. It also provides an insight on the role of microbial biopesticides in carrying sustainable agricultural practices and generating livelihood opportunities.

Keywords: *Biopesticides, Livelihood generation, Sustainable agriculture, Mechanism, Microbial biopesticides.*

INTRODUCTION

Interest in biopesticide is gaining because of its advantages associated with environmental safety, target-specificity and efficacy in very small quantity; natural decomposition and suitability in sustainable agriculture. They can substitute some of the harmful chemical pesticides. Biopesticide has enhanced the growing demands for safe and healthy organic food. Although the use of agrochemicals is indispensable to meet the ever-growing demands of food, feed and fodder, opportunities do exist in selected crops and niche areas where biopesticides can be used as a component in organic and sustainable agricultural practices.

Over some past few years, an increase in demands for residue-free crop products; growth in the organic food market and easier registration procedure in comparison to chemical pesticides are some of the key drivers of the biopesticide usage and its commercialization. There is an increasing amount of apprehension for the conservation of biodiversity and to reduce the number of threats to some of the endangered species, being faced due to the overuse and rampant use of agrochemicals including chemical pesticides. Hence, there is an urgent need to replace chemical agrochemicals with natural and organic constituents.

While there is commercial pressure from the manufacturing and profit-making side to develop products, based on a single strain that holds broad spectrum to control a wide range of pests infecting different crops and may not be endemic to the areas of application; the environmentalists want narrow-spectrum products based on strains from the area of use. To reconcile these contradictory demands, biopesticides in the market have been maintained at minimal negative impact, if any, on the environment

Biopesticides are prepared from naturally occurring materials that control pests by non-toxic mechanisms and in an eco-friendly approach. They may be derived from animal sources (*e.g.*nematodes), plant sources (*Chrysanthemum, Azadirachta*) and microbial sources (*e.g. Bacillus thuringiensis, Trichoderma, Nucleopolyhedrosis virus*) (Kumar, 2015).

Biopesticides are target-specific and have little or no residual effects and are thus acceptable in organic farming. On a global scale, the use of biopesticides is increasing progressively by 10 per cent each year (Kumar and Singh, 2015). About 175 biopesticides active ingredients and 700 products have been registered worldwide.

Out of all currently used biopesticides; microbial biopesticides constitute the largest group of broad-spectrum biopesticides, which are pest-specific (*i.e.*, do not target non-pest species and are environmentally benevolent). There are at least 1500 naturally occurring insect-specific microorganisms, a hundred out of which possess insecticidal properties (Koul, 2011). Microbial biopesticides are based on naturally occurring bacteria, fungi, protozoa, viruses and nematodes. They have offered some boosting alternatives to chemical pesticides. There are many reasons for the recent increased interest in microbial biopesticides, including the increase of resistance to traditional synthetic pesticides, rise in public concern over the health-related threats associated with synthetic pesticides, host-specificity of microbial pesticides and improvement in the production and formulation technology of microbial biopesticides. Therefore, soon, as the demand for organic products is increasing there is a possibility of expansion of natural and organic agrochemical products like microbial biopesticides in the global market. Moreover, the environmentalists, policymakers of different countries, and other stakeholders are looking forward to training and urge people to go for organic and sustainable farming and to help the production of such organic chemicals on large scale. Consequently, these will open up greater opportunities and avenues to earn livelihoods.

This chapter focuses on the classification, mode of action and mechanism of pest killing activity of biopesticides. It also provides an insight into the role of microbial biopesticides in sustainable agricultural practices and livelihood generation.

Current Status of Biopesticides and Microbial Biopesticides

Currently, biopesticides hold only 2 per cent of the plant protectants at an international level; however, its growth rate shows an increasing trend from a few decades. Globally, the use of biopesticides is increasing steadily by 10 per cent every year. About 90 per cent of the microbial biopesticides are derived from just one entomopathogenic bacterium (entomopathogenic bacteria means the bacteria which bear the capacity to infect and kill insect – pests), *Bacillus thuringiensis*. On an international scale, the overall production of biopesticides has been estimated to be over 3,000 tons per year and is rising very fast. A rise in demand for residue-free agricultural products, an upward trend of an organic food market and an easier registration procedure in comparison to chemical pesticides are some of the key drivers of the biopesticide market. The above 200 such products are being sold in the US market, compared to only 60 comparable products in the European Union. However, Asia lags with the usage to only 5 per cent of biopesticides sold the world over (Bailey *et al.*, 2010). Most of the countries have amended their policies to reduce the use of chemical pesticides and endorse the utilization of biopesticides.

Even though for successful utilization of biopesticides several technological and policy gaps have been identified, yet they need to be dealt with properly by each country separately maintain coherency with other countries and for a hazard-free environment. Policy measures need to be strengthened to conserve biodiversity also as day by day the numbers of species are declining because of the rampant use of different physical and chemical agents for different purposes which also include agriculture.

Although microbial biopesticides offer a good opportunity for sustainable agricultural practices yet, a scrupulous insight on the mode of action, their efficacy in the control of pest and contribution towards their sustainability should be carefully investigated; as environmental safety, is a global concern and the prime need of the present time.

Microbial Biopesticides

The active ingredient of a microbial biopesticide is a microorganism (bacterium, fungus, virus, protozoan, alga, rickettsia, mycoplasma and nematodes) that either occurs naturally or is genetically modified through genetic engineering. The pesticidal action may be from the organism itself or from a substance it produces. They kill or suppress the growth of pests either by producing toxic metabolites specific to the pest, causing diseases in them or by impairing their reproductive systems and infecting their larva and through several other modes of action (Clemson, 2007).

Basing on the organism used for the pesticidal activity, microbial pesticides are classified. Thus, microbial biopesticides can be categorized into **bacterial, fungal, viral, protozoan and nematode biopesticides**.

a. Bacterial Biopesticides

To generate bacterial biopesticides both spore-forming and non-spore-forming bacteria can be used. Most of the insect pathogenic bacteria belonging to the families of Pseudomonadaceae, Bacillaceae, Streptococcaceae, Enterobacteriaceae and Micrococcaceae. However, the *Bacillus* spp of Bacillaceae, have gained much more interest than others because of their importance as microbial control agents (Tanda and Kaya, 1993). Bacterial biopesticides are the most widespread microbial biopesticide. As insecticide, they specifically act on species of moths, butterflies, beetles, flies, and mosquitoes.

Mode of Action of Bacterial Biopesticides

The pathogenic bacteria first grow on the plants and then come in contact with the pest attacking the plant. These bacteria after making an entry into the pest disrupt their digestive system by producing endotoxins, which are species specific, and hence kill a particular pest. Similarly, bacteria with antibacterial and antifungal activity, control pathogenic bacteria or fungus, by colonizing the plant and subsequently pest and then crowds out the pathogenic species from the plant (Brein *et al.*, 2009).

The most widely exploited subspecies and strains of bacteria to be used as microbial pesticide is *Bacillus. thuringiensis* (Bt), accounting for approximately 90 per cent of the biopesticide market in USA(Chattopadhyay *et al.*, 2004). Presently around 40,000 species of *Bacillus thuringiensis* are found to be effective against *Lepidoptera* or *Dipterans* pest. At the time of spore formation Bt synthesizes Δ endotoxins or Cry proteins, the crystalline inclusions of Bt bacteria. These endotoxins bear insecticidal properties.

The Bt δ endotoxins or Cry proteins produced by Bt bacteria binds to cellular lining of the pest digestive tract and results in termination of feeding activity and lead to death of the pest. This mode of pest inhibition in agricultural crops, trees and plants of forestry and of medicinal crops makes the inclusion protein of Bt efficient and safe. Thus, killer proteins of Bt provides a sustainable alternative to chemical pesticides for the control of insect pests (Roy *et al.* (2007). Apart from *Bacillus thuringiensis,* several other bacterial species like *B. subtilis* and *Pseudomonas aureofaciens* have also been used to control and elimination of causative pathogens of damping off and soft rots diseases (Haas and Defago, 2005; Berg, 2009).

b. Fungal Biopesticides

Entomopathogenic fungi are used as mycoinsecticides against many insect pests (Dutta *et al.*, 2015). Fungal biopesticides are effective against numerous insects as well as other plant diseases. In addition, biofungicidal products of fungi stimulate plant defense system enabling plants more resistant to a variety biotic and abiotic stress (Srinivasa *et al.*, 2008).

There are about 90 genera and above 700 species of entomopathogenic fungi belonging to all major classes of fungi (Moorhouse *et al.*, 1992). However, less than ten species of fungi have been exploited for commercial biofungicides all across the world (Copping and Menn, 2000).

Trichoderma spp. and *Beauveria bassiana* both are used as fungal biopesticides mainly applied in the nursery, field crop, ornamental and vegetable crops. *Trichoderma* is also employed as fungicide against various foliar and soil-borne plant pathogens like *Rhizoctonia, Sclerotium, Fusarium, Pythium, Ceratobasidium, Macrophomina* and *Phytophthora* spp (Anand and Reddy, 2009). *Beauveria bassiana* has used against asaphids, thrips, whitefly and Q-Biotype White fly.

Mode of Action of Fungal Biopesticides

Entomopathogenic fungi make an entry into pest body through their integument. They also enter during ingestion or through wounds and/or trachea (Holder and Keyhani, 2005). Fungi begin to synthesize germination tube on the integument of the pest with the help of enzymes like proteases, chitinases, quitobiases, upases and lipoxygenases secreted from their body. These enzyme help in the disintegration of the cuticle layer and form an appressorium (a specialized structure in the germinative tube) and through it make an entry to the pest body. Inside the pest body, the fungi develop hyphal bodies that spread through the haemocoel and attack diverse muscle tissues, fatty bodies, malpighian tubes, mitochondria and haemocytes of the pest leading to its death within 3 to 14 days of infection. After the death of the insect, hyphae of fungi pierce the cuticle from the interior of the insect and appear at the surface, where they again begin spore formation under suitable environmental conditions (Kachhawa, 2017).

c. Viral Pesticides

A virus is a set of one or more nucleic acid template molecule, normally encased in a protective coat of protein or lipoprotein that is able to organize its own replication only within suitable host cells. Viruses have been isolated from more than 1000 species of insects from at least 13 different insects order (Srivastava and Dhaliwal (2010). Entomogenous viruses (viruses infecting insects) can be differentiated into two groups. The first group is inclusion viruses (IV) which produces inclusion bodies in the host cells and the second group are noninclusion viruses (NIV) which do not produce inclusion bodies.

The inclusion viruses are further subdivided into polyhedron viruses (PV) or polyhedroses, which produce polyhedral bodies and granulosis virus which produce granular bodies. Polyhedroses could inhabit the nucleus and are called nuclear polyhedrosis viruses (NPV) or the cytoplasm and are called cytoplasmic polyhedrosisvirus (CPV) (Rohrmann, 2013).

Beside IV and NIV, viruses belonging to family Baculoviridae also infect insects. Hence, these are employed in the development of most commercial viral biopesticides (Moore *et al.*, 1987). Members of this family are regarded as safe for vertebrates and, to date, no cases of pathogenicity of a baculovirus to a vertebrate have been reported. Baculoviruses are characterized by rod-shaped enveloped virions and a circular, dsDNA genome of 80–180 kbp. Their host range is restricted to arthropods and most particularly to insects and hence considered to be safe to be used in agricultural practices (Rohrmann, 2013).

Mode of Action of Viral Biopesticides

Baculoviruses infect midgut epithelial cells after getting ingested and dissolve the ocular bodies (OB) in the alkaline environment of the midgut. Typically, two virion phenotypes occur in baculovirus infections, occlusion-derived virions (ODV) and budded virions (BV) that spread the infection to tissues throughout the host. Thus, the ODV are responsible for horizontal transmission between insect hosts and BV for systemic spread of infection within a host.

Baculoviruses kill their hosts by virtue of a complex replicative cycle that usurps the metabolic machinery of the host cell and diverts it to the production of baculovirus nucleic acids and structural proteins for virus production. Baculoviruses encode a number of proteins and enzymes that enhance their ability to specifically infect and replicate to high levels in insects and to increase the efficiency of spread and persistence of viruses within the host insects' environment. All baculoviruses encode an occlusion body protein (polyhedrin or granulin) which forms the bulk of the proteinaceous crystal or OB that occludes the virions in the later stages of virus replication. The OBs are responsible for the highly stable nature of these viruses in the environment. While the occluded form of the virus is susceptible to UV inactivation, in the soil the occluded virions can remain viable for several years.

Upon ingestion of baculovirus OB by a host insect larva, the OB structure and polyhedrin protein is degraded in the alkaline environment of the insect midgut which also includes a diverse mixture of digestive proteases. The process releases the infectious occluded virions into the midgut lumen. However, to infect the midgut epithelial cells the virions must transit the peritrophic matrix (PM), a chitin and glycoprotein layer which lines the midgut lumen of most insect groups. The PM is a physical barrier that protects the underlying midgut epithelial cells from abrasive food particles, digestive enzymes and pathogens that are infectious after entering through mouth. A number of baculoviruses encode a family of metalloproteases referred to as "viral enhancing factor" or "enhancins". These metalloproteases are associated with the OB or are incorporated into the envelope of the occluded virion. Enhancins have been demonstrated to increase the oral infectivity of baculoviruses from 2 to 15 fold when present at normal levels in OB or ODV (Popham *et al.*, 2001; Li *et al.*, 2003).

Similarly, a bacteriophage (virus that infects bacterial cell) can be used as a biopesticide against pest those contain bacteria in their gut. Bacteriophages have been helpful for the control the lepidopterous pests of vegetables, cotton and rice. When ingested by the host insect, infectious virus particles are liberated internally and become active. Once in the larval gut, the virus's protein overcoat quickly disintegrates, and the viral DNA proceeds to infect digestive cells. Within a few days, the host larvae are unable to digest food and so weaken and die (Thakore, 2006). In 1975, Elcar™ (first viral insecticide) was proposed, by Sandoz Inc. (Ignoffo and Couch, 1981). Elcar™ was prepared by *Heliothiszea* NPV. HzSNPV not only are effective against cotton boll worm, but are also effective against the pests of tomato, maize, beans, soybean and sorghum.

d. Protozoan Biopesticides

Protozoan pathogens are diverse group of organisms comprising about 1000 species attacking invertebrates including pest and are commonly referred as microsporidians. These pathogenic protozoans are generally host specific and slow acting. They produce chronic infections with deadly illness of the host (pest).

Mode of Action of Protozoan Biopesticides

The pests get infected when they ingest the protozoan spores. Spore germinates in the midgut and sporoplasm is released which invade target cells causing infection in the host. The infection results in reduction in feeding, vigour, fecundity and longevity of the insect host as generally observed when there is an application of microbial control agents. Till today only a few protozoan species have been moderately successful to act as biopesticides.

e. Nematode Biopesticides

Entomopathogenic nematodes infect insects. These nematodes are soft bodied, non-segmented roundworms that are obligate or sometimes facultative parasites. Entomopathogenic nematodes occur naturally in soil environments and locate their host in response to carbon dioxide, vibration, and other chemical clues. Species belonging to families Heterorhabditidae and Steinernematidae have been effectively used as bioinsecticides in pest management programs (Grewal and Peters, 2005).

Entomopathogenic nematodes fit nicely into integrated pest management (IPM) programs as they are considered to be non-toxic to human, specific to their target pests, and can be applied with standard pesticide equipment.

Mode of Action of Nematode Biopesticides

Nematode in their juvenile stage makes entry into the insect- pests. It is the only free living stage of entomopathogenic nematodes. The juvenile stage penetrates the host insect via spiracles, mouth, anus, or in some species through inter-segmental membranes of the cuticle. Finally, they make their entry into the hemocoel. Both Heterorhabditis and Steinernema remain in mutualistic association with *Photorhabdus* and *Xenorhabdus,* genera of bacteria respectively (Ferreira and Malan 2014).

The juvenile nematodes carrying intestinal bacteria (bacteria remain in a symbiotic relationship with nematodes) release them into the pest hemocoel. The bacteria multiply in the insect hemolymph, and the infected host usually dies within 24 to 48 hours. After the death of the host, nematodes continue to feed on the host tissue, get matured and reproduce. The progeny of nematodes through four juvenile stages develop into adult nematodes. Depending upon the availability of resources, one or more generations of nematodes may occur within the host cadaver, and a large number of infective juveniles are eventually released into the environment to infect fresh hosts continue their life cycle (Kachhawa, 2017).

Microbial Biopesticides in Biodiversity Conservation and Sustainable Agricultural Practices

The science of microbial biopesticides is still considered to be young and evolving. Microbial biopesticides are target specific and hence do not affect the beneficial organisms such as the natural enemies and thus helpful in biodiversity conservation. These are effective in small quantities and their usages promote sustainable pest management and so, contribute towards sustainable agricultural practices (Nawaz *et al.,* 2016). Furthermore, microbial biopesticides do not create any form of resistance in pests

In addition, recombinant DNA technology is also being deployed for enhancing efficacy of microbial biopesticides in an indirect form. In this technique novel fusion proteins from microbial source are being designed to develop next-generation microbial biopesticides. The technology allows a toxin (not toxic to higher animals) to be combined with a carrier protein which makes it toxic to insect pests when consumed orally, while it was toxic only when injected into a target prey by a predator. Further, the fusion protein may be produced as a recombinant protein in microbial system, which can be scaled up for industrial production and commercial formulations. Several other innovative approaches are also being applied to develop microbial biopesticides or recombinant biopesticides using microbes as carrier agents or vectors for effective, efficient and acceptable means for sustainable pest control measures and conservation of non targeted plants and animals those get killed with the use of chemical pesticides.

But for the use of microbial biopesticides on a large and effective scale, an in-depth research and hands on training is needed in many areas such as culture and production of microorganisms, inoculation of the microbes in agricultural fields, effective and timely delivery and commercialization of the products. Furthermore, use of microbial biopesticides in agricultural practices can add on to conservation of biodiversity, as they do not kill any non targeted pest, animal or plants and can achieve another milestone in sustainable agricultural practices.

Microbial Biopesticides an Opportunity for Improvement of Livelihood Generation

Agriculture and allied activities maintain livelihoods of nearly 70 per cent of India's rural population (Hiremath 2007). The modern world is persuaded by rapid technological developments. Capability to generate, enhance and manage change seems the key factor in formulating the livelihood strategies of poor people involved in agriculture and allied areas in India. In livelihood analysis, technology plays a greater role in four interrelated constituents *viz.* infrastructure (machines and equipment), knowledge and skills (know-how and skills), organization (systems, procedures, practices and support structures), and art of product synthesis and generation (design, specification) (Pauli, 1999).

Agriculture and its allied field enhance rural livelihood systems through the locally available technological backstop. In principle, poor people stand to gain from technological change – generating easier access to information, higher productivity, at lower inputs costs, less wastage and better sustainable practices and management. However, speed and unpredictable change are some practical and vulnerable problems, particularly when allied activities are left behind agriculture development or forced to take greater risks to keep pace with increasing risk factors. As a result, the livelihood outcomes that allied practices owners practice is likely to get increased which is in turn determined by the activities that generate and manage technological change. In the long run, effective analysis of the factors that influence technological change in and around agriculture and allied enterprises is important for understanding the livelihood strategies and options for poor people who work in these areas. An approach in understanding the livelihood opportunities with such activities that involve the adoption of changes in the traditional system is presented in Figure 32.1.

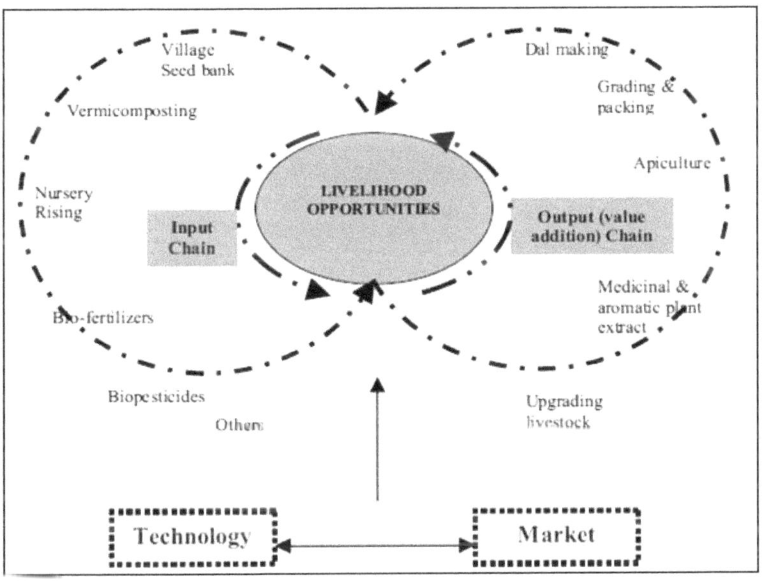

Figure 32.1. Input-Output Chain in Livelihood Promotion in Agriculture and its Allied Practices (Wani *et al.*, 2009).

This approach clearly link between developments of rural livelihood through organic farming. Rural livelihood system is dependent on input and output chains which are centered on utilization of natural resources or nature derived resources. In this model the input chain is mainly supplying support to reach higher growth and larger income flows to different category of people who are depending on these activities. These are the keys to value addition to their income activities that are dependent on market and technology, like marketing and distribution of organic seeds, manufacturing biopesticides including microbial biopesticides, organic and biofertilizers *etc.*

Here, the figure represents the input and output chain in organic farming practice. Input chain involves constitution of village seed bank, preparation of vermicompost and other organic and bio-fertilizers *etc.*, which can provide ample opportunity to the value addition in the household economy (Wani *et al.*, 2009). The output chain involves the production of value added products like cultivating and marketing of organic pulses, cultivation of organic medicinal crop plants. Besides, these livelihoods can also be generated by apiculture, poultry and others like livestock maintenance and their marketing.

Sustainable development with organic farming provides better training and development to farming communities to form opportunities to reduce migration to urban areas for seeking employment during off-farm season. Selection of micro-enterprises like making of biopesticides of different types including microbial biopesticides, biofertilizers, organic fertilizers, proper preservation, packing, disbursement and distribution of organic seeds along with biopesticides, biofertilizers, organic fertilizers, offering training on them can be helpful in getting better livelihood sources in their native places. The selection of such activities can also help women and landless people to attain their livelihoods and to improve the economic condition by using locally available resources. Several such activities can provide an opportunity to diversify livelihood activities and to improve the crop productivity with methods of organic farming (Wani *et al.*, 2002; 2009). These activities avail market facilities at the nearest places to sell their products.

In this regard we can consider use and production of microbial pesticides as a source of livelihood generation. People can get training on the microbial culture techniques, making of inoculums and inoculating them in agricultural fields and further providing hands on training to other untrained farmers. It will thus boost up their income sum through sustainable agricultural practices as the demand for organic products and their marketing are booming up all across the globe. Therefore, timely adoption of new skill and having proper knowledge on own working sector will definitely heighten the livelihood in rural India.

Therefore, provision are to be made by both government and non government organizations to encourage organic and sustainable agricultural practices and promotion of their allied field like manufacturing organic compost, fertilizers, biopesticides, organic seeds *etc.* so as to offer greater opportunities to strengthen rural livelihoods.

CONCLUSION

Microbial biopesticides offer a viable alternative to chemical pesticides in sustainable pest management programs. Baculoviruses such as nucleopolyhedrovirus (NPV) and granulovirus (GV) have been exploited as successful biological pesticides in agriculture, horticulture and forestry. *Marucavitrata* multiple nucleocapsid NPV

(MaviMNPV) was found to be a unique baculovirus specifically infecting pod borer on food legumes, and it has been successfully developed as a biopesticide in Asia and Africa. Entomopathogenic fungi also offer sustainable pest management options. Several strains of *Metarhizium anisoplia e*and *Beauveria bassiana* have been tested and developed as biopesticides in Asia and Africa. Hence, proper training on handling and inoculating microbes in crop plants to kill pest needs to be done at a faster rate to get maximum advantage from sustainable farming.

Nevertheless, further research is recommended to close the gaps in formulation of microbial biopesticides and techniques of their usage. Stable products under field conditions should provide guarantee of absolute effectiveness of microbial biopesticides in crop pest management. Researchers should therefore work together with officers and other official members in the government and industry as well as farmers to provide stable, durable formulations of microbial biopesticides to achieve livelihood through sustainable agricultural practices and conserving biodiversity by giving protection to the non- target pests, animals and plants those get usually killed by the application of inorganic agro-chemicals.

REFERENCES

Anand, S., and Reddy, J. (2009). Biocontrol potential of *Trichoderma* sp against plant pathogens. Inter. *J. Agri. Sci*, 2, 30-39.

Bailey, K. L., Boyetchko, S. M., and Längle, T. (2010). Social and economic drivers shaping the future of biological control: a Canadian perspective on the factors affecting the development and use of microbial biopesticides. *Biol Control,* 52, 221-229.

Berg, G. (2009). Plant-microbe interactions promoting plant growth and health: perspectives for controlled use of microorganisms in agriculture. *Appl. Microbiol. Biotechnol,* 84(1), 11-18.

Clemson, H. G. I. C. (2007). *Organic pesticides and biopesticides, Clemson extension, home and garden information center*. Clemson University, Clemson

Chattopadhyay, A., Bhatnagar, N. B., and Bhatnagar, R. (2004). Bacterialinsecticidal toxins. *Crit Rev Microbiol*, 30, 33-54.

Copping, L. G. and Menn, J. J. (2000). Biopesticides: a review of their action, applications and efficacy. *Pest Manag Sci*, 56, 651-676.

Dutta, S. (2015). Biopesticides: An Ecofriendly approaches for pest control. *World Journal of Phamacy and Pharmaceuticsl Science*, 4(6), 250- 265.

Ferreira, T. and Malan, A. P. (2014). *Xenorhabdus* and *Photorhabdus*, bacterial symbionts of the entomopathogenic nematodes Steinernema and Heterorhabditis and their in vitro liquid mass culture: a review. *African Entomolog,*. 22, 1-14.

Grewal, P. S. and Peters, A. (2005). Formulation and quality. In. Grewal, P. S., Ehlers, R. U, and D. I Shapiro-Ilan (Eds.), *Nematodes as biocontrol agents.* (pp. 79-90) Wallingford, UK: CAB International.

Haas, D. and De'fago, G. (2005). Biological control of soil borne pathogens by fluorescent pseudomonads. *Nat. Rev.Microbiol*, 3, 307-319.

Hiremath, B. N. (2007). The Changing faces of Rural Livelihoods in India. Paper presented at the National Civil society Conference on 'What it takes to eradicate poverty' held during December 4-6, 2007. Institute of Rural Management, Anand, Gujarat. pp. 1-10.

Holder, D. J., and Keyhani, N. O. (2005). Adhesion of the entomopathogenic fungus *Beauveria* (*Cordyceps*) *bassiana* to substrata. *Applied and environmental microbiology*, 71(9), 5260-5266.

Ignoffo, C. M. and Couch, T. L. (1981). The nucleopolyhedrosisvirus of *Heliothis* species as a microbial pesticide. In. H. D. Burges (Ed), *Microbial Control of Pests and Plant Diseases*. (pp. 329-362). Academic Press, London.

Kachhawa, D. (2017). Microorganisms as a biopesticides. *Journal of Entomology and Zoology Studies*, 5(3), 468- 473.

Kumar, S. (2015). Biopesticide: an environment friendly pest management strategy. *J Biofertilizer, Biopesticid.*

Kumar, S. and Singh, A. (2015).Present Status and the Future Prospects. *J Fertil Pestic Biopesticide*, 6.

Li, Q., Li, L., Moore, K., Donly, C., Theilmann, D. A., and Erlandson, M. A. (2003). Characterization of Mamestra configurata nucleopolyhedrovirusenhancin and its functional analysis via expression in an Autographacalifornica M nucleopolyhedrovirus recombinant. *J Gen Virol*, 84, 123-132.

Moore, N. F., King, L. A., and Possee, R. D. (1987). Viruses of insects. *Insect Sci. Appl*, 3, 275-289.

Moorhouse, E. R., Gillespie, A. T., Sellers, E. K., and Charnley, A. K. (1992). Influence of fungicides and insecticides on the entomogenous fungus *Metarhizium anisopliae* a pathogen of the vine weevil, *Otiorhynchus sulcatus*. *Biocontrol Science and Technology*, 2(1), 49-58.

Nawaz, M., Mabubu, J. I., and Hua, H. (2016). Current Status and Advancement of Biopesticides: Microbial and Botanical Pesticides. *Journal of Entomology and Zoology Studies*, 2241-246.

O'Brien, K. P., Franjevic, S., and Jones, J. (2009). Green chemistry and sustainable agriculture: the role of biopesticides, advancing green chemistry. http: // advancinggreenchemistryorg/wpcontent/uploads/Green-Chemand- Sus.-Ag.-the-Role-of-Biopesticides.

Pauli, G. (1999). Towards technology strategy for sustainable livelihoods, Prepared for the sustainable livelihoods unit of the UNDP.

Popham, H. J. R., Bischoff, D. S., and Slavicek, J. M. (2001). Both Lymantria dispar Nucleopolyhedrovirus Enhancin genes contribute to viral potency. *J Viro*, 75, 8639-8648.

Rohrmann, G. F. (2013). Baculovirus Molecular Biology: Third Edition (Internet). Bethesda (MD): National Center for Biotechnology Information (US), Available online: http: //www.ncbi.nlm.nih.gov/books/NBK114593.

Roy, A., Moktan, B. and Sarkar, P. K. (2007). Characteristics of *Bacillus cereus* isolates from legume-based Indian fermented foods. *Food Contr*, 18, 1555-1564.

Srinivasa, M., Babu, J. and Anitha G. (2008). Girish Laboratory evaluation of available commercial formulationsof HaNPV against *Helicoverpa armigera* (Hub.). *Journal of Biopesticides*, 1, 138-139.

Tanda Y. and Kaya, H.K. (1993). Insect pathology. Academic Press Inc., Harcourt Brace Jovanovich Publishers, San Diego.

Thakore, Y. (2006). The biopesticide market for global agricultural use. *Industrial Biotechnology*, 194-208.

Wani, S. P., Pathak, P., Tam, H. M., Ramakrishna, A., Singh, P., and Sreeedevi, T. K. (2002). Integrated Watershed Management for Minimizing Land Degradation and Sustaining Productivity in Asia. In *Integrated Land Management in Dry Areas. Proceedings of a Joint UNU-CAS International Workshop* Zafar Adeel (Ed.), (pp. 207-30). 8-13 September 2001, Beijing, China.

Wani, S. P., Venkateswarlu, B., Sahrawat, K. L., Rao, K. V., and Ramakrishna YS (eds.). (2009). Best-bet Options for Integrated Watershed Management. Proceedings of the Comprehensive Assessment of Watershed Programs in India, 25-27 July 2007, ICRISAT, Patancheru 502 324, Andhra Pradesh, India. Patancheru 502 324, Andhra Pradesh, India: International Crops Research Institute for the Semi-Arid Tropics. 312 pp. ISBN 978-92-9066-526-7: Order code: CPE 167.

Chapter 33

Changing Socio-Cultural Tradition of Santal Tribes: A Case Study of Mayurbhanj District in Odisha, India

Suvendu Kumar Satapathy

North Orissa University, Takatpur,
Baripada – 757 003, Odisha
e-mail: ssuvendukumar51@gmail.com

Modernisation is based upon the social structure prevailing in highly advanced countries mainly characterized by rapid urbanization, growing literacy, a meaningful concourse of idea among people through new media, radio, television and books, A major issue that the study reveals is the large-scale impact of modernisation on tribal areas of Mayurbhanj, that is rapidly eroding the age-old customs and traditions of the Santal tribes. The first and major impact which tribal population faces is in the shape of loss of tribal identity. Tribal customs and traditions come under pressure, due to contact with the urban culture and a consequent change in the attitudes of the tribals in Mayurbhanj can be seen. Tribal crafts and cultivation have also steadily declined. This is the need of the hour for discussion about preservation of the indigenous tradition and cultural practices. This paper tries to analyse the impact of changing socio-cultural tradition as well as religious aspects in particular to Santal tribes in Mayurbhanj district of Odisha.

Keywords: *Customs, Modernisation, Traditions, Culture, Santal tribe.*

INTRODUCTION

In India, 90 million people belong to the indigenous communities known as Adivasis or tribal. According to the 2011 census, India has 8.10 per cent of tribal population. Odisha claims to be the second-largest leading tribal state which comprises 62 number of tribes representing an impressive 22.13 per cent population of the state. Santals are considered to be the largest homogeneous Scheduled Tribe in India and can be found mainly in the states of Jharkhand, Bihar, West Bengal, Madhya Pradesh, Assam, Tripura, and Odisha. in Odisha Santals are predominantly found in the areas of Mayurbhanj and Keonjhar district. The overall population of Mayurbhanj is 25,19,738 out of which 14,79,576 are Tribals counting to 58.71 per cent of the population. Regionally called, Santal or Santhal. But they are locally known as Majhi found by the researcher. It is being widely seen today that the traditional features of tribal life are gradually changing from being deeply ingrained in tribal customs and traditions to something that is more modernised, in a developmental sense, due to adaptation of modern ways of living and altered lifestyle pattern. According to M.D. Jenson, "Social change may be defined as modification in ways of doing and thinking of people. According to Jones "Social change is defined variations in or modifications of any aspect of a social system, processes, social patterns, social interaction or social organization". Concerning recent changes in social structure and cultural patterns of tribal societies in the process of mainstreaming is the formation of a new identity or in a sense, acceptance to loss of indigenous identity. It is a more complex process. In this process of mainstreaming, many tribal people both men and women are coming across the non-tribal influence. The cultural discontent is heading towards a total cultural diffusion like situation rather than balanced reciprocity of cultural exchanges. (Acharya K *et al.*, 2014). This research is an attempt to analyse the socio-cultural crises on the religious customs and traditions of the Santal tribes living in Mayurbhanj district of Odisha. It shows how the modern practice which has been created by modernisation has brought major changes in Culture and Tradition.

Significance of the Study

The significance of the study arises from the fact that tribal cultures and indigenous traditions need to be preserved. These are our indigenous cultures and traditions and if we do not take the effort to preserve them, then we would lose a life that was once an integral part of our everyday existence. The goals should be to let people develop along the lines of their ingenuity and avoid imposing anything on them and to encourage their traditional art and culture. Tribal people should be allowed to administer themselves and avoid entry of too many outsiders into the tribal area in the name of administration, without overwhelming them with too many schemes and work in cooperation with their own social and cultural institutions. The development factor should be judged according to the perceived indices of actual level of human development seen in the context.

Materials and Methods

The study is based on both primary and secondary sources of data, but more emphasis is given on primary data which are collected by authors through field surveys. The data are collected on basis of a stratified random sample, from Thakurmunda block of Mayurbhanj district where the majority of tribals are belonging to the Santal Community. In the next phase, two gram panchayat were selected from the Thakurmunda block which is Kendujiani and Hatigoda. And in the third phase, 20 households were selected randomly from each village which includes, Sunariposi and Sarubil from Kendujiani panchayat and Kadapani and Jamudiha from Hatigoda panchayat. As a whole data were collected from a total80 respondents in the study area The secondary data have been collected published sources. The data were arranged in a suitable table form presented a chart, diagram for better understanding through analysis and interpretation. At last, all the data were represented through the graphical method and a statistical method were used according to the need of the present study.

Study Area

The present research work was undertaken among the Santals tribes of Thakurmunda block, Mayurbhanj district of Odisha. The village is about 130 km from the district headquarter town of Baripada and about 20 km from Thakurmunda block. The tribal Village is a multi-ethnic village but it is dominated by the Santal tribe. The tribals of the village speak Odia language (regional language) but they speak their own dialect Santali also.

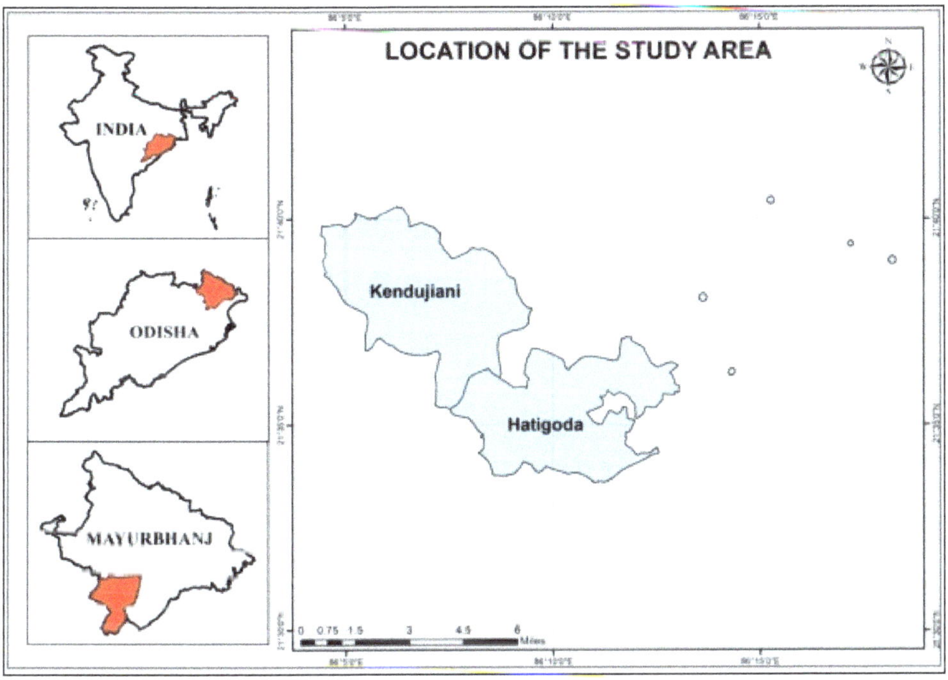

Figure 33.1. Map of Kendujiani and Hatigada GP in Thakurmunda Block.

Results and Discussion

Characteristics of Responses

Description of the characteristics of the target population gives some basic information about the sample population involved in the study, thus the following tables contain about the general, educational, economic status of the santal tribe.

Table 33.1. Distribution of Respondents on the Basis of Age, Marital Status and Education

Sl.No.	*Characteristics of Responses*	*No. of Respondents*	*Percentage*
1	**Age**		
A	Below 20	4	6
B	20-30	8	12
C	30-40	30	38
D	40-50	28	32
E	Above 50	10	12
	Total	**80**	**100**
2	**Marital Status**		
A	Married	66	68
B	Unmarried	8	18
C	Widow	6	14
	Total	**80**	**100**
3	**Education Status**		
A	Elementary	21	26.25 per cent
B	Secondary	36	45 per cent
C	Intermediate,+2	15	18.75 per cent
D	Graduation	6	7.5 per cent
E	Post-graduation and Above	2	2.5 per cent
	Total	**80**	**100 per cent**

Source: Primary data.

Age of the Respondents

The Table 33.1 show that 38 per cent of respondents belonged to the age group of 30-40 year old, whereas 6 per cent of them come under the age group of below 20. While 32 per cent of them belong to the age group of 40-50 years old and 12 per cent of them come under in two ages groups 20-30 and above 50. Thus, from the above analysis, it is learned that most of them belong to the age group above 30-40.is near to middle age who are deeply involved in their culture and tradition.

Marital Status of the Tribal

The Table 33.1 also indicates that 68 per cent of respondents are married, whereas 14 per cent of them are widows While 18 per cent of the respondents are unmarried. Thus from the above analysis, it is found that the majority of the respondents (68 percent) were married and involved in cultural and traditional beliefs.

Education and Literacy among Tribal

Modernization has its positive aspects too, modernization has brought along with education and literacy which has helped the tribes to get themselves better living conditions and has also aided them from being exploited by the non-tribal groups. It can be seen from the results that most of the tribal encouraged their children to pursue education and send them to schools and colleges for higher studies. It is also seen from the study that in many cases the parents are not well educated or sometimes not at all educated but they prefer to send their children to school. On the other hand, the Government has also been actively participating in providing education to the tribal people by implementing various schemes and facilities that help the tribals to get educated.

From the Table 33.1, it can be interpreted that 26.25 per cent of respondents completed their Elementary education. A majority of respondents *i.e.* 45 per cent have completed their secondary education and 18.75 per cent have completed their intermediate studies. Only 7.5 per cent and 2.5 per cent have completed their graduation and post-graduation studies respectively.

Language Spoken by Santal Tribes

Language is a vocal identity of the tribal society. Santali language is written Al-Chiki script. It is their mother tongue. All Santali people in this village speak in this language. However, they are used to make conversation in Odia language with their Odia neighbours.

From the Figure 33.2, it can be analysed that the majority of the respondents *i.e.* 66.25 per cent spoke both the languages (Odia and Al-Chiki) whereas 33.75 per cent of the respondents spoke only Al-Chiki. Thus the Santhal people have not only kept the knowledge of their dialect but also adopted the local language to maintain proper communication with the mainstream society. As a result, peoples are using Odia language during communication due to which their language has reached a point of deterioration.

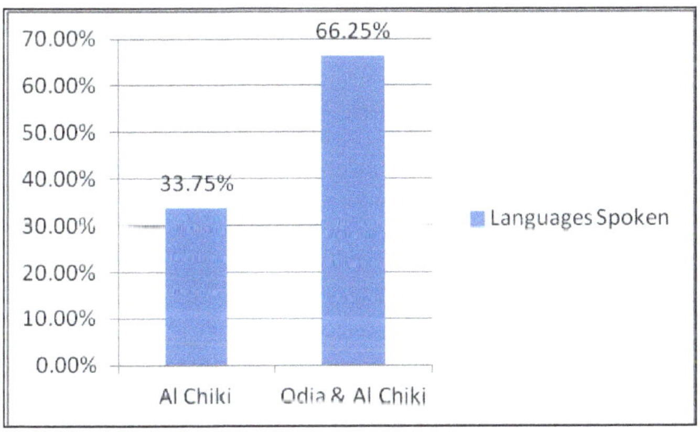

Figure 33.2. Language is Spoken by Santal Tribes.

Dress of Santal People and Desire for Wearing Traditional Dress

Indian tribes present the heritage of traditional culture because most of the tribal people still living under primitive belief. Santal people wearing typical dresses that are different from modern culture and still they are strongly following their traditional dresses. Normally male members wear a medium piece of cloth around his waist which is commonly known as Dhoti and during the time of working, they usually wear a small piece of cloth which is called kopni. Generally, Santali women are dressed with a larger piece of cloth or saree, the white Saree with a red border is of common use during the time of the festival. But their traditional dress is Panchi Parhat. All these clothes are made of cotton. The Santali women like to wear various types of ornaments; these include necklace, earring, bangles, wristlets, ring, girdles *etc.* these are commonly made of (brass) silver nickel. An interview taken with the tribal people about the choice of their traditional dress has been presented below in tabular as well as in diagrammatic forms.

From the Figure 33.3, it can be interpreted that 47.5 per cent of respondents preferred to wear modern dresses usually while 30 per cent of the respondents prefer to wear traditional dresses all the time. Other than that 22.5 per cent of the respondents like to wear traditional attires on the festive occasions only.

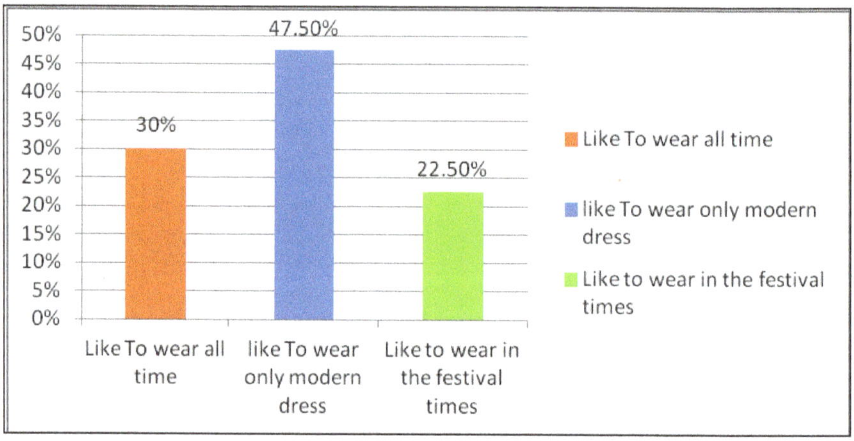

Figure 33.3. Distribution of Respondents based on a Desire for Wearing the Dress.

Knowledge about the Festivals

Nowadays, due to the impacts of Modernization Santal people are also changing their traditional festivals and adapting a new socio-modern culture. From the study area, it was found that more than 52 per cent of Santals practice their ancestral festival due to the level of modern influences. More than 48 per cent of Santal people are strongly participating and have good knowledge of their festivals.

Table 33.2. Distribution of Respondents based on Knowledge about the Festival

Knowledge about Festivals and Rituals	*Per cent of Respondents*
Know all about them	48 per cent
Know some of them	52 per cent
Do not know	0
Total	**100 per cent**

Source: Primary data.

Participation in Cultural Activities

The Santal Tribe has a profound cultural background which is quite renowned. They follow their culture as much as possible with ample dedication and devotion. The Santhal tribal culture and heritage include various constituents like Dance, Prayers, Sacrificial Ceremonies, and Festivals. These are passed down from generation to generation for their continuation in observation. Santal Tribes enjoy and live dancing. It forms an important part of the Santal fairs and festivals. Apart from dance, Santals rejoice and enjoy playing music. The following table illustrates the respondent's interests in various cultural activities.

The Figure 32.4 indicates that the maximum percentage of respondents *i.e.* 65 per cent prefer song and dance as their primary cultural activities. The tribals enjoy by observing different festivals like Dal Puja, Baha Bhanga in which the maximum younger generations participated. 23 per cent of respondents prefer offering prayers and fasting. The male members praying at Jahira place his almighty nature 12 per cent of the respondents' belief that if they sacrifice hens and goats as an offering to their deities.

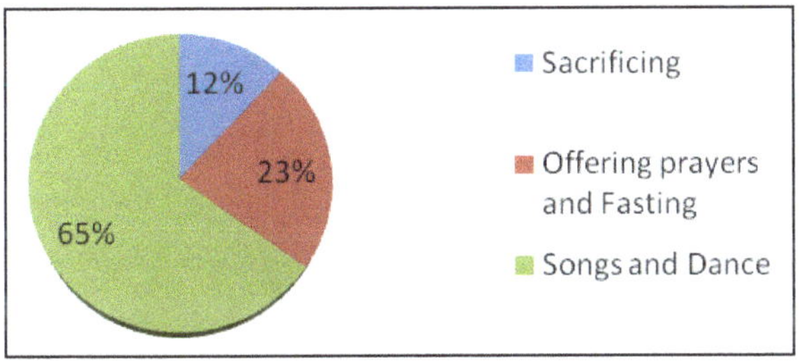

Figure 33.4. Distribution of Respondents based on Participation in Cultural Activities.

Views on Modernisation about their Culture and Tradition

Modernisation is an integral part of any society for its overall development. Now the process of modernisation and its impact on the santal tribes is the moot able part. To unveil the acceptance level of modernisation among the tribals, the following questions were asked to the respondents and their responses were recorded (Figure 33.5).

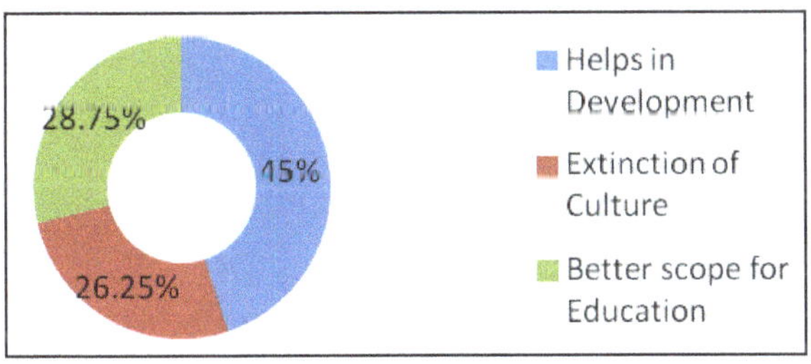

Figure 33.5. Views on Modernisation about their Culture and Tradition.

From the Figure 33.5, it can be concluded that 45 per cent of respondents believed that modernisation helps in the development and 28.75 per cent of the respondents viewed that modernisation provides better scope for education. On the other hand, 26.25 per cent of respondents still believe that modernisation can lead to depletion or extinction of a culture.

Views on Preservation of Culture, Tradition, and Religion

Cultural change among santal appears to follow a pattern similar to that of other ethnic groups in India. The process of change in culture among the Santal is described in terms of a cultural focus consisting of complexes centering about hunting, cultivation, annual festivals, and the village community organization (Datta Nabendu, 1958). Tribal society at the present moment is going through the critical stage of transformation. The forces of modernization have brought a rapid transformation in their socio-cultural milieu. Due to the rise of Cultural intermixture, a perceptible change in the lifestyle of tribal is seen in the study area. Therefore, attention should be paid to preserve their traditional systems. The following table shows the number of respondents who wanted to preserve their culture and tradition as well as the adoption of the modern culture.

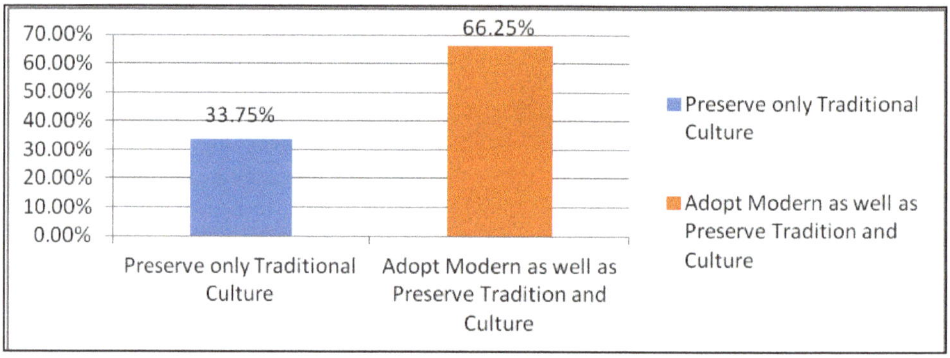

Figure 33.6. Views on Preservation of Culture, Tradition, and Religion.

From the Figure 33.6, it could be analysed that the tribals though are coping with the modernisation incessantly, wanted to preserve their age-old traditions. It can be seen that 33.75 per cent of respondents wanted to preserve their culture and traditions. On the other hand, 66.25 per cent of the respondents not only wanted to preserve their culture and tradition but also were in support to adopt the modern outlook to maintain parallelism between the age-old traditions and the modern traditions.

CONCLUSION

Santhal-life and their cultural panorama is a resource to our 'Cultural-Bank'. So we have to manage, preserve the traditional cultural life of Santhal's community. Cultural resource must be managed because they are threatened. Primary threats include development, looting, erosion, and inadvertent impact from recreationists (Stapp and Buurney, 2002). It is concluded from the study undertaken that due to the threat of modernisation in these tribal villages, the tribals' perceptions, thoughts, food habits, daily lifestyles, dresses, festivals, rule and rituals, the eco-friendly, nature-related culture, *etc.*, are all deteriorating due to the impact of modernisation. As modernisation has entered the tribal lifestyle, there is a chance to lose the traditions and tribal uniqueness in the long run, in the changing scenario (Chaudhuri, 1965). The study also reveals that the santal tribes constantly face trauma and pressure from the society that they live in. To be a part of the society has resulted in linguistic assimilation and many tribal communities have lost their original mother tongues and speak Odia one of the main languages of Odisha. Recently education and related awareness have aided the tribals from being seriously exploited by the non-tribals. It is paradoxical that in many areas the tribals are exposed to the influence of the society they will be exploited, thus to be competent and to be a part of the society the tribals have undergone a major change in their lifestyle. Thus, modernisation is seen to be slowly changing the tribal life as it is gradually disintegrating the cultural resources of the tribal community. To conserve the tribal-culture many steps are taken by the government but it needs to be preserved at the community level.

REFERENCES

Acharya, S.K., and Gautam, K.K. (2014). Social Transformation, Identity of Indian Tribes in Recent Time: An Anthropological prospective. *Afro Asian Journal of Anthropology and Social Policy*, 5(2), 73-88

Chaudhuri, N.C. (1965). The Tribal Problem in All-India Perspective. *Education*, 31(1): 40-47.

Datta-Majundar, N. (1956). *The Santal: A study in Culture-change*, Manager of Publications, 2, pp.150.

Subrata, G. and Ismail Md. (2015). Socio-Cultural changes of Tribes and their Impacts on Environment with Special Reference to Santhal in West Bengal, *Global Journal of Interdisciplinary Social Sciences*, 4(3), 148-156.

Malyadri, P. (2012). Education for tribal children: An engine for human development. *International Journal of Research Studies in Education*, 1 (1), 99-106.

Pal, T. (2011). Changing Tribal Culture: A Photo-Geographical Explanation. *Current Research Journal of Social Science*, 3(6), 483-489.

Rath, R. (2011). Development and Cultural Change Among the Kandh Tribals of Kandhamal, Odisha. *Review*.5, 34-40

Stapp, D.C. and Buurney, M.S. (2002). Tribal Culture Resource Management: The Full Circle to Stewardship. Rowman and Littlefield Publishers, England, pp: 6.

Singh, K.S. (1982). Transform of Tribal Society. *Economic and Political Weekly* (17) 34, 1376-1384.

Index

G

H

I

K

L

M

R

S

T

U

COLOUR PLATES

Plate 5.1: Orchids of SBR. (Pg 67)

Aloe barbadensis Aloe Vera

Hemidesmus indicus Anantmool

Withania somnifera Ashwagandha

Bacoppa monerii Brahmi

Sweretia chirayta Chiraita

Datura somnifera Dhatura

Plate 7.1. Some Important Useful Medicinal Plants of Vindhyan Region, Uttar Pradesh. (Pg 94)

Abrus precatorius Ghumchi

Tinospora cardifolia Giloe

Gymnema sylvestre Gudmar

Commiphora wightii Guggal

Andrographis panniculata Kalmegh

Centella asiatica Mandookparni

(Pg 95)

Helicteres isora Marorphali

Glycyrrhiza glabra Mulethi

Cyprus rotendus Nagarmotha

Evolvulus alsinoides Sankhpuspi

Rauwolfia serpentina Sarpgandha

Asparagus racemosus Satawar

Acorus calamus Vach (Pg. 97)

Figure 16.1. (Pg. 201)

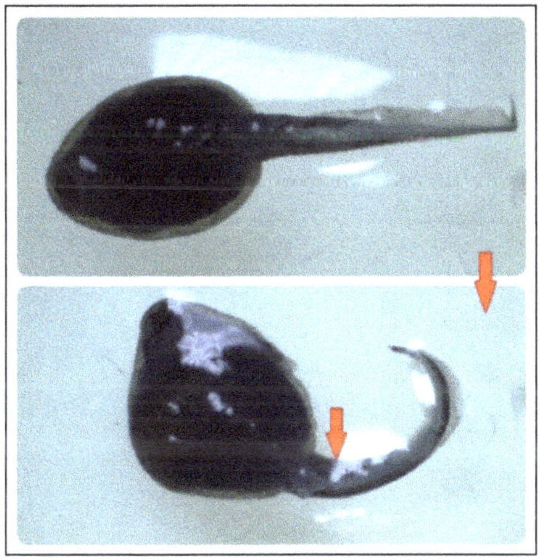

Figure 16.2. (Pg. 201)

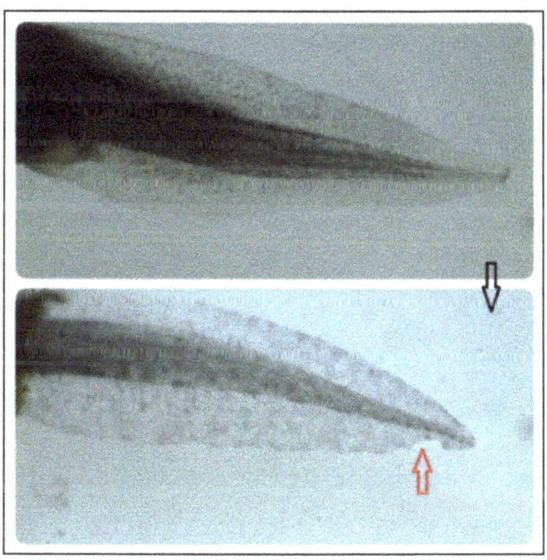

Figure 16.3. (Pg. 202)

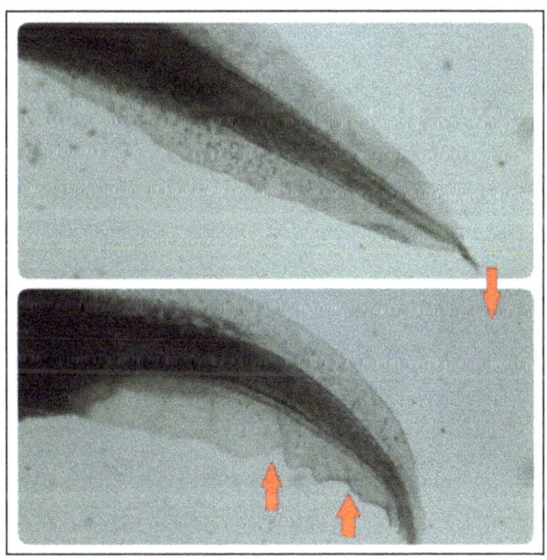

Figure 16.4 (Pg. 202)

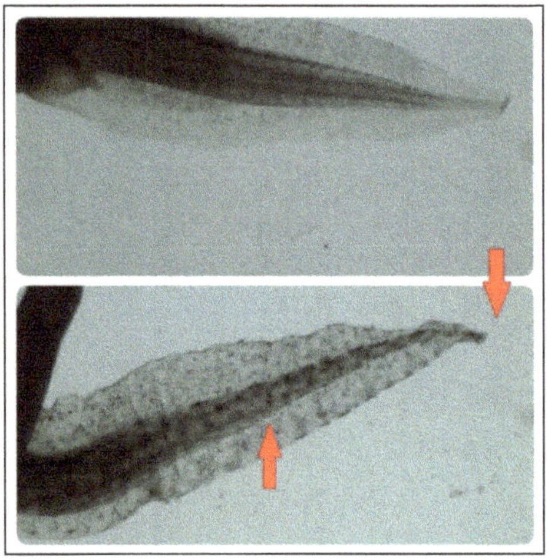

Figure 16.5 (Pg. 202)

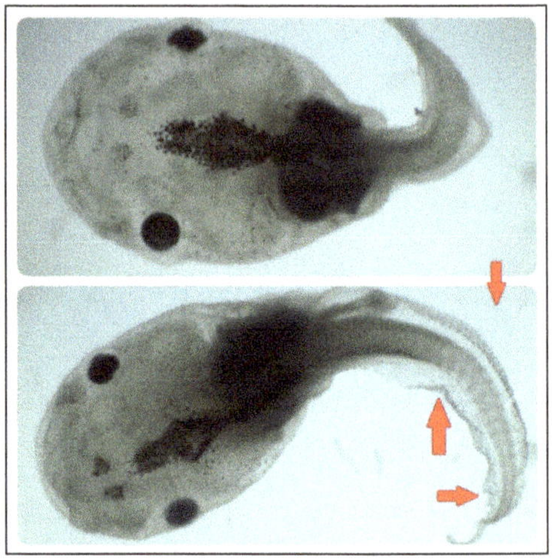

Figure 16.6. (Pg. 202)

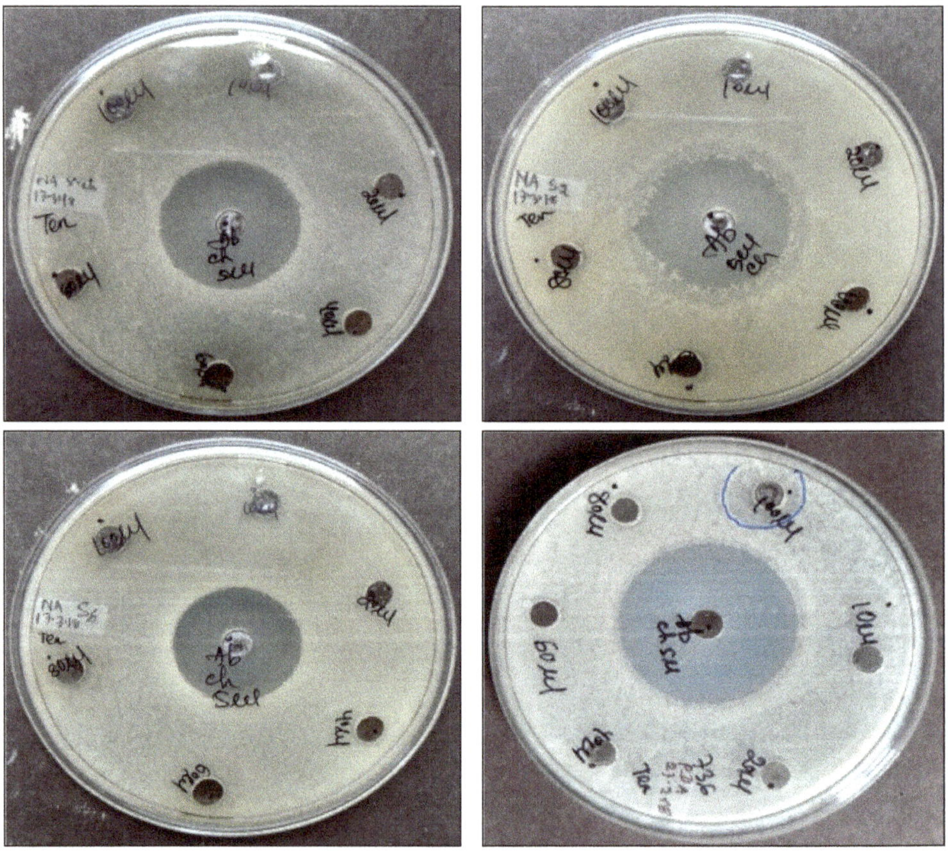

Figure 17.4: Agar Cup Well Diffusion of Termite Extract against Bacterial Strains. (Pg. 210)

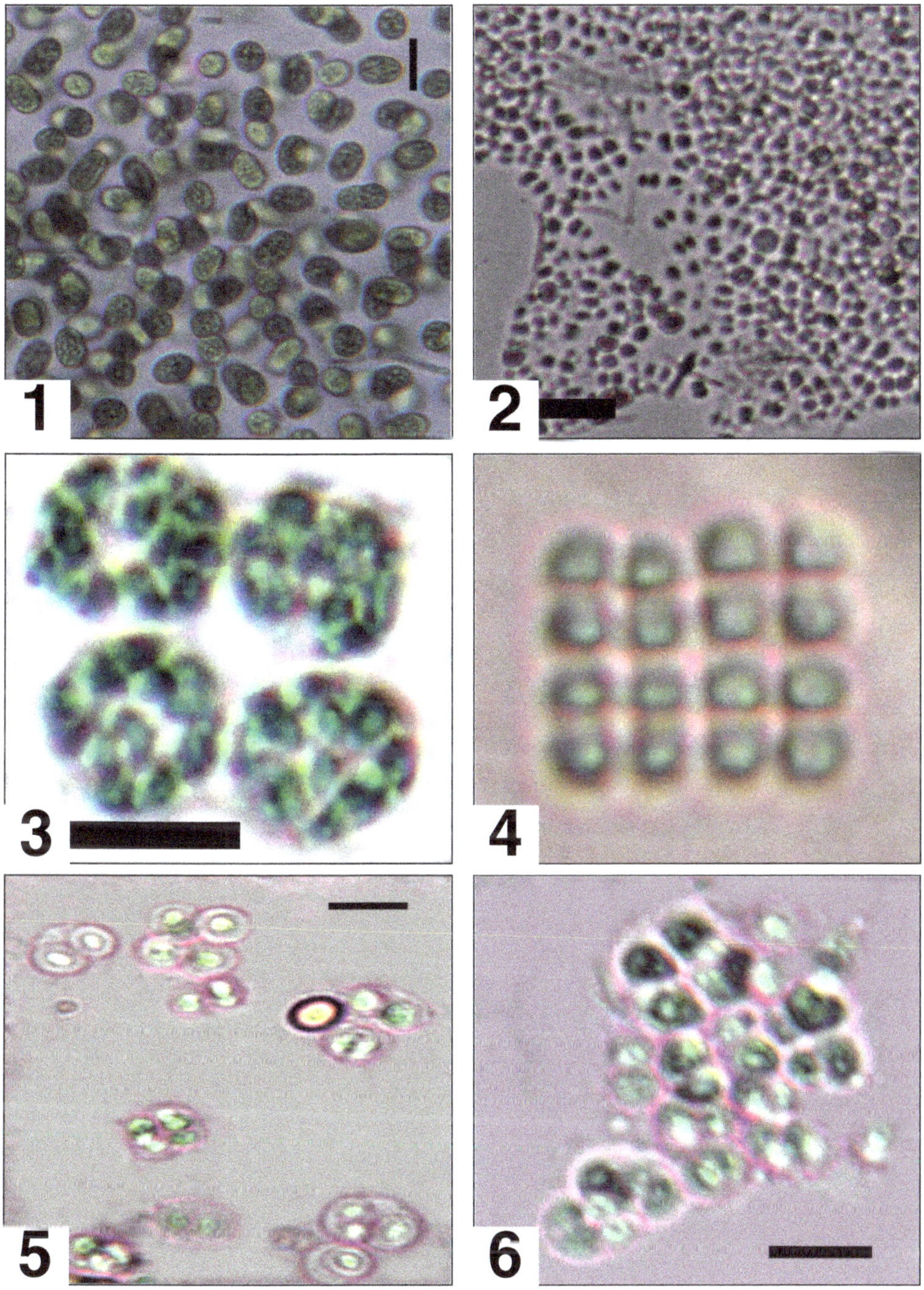

Scale Bar: -1, 3, 4, 5, 6 = 10 µm, 2= 40 µm.

Plate 18.2. Photomicrographs of Cyanobacteria Species in Similipal Biosphere Reserve.

1**:** ***Aphanothece microscopica***; 2: ***Microcystis*** sp.; 3: ***Gomphosphaeria*** sp.; 4: ***Merismopedia glauca;*** 5: ***Gloeocapsa*** sp.; 6: ***Gloeocapsa magma (Pg. 219)***

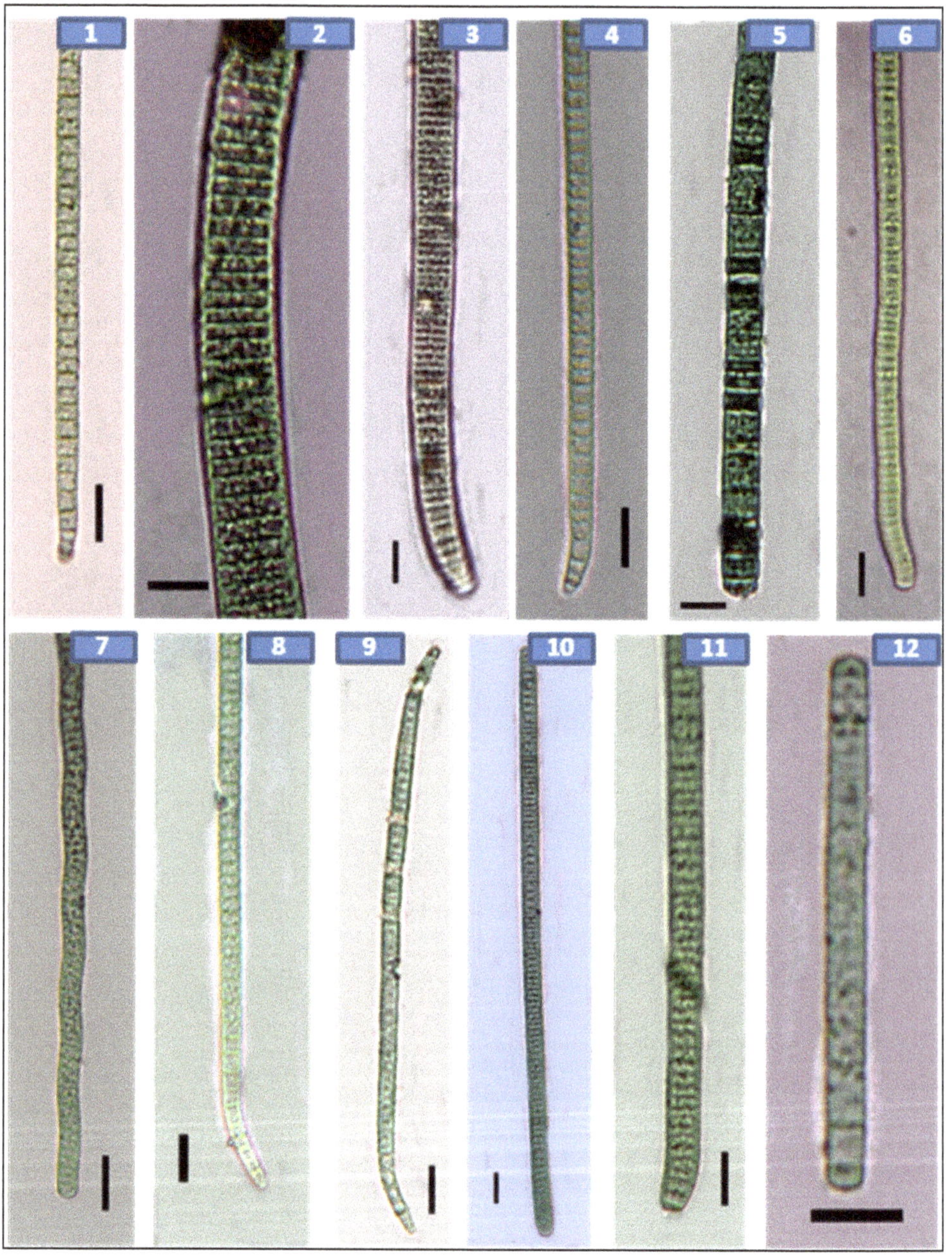

Scale Bar: - 1, 2, 3, 4, 5, 6, 7, 8, 9, 10, 11 and 12 = 10 μm

Plate 18.3: Photomicrographs of Cyanobacteria Species in Similipal Biosphere Reserve.

1: ***Oscillatoria amoena***; 2: ***Oscillatoria perornata***; 3: ***Oscillatoria formosa;*** 4, 8 and 9: ***Oscillatoria acuta***; 5: ***Oscillatoria vizagapatensis***; 6: ***Oscillatoria chalybea;*** 7: ***Oscillatoria curviceps***; 10: ***Oscillatoria subbrevis***; 11: ***Oscillatoria proteus;*** 12: ***Oscillatoria limosa***. (Pg. 220)

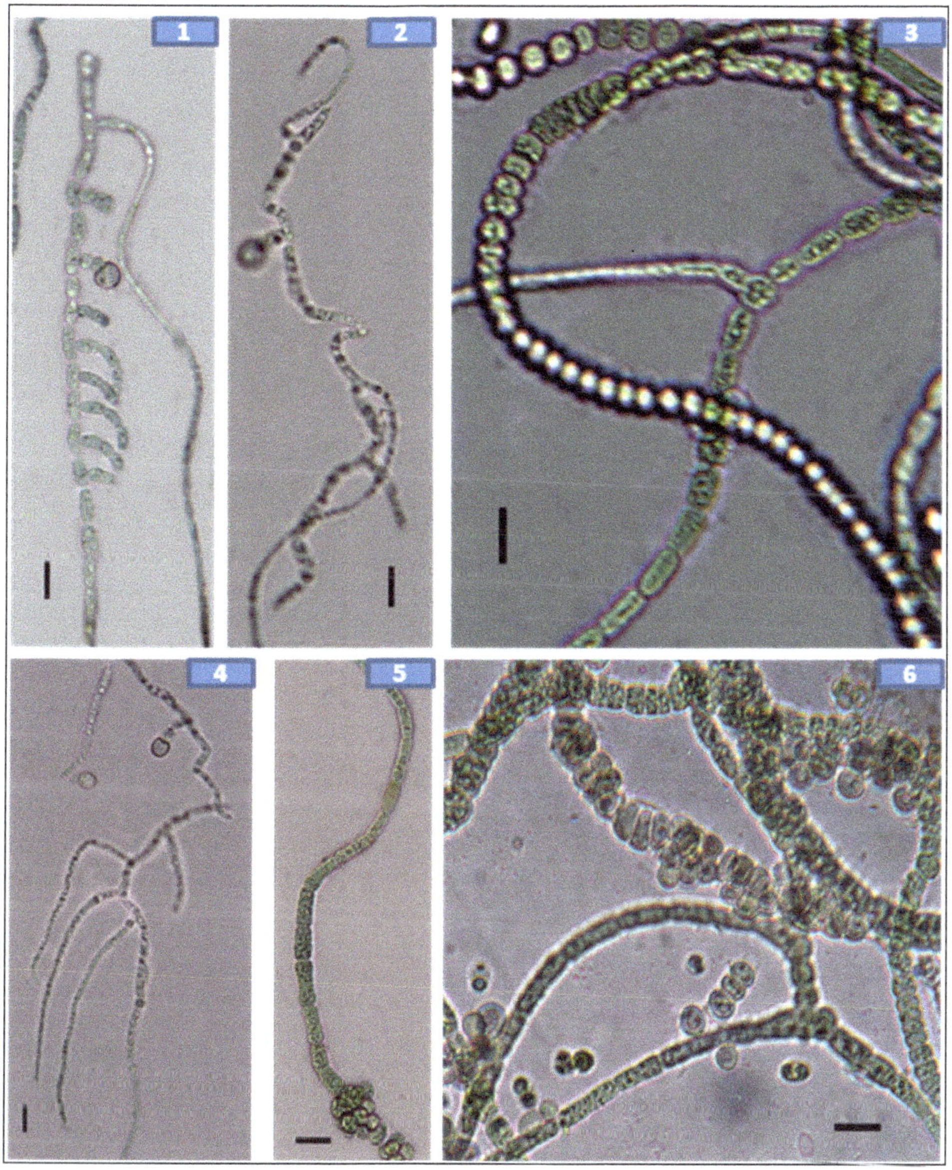

Scale Bar: - 1, 2, 3, 4, 5, 6 = 10 µm

Plate 18.4: Photomicrographs of Cyanobacteria Species in Similipal Biosphere Reserve.

1: ***Nostochopsis lobatus***; 2 and 4: ***Nostochopsis hansgirgi***; 3: ***Hapalosiphon welwitschii;*** 5 and 6: ***Westiellopsis prolifica***. (Pg. 221)

Parmelia caperata — *Trapelia* sp. — *Diorygma hieroglyphicum* — *Parmotrema ravum* — *Parmelia* sp. — *Flavoparmelia* sp.

Figure 21.1. Photographs of different Lichen Thalli Used for Isolation of Endolichenic Fungi. (Pg. 260, 261, 262)

Cryptothecia sp.

Lecanora sp.

Parmotrema ravum

Parmotrema tinctorum

Cryptothecia sp.

Parmelia sp.

(Pg. 261)

Graphis sp.

Cryptothecia sp.

Xanthoparmelia sp.

Cryptothecia sp.

(Pg. 262)

Figure 28.2. Wild Edible Mushrooms of the Study Area. (Pg. 405)

(Pg. 405)

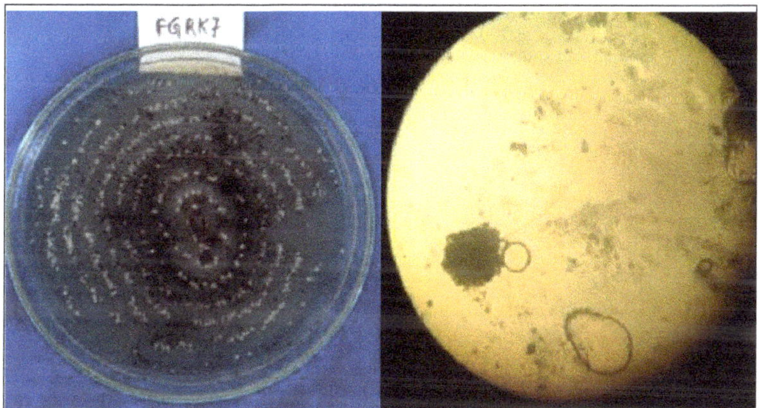

D

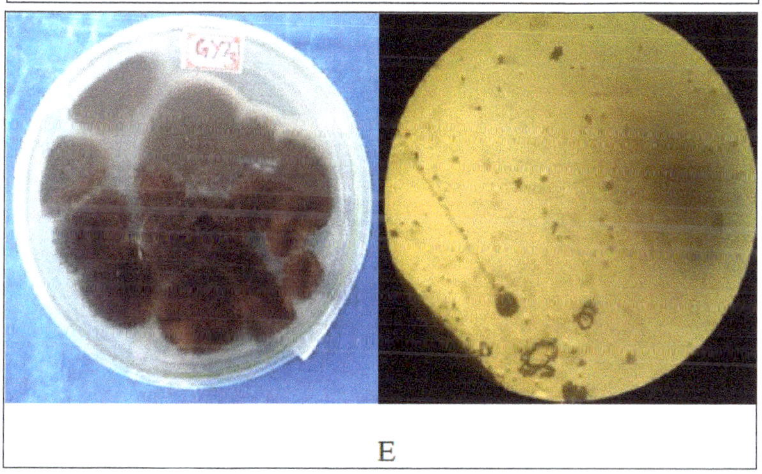

E

(Pg. 266)

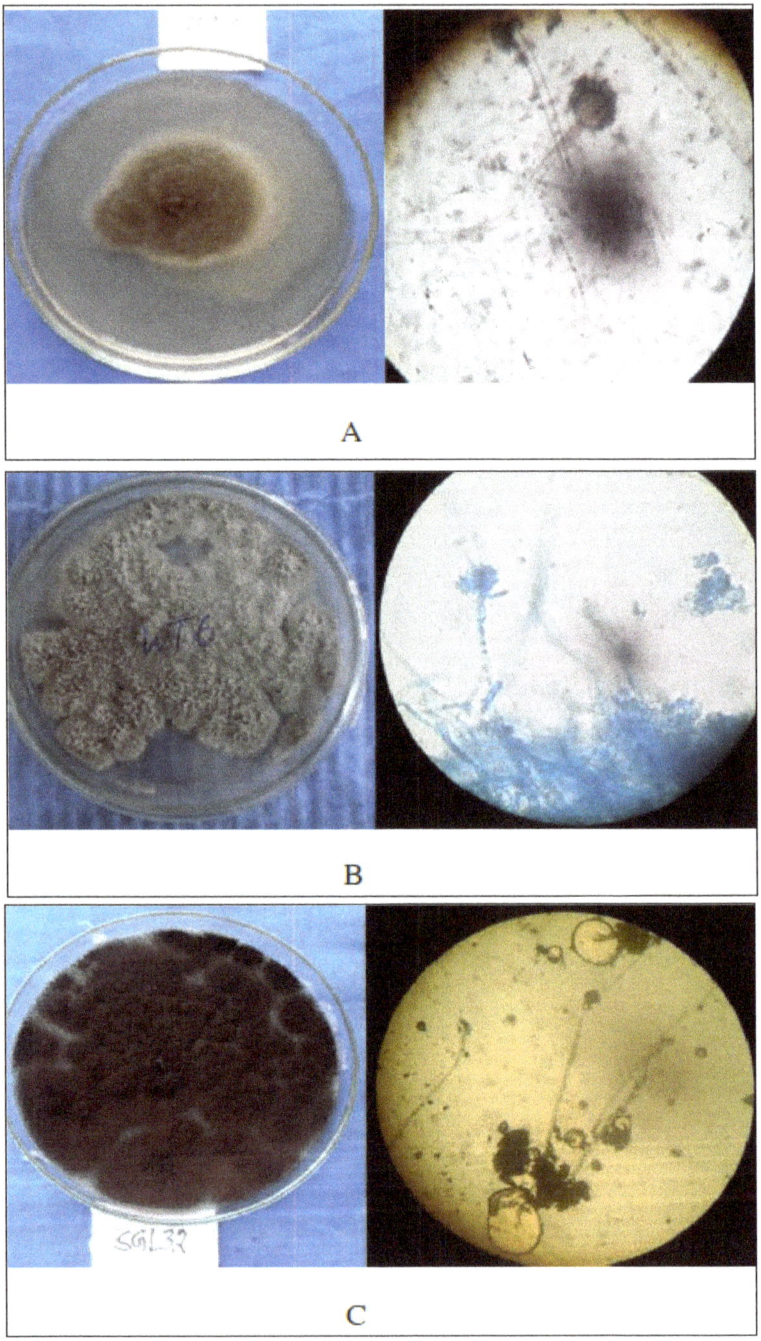

Figure 21.3. Photographs Showing Colonial Morphology and Microscopic View of Selected Isolates Described in this Study. (Pg. 265)

Plate 31.1. Wild Foods Consumed by Local Community of Imphal, Manipur.
(P.g 413)

(Pg. 414)

www.ingramcontent.com/pod-product-compliance
Ingram Content Group UK Ltd.
Pitfield, Milton Keynes, MK11 3LW, UK
UKHW060117010826
14089UKWH00003B/27

9 789359 190983